Internet of Things A to Z

IEEE Press
445 Hoes Lane
Piscataway, NJ 08854

IEEE Press Editorial Board
Ekram Hossain, *Editor in Chief*

Giancarlo Fortino
David Alan Grier
Donald Heirman
Xiaoou Li
Andreas Molisch
Saeid Nahavandi
Ray Perez
Jeffrey Reed
Linda Shafer
Mohammad Shahidehpour
Sarah Spurgeon
Ahmet Murat Tekalp

Internet of Things A to Z

Technologies and Applications

Edited by Qusay F. Hassan

Copyright © 2018 by The Institute of Electrical and Electronics Engineers, Inc. All rights reserved.

Published by John Wiley & Sons, Inc., Hoboken, New Jersey.
Published simultaneously in Canada.

No part of this publication may be reproduced, stored in a retrieval system, or transmitted in any form or by any means, electronic, mechanical, photocopying, recording, scanning, or otherwise, except as permitted under Section 107 or 108 of the 1976 United States Copyright Act, without either the prior written permission of the Publisher, or authorization through payment of the appropriate per-copy fee to the Copyright Clearance Center, Inc., 222 Rosewood Drive, Danvers, MA 01923, (978) 750-8400, fax (978) 750-4470, or on the web at www.copyright.com. Requests to the Publisher for permission should be addressed to the Permissions Department, John Wiley & Sons, Inc., 111 River Street, Hoboken, NJ 07030, (201) 748-6011, fax (201) 748-6008, or online at http://www.wiley.com/go/permission.

Limit of Liability/Disclaimer of Warranty: While the publisher and author have used their best efforts in preparing this book, they make no representations or warranties with respect to the accuracy or completeness of the contents of this book and specifically disclaim any implied warranties of merchantability or fitness for a particular purpose. No warranty may be created or extended by sales representatives or written sales materials. The advice and strategies contained herein may not be suitable for your situation. You should consult with a professional where appropriate. Neither the publisher nor author shall be liable for any loss of profit or any other commercial damages, including but not limited to special, incidental, consequential, or other damages.

For general information on our other products and services or for technical support, please contact our Customer Care Department within the United States at (800) 762-2974, outside the United States at (317) 572-3993 or fax (317) 572-4002.

Wiley also publishes its books in a variety of electronic formats. Some content that appears in print may not be available in electronic formats. For more information about Wiley products, visit our web site at www.wiley.com.

Library of Congress Cataloging-in-Publication Data is available.

ISBN: 978-1-111-945674-2

Printed in the United States of America.

10 9 8 7 6 5 4 3 2 1

Table of Contents

Preface

Information and communication technology (ICT) has always been dynamic and evolutionary in nature, leading to the continuous emergence of new technologies and business models. The recent advances in terms of available computing resources, software systems and communication networks, and the continuing miniaturization of hardware components have made it possible to integrate ICT into virtually anything, thus leading to the rise of a new computing paradigm known as the Internet of Things (IoT). The IoT aims at realizing an old dream of turning everyday objects into smart ones that are interconnected via the Internet and able to collect and exchange data and to make decisions autonomously. This additional "smartness" covers both the communication infrastructure and applications, including monitoring systems, industrial automation, and ultimately smart cities. In a recent report, Gartner estimates that there will be 11.2 billion connected devices in use in 2018 and 20.1 billion in 2020.[1] This clearly demonstrates the great potential and importance of this model and is also the motivation behind this book.

The potential of IoT is great and the possible applications of this model are countless. Therefore, uncovering the ins and outs of the IoT is crucial to both technology and business communities. This book seeks to provide a holistic coverage of the IoT model by presenting its principles, enabling technologies, and some of its numerous application domains. Important aspects such as the need for standardization, as well as security issues are also highlighted. The book also presents two sample applications showcasing how the discussed concepts and technologies can be leveraged and put to practical use to solve some real-world problems.

The book is a cohesive material that is composed of 21 chapters authored by several internationally renowned researchers and industry experts. Each chapter focuses on a specific subject and also provides the reader with the necessary background information, thus improving understandability and encouraging the reader to think further.

---

1 https://www.gartner.com/newsroom/id/3598917.

The book may be used as a textbook for both undergraduate and graduate students. It also comes in handy as a reference for researchers and IT professionals who are interested in IoT concepts, technologies, and possible applications. I hope the readers will enjoy reading this book as much as I enjoyed reviewing and editing the submissions.

Organization of the Book

The book is organized in a way that helps the reader to first grasp the concepts and then learn about key enabling technologies before moving to some potential applications. Although I would advise to read the entire book, most chapters are self-contained allowing the reader to focus on the topics they are interested in. Cross references between chapters are provided to help the reader to navigate between them.

This book is divided into five parts, each of which is devoted to a distinctive area.

Part I: Concepts and Perspectives

This part is composed of two chapters that cover the core concepts underlying the IoT, as well as its evolution and impacts.

Chapter 1 provides an introductory overview of the IoT, including its core conceptual ideas. The chapter also covers closely related concepts and paradigms, as well as a list of initiatives and organizations that contribute to their further development. In addition, the chapter structures the broad range of technical as well nontechnical aspects by presenting a four-layer framework that addresses the IoT's enabling technologies, derived qualities of modern ICT and how they are supporting the IoT, potential for new innovations based on the IoT ecosystem, and finally IoT implications and challenges.

Chapter 2 explores some very important aspects that are usually ignored or forgotten when talking about IoT. Specifically, the chapter addresses issues like the heterogeneity of objects utilized as well as the diversity of environments within which IoT systems run, how time is critical to IoT systems in terms of development and support, people involved in the IoT ecosystem, and new security challenges the IoT poses. It also looks at big data, and both the technical and the moral challenges that result from the adoption of such intelligent objects.

Part II: Enablers

This part comprises five chapters about various IoT enabling technologies that range from hardware items and communication technologies to data processing and storage to the emerging standards.

Chapter 3 gives a general overview of various enabling technologies of the IoT. The discussion of these technologies is based on their application and functionality in the IoT five-layer model. This includes hardware components, network technologies, middleware technologies, application services, and business-related technologies. Moreover, the chapter provides a brief overview of some of the key platforms and operating systems that are widely used in IoT environments.

Chapter 4 provides an introduction to cloud computing and fog computing, two of the key back-end technologies in IoT systems. The chapter highlights the importance of these two models and shows how they complement each other and work together. The chapter presents their advantages and disadvantages, as well as some examples of IoT applications where they can be used.

Chapter 5 introduces RFID, one of the core technologies in IoT systems, and describes the important role it plays. The chapter gives a brief history of RFID showing how it is linked with the emergence of the IoT and highlights some of its possible applications and implementation challenges.

Chapter 6 provides a tutorial that explains the design and development of IoT prototypes using Arduino and Raspberry Pi platforms. The chapter offers a guide that covers both the hardware and software aspects of these platforms for beginners who wish to learn about developing IoT applications. Detailed examples are provided to demonstrate how to implement IoT projects using these two platforms.

Chapter 7 looks at the development over the past 20 years of the standardization efforts for the IoT and four of its applications (and their respective predecessor technologies, if any) and at the links between them. The chapter speculates about the future of standardization in those domains based on what has been accomplished so far.

Part III: Security Issues and Solutions

This part provides three chapters about various security issues, technologies, and considerations in IoT environments.

Chapter 8 explores the main security protocols and technologies currently used in a typical IoT communication stack. This technically oriented chapter also discusses security issues in IoT environments and presents solutions enabled by research and standardization efforts to address them.

Chapter 9 introduces blockchain technology and how it can be leveraged to secure IoT systems and protect their data. The chapter provides some examples where blockchain-based solutions were proposed to secure IoT and gives an idea about integration challenges and current research efforts in this area.

Chapter 10 highlights the importance of IT auditing for organizations adopting the IoT. The chapter discusses the risks associated with the IoT and how routine and thorough IT auditing can prevent them. Risk identification and assessment, as well as audit considerations and policies are presented.

Part IV: Domains

This part is composed of nine chapters presenting various application domains where IoT technologies can be utilized. These chapters present the concepts, underlying technologies, implementation details, and advantages and challenges of such integration.

Chapter 11 represents a foundational chapter for this part as it gives an introduction to the use of IoT in several domains with focus on the Industrial Internet of Things (IIoT) showing how IoT can be leveraged in industrial fields of application. Two of the main initiatives in the IIoT are highlighted, namely, the Industrial Internet Consortium (IIC) and the Plattform Industrie 4.0.

Chapter 12 provides an overview of IoT applications for smart cities and how it can help in improving resource management. The technical aspects and general requirements of such solutions as well as the challenges the broad adoption of IoT faces in smart cities are presented.

Chapter 13 provides a contemporary overview of the IoT applications in smart homes, or what is called in the chapter as smart connected homes. The chapter presents the underlying technologies and architectures of smart connected homes, as well as the services they offer to householders. Both the technical and social challenges are also highlighted.

Chapter 14 addresses the integration of the IoT in the energy domain. The chapter provides a broad discussion of the motivations, approaches, and challenges of this integration and presents some specific applications including smart grid, green IoT, and smart lighting.

Chapter 15 continues the discussion made in Chapter 14 by highlighting various essential developments that are required for deploying and managing smart grids and renewable generation sources using IoT. The technical requirements, industry standards, and security concerns are addressed in this chapter.

Chapter 16 provides a comprehensive discussion on the integration of the IoT in patient-focused health applications. The chapter first describes the key elements of IoT-based health care ecosystems, and then explores the different types of applications that utilize this model. Challenges and expectations of future developments are highlighted as well.

Chapter 17 discusses how paramedics can use the IoT for emergency support. The chapter starts with an overview of how IoT enables the realization of smart ambulance, and then it provides a case study that assesses the adoption of this model for diagnosis and prognosis of chronic obstructive pulmonary disease. Challenges for the global deployment of smart ambulance are highlighted.

Chapter 18 reviews the various applications the IoT offers for precision agriculture (PA). The chapter starts with an introduction to PA, and then illustrates how the integration of IoT technologies into its different sections can be revolutionary in terms of quantity, quality, efficiency, sustainability, and cost-effectiveness.

Chapter 19 provides a broad overview of the integration of the IoT into Unmanned Aerial Vehicles (UAVs; commonly known as drones). The chapter describes this new model and presents its underlying technologies as well as several applications where it can be utilized. The chapter also highlights the challenges of this model with special focus on the security and safety issues.

Part V: Relevant Sample Applications

This part contains two chapters that offer two exemplary applications of IoT. Both chapters are implemented in lab scale using smart objects with the aim of tackling some real-life issues. Both technical and social details of these applications are discussed. The objective of providing these examples is to present the implementation details and show the reader how the IoT can actually be implemented.

Chapter 20 demonstrates the use of IoT technologies in the behavioral and psychological research field by capturing the emotions of children with autism spectrum disorder (ASD) using smart objects integrated into their play environment. First, ASD is explained and a literature review of related work is provided. Then, the technical details of the presented application are provided.

Chapter 21 presents a low-cost IoT framework for detecting and reporting landslides in landslide-prone areas. The chapter discusses the technical details of this framework, including the system design, used hardware components, and test results. The proposed framework has only been tested in lab, but it clearly shows how IoT technologies can be effectively integrated into such critical scenarios.

USAID

Qusay F. Hassan
Cairo, Egypt

Acknowledgments

I would like to express my gratitude to each and every individual who participated in this project. In particular, I would like to acknowledge the hard work of authors and their patience during the revisions of their chapters.

I also wish to thank Daniel Minoli, Marta Vos, Qiang Yang, Supriya Mitra, and Xiaojun Zhai for their comments that helped improve the quality and organization of this book.

Finally, I am very grateful to the editorial team at Wiley-IEEE Press for their support through the stages of preparation and production. Special thanks to my editor Mary Hatcher for her great support throughout the entire process. Thanks to Vishnu Narayanan for overseeing the editorial phase. Also, thanks to the anonymous reviewers provided by Wiley-IEEE Press whose comments helped improve the quality of the chapters.

Qusay F. Hassan

Acknowledgments

[illegible]

Contributors

Abdullah Abuhussein
Department of Information Systems, St. Cloud State University, St. Cloud, MN, USA

Faisal Alsubaei
Computer Science Department, University of Memphis, Memphis, TN, USA

and

Faculty of Computing and Information Technology, University of Jeddah, Jeddah, Saudi Arabia

Kalinka Regina Lucas Jaquie Castelo Branco
Institute of Mathematics and Computer Sciences (ICMC), University of São Paulo (USP), São Carlos, São Paulo, Brazil

Willie L. Brown, Jr.
Department of Engineering and Aviation Sciences, University of Maryland Eastern Shore, Princess Anne, MD, USA

Joseph Bugeja,
Internet of Things and People Research Center and Department of Computer Science and Media Technology, Malmö University, Malmö, Sweden

Pratik Chaturvedi
Defence Terrain Research Laboratory, Defence Research and Development Organization, New Delhi, India

and

Applied Cognitive Science Laboratory, Indian Institute of Technology Mandi, Kamand, India

Ibibia K. Dabipi
Department of Engineering and Aviation Sciences, University of Maryland Eastern Shore, Princess Anne, MD, USA

Paul Davidsson
Internet of Things and People Research Center and Department of Computer Science and Media Technology, Malmö University, Malmö, Sweden

João Vitor de Carvalho Fontes
São Carlos School of Engineering (EESC), University of São Paulo (USP), São Carlos, São Paulo, Brazil

Varun Dutt
School of Computing and Electrical Engineering; School of Humanities and Social Sciences, Indian Institute of Technology Mandi, Himachal Pradesh, India

Jeanette Eriksson
Department of Computer Science and Media Technology, Internet of Things and People (IoTaP) Research Center, Malmö University, Malmö, Sweden

Akaa Agbaeze Eteng
Department of Electronic and Computer Engineering, University of Port Harcourt, Port Harcourt, Nigeria

Lucas Finco
Principal Consultant, Strategain, New York, NY, USA

Bernard Fong
School of Public Health, Auckland University of Technology, New Zealand

A. C. M. Fong
Department of Computer Science, Western Michigan University, Kalamazoo MI, USA

Virginia N. L. Franqueira
Department of Electronics, Computing & Mathematics, University of Derby, Derby, UK

Mário Marques Freire
Instituto de Telecomunicações and Department of Computer Science, Universidade da Beira Interior, Covilhã, Portugal

Daniel Happ
Technische Universität Berlin, Telecommunication Networks Group (TKN), Berlin, Germany

Pedro Ricardo Morais Inácio
Instituto de Telecomunicações and Department of Computer Science, Universidade da Beira Interior, Covilhã, Portugal

Andreas Jacobsson
Internet of Things and People Research Center and Department of Computer Science and Media Technology, Malmö University, Malmö, Sweden

Kai Jakobs
Computer Science Department, RWTH Aachen University, Aachen, Germany

Venkata Uday Kala
School of Engineering, Indian Institute of Technology Mandi, Himachal Pradesh, India

Jonny Karlsson
Department of Business Management and Analytics, Arcada University of Applied Sciences, Helsinki, Finland

Sudhakar Kumar
School of Computing and Electrical Engineering, Indian Institute of Technology Mandi, Himachal Pradesh, India

Chee Yen Leow
Wireless Communication Centre, Universiti Teknologi Malaysia, Johor, Malaysia

C. K. Li
Add-Care Ltd., Hong Kong

Naresh Mali
School of Engineering, Indian Institute of Technology Mandi, Himachal Pradesh, India

Manuel Meruje
Instituto de Telecomunicações and Department of Computer Science, Universidade da Beira Interior, Covilhã, Portugal

Daniel Minoli
IoT Division, DVI Communications, New York, NY, USA

Benedict Occhiogrosso
Intellectual Property Division, DVI Communications, New York, NY, USA

Daniel Fernando Pigatto
Graduate Program in Electrical and Computer Engineering (CPGEI), Federal University of Technology Paraná (UTFPR), Curitiba, Paraná, Brazil

Alex Sandro Roschildt Pinto
Federal University of Santa Catarina (UFSC), Blumenau, Santa Catarina, Brazil

Göran Pulkkis
Department of Business Management and Analytics, Arcada University of Applied Sciences, Helsinki, Finland

Sharul Kamal Abdul Rahim
Wireless Communication Centre, Universiti Teknologi Malaysia, Johor, Malaysia

Hejamadi Raghav Rao
Department of Information Systems and Cyber Security, University of Texas San Antonio, San Antonio, TX, USA

Shahid Raza
RISE SICS Security Lab, Kista, Stockholm, Sweden

Mariana Rodrigues
Institute of Mathematics and Computer Sciences (ICMC), University of São Paulo (USP), São Carlos, São Paulo, Brazil

Jason M. Rosenberg
University at Buffalo School of Management, The State University of New York at Buffalo, Buffalo, NY, USA

Nancy L. Russo
Department of Computer Science and Media Technology, Internet of Things and People (IoTaP) Research Center, Malmö University, Malmö, Sweden

Musa Gwani Samaila
Instituto de Telecomunicações and Department of Computer Science, Universidade da Beira Interior, Covilhã, Portugal

and

Centre for Geodesy and Geodynamics, National Space Research and Development Agency, Toro, Bauchi State, Nigeria

Detlef Schoder
Department of Information Systems and Information Management, University of Cologne, Köln, Germany

Sajjan Shiva
Computer Science Department, The University of Memphis, Memphis, TN, USA

John Shu
Department of Information Systems and Cyber Security, University of Texas San Antonio, San Antonio, TX, USA

Jan Sliwa
Department of Engineering and Information Technology, Bern University of Applied Sciences, Bern, Switzerland

James Smith
Computer Science and Creative Technologies (FET), University of the West of England (UWE), Bristol, England, United Kingdom

Tiffany Y. Tang
Wenzhou-Kean Autism Research Network, Assistive Technology Research and Development Center, Department of Computer Science, Wenzhou-Kean University, Zhejiang Province, China

Kamal Kishore Thakur
Computer Science and Engineering Department, Thapar Institute of Engineering and Technology, Patiala, India

Marco Tiloca
RISE SICS Security Lab, Kista, Stockholm, Sweden

Shambhu Upadhyaya
University at Buffalo, The State University of New York
Department of Computer Science and Engineering, Buffalo, NY, USA

Magnus Westerlund
Department of Business Management and Analytics, Arcada University of Applied Sciences, Helsinki, Finland

Alexander Willner
Fraunhofer FOKUS, Software-based Networks (NGNI), Berlin, Germany

and

Technische Universität Berlin, Next Generation Networks (AV), Berlin, Germany

Pinata Winoto
Wenzhou-Kean Autism Research Network, Assistive Technology Research and Development Center, Department of Computer Science, Wenzhou-Kean University, Zhejiang Province, China

Srishti Yadav
School of Computing and Electrical Engineering, Indian Institute of Technology Mandi, Himachal Pradesh, India

Lei Zhang
Department of Engineering and Aviation Sciences, University of Maryland Eastern Shore, Princess Anne, MD, USA

Part I

Concepts and Perspectives

1

Introduction to the Internet of Things

Detlef Schoder

Department of Information Systems and Information Management, University of Cologne, Köln, Germany

1.1 Introduction

Early in 1926, Nikola Tesla envisioned a "connected world." He told *Colliers Magazine* in an interview (Kennedy, 1926):

> "When wireless is perfectly applied, the whole Earth will be converted into a huge brain, which in fact it is, all things being particles of a real and rhythmic whole [. . .] and the instruments through which we shall be able to do this will be amazingly simple compared with our present telephone. A man will be able to carry one in his vest pocket."

Kevin Ashton was the first to use the term Internet of Things (IoT) in 1999 (Ashton, 2009) in the context of supply chain management with radio frequency identification (RFID)-tagged or barcoded items (things) offering greater efficiency and accountability to businesses. As Ashton wrote in the *RFID Journal* (June 22, 2009):

> "If we had computers that knew everything there was to know about things – using data they gathered without any help from us – we would be able to track and count everything, and greatly reduce waste, loss and cost. We would know when things needed replacing, repairing or recalling, and whether they were fresh or past their best."

In the same year, Gershenfeld (1999) published his work "When Things Start to Think," in which he envisioned the evolution of the World Wide Web as being a state in which "things start to use the Net so that people don't need to." ATMs could be considered as one of the first smart objects, which went online as early

Internet of Things A to Z: Technologies and Applications, First Edition. Edited by Qusay F. Hassan.
© 2018 by The Institute of Electrical and Electronics Engineers, Inc. Published 2018 by John Wiley & Sons, Inc.

as 1974. In addition, early examples of various prototype devices include vending machines in the 1980s performed by the Computer Science Department of Carnegie Mellon University. Since then, understanding of the possible breadth of IoT has become much more inclusive, comprising a wide range of application domains, including health care, utilities, transportation, and so on, as well as personal, home, and mobile application scenarios (Gubbi et al., 2013; Sundmaeker et al., 2010). More recently, the "Industrial Internet of Things" (IIoT) has further expanded the scope of IoT (see Section 1.2.2 and Chapter 11). With IoT, a world of networked, "intelligent," or "smart" objects (Ashton, 2009; Weiser, 1991; Weiser and Brown, 1996; Lyytinen and Yoo, 2002; Aggarwal et al., 2013; Gubbi et al., 2013; Mattern and Flörkemeier, 2010; Atzori et al., 2014; Chui et al., 2010) is envisioned. Recently, novel extensions of IoT have emerged, which include not only physical objects but also virtual objects[1] (which may blur the core concept of IoT that predominately focuses on physical things and objects). The common denominator of these varied conceptions of IoT is that "things" are expected to become active elements in business, information, and social processes.

If one recognizes the broad spectrum of application scenarios, the more general term "Net" would be more adequate than "Internet," since not all communication occurs via the Internet. Communication also does not exclusively occur between things/devices, but also between things and people. So, it would be more appropriate to use the terms the "Internet of Everything"[2] or "Net of Everything" instead of "Internet of Things."

As the most well-known visionary of the computerized and interlinked physical world, Mark Weiser asserts that a connected world of things is designed to help people with their activities in an unobtrusive manner. Interaction occurs with everyday—but computationally augmented—artifacts through natural interactions, our senses, and the spoken word (Weiser, 1991). In the course of miniaturization, the increasingly smaller technical components will be embedded into physical components, with as little intrusiveness for users as possible or without attracting attention at all. For example, miniaturized computers (or components thereof) and wearables with sensors are directly

1 Increasing numbers of physical objects/things are beginning to be seen in digital format or even only in digital format. Examples of this include books, maps, e-tickets of any sort, business cards, electronic purses, and so on. Consequently, not all "virtual objects" that are currently used are digital models of physical objects (pendants), but rather these objects "stand on their own" with no physical counterpart. Virtual objects can be defined as a digital element with a specific purpose, comprised of data and capable of performing actions (Espada et al., 2011).

2 The term "Internet of Everything" was coined by Cisco Systems and basically refers to applying the IoT model to everything, thus creating new capacities and smart processes in virtually every field. Cisco calls it the connection of "people, process, data and things." The Internet of Everything may be perceived as a variation or extension of IoT, subtly distinguishing itself by emphasizing the connection of people to things.

incorporated into pieces of clothing. In his essay in 1991, "The Computer for the 21st Century," Mark Weiser first expressed this vision while he was a Chief Technologist at the Xerox Palo Alto Research Center in the late 1980s (Weiser, 1991, 1993; Weiser and Brown, 1996; Weiser et al., 1999). Since then, this work ranks among the most cited academic papers in related academic disciplines that envision a connected world of everyday things. This vision and the related developments are referred to by Weiser as "Ubiquitous Computing" (also known as "Ubicomp"). Since its conceptual inception more than 25 years ago, many more related and modified concepts have emerged, including pervasive, nomadic, calm, invisible, universal, and sentinel computing, as well as ambient intelligence.[3] The Cluster of European Research Projects on the Internet of Things (CERP-IoT) blend together building blocks that derive from the aforementioned concepts and emphasize the symbiotic interaction of the real and physical with the digital and virtual world. From their perspective, physical objects have virtual counterparts representing them, which translate them into computable parts of the physical world. The CERP-IoT vision has recently become even more comprehensive by incorporating issues of Social Media, anticipating massive user interaction with things and linking to additional information regarding identity, status, location, or any other business, social, or privately relevant information (Chapter 1 of Uckelmann et al., 2011). Essentially, ITU (2005) defines IoT as a concept that allows people and things to be connected anytime, anyplace, with anything and anyone (and adding—according to CERP-IoT, 2009—ideally using any path/network and any service). Another line popularized by CISCO asserts a simple concept: The IoT is born when more things are connected via the Internet as human beings. As such, the advent of IoT may be dated around 2008/2009 (Evans, 2011) or 2011 (Gubbi et al., 2013). According to the International Data Corporation (IDC)'s Worldwide Internet of Things Forecast, 2015–2020, 30 billion connected (autonomous) things are predicted to be part of the IoT by 2020. Another estimate anticipates approximately 1000 devices per person by 2025 (Sangiovanni-Vincentelli, 2014).

IoT is at the center of overlapping Internet-oriented (middleware), things-oriented (sensors), and semantic-oriented (knowledge) visions (Atzori et al., 2010). Specifically, (i) Internet-oriented, which emphasizes the networking paradigm and exploiting the established IP-based networking infrastructure, in order to achieve an efficient connection between devices, and on developing lightweight protocols in order to meet IoT specifics (see Section 1.5.2); (ii) things-oriented, which focuses on physical objects and on finding means that are able to identify and integrate them with the virtual (cyber) world; and

3 For a discussion on similarities and differences, see, for example, Aarts et al. (2001); for a summary of more than 20 years of the "Ubicomp" vision, see, for example, Cáceres and Friday (2012).

(iii) semantic-oriented, which aims to utilize semantic technologies, making sense of objects and their data to represent, store, interconnect, and manage the enormous amount of information provided by the increasing number of IoT objects (Atzori et al., 2010; Borgia, 2014).

As IoT continues to evolve, its comprehensive definition is also likely to develop.[4] Accordingly, the IEEE IoT initiative gives its community members an opportunity to contribute to the definition of the IoT (IEEE, 2015, 2017). The document presents two definitions, one for small-scale scenarios: "An IoT is a network that connects uniquely identifiable 'Things' to the Internet. The 'Things' have sensing/actuation and potential programmability capabilities. Through the exploitation of unique identification and sensing, information about the 'Thing' can be collected and the state of the 'Thing' can be changed from anywhere, anytime, by anything." The second definition is for large-scale scenarios: "Internet of Things envisions a self-configuring, adaptive, complex network that interconnects 'Things' to the Internet through the utilization of standard communication protocols. The interconnected things have physical or virtual representation in the digital world, sensing/actuation capability, a programmability feature and are uniquely identifiable. The representation contains information including the thing's identity, status, location or any other business, social or privately relevant information. The things offer services, with or without human intervention, through the exploitation of unique identification, data capture and communication, and actuation capability. The service is exploited through the use of intelligent interfaces and is made available anywhere, anytime, and for anything taking security into consideration."

Incorporating various perspectives while revealing its nucleus, we may consolidate and define:

> IoT is a world of interconnected things which are capable of sensing, actuating and communicating among themselves and with the environment (i.e., smart things or smart objects) while providing the ability to share information and act in parts autonomously to real/physical world events and by triggering processes and creating services with or without direct human intervention.

We intentionally leave out whether this "big plot" will necessarily be realized on standard communication protocols or a unified framework. Although a unified framework would certainly be optimal, it may not be necessary or even achievable given the dimensionalities and complexities of a likely very highly heterogeneous computerized world of interconnected things.

In order to better structure the scale and scope of IoT, this chapter provides an introductory overview and briefly outlines the conceptual core ideas as laid out

---

4 A broad range of IoT definitions can be found in Minoli (2013).

prior to IoT with "Ubiquitous Computing." The chapter covers not only technical but also nontechnical issues of IoT.

1.2 Internet of Things Concepts

With technical advancements, our interaction with information systems is changing, both at work and during leisure time. Information, sensor, and network technology are becoming increasingly small, more powerful, and more frequently used. People no longer only encounter information technology at common points in their lives, such as in offices or at desks, but as information and communication infrastructures, which are present in increasing areas of everyday life. These infrastructures are characterized by the fact that they not only include classic devices, for example, PCs and mobile phones, but that information and communication technology is also embedded in objects and environments.

The Ubiquitous Computing vision of Mark Weiser implies that computers, as we currently know them, "disappear," or, more precisely, move into the background. Everyday objects and our immediate environment then assume the tasks and abilities of computers (Weiser and Brown, 1996). In his seminal paper, Weiser describes this as follows: "The most profound technologies are those that disappear. They weave themselves into the fabric of everyday life until they are indistinguishable from it" (Weiser, 1991). Through the physical embedding of IT, everyday objects and our everyday environment become "smart," that is, capable of processing and providing information, but not necessarily intelligent in the sense of human cognitive intelligence. In another highly regarded article, Weiser together with Brown introduced the notion of "Calm Computing." They also refer to a connected world full of computers. However, only in cases of service provision or when a need exists for interaction do those computers or their respective services become "visible"; at other times, those capabilities are "calm" in the background, and not intrusive or even visible to the users (Weiser and Brown, 1996).

The core concepts comprising IoT, as well as related concepts and models, will be presented in the following sections.

1.2.1 Core Concepts: Smart Objects and Smart Environments

A smart object is a physical object in which a processor, data storage system, sensor system, and network technology are embedded (Poslad, 2009; Kortuem et al., 2010; Sánchez López et al., 2011). Some smart objects can also affect their environment by means of actuators. In principle, all physical objects can be turned into smart objects, for example, conventional everyday objects such as pens,[5] wristwatches (there are numerous wristwatch models with sensors and

5 A well-known example of a computerized version of a pen is the Anoto Pen, see www.anoto.com.

processors, for example, to measure the heart rate or to determine geographic position), or automobiles (more recently, autonomous automobiles). In an industrial context, it could be a machine or the product to be manipulated. Smart objects may also be anywhere. In fact, there are almost no restrictions regarding domains: consumer electronic devices, home appliances, medical devices, cameras, and all sorts of sensors and data-generating devices. Most smart objects have a user interface and interaction capabilities to communicate with the environment or other devices (e.g., displays). The capability of smart objects to communicate with other objects and with their environment is a core component of IoT. In line with this is the idea that specific information can be retrieved via any networked smart object, which is uniquely identified and localized, and may have its "own home page," that is, unique address. Today, one can take advantage of a broad range of fairly inexpensive, tiny, and relatively powerful components, including sensors, actuators, and single board computers (SBC), to enrich physical things and connect them to the Internet. SBCs, such as Raspberry Pi, BeagleBone Black, and Intel Edison Open, as well as open-source electronics, such as Arduino, which entered the market between 2005 and 2008, catalyzed millions of new ideas and projects. Creating and collecting data about the status of physical objects may establish the basis for interesting home and office automation projects, education, and leisure activities with real-time visualizations of information generated from data "on the go" (Baras and Brito, 2017). Moreover, one can utilize the remote networks of intelligent devices deployed somewhere else.

Tightly coupled to "smart objects" is the concept of "smart environments." One definition emphasizes the physical extent to which smart objects are deployed and interacting. A compilation of smart objects within a given space, such as a closed space (automobile, house, room) or an outside area, for example, a district or an entire city (i.e., a smart city; see Chapter 12), turns a common environment into a smart one. Another definition asserts that sensors are the key factor in a smart environment. Essential for a smart environment is the context information gathered by sensors in order to provide adapted applications and services. Weiser et al. (1999) defined a smart environment as "the physical world that is richly and invisibly interwoven with sensors, actuators, displays, and computational elements, embedded seamlessly in the everyday objects of our lives, and connected through a continuous network."

1.2.2 Related Concepts: Machine-to-Machine Communications, Industrial Internet of Things, and Industry 4.0

IoT is not a construct that has appeared suddenly or without precursors. Technological forerunners and various conceptualizations exist prior to the relatively new "IoT" label, for example, machine-to-machine (M2M) communications. In addition, recent derivatives exist, for example, the Industrial

Internet of Things and Industry 4.0. The subsequent sections attempt to discern their similarities and differences, and how these concepts relate to each other.

1.2.2.1 Machine-to-Machine Communications

M2M communications refers to direct wired or wireless communication between devices using any communications channel that does not necessarily require direct human intervention (ETSI, 2010). As such, M2M can be viewed as the forerunner of IoT. M2M communication can include industrial production facilities, enabling a sensor or meter to communicate the data that it records (e.g., temperature, throughput, and inventory level) to application software that can further process them (e.g., adjusting an industrial process based on technical parameters, such as temperature or triggering new processes, such as placing orders to replenish inventory). Such communication was aimed at monitoring remote machines from which data were received, processed at some central station, and eventually relayed back to those machines with adjusted parameters, if necessary. A core motivation for many organizations is to reduce service management costs through remote diagnostics, remote troubleshooting, remote updates, and other remote capabilities that reduce the need to deploy field service personnel (Polsonetti, 2014). IoT accommodates the same devices/assets/machines as M2M applications, but also very small (low-power), personal, and inexpensive devices with sometimes very limited functionality that might not be able to justify a dedicated M2M hardware module. Although IoT and M2M communications have remote access to machines, or in more general terms "devices," in common, there are no other major similarities. For example, traditional M2M solutions typically rely on point-to-point communications using embedded hardware modules and dedicated protocols. In contrast, IoT solutions depend predominantly on IP-based networks to interface device data to a cloud or middleware platform primarily using common/open protocols (in order to ensure maximum interoperability, in the sense of a remote device connected to some central hub, as well as particular interoperability among the devices themselves). Another difference is that M2M solutions offer remote access to machine data that are traditionally targeted at point solutions in service management applications. In the past, these data are rarely, if ever, integrated with enterprise applications to help improve overall business performance. Finally, IoT-based data delivery increasingly involves cloud services enabling access by any sanctioned enterprise application, whereas M2M typically employs direct point-to-point communication. The cloud-based architecture also makes IoT inherently more scalable, eliminating the need for incremental hard-wired connections or SIM card installations. M2M is often referred to as "plumbing," while IoT is viewed as a universal enabler (Polsonetti, 2014). It could be argued that the conceptual boundaries and visions of IoT and M2M have become increasingly overlapping. Indications of this include that more recent M2M communications have evolved into a system of networks that transmits data to personal appliances. In this sense, M2M communication is taking increasingly advantage of the expansion of IP networks globally by switching from point-to-point proprietary style connections to IP-based multipoint communications. We may conclude that the focus of M2M issues tends to be more on the technical infrastructure layer. In contrast, the emerging IoT possesses much greater scope. IoT calls for the integration of device and sensor data with business intelligence, analytics, and other enterprise applications in order to achieve numerous benefits throughout manufacturing enterprises with a strong emphasis on improving products, processes, and business models.

1.2.2.2 Industrial Internet and Industry 4.0

A broad segmentation of IoT comprises (i) a consumer-oriented perspective, including smart phones, connected automobiles, smart TVs, and wearables, and (ii) an industrial perspective. The latter includes, for example, power grids and power plants, transportation, wind turbines, and industrial equipment (Jeschke et al., 2016). The straightforward analogy is to translate objects within an industrial (production) context into smart objects. Production facilities, such as tools, conveyors, and even the products to be manipulated or built will become smart objects as conceptually defined here. In line with this perception, a "common factory" turns into a smart factory. It could be asserted that this may constitute the foundation for a fundamental new way of coordinating and producing goods. These expectations coalesce in labeling the upcoming era the "Fourth Industrial Revolution." Accordingly, the term Industrial Internet of Things or just Industrial Internet[6] is typically used. Moreover, in the context of IIoT, the term is often employed synonymously with Industry 4.0 or the original German term "Industrie 4.0,"[7] which is a label for various government initiatives in Germany (World Economic Forum, 2015). The differences between the terms or initiatives mainly concern stakeholders, geographical focus, and representation (Bledowski, 2015). IIoT also semantically describes a technological movement, while Industry 4.0 is more associated with expected economic impacts. The Industry 4.0 vision is anticipated to be realized by IIoT (Jeschke et al., 2016).

The proclaimed implications and benefits of IIoT are manifold and are rooted—as outlined in Section 1.6—in "derived qualities" from modern ICT, in particular context sensitivity, adaptability, proactivity, and increased data quality. Eventually, these derived qualities may help to achieve greater resource efficiency, shorter time-to-market, higher value products, and new services (Jeschke et al., 2016).

---

6 General Electric, which coined the term "Industrial Internet" as a holistic new application, joined with AT&T, Cisco, Intel, and IBM in 2014 to set up the Industrial Internet Consortium http://www.iiconsortium.org/.

7 The term "Industrie 4.0" is considered to have first been used at the Hannover Messe, Germany, in 2011.

1.3 Who Works on the Internet of Things?

A truly connected world in terms of IoT has not yet been fully achieved. However, a large number of organizations and alliances across industry, academia, and various levels of government are working on IoT and closely related streams, often in parts under different labels. This section compiles prominent national and international representatives from governmental bodies, academia, and industry (Rose et al., 2015; Gubbi et al., 2013; for standardizing, see Chapter 7 of this book). Supported by the European Commission 7th Framework program, a significant number of initiatives and projects have been funded, such as the Internet of Things Architecture (IoT-A) project and the Internet of Things-Initiative (IoT-i). The Internet of Things European Research Cluster (IERC), which coordinates ongoing activities in the area of IoT across Europe,[8] is a major organization in this area. The online companion website of the special issue on interoperability of IoT lists, among other collections, EU-funded projects started from January 1, 2016 and international IoT-projects.[9] The Alliance for Internet of Things Innovation (AIOTI) was launched by the European Commission to support the development of a European IoT ecosystem, including standardization policies.[10] Their working groups correspond to application areas of IoT, including smart living environments for aging well, smart farming and food safety, wearables, smart cities, smart mobility, smart water management, smart energy, and smart buildings and architecture. ETSI with its Connecting Things program is developing standards for data security, data management, data transport, and data processing related to potentially connecting billions of smart objects into a single communications network.[11] IEEE has a dedicated IoT initiative and clearinghouse of information for the technical community involved in research, implementation, application, and usage of IoT technologies.[12] The Internet Engineering Task Force (IETF) is the Internet's premier standards-setting body, and has an IoT directorate that coordinates related efforts across its working groups, reviews specifications for consistency, and monitors IoT-related activities in other standards groups.[13] As a major need exists for consensus around IoT technical issues, the Internet Protocol for Smart Objects (IPSO) Alliance was created. It has more than 60 member companies from leading technology communications and energy companies working together with standards bodies, such as IETF, IEEE, and ITU. China has prominently set IoT on its strategic agenda, including

---

8 http://www.internet-of-things-research.eu/.

9 https://www.computer.org/web/ computingnow/archive/interoperability-in-the-internet-of-things-december-2016-introduction?lf1=8161623904b709516091377b61283885.

10 https://ec.europa.eu/digital-agenda/en/alliance-internet-thingsinnovation-aioti.

11 http://www.etsi.org/technologies-clusters/clusters/connectingthings.

12 http://iot.ieee.org/.

13 https://trac.tools.ietf.org/area/int/trac/wiki/.

state-based and industry-funded initiatives (e.g., "Internet of Things Union Sensing China" in Wuxi). In addition, the Industrial Internet Consortium (IIC) works on an industrial grade IoT architectural framework and released a reference architecture for IoT in 2015.[14] In addition, literally all major national and supranational standardization bodies work on IoT issues, including ISO/IECJTC-1.[15] For example, the ITU has set up a "Study Group 20."[16] The Manufacturers Alliance for Productivity and Innovation (MAPI) is developing Industry 4.0 for industrial applications of IoT.[17] OASIS is developing open protocols to ensure interoperability for IoT, especially based on Message Queuing Telemetry Transport (MQTT) as its messaging protocol of choice for IoT.[18] The Online Trust Alliance, a group of security vendors, has developed a draft trust framework for IoT applications, focused on security, privacy, and sustainability.[19] The Open Management Group is a technical standards consortium that is developing several IoT standards, including Data Distribution Service (DDS) and Interaction Flow Modeling Language (IFML) along with dependability frameworks, threat modeling, and a unified component model for real-time and embedded systems.[20] At the same time, large-scale initiatives are underway in Japan, Korea, the United States, and Australia where industry, associated organizations, and government agencies are collaborating on various programs, such as smart city initiatives, smart grid programs incorporating smart metering technologies (in some European countries, smart metering has become legally mandated for new buildings), and the implementation of high-speed broadband infrastructures (e.g., in Germany).

1.4 Internet of Things Framework

The brief discussion in the following paragraph of technical, economic, and social issues reveals that IoT encompasses a wide area of topics and disciplines. Aimed at structuring the field, we propose the following four-layer "Internet of Things Framework" (Figure 1.1).

At the core, modern information and communication technologies form the technical foundation of IoT (covered in layer 1). IoT generates a network of unambiguously identifiable physical objects (things). Networking, and thus also the ability to communicate, does not only refer to human participants but also to the objects (or things) involved. These things are equipped with miniaturized

14 http://www.industrialinternetconsortium.org/.
15 http://www.iso.org/iso/internet_of_things_report-jtc1.pdf.
16 http://www.itu.int/en/ITU-T/studygroups/2013-2016/20/Pages/default.aspx.
17 https://www.mapi.net/research/publications/industrie-4-0-vsindustrial-internet.
18 https://www.oasis-open.org/committees/tc_cat.php?cat=iot.
19 https://otalliance.org/initiatives/internet-things.
20 http://www.omg.org/hottopics/iot-standards.htm.

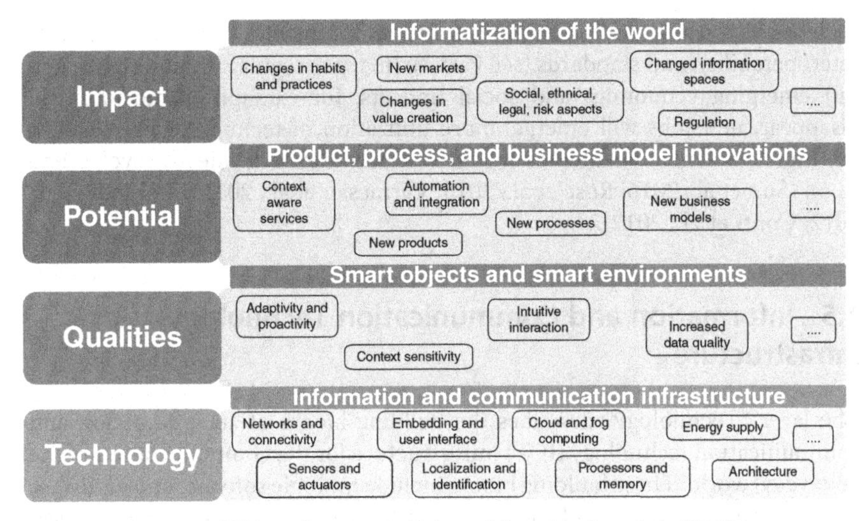

Figure 1.1 Internet of Things framework. (Adapted from Laudon et al., 2016.)

processors and actuators, for example, mechanical elements, temperature controllers, and audio or video output devices that can be utilized to control the objects and the environment. This allows for adapting objects and environments to our needs, interacting with the situation, and the provisioning of information and services according to specific situational requirements, that is, they become "smart objects" and "smart environments" (covered in layer 2). The automatic identification via RFID is often regarded as the basis for IoT. Sensors and actuators expand functionality by capturing states and the execution of actions or effects on reality.

This results in potential for new services, including consumer products as well as new business processes and business models (covered in layer 3). Products for consumers, for example, may hold and provide a large amount of information and can also offer customers with additional context-specific services during the post-sales phase. IoT also provides a higher level of data quality for business processes, enabling organizations to respond more rapidly and properly to events, and may contribute to improved efficiency, accuracy, and economic benefits (Sun et al., 2016). These potentials will lead to various product, process, and business model innovations. As these innovations affect our everyday lives, they have a wide impact on individuals, society, markets, and companies (covered in layer 4). On the one hand, companies are under pressure to adapt to changed value creation and market structures, as well as changing customer needs. On the other hand, innovative companies are given the opportunity to develop new products, processes, and business models that enable them to better meet the needs of their customers, and thus participate in the design of a computerized world. The effects are manifold and not always solely positive for everyone. Indeed, IoT poses severe challenges to companies, individuals, and societies as a whole. Major challenges and issues include (i) security, privacy, interoperability, and standards (see Part 3); (ii) legal, regulatory, and rights; and (iii) emerging economies and social impacts, for example, some jobs will disappear, new jobs will emerge, more utilization of technology may lead to less human and manual interaction, different forms of social life may evolve, and so on (Sun et al., 2016; Rose et al., 2015; Vermesan et al., 2011; Miorandi et al., 2012; Conti et al., 2012).

1.5 Information and Communication Technology Infrastructure

The layer "Technology" describes the building blocks of an information and communication technology (ICT) infrastructure for the computerization of the (everyday) world. These building blocks include multiple software and hardware components, as well as highly developed and novel technologies. They are used to connect virtual information about or from things to the physical real world. These include technologies for computing, storage, embedding, and mobile and wireless networking, as well as sensors and actuators. Furthermore, improved methods for energy supply, identification, and localization constitute basic elements of IoT. Typically, in order to deal with the enormous resultant complexity, a layered approach is taken.

The next sections describe the building blocks of the technology layer, which are a foundational dimension of IoT.

1.5.1 Architecture and Reference Models

Especially for the technology layer, the extant literature covers a multitude of architectural proposals, reference models, and technical descriptions of the current or envisioned state of IoT.[21] Figure 1.2 presents a high-level view in terms of a three-layered stack of IoT-relevant technologies: (i) the thing or device layer, (ii) the connectivity layer, and (iii) the application layer. At the device layer, IoT-specific hardware such as sensors, actuators, memory, and processors are added to existing core hardware components, and embedded software is intended to manage and operate the functionality of the particular physical thing. At the connectivity layer, communication protocols enable communication between things and connected infrastructure, for example,

21 A vast literature exists on the architecture of IoT. Overviews (and alternative perceptions of such architectures) are provided, for example, Porter and Heppelmann (2015), Gubbi et al. (2013), Borgia (2014), Karzel et al. (2016), Wortmann and Flüchter (2015), and Weyrich and Ebert (2016). Edited multipaper volumes have several additional illustrations, for example, Vermesan and Friess (2016).

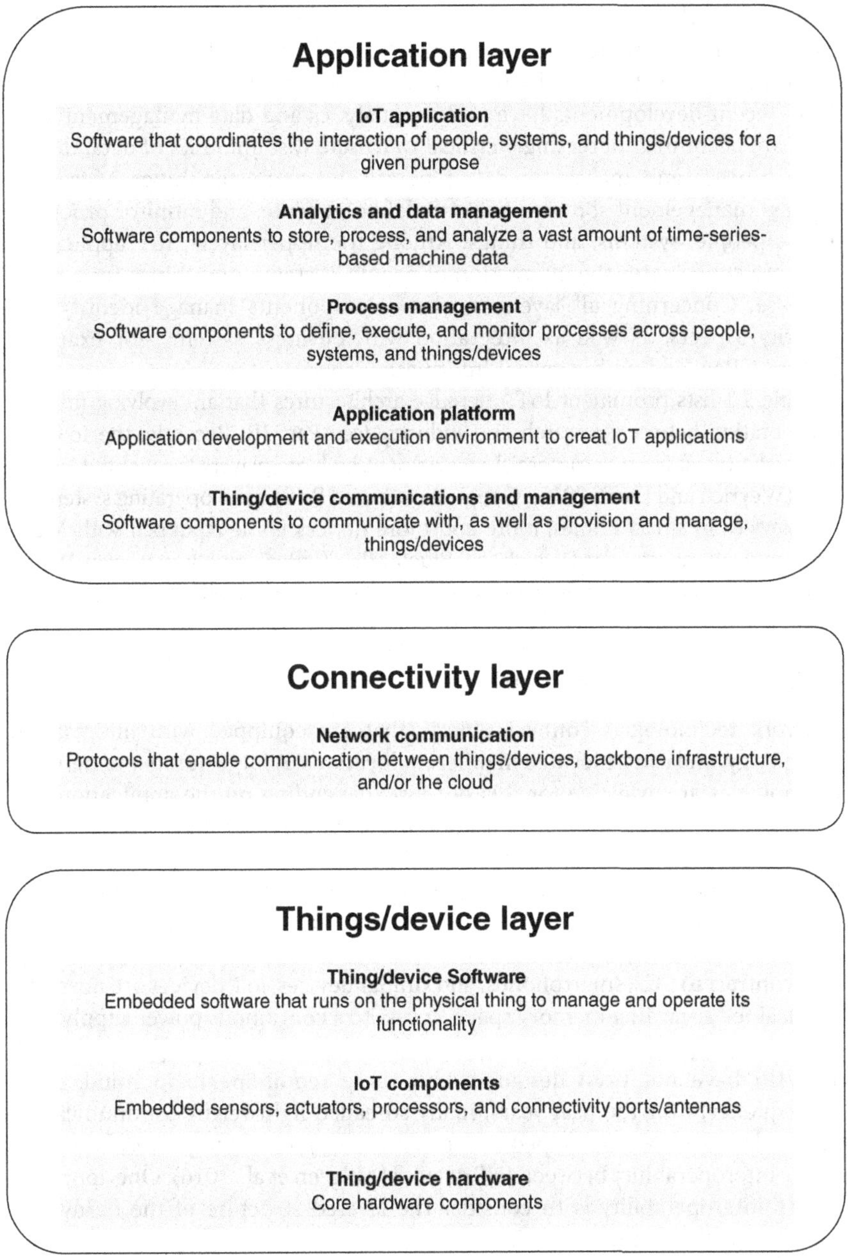

Application layer

IoT application
Software that coordinates the interaction of people, systems, and things/devices for a given purpose

Analytics and data management
Software components to store, process, and analyze a vast amount of time-series-based machine data

Process management
Software components to define, execute, and monitor processes across people, systems, and things/devices

Application platform
Application development and execution environment to creat IoT applications

Thing/device communications and management
Software components to communicate with, as well as provision and manage, things/devices

Connectivity layer

Network communication
Protocols that enable communication between things/devices, backbone infrastructure, and/or the cloud

Things/device layer

Thing/device Software
Embedded software that runs on the physical thing to manage and operate its functionality

IoT components
Embedded sensors, actuators, processors, and connectivity ports/antennas

Thing/device hardware
Core hardware components

Figure 1.2 High-level view of an IoT architecture.

through cloud services. Accordingly, at the IoT application layer, device communication and related functionality is provided, while an application platform enables the development and execution of IoT applications. As more recent developments have proven, analytics and data management software are becoming increasingly critical to handle vast amounts of data, that is, store, process, and analyze the data generated by connected things. Moreover, process management software helps to define, execute, and monitor processes across people, systems, and things. Among the upper layers, IoT application software coordinates the interaction of people, systems, and things for a given purpose. Concerning all layers, software components manage identity and security aspects, as well as integration with business systems, for example, ERP or CRM, and with external information sources.

Table 1.1 lists prominent IoT reference architectures that are evolving in close collaboration between research and industry (see Part II). Recently, the IoT has received a boost from commercial engagement by large players throughout industries (Weyrich and Ebert, 2016): Google announced Brillo as an operating system for IoT devices in smart homes; more and more devices come equipped with M2M communications standards such as Bluetooth, ZigBee, and low-power Wi-FI; Microsoft has announced that Windows will support embedded systems and so on.

1.5.2 Networks and Connectivity

Network technologies connect objects that are equipped with information technology, and can be located in different locations. A large number of network technologies are available for this purpose, depending on the application. An application-related distinction feature is the scaling of the range. It ranges from global networks (satellites) over regional and local networks to so-called personal, body, and intrabody area networks. Personal area networks (PANs) can, for example, network via WLAN devices, typically in an area of up to 10 m^2 around one or two people.

In contrast to PCs, smartphones, and similar devices, loT devices are normally constrained regarding memory space, access to a continuous power supply, and processing capacity. Traditional protocols (in particular, the protocol stack TCP/IP) have not been designed with these requirements in mind. As a consequence, over the past years, many so-called lightweight communication protocols have been developed on virtually all layers of the protocol stack to create interoperability between IoT devices (Ahlgren et al., 2016). One approach to IoT interoperability is to consider the layered structure of the hardware/software stack (Fortino et al., 2016):

- The lower layers (according to the OSI model, the physical and data link layers; in the non-OSI context, sometimes labeled as the device layer) are aimed at seamlessly integrating new devices into the existing IoT ecosystem.

Table 1.1 Examples for IoT reference architectures.

Reference model	Founders	Latest release	IoT domain(s)	Viewpoints[a]	Brief description
Internet of Things—Architecture (IoT-A)	NEC, CFR, ALBLF, SAP UniS, HEU, HSG, CEA, SIEMENS, ALUBE, FhG IML, and CATTID,	July 2012	Any	Functional and information	The "Internet of Things Architecture" (IoT-A) is an EU project. Based on a system requirement process, the outcomes cover a detailed architecture including the definition of a range of key components. It centers on a functional and an information perspective. http://www.meet-iot.eu/deliverables-IOTA/D1_3.pdf
Industrial Internet Reference Architecture (IIRA)	AT&T, Cisco, General Electric, IBM, and Intel	January 2017	Manufacturing	Business, usage, implementation, and functional	The IIRA is a standards-based architectural template and methodology. It is meant to enable Industrial Internet of Things system architects to design their own systems based on a common framework and concepts. http://www.iiconsortium.org/IIC_PUB_G1_V1.80_2017-01-31.pdf
Reference Architecture Model Industrie 4.0 (RAMI 4.0)	The German Electrical and Electronic Manufacturers' Association, and its partners	July 2015	Manufacturing and Logistics	Business, functional, information, communication, integration, asset, lifecycle/value chain, and hierarchy	The RAMI 4.0 is a reference architecture taking into account particularities of Industrie 4.0/smart factories, which started in Germany and today is driven by all major companies and foundations in a large number of industry sectors. The RAMI 4.0 consists of a three-dimensional coordinate system that describes aspects of Industrie 4.0. www.zvei.org/en/association/specialist-divisions/automation/Pages/default.aspx
Cisco's Internet of Things Reference Model	Cisco	June 2014 (Draft)	Any	Any	The proposed IoT reference model is comprised of seven levels standardizing the concept and terminology surrounding IoT. From physical devices and controllers at level 1 to the collaboration and processes at level 7, the reference model sets out the functionalities required and concerns. http://cdn.iotwf.com/resources/71/IoT_Reference_Model_White_Paper_June_4_2014.pdf

a) A viewpoint is a set of conventions, templates, and patterns for creating a kind of view. It specifies the stakeholders whose interests are reflected in the viewpoint and the principles, guidelines, and template models to construct its views.

- The networking layer handles object mobility and information routing.
- The middleware layer facilitates seamless service discovery and management of smart objects.
- The application layer reuses heterogeneous application services from heterogeneous platforms.
- The data and semantics layer introduce common understandings of data and information.

The following are the prominent examples of standardized IP-based communication protocols for IoT devices: (i) on the application layer, IETF Constrained Application Protocol (CoAP)/REST engine and Message Queuing Telemetry Transport (MQTT); (ii) on the networking layer, IPv6 and RPL (and a derivative for low-power wireless personal area networks “6LoWPAN”); (iii) on the physical layer, IEEE 802.15.4 (Ahlgren et al., 2016). Examples for semantic oriented protocols include OPC UA (OPC Unified Architecture), UPnP (Universal Plug and Play), DPWS (Devices Profile for Web Services), CoAP (Constrained Application Protocol), and EXI (Efficient XML Interchange) (Weyrich and Ebert, 2016).

Interoperability has several dimensions. It is worth noting that even a high degree of standardization of protocols does not imply a high degree of standardization of data formats or device compatibility. In fact, interoperability is currently hampered by this condition. In an ideal situation, communication must be independent of the creator of a given fragment of the infrastructure. In reality, however, various players (including vendors) have their own IoT solutions that are more or less incompatible with other solutions, thus creating local “IoT silos” (Fortino et al., 2016). A large body of recent research into IoT is thus devoted to interoperability. The EU’s Unify-IoT project may serve as an indication of this: They estimate more than 360 available IoT platform providers and determine that approximately 20 are somewhat popular. This indicates clearly that massive research efforts do not necessarily converge (Unify-IoT, 2016).[22]

For exchanging data between applications, devices, and objects, well-known communication standards exist, including Bluetooth, Wi-Fi,[23] and various mobile communication standards, such as GSM. Based on the use cases of

---

22 For various reports (deliverables) on IoT platform research of the EU-funded project Unify-IoT, refer to http://www.unify-iot.eu/.

23 Wi-Fi is a trademark of the Wi-Fi Alliance. It is a technology for wireless local area networking with devices based on the IEEE 802.11 standards. IEEE 802.11 is the radio frequency needed to transmit Wi-Fi. Devices that can use Wi-Fi technology include personal computers, video-game consoles, smartphones, digital cameras, tablet computers, digital audio players, and modern printers. Wi-Fi compatible devices can connect to the Internet via a WLAN network and a wireless access point, usually within a range of approximately 20 m indoors and a greater range outdoors.

mobile communications, major technological progress was achieved in terms of higher bandwidth (and accordingly, higher bit rates), multimedia streaming capabilities, and so on. However, as previously mentioned, most IoT use cases involve resource-constrained devices. Consequently, the goal of "Low Power, Wide Area Networks" (LPWAN) has become a core topic in IoT over the last few years. LPWAN is a broad term for a variety of technologies used to connect sensors and controllers to the Internet without the use of traditional Wi-Fi or cellular networks. At the same time, however, major players in cellular network industries are also further developing cellular-based networking standards, for example, LTE-M and NB-IoT. The latter is backed by leading manufacturers and by the world's 20 largest mobile operators. Further examples of activities forming new standards better suited for IoT use cases include LoRa and N-Wave, and Sigbox. The predominant design considerations are low energy consumption (up to more than 10 years of autonomy), strong penetration in indoor environments, and connecting a large number of sensors and devices with low bandwidth requirements. Table 1.2 summarizes selected communication protocols and standards currently under investigation or in use.[24]

1.5.3 Embedding

The anticipated omnipresence of a computerized world is, however, not to be implemented by setting up computers on the corners of every street. Instead, functionalities are embedded in objects and spaces. For example, conductive materials are woven into or printed on textiles. Objects are then computerized in this way, allowing us to immediately receive information about them and process them. The miniaturization of hardware is an essential prerequisite for the embedding of IT into objects. According to the still valid Moore's law,[25] miniaturization is accompanied by the improved performance of processors and increased storage capacities, with the cost of manufacturing the components remaining the same or even decreasing. These developments promote the general diffusion of information and communication technologies and allow them to be embedded in any, even small and short-lived, objects. This does not always concern increased performance, but can include other factors, for example, the energy efficiency of components. While embedding computers or components in physical things, novel challenges for the user interface often arise. For example, how does one communicate with "disappearing" computers? Displays, keyboards, and other commonly used input and output

24 Compilation taken from Baras and Brito (2017), with additions. Data sources are Postscapes (2017), Opensensors (2017), and ETSI (2016).

25 Whether and how long Moore's law may be still valid remains controversial (see several papers in *IEEE Spectrum*, 04/15).

Table 1.2 Overview of communication technologies and standards for IoT.

Name	Frequency	Range	Examples	Standards
Bluetooth BLE	2.4 GHz	1–100 m >100 m	Headsets, wearables, sports and fitness, health care, proximity, automotive	IEEE 802.15.1[a)] Bluetooth SIG [b)]
EnOcean	315 MHz, 868 MHz, 902 MHz	300 m outdoor, 30 m indoors	Monitoring and control systems, building automation, transportation, logistics	ISO/IEC 14543-3-10[c)]
GSM, LTE, LTE-M	Europe: 900 MHz and 1.8 GHz, USA: 1.9 GHz and 850 MHz		Mobile phones, asset tracking, smart meters, M2M	3GPP[d)]
6LoWPAN	2.4 GHz	10–30 m	Automation and entertainment applications in home, office, and factory environments	Adaption layer for Ipv6 over IEEE802.15.4[e)]
LoRa	Sub 1 GHz ISM band	2–5 km urban; 15 km suburban; 45 km rural	Smart city, long-range M2M	LoRaWAN[f)]
NB-IoT (narrow-band-IoT)	700–900 MHz	10–15 km rural deep indoor penetration	Smart meters, event detectors, smart cities, smart homes, industrial monitoring	3GPP LTE Release 13[g)]
NFC	13.56 MHz	Under 0.2 m	Smart wallets, smart cards, action tags, access control	ISO/IEC 18092[h)] ISO/IEC 14443-2,-3,-4[i)]
NWave	Sub 1 GHz ISM band	Up to 10 km	Agriculture, smart cities, smart meters, logistics, environmental	Weightless[j)]
RFID	120–150 kHz (LF), 13.56 MHz (HF), 2450–5800 MHz (microwave), 3.1–10 GHz (microwave)	10 cm to 200 m	Road tolls, building access, inventory, goods tracking	ISO 18000[k)]

DASH7	433 MHz (UHF), 865–868 MHz (Europe), 902–928 MHz (North America) UHF	0–5 km	Building automation, smart energy, smart city logistics	
SigFox[l)]	900 MHz	3–10 km urban 30–50 km rural	Smart meters, remote monitoring, security	
Weightless	470–790 MHz	Up to 10 km	Smart meters, traffic sensors, industrial monitoring	Weightless[m)]
Wi-Fi	2.4 GHz, 3.6 GHz, 4.9–5 GHz	Up to 100 m	Routers, tablets, smartphones, laptops	IEEE 802.11[n)]
Z-Wave	ISM band 865–926 MHz	100 m	Monitoring and control for homes and light commercial environments	Z-Wave[o)]; recommendation ITU G.9959[p)]
ZigBee	2.4 GHz; 784 MHz in China, 868 MHz in Europe, and 915 MHz in USA and Australia	10–20 m	Home and building automation, WSN, industrial control	IEEE 802.15.4[q)]

a) http://www.ieee802.org/15/.
b) https://www.bluetooth.com/.
c) https://www.iso.org/standard/59865.html.
d) http://www.3gpp.org/.
e) https://standards.ieee.org/about/get/802/802.15.html.
f) https://www.lora-alliance.org/What-Is-LoRa/Technology.
g) http://www.3gpp.org/release-13.
h) https://www.iso.org/standard/56692.html.
i) https://www.iso.org/standard/50941.html; https://www.iso.org/standard/50942.html; https://www.iso.org/standard/50648.html; JIS X 6319-4 "FeliCa", http://nfc-forum.org/.
j) http://www.nwave.io/; http://www.weightless.org/.
k) https://www.iso.org/standard/46145.html.
l) https://www.sigfox.com/.
m) http://www.weightless.org/.
n) http://www.ieee802.org/11/.
o) http://www.zwave.de.
p) https://www.itu.int/rec/T-REC-G.9959.
q) http://standards.ieee.org/getieee802/download/802.15.4-2015.pdf.

devices may not always constitute the optimal solution. A need for new metaphors and user interfaces exists, in particular those suited for intuitive interaction (see Section 1.6.2.)

1.5.4 Sensors

Sensors are technical components for the qualitative or quantitative measurement of certain chemical or physical variables and properties, for example, temperature, light (intensity and color), acceleration, electricity, and so on. The recorded measured values are usually converted into electronic signals. Currently, we are already surrounded by sensors in many places. For example, modern automobiles contain hundreds of sensors, for example, rain sensors for windshield wiper systems, crash sensors for air bag release systems, and lane and parking-assist sensors. Indeed, modern automobiles, some with far more than 200 sensors and a few dozen microprocessors (Economist, 2009), constitute a good example of this. In fact, the ordinary automobile is increasingly becoming one unified computerized object. In addition, when a sensor is employed together with a processor (controller), a power supply, and a unit for data transmission, this is referred to as a *sensor node.*

A sensor node's primary function is to collect, preprocess, and transmit sensor data from its environment to other sensor nodes or a base station. Examples of sensor categories include (Baras and Brito, 2017) the following:

- *Location:* GPS, GLONASS, Galileo
- *Biometric:* fingerprint, iris, face
- *Acoustic:* microphone
- *Environmental:* temperature, humidity, pressure
- *Motion:* accelerometer, gyroscope

Sensor nodes can form Wireless Sensor Networks (WSN) by means of their transmission unit. For example, these are utilized to (i) detect earthquakes, forest fires, avalanches, as well as terrorist attacks; (ii) monitor vehicle traffic, particularly in tunnels; (iii) track the movements of wild animals; (iv) protect property; (v) operate and manage machines and vehicles efficiently; (vi) establish security areas; (vii) monitor supply chain management; and (viii) discover chemical, biological, and radiological material. For the operation of sensor networks, special software is required, which ensures a dynamic and robust self-organization of the sensor network that functions in a safe and scalable manner. This is because sensor nodes can fail, change their position, or be only online intermittently. WSN can consist of several hundred or hundreds of thousands of sensor nodes, which are deployed either inside of the phenomenon or very close to it.[26] Sensor nodes are connected to an intermediary network that forward the data

26 For an excellent overview on sensor networks, see Akyildiz et al. (2002).

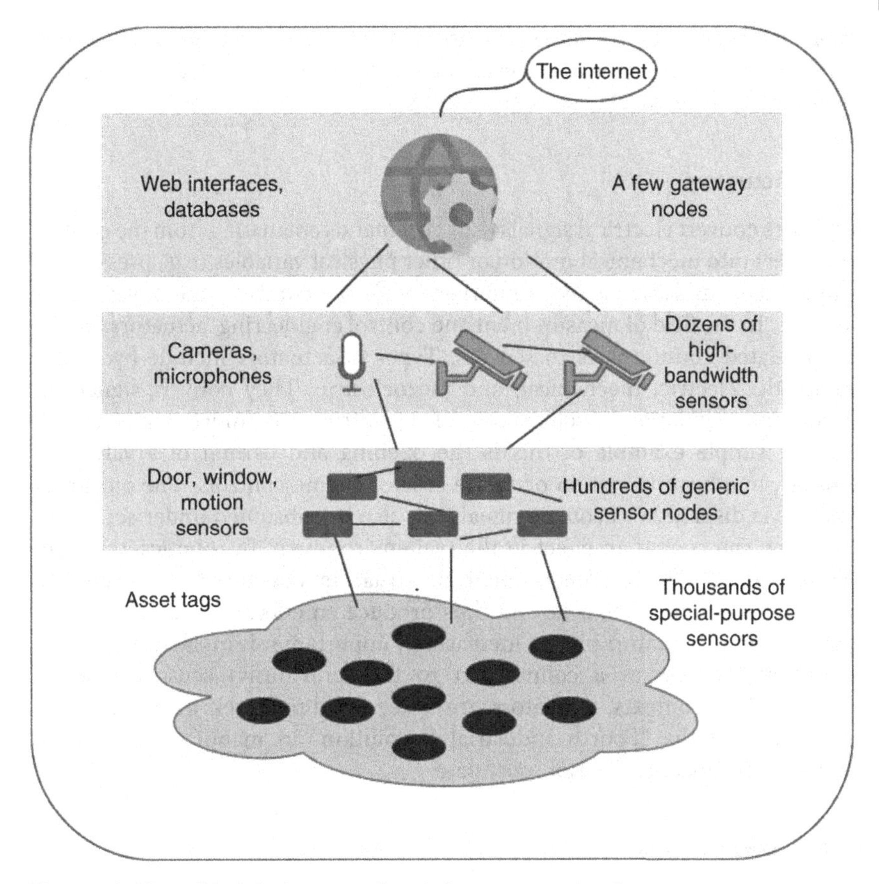

Figure 1.3 Hierarchical deployment of a wireless sensor network.

that they collect to a computer for analysis. Sensor nodes are installed in their workspace to function for years, preferably without requiring any maintenance or human intervention. They must therefore have a low energy requirement and have batteries that are functional over several years. The construction of a typical WSN is layered (see Figure 1.3) (Hill et al., 2004). Specifically, it begins with sensors on the lower level and continues up to the top-level nodes for data collection, analysis, and storage. Simple and complex data are routed through a network to an automated facility that provides continuous monitoring and control of the dedicated environment. WSNs do not necessarily operate on all layers with the common TCP/IP stack and may use dedicated lightweight protocols instead.

Each platform class handles different types of sensing (according to Hill et al., 2004, p. 42). As sensors are foundational to both smart objects and sensor nodes, they are a crucial component of an IoT world. In fact, WSNs will facilitate the proliferation of many applications. The small, robust, inexpensive, and low-powered WSN sensors will bring the IoT to even the smallest objects installed in any kind of environment, at reasonable costs (IEC, 2014).

1.5.5 Actuators

Actuators convert electrical signals (e.g., commands emanating from the control computer) into mechanical motion or other physical variables (e.g., pressure or temperature), and thus actively intervene with the control system and/or set variables. In the field of measurement and control engineering, actuators are the signal-related counterparts to sensors. Types of actuators include hydraulic, pneumatic, electric, mechanical, and piezoelectric. They convert signals or setting and regulation specifications of a control into (mostly) mechanical work. A simple example of this is the opening and closing of a valve, for example, in a heating system or in the case of engine controls. The output of optical (via displays) or acoustic signals can also be subsumed under actuators, since they can trigger an effect in the real environment. In robotics, the term *effector* is often used as an equivalent for actuators. Effectors allow a robot to grasp and manipulate objects, and thus produce an effect. In a computerized world of things, actuators play an increasingly important role in the realization of actions and effects as a counterpart to the (previously) sensory-detected corresponding contexts. Actuators are a key building block in more recent perceptions of the "Fourth Industrial Revolution" in manufacturing as an Industry 4.0 conceptualization postulate.

1.5.6 Power Supply

While many technologies are already available on the market or at least have been tested in research contexts, unsolved technical problems remain. A very limiting factor of the mobility of smart objects is their energy supply. Although batteries are becoming smaller and more powerful, today's mobile devices still have very limited battery capacities. The heavy research on improved battery technologies has only produced relatively mild progress in battery performance. In fact, it is continually falling behind other relevant technological developments. Some argue that there will soon be (or even exist today, as initial reports on burning smartphone devices may prove) a limit reached in which energy density becomes so high that the respective devices become a serious threat to safety. To counteract these challenges, several avenues of research are being pursued, including intelligent designs that require less battery power. This can be achieved by departing from the idea that everything has to be online all of the time. Sometimes, it is sufficient to only occasionally know about a status shift of an object. This can be communicated with much less relative effort and demand on bandwidth and energy. Another strategy is to harvest energy "on the fly." The development of technologies for the utilization of alternative sources of energy, such as the sun, wind, and water is progressing rapidly, partly due to political pressure. We have already witnessed this type of integration into portable devices, for example, smartphones with solar cells. Moreover, approaches to extracting energy from the external sources of solar, thermal, piezoelectric, mechanical, and kinetic energy are already established, referred to as *energy harvesting* (Anton and Sodano, 2007; Sudevalayam and Kulkarni, 2011). These approaches are particularly suitable (because of their independence of fixed infrastructures) for the power supply of mobile and autonomous devices, such as sensor nodes. A promising idea for personally used mobile devices is the tapping of the energy that a person naturally produces and emits. Through movement and metabolism (warmth), a person expends several kilowatt hours (heat and movement power). At the same time, several hundreds and up to 1000 W can be generated and stored, which could theoretically generate sufficient power for the operation of a notebook computer. Energy generation from blood glucose or other energy potentials, such as the pH level of body fluids, is also conceivable. Effectively, however, only a fraction of this can currently be accessed, if at all, and the impairment imposed by the required devices on the user may, in some instances, still be too great. Other innovative approaches are biofuel cells that work with bacteria. Through the decomposition products of bacteria, energy can be generated from organic substances. An application of this is to install biofuel cells in wastewater treatment plants and sewage treatment plants, where large quantities of energy-rich organic substances are present.

1.5.7 Identification

An important prerequisite for the linking of information with real entities in our environment is an unambiguous identification of things and people. The umbrella term "Automatic Identification (Auto-ID) and Mobility (AIM) technologies" describes a diverse family of technologies that share the common purpose of identifying, tracking, recording, storing, and communicating essential business, personal, and product data. Several identification technologies exist, for example, biometric, barcodes, and RFID. Applications of RFID, which have been known since the 1960s, have especially become a catalyzer for IoT scenarios.

1.5.7.1 Radio Frequency Identification

Radio frequency identification systems use tiny, so-called *tags* with embedded microchips that typically contain a small amount of computer memory and transmit their content via radio signals over a short distance to specific RFID readers (see Chapter 5). The reader captures these data, decodes them, and sends them to a host computer for further processing via a wired or wireless network. In fact, RFID tags can be considered as electronic barcodes (Welbourne et al., 2009). However, in contrast to barcodes, RFID tags do not require visual contact in order to be read. The RFID reader consists of an antenna and a radio transmitter with a decoding function, and is attached to a stationary or handheld device. Depending on output power, radio frequency, and ambient conditions, the reader emits radio waves in ranges between 2.5 cm and 30 m. If a passive RFID tag reaches the range of the reader, the tag is activated and begins sending data, that is, the prerecorded number(s) in the tag. In the case of active tags, which are battery powered, the tag itself is capable of sending data. As RFID tags can store a (unique) number and can be physically attached to an object, the object becomes automatically and contactless identified. Due to these major functionalities, RFID is considered to constitute a key technology as it bridges the physical world and virtual world, that is, physical objects become uniquely identifiable. In materials management and supply chain management, RFID systems can record and manage more detailed information about specific items in warehouses or in production much better than do barcode systems. When a large number of items are shipped together, RFID systems track each pallet, batch, or individual item in the delivery. Moreover, the number of reading points is technically unlimited. When there are more reading points in place, manufacturers can better follow the life cycle of each product, aimed at understanding product deficiencies and successes. Another example is books in libraries that use an RFID chip to allow users, by means of RFID reading systems, to borrow and return books without other assistance. In this way, hours of operation restrictions and waiting in lines are avoided. RFID has been available for decades, but the widespread use of tags was delayed as long as the cost of each tag ranged between €1 and 20. Currently, the simplest tags—purchased in large quantities—cost less than €0.10, and probably in only few years will cost less than €0.01. With this dramatic reduction in the cost of tags, RFID has become profitable for many more applications. In particular, the deployment of a large number of tags has become economically feasible, even for low-value items. Cost drivers for an RFID system, however, also include the installation of RFID readers and tagging systems. In addition, companies are likely to have to upgrade their hardware and software systems to process the enormous amounts of data produced by RFID systems. In fact, the monitored transactions could easily add up to hundreds of terabytes. In order to filter, collect, and prevent RFID data from overloading corporate networks and system applications, special middleware is required. The applications need to be redesigned to accommodate the massive volumes of RFID-generated data, as well as to share data with other applications. Large enterprise software vendors, including SAP and Oracle, offer RFID-enabled versions of their supply chain management applications. The power of RFID for IoT is amplified when used together with addressing schemes, in particular the Electronic Product Code (see the next section).

1.5.7.2 Addressing Schemes Based on IPv6 and Electronic Product Code

Addressing schemes has become a crucial task in identifying things. The challenge in an IoT scenario is to uniquely identify billions of devices and, for many application scenarios, to also control them. The top technical challenges are uniqueness, reliability, persistence, and scalability. Internet Protocol version 6 (IPv6) and the Electronic Product Code are important building blocks for IoT.[27]

IPv6 is the most recent version of the Internet Protocol (IP), which is the communications protocol that provides an identification and location system for computers on networks and helps to route traffic across the Internet. The idea of IP is to connect every device to a network while assigning a unique IP address for identification and location definition. With the rapid growth of the Internet after commercialization in the 1990s, it became evident that far more addresses would be needed to connect all devices than its predecessor—the IPv4—had available. By the late 1990s, the Internet Engineering Task Force (IETF) formalized the successor protocol, that is, IPv6. IPv6 uses a 128-bit address, theoretically allowing 2^{128}, or approximately 3.4×10^{38} addresses. In other words, the total number of possible IPv6 addresses is more than 7.9×10^{28} times as many as IPv4, which uses 32-bit addresses and provides approximately 4.3 billion addresses. This seemed to be more than sufficient to assign a unique address to any number of man-made objects present or to be built. IPv6 has incorporated both a rich address scheme and a great deal of sophisticated functionality (for dynamic address management, intelligent routing, etc.), which adds to the so-called protocol overhead and renders IPv6 a relatively heavy protocol. In addition, IPv6 does not fit well, especially regarding the application scenarios of WSNs, which may coordinate extreme large numbers of networked sensors and would not need all of the networking functionalities that come with IP. However, not all layers of a typical WSN usually operate within the established IP stack and can therefore not take advantage of the address scheme provided by IPv6. This calls for an additional subnet layer or for the development of a lightweight form of IPv6 (e.g., 6LoWPAN) that are better suited for IoT scenarios.

The Auto-ID Center at MIT (now Auto-ID Labs, an international research network)[28] and the development community around RFID played a crucial role

---

27 Many more addressing schemes exist. An early concept that may gain more interest again in terms of IoT is the so-called MAC address. "MAC" reads media access control. It is a unique identifier assigned to network interfaces for communications at the data link layer of a network segment. MAC addresses are used as a network address for most IEEE 802 network technologies, including Ethernet and Wi-Fi. MAC addresses are most often assigned by the manufacturer to devices that will be connected to a network. Usually the unique number is stored in the device, such as the card's read-only memory or some other firmware mechanism. This unique number could also be defined as an Ethernet hardware address, hardware address, or physical address. Accordingly, things that have a MAC address become uniquely addressable.

28 http://autoidlabs.org/wordpress_website/.

in the conceptualization and identification of the standardization efforts needed. The core idea is to discover information about a (RFID-) tagged object by browsing an Internet address or a database entry corresponding to a particular code stored within an RFID tag. They worked on the development of the Electronic Product Code (EPC) (EPCglobal Inc., 2014), that is, a universal identifier that provides a unique identity for every physical object, for all time.[29] Today, the concepts are more general and are not limited to RFID only. A thing can be any real/physical object, but also a virtual/digital entity, which moves in time and space and can be uniquely identified by assigned identification numbers, names, and/or location addresses. For virtual objects, corresponding concepts are Uniform Resource Identifiers (URI) and IP addresses, which allow identifying and discovering an object's presence on the Web.[30] Based on the well-established Domain Name System (DNS),[31] in an IoT context, IP addresses can also be utilized as identifiers for networked objects together with name labels. The core idea is to extend the already existing DNS programming interfaces and formats to small networks where there are no name servers available. One key concept is the multicast Domain Name System (mDNS), which resolves host names to IP addresses within small networks that do not include a local name server (Cheshire, 2017).

1.5.8 Localization

In addition to identification, the position of an object or a human being is essential contextual information. Localization techniques can be employed for determining position, which either localize an object externally or with which an object determines its position itself. Examples of "global" positioning systems are the Global Positioning System (GPS) of the Unites States, GLONASS (Russia), Galileo (European Union), and BeiDou (China). A distinction is made between four types. In trilateration, distances are measured to at least three points, the position of which is known, and the geometrical intersection is used to determine the position. This can be carried out in networks simply by means of propagation times of transmitted signals. Similarly, triangulation, in which angles or directional dimensions are used for the calculation of distance and

29 Its structure is defined in the EPCglobal Tag Data Standard, which is an open standard freely available for download from the website of EPCglobal, Inc.; http://www.gs1.org/epcglobal.
30 A corresponding standardization initiative is organized by the Hypercat Alliance, http://www.hypercat.io/.
31 The Domain Name System (DNS) is a hierarchical decentralized naming system for resources connected to the Internet or a private network. Most prominently, it translates domain names (which are more suited for human readability) to the numerical IP addresses needed for locating and identifying computer services and devices with the underlying network protocols. Since its inception in the 1980s, it has become a worldwide, distributed directory service and an essential component of the functionality of the Internet.

position, also exists. On the other hand, position is measured with the ambient determination by means of the next known point. This method is already utilized today in mobile radio localization in GSM networks by assignment to a mobile radio cell. Another technique, scene analysis, determines position based on specific features of the point of view (called a "footprint"). These features can be actual images of the landscape from the corresponding viewing angle or can be stored beforehand in a table with specific measured values of a point of view, for example, electromagnetic values or radiation specifications in one or several present WLANs. Challenges in localization procedures are the tracking of moving objects and the handling of covered or indoor objects (problematic with GPS positioning) or radiation and falsification of radio waves. However, in recent years, much effort has been invested in indoor localization technologies (Koyuncu and Yang, 2010).

1.5.9 Cloud Computing and Fog Computing

The large and increasing numbers of IoT devices will lead to rapid growth of collected data. Often, such data will have a device–time–space relationship (i.e., time and position data that tightly relate to a specific device). In IoT scenarios, it is likely that such data are shared among several applications, necessitating greater interoperability. Moreover, additional dimensions of objects might be of interest, including different types of sensor data or meta-data, about the object. This creates new data management issues and may change the predominant way of processing. Specifically, processing may move away from a formerly "offline" or batch mode, in which storage and query as well as processing and transactions might have occurred with some delay without negatively affecting applications or services toward a more "online" or real-time world, where collecting, processing, and acting upon data may not allow major delays (Borgia, 2014). Apart from "real-time processing" needs, data archiving with intelligent policies to distill, index, and intentionally delete data in efficient ways is still a major challenge. Several alternative solutions exist, including central approaches, decentralized or data-centric storages that are as near as possible to its production points or – as a kind of mixture – dynamically adjust the data storage position according to specific conditions (see Borgia, 2014, "Data management" for a set of references and Chapter 4 of this book). In order to meet data management challenges,[32] cloud and fog computing are among the most important approaches to cope with IoT data management issues.

Cloud computing is a concept in which computing performance, storage, software, and other services are provided as a group of virtualized resources over

32 Borgia (2014) lists IoT general requirements, such as heterogeneity, scalability, cost minimization, self-service, flexibility, quality of service, and secure environment for which cloud computing and fog computing may contribute answers.

a network, primarily the Internet. In addition to this, the "Cloud" of resources can be accessed at any time from any connected device and site (Zhang et al., 2010; Weinhardt et al., 2009; Armbrust et al., 2010). Typically, users automatically receive cloud resources, such as server time or network storage, without a need for further negotiation with the service provider in an "on-demand self-service" and in an "elastic" manner. This is of tremendous value to the user, as he or she need not to hold available such resources even in the case of large demand. Those resources, and in particular the management of up- and down-scaling, are delegated to the cloud service provider. Most often, cloud services come as a measured service: Cloud resource charges are based on the resources actually used (Mell and Grance, 2011). Cloud computing is seen as a major building block of Ubicomp scenarios (Gubbi et al., 2013; Cáceres and Friday, 2012) in order to cope with the challenges of efficient, secure, scalable, and market-oriented computing and storage. In principle, Cloud computing achieves excellent results in terms of networking resources and storing and accessing data related to or derived from connected things. However, regarding latency-sensitive applications, which require nodes in the vicinity to meet their delay requirements, cloud computing may possess some limitations—especially when millions of devices are to be handled in a time-critical manner. New use cases may arise that call for tight control of physically dispersed, yet specifically located, sensors or actuators (e.g., a plant with machines that have to react to sudden changes in the environment or production process). In response to these challenges, the fog computing paradigm (also referred to as "edge computing") is proposed (Bonomi, 2011; Bonomi et al., 2012), which should not replace the cloud computing paradigm, but extend it.[33] Fog computing, as a highly virtualized platform, provides computing, storage, and networking services between end devices and traditional cloud computing data centers that are typically, but not exclusively, located at the edge of a network. Focusing more on the "edge of the network," however, implies a number of characteristics that make fog computing a nontrivial extension of cloud computing. Fog computing is expected, for example, to deal with widely distributed and mobile deployments in which very large numbers of nodes are involved, for example, fast-moving and large groups of vehicles along highways or large-scale sensor networks to monitor the environment). Since its conceptual inception just some years ago, fog computing has achieved remarkable interest in academia (Dastjerdi and Buyya, 2016) and industrial research (see, for example, the Open Fog Consortium, founded in 2015).[34]

---

33 There is a significant increase in extant literature on fog computing. *IEEE Internet Computing* devotes a special issue to fog computing in its March 2017 issue.
34 https://www.openfogconsortium.org/.

1.6 Derived Qualities of Modern ICT

Modern information and communication technology infrastructures enable the following qualities: context awareness, adaptability, proactivity, high data quality, and intuitive interaction.

1.6.1 Context Awareness, Adaptability, and Proactivity

Context awareness (also context dependency) is a behavior that depends on information about the context of any entity (programs, people, objects). Information about contexts can be obtained from a wide variety of sources, in particular, via sensors. This information is used to draw conclusions about the context and to adapt the behavior adequately. The utilization of contextual information is most frequently associated with time and location aspects, in the latter case referenced as location-based services. However, any further aspects can be included in a context model if corresponding information sources or sensors exist (Perera et al., 2015). This can be, for example, archive data or biometric data, the temperature in an environment, or relationships between people (Dey, 2001; Dourish, 2004; Coutaz et al., 2005).[35]

Context sensitivity allows for adaptability and proactivity. It is even less intrusive and disruptive when services and functionalities provided by smart environments adapt to the context and are proactively offered outside of a smart environment. Currently, the degree of customization of conventional computers and mobile phones is very low. Adaptations to regional conditions, such as language and time settings, are customary. It is expected that more contextual information will be utilized in the future, and device settings and services will automatically adapt accordingly, such as the position of the user, his or her health or emotional state, his or her plans, tasks to be done, and other factors in the environment that affect the user. Proactivity unites the adaptability of applications in the background and an anticipated interaction of a designated user with the offered service. Services are automatically offered to a user in the ideal case wherever and whenever they are needed. The initiator is the smart environment itself, and not the potential user. This quality entails a major requirement: The smart environment must be able to correctly recognize the context and the intentions of the user. It is questionable whether this can also be achieved reliably in complex situations. A simple example shows only one of the difficulties for a reliable implementation: If a person falls unconsciously to the ground, the automated sending of an emergency call is useful, but this case is different from "similar" occurrences, for example, if the person drops suddenly and deliberately on the sofa to rest. The recognition of the situation and the

35 For an in-depth recapitulation of what context is and whether context can be computerized, consult Dourish (2004).

"right" context (context awareness) is one of the core challenges of the realization of a computerized world.

1.6.2 Increased Data Quality

The improved availability of data in terms of quantity ("we know more about the status of a thing or related process") and quality ("we know more details about it") may constitute the most obvious change resulting from omnipresent information-gathering mainly through sensors. Obtaining better data about things in general is fundamental for any improvement related to products, processes, and business models. The subsequent sections will differentiate four dimensions of (improved) data quality and its effects, that is, the substitution and elasticity effect.

1.6.2.1 Dimensions of Data Quality

IoT platforms allow for an increase of data quality at approximately the same cost and in a simpler way than previously. These improvements can be described by four dimensions of data quality.

1) *Object Granularity and Type.* Falling hardware costs and miniaturization simplify the use of technical components on individual objects at lower costs. Granularity refers to the number of objects of a group or class, through which the information is aggregated. Due to certain concepts, such as ubiquitous computing, fine-grained data can be acquired for individuals, and even very small objects. Today, containers and pallets are tracked on their delivery routes using RFID and GPS. Soon, the data acquisition for each of the products on the pallets, including a small item, such as a yogurt cup, becomes affordable. This means that all object types, including products of low value and with a short lifetime, are also recorded within economic boundaries.
2) *Time Granularity.* Efficient data transmission and wireless networks in smart environments enable simple, continuous data collection in real time. Although inventory is still carried out periodically and manually in many companies, it can run continuously and automatically with an RFID system. This means that current inventory data can be called up at any time, and changes can be viewed in real time or very promptly. However, real-time data collection is problematic, for example, on flights in which data transmissions can interfere with air traffic, and when objects move very quickly or relevant features of the environment change very rapidly.
3) *Data Content.* RFID is a cost-effective, highly tested technology for contactless individual object identification. It offers several advantages over the conventional barcode and serial IDs, which can only be read by visual contact. Through RFID, an individual ID can be linked to the object both physically and digitally simultaneously. The EPC is such a unique ID. Depending on the tag type, additional data, such as the date of manufacture and the production location, can be stored on an RFID tag. However, the

storage space is typically limited to a few kilobytes. Only the utilization of additional data stores and sensors at the object and in the environment allows for more comprehensive object or context data.

4) *Reach.* The dimension reach is less dependent on technologies than on application concepts. Through networking, the integration of applications and information systems is generally possible throughout a company or in an interorganizational manner. However, cooperative agreements and agreements on standards are crucial for the success of implementations. In supply chain management, data standards, such as EAN and EPC from GS1, are particularly widespread. Other standards, such as XML, Semantic Web standards, and Web Services, make it easier to implement these applications.

1.6.2.2 Effects of Increased Data Quality

Equipping the infrastructure with sensors and actuators has two effects. First, there is a substitution effect. Conventional data collection and retrieval (e.g., manual or barcode) are automated, and media discontinuities are avoided. Second, an elasticity effect exists (Fleisch and Tellkamp, 2006). In addition, new data can now be collected and utilized. As a result, companies can map real-world information in real time, and thus use it to directly control processes and activities. This allows digitization of management regimes and leads to better decision-making. Business can easily collect more data and enrich existing collections with new data quality. The data may also be employed for triggers and alarm functions for certain events, for example, if a delivery transport is stuck in heavy traffic. If this concept is implemented together with business partners and transferred into an integrated information system, the so-called event-driven supply chain management can be implemented. Furthermore, automated processes lead to independent monitoring and control, for example, in production processes. With very high data quality, in particular with high time granularity, a real-time process control of the company can be implemented on the basis of the automatically recorded data, which are directly available for management via fast network connections, regardless of where the decision-makers would like to retrieve them. It is critical to consider whether real-time data are actually required for all processes and tasks, or whether summarizing the data in larger reporting cycles is already sufficiently appropriate.

1.6.3 Intuitive Interaction

Technology disappears by embedding it in the physical environment so that it is no longer perceptible. This makes it even more necessary that functionality and operability remain recognizable to the user. This can be termed the "invisibility dilemma." The solution to this dilemma constitutes the design of an intuitive human–computer interaction. A key concept is the implicit use of information systems (Kranz et al., 2010). It works like automatic sliding doors, which open as soon as a person approaches, without an explicit command. For example, the

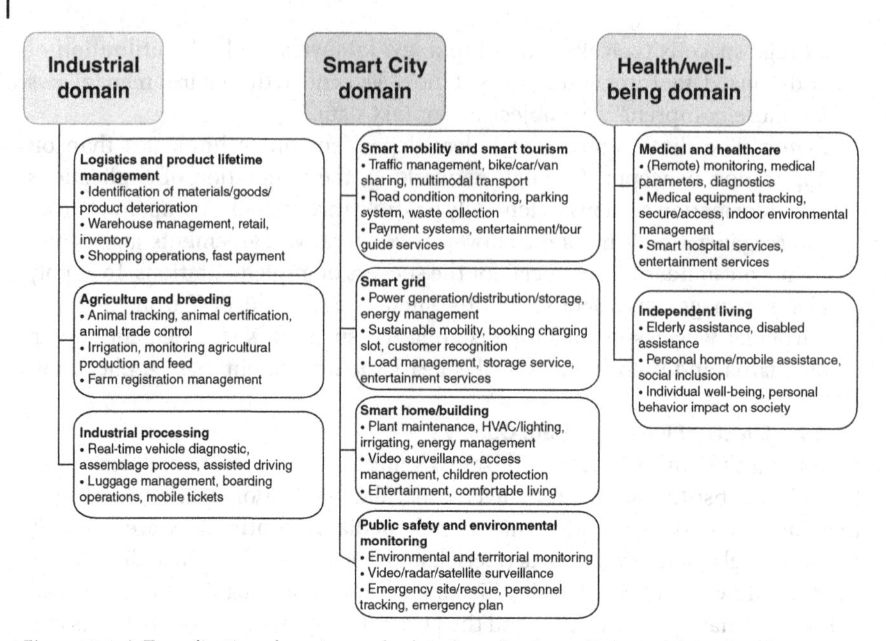

Figure 1.4 IoT application domains and related applications. (Adapted from Borgia, 2014, p. 9.)

natural behaviors of people are used, which are recognized, for example, by language, glances, facial expressions, and movements.

1.7 Potential for Product, Process, and Business Model Innovations

Opportunities for product, process, and business model innovations reside in two fields: (i) innovating within the IoT ecosystem; and (ii) innovating based on the IoT ecosystem. The focus here will be on the latter.[36] The presented qualities of modern information and communication infrastructures, as well as the congruency of smart objects and smart environments, offer great potential for innovation in nearly every field (see Figure 1.4). This is mainly due to the new or enriched "qualities" that an informatized infrastructure provides (see Section 1.6).[37]

36 A good overview of aspects of innovating within the IoT ecosystem is provided by the EU-funded H2020 UNIFY-IoT project: Supporting Internet of Things Activities on Innovation Ecosystems, Deliverable 02.01, IoT Business Models Framework, http://www.unify-iot.eu/wp-content/uploads/2016/10/D02_01_WP02_H2020_UNIFY-IoT_Final.pdf (retrieved January 29, 2017).
37 Borgia (2014) attempts to enumerate the domains and perceivable applications; see Parts IV and V for detailed examples.

Companies can develop new or improved processes or products (which include services here) in order to gain an advantage against their competitors. Existing business models may also be changed (Iansiti and Lakhani, 2014). However, how can companies create such innovations? For the development of applications, two approaches can be determined: problem-initiated innovation and the technology-driven innovation.

In the case of problem-initiated innovation, new technologies are developed or utilized in a targeted manner to solve a specific problem. This often leads to incremental innovations that initially increase the efficiency of existing business processes or products or services. In his seminal article, March (1991) speaks of "exploitation." These innovations are usually triggered by the user, who expresses a desire for improvement. Through IoT, control and information-intensive processes can be improved. By using RFID, sensors, and localization procedures, supply chains can be automated and controlled in real time. This avoids or reduces costs due to unexpected disturbances. In addition, antitheft protection can be improved and anticounterfeiting measures can be increased.

Technology-driven innovations sometimes exhibit a radical character since they help solve existing problems in a completely new way. In terms of March (1991), a highly cited technology management researcher, this is labeled "Exploration." In a typical case, the developer (inventor) or an expert in the corresponding technology has an idea of how to use it in a valuable way. He or she focuses on the special features of the technology. The features of smart environments have already been presented. As a result, new services and products can be developed that offer customers added value over old and comparable products. With IoT, computerized products and context-based services can be offered. As the technology-driven innovation does not originate from the user, a danger exists that it will not fulfill user needs. Therefore, users should be integrated into the innovation process as early as possible. If innovations are aligned with the actual needs of their users, business processes and products can not only be improved but also be fundamentally innovated.

The power for innovation may be illustrated by three categories: (i) new products; (ii) new processes; and (iii) new business models.

1.7.1 Product Innovation

Most traditional products can become smart objects by enriching them with information technology. Then, the products can store information about their entire product life cycle from manufacture to disposal, and possibly exchange it with other products, smart environments, or users. Equipped with appropriate processors and a control program, they can even adapt their behavior to specific contexts or trigger autonomous actions. A real example is pans, which read in recipes via RFID and prepare the food with the stated temperature and cooking time. For this purpose, they can communicate with the stove (which must have appropriate, coordinated communication standards) and regulate the degree of heat. New products and related value-added services benefit from data present in higher granularity (Fano and Gershman, 2002; Ferguson, 2002; Allmendinger and Lombreglia, 2005; Iansiti and Lakhani, 2014).

1.7.2 Process Innovation

In combination with novel information and communication technology infrastructures that achieve a previously unprecedented level of data quality, processes can be more precisely captured and assessed, as well as processed faster, and in a more integrated and automated manner. In addition, these achievements can be obtained in extreme cases in near-real time or in real time. Many processes benefit from context-based information.

The core factor for improved processes is improved data, or data that have been distilled to more meaningful information. The ubiquity of information gathering and presentation is accompanied by the fact that the number and size of media discontinuities between the virtual and the real world are reduced. This closes the gap between the real and the virtual world. This also opens the path to better automation and integration (Chui et al., 2010). When data are entered manually into the system via the keyboard, errors can occur at every media discontinuity; apart from another problem, that time will elapse before data are recorded and ready for further processing. A technical approach is the encoding of data using barcodes. This idea first appeared in the 1930s. Successors of the first, one-dimensional barcodes are two-dimensional codes, also called *2D codes*. The information is stored not only on one axis, but vertically and horizontally. There are many coding schemes, of which one of the best known is the "QR code"—quick response code. The acceptance of further dimensions (color, time) results in 3D or 4D code, which can also store more information in a compact manner. With RFID (see Section 1.5.7.1), media discontinuities are greatly reduced, and the data are immediately transferred to a connected back end system after contactless detection. The same applies to data from wireless sensor nodes. Data acquisition, processing, and distribution are automated in the computerized world, that is, human intervention is no longer required. However, intervention points, for example, for configuration, subsequent control, or in the event of a malfunction, should still be available. Through automatic data transmission between networked objects and environments, a media-free integration of applications and enterprise systems can be implemented. This means that data are forwarded to authorized systems according to defined rules, and processed there according to the application. The prerequisites for this are uniform data formats and communication rules (protocols). In other words, the systems must be capable of mutual understanding. For example, which context data belong to which object and how to interpret special sensor measurement values must be known. If smart objects are equipped with artificial intelligence, self-controlling processes can be realized. In this context, for example, delivery packages or products "take their own way to the destination" and pass on production information to machinery or transport vehicles. These intelligent objects make autonomous decisions and organize themselves in a decentralized manner. One way of embedding these skills into objects is software agents, that is, a self-executing software program that makes decisions based on rules and learned knowledge, which, in some way, control or influence their environment through actuators, adapt to changes, and react to expected and unexpected events.

1.7.3 Business Model Innovation

Business models are also affected or altered by computerized worlds or can only be realized through them (Chan, 2015). For example: (i) Companies have the opportunity to redesign their pricing through the improved information base. In this way, customers' different payment options could be better recognized by means of price discrimination. For example, in the course of exploiting individual contexts, corresponding pricing can be made. An actual implementation of such price models is the "pay as you drive" tariffs for automobile insurance. (ii) Enterprises can redefine existing value chains. One example of this is the Zipcar, one of the world's largest car-sharing companies. The available automobiles or their positional data are automatically transmitted to the control center so that car-sharing members can quickly identify driving opportunities via a web interface. The company views itself less as a car rental company than as a flexible mobility service provider. (iii) The computerization of the everyday world could lead to new care services, for example, in the health sector. Together with the presented "smart home," people that require intensive care could live better and longer in an environment that is familiar to them.

Particularly in the field of mobile communications, location-based services are already being used, which consider the position of the user and, for example, display restaurants in the current environment of the user. Context-based services include not only location information but also other relevant information about the environment and the user. In smart environments, context data can be utilized to provide services that are adapted to the situation, the user, his or her tasks, wishes, plans, and other factors, or react to a specific context with meaningful actions or suggestions. Navigation systems that receive information about road conditions and traffic on the target route in real time are able to reconcile this context information with the user's target data and then make flexible route adjustments. This could also warn the driver of any short-term accidents coming up or imminent tire damage (if sensors are installed on the tire/wheel system of the automobile). In addition, context-based marketing is tuned to customers, their whereabouts, and other context factors, so that as little randomness as possible is caused by unsuitable advertising campaigns, for example, offers of umbrellas, which can be bought in the surrounding area during rainy weather. Personal customer data can also be used to differentiate customer groups. In the case of scarce resources, service differentiation can be carried out. Important customers are treated preferentially. Product and information individualization also create added value. Information is individually tailored and adjusted, and product properties adapted to individual preferences, so that the customer can achieve a higher level of satisfaction.

The mentioned examples and the major trend that increasing numbers of things are creating more data have produced conceptualizations, including "data centricity," "competing on analytics," "Big Data-based business models," and so on. IoT and its implications for sensors and creating ever-increasing amounts of data constitute a new opportunity for creativity aimed at transforming data into value-creation activities.

1.8 Implications and Challenges

The computerization of the (everyday) world is accompanied by major implications and challenges, which can be characterized as (i) new markets; (ii) changed value creation; (iii) increased awareness of information spaces; and (iv) and social, ethical, legal, and risk aspects.

1.8.1 New Markets

A computerized world of connected things opens the door to innovations that facilitate new interactions among things and humans, and allows the realization of smart cities, infrastructures, and services that promise an enhancement of quality of life. By 2025, IoT could have an economic impact of US$11 trillion per year, which would represent approximately 11% of the world economy; and that users will deploy 1 trillion IoT devices (Manyika et al., 2015; Buyya and Dastjerdi, 2016).

Many reports and white papers (Ducatel et al., 2001) provide scenarios for impacts on hospitals, transportation systems, parcel services, supermarkets, offices, and other areas of everyday life. An illustrative example of the impact of computerized worlds on our everyday lives is the "smart home." In the smart home, devices, objects, and rooms are computerized and networked. The inhabitants can control furnishings, such as lights, doors, refrigerators, curtains, and so on, via remote control, voice control, or hand movements. They are also able to use the Internet to check whether everything is working well and acceptable in the house. The smart home also recognizes sensors that indicate when someone is in the house, and can turn on the lights automatically when an occupant enters a dark room. It can also recognize and store the preferences of residents. For example, in the case of a resident who watches his or her favorite television series every Saturday afternoon, the television is turned on with a corresponding transmitter or, if the resident is not at home, the sequence is automatically recorded. In addition, when food and related supplies are used daily, a sensor, after checking the contents of the refrigerator, sends a message to the digital notepad in the kitchen and places the products on a shopping list, which each of the residents can access by smartphone, for example, while they are at the supermarket. More integrated scenarios might trigger autonomous replenishment systems consisting of third-party-provided robots physically refilling, for example, a refrigerator. This and other scenarios can be developed much further. The essential point, however, is that the actors in a computerized world are aware of the potential impact on value-added features and markets. These are particularly the result of the fact that, as the example scenario shows, many more actors are involved in value creation for a customer.

1.8.2 Changed Value Creation

Together with higher data quality (as shown above), the importance of data and information as a resource for value creation is clear. This can be seen simply by observing the effects of a computerized world on value creation at different levels. On the individual level, consumers and producers are living in a computerized world. On the one hand, consumers are provided with information as consumer goods (either in the form of information services or in combination with computerized products) and, on the other hand, as input for decisions. Information can reduce search costs and facilitate rational action, as decisions can be weighed more accurately with more relevant information. On the other hand, for the convenience of context-based offers, the disclosure of personal preferences, personal data, and payment needs is required. In addition to efficiency improvements and cost advantages, producers can also benefit from differentiation, price discrimination, and bundling strategies by improving the information base. This creates great potential for the optimized elimination of the consumer's willingness-to-pay. For groups of individuals and organizations, the coordination and control of certain processes is facilitated. This makes it easier to ascertain the location and activities of the employees. Members of organizations can be brought to the same level of information due to better networking. This creates starting points for the analysis and improvement of group coordination. Contracts in the field of risk distribution and incentives can be made more equitable by capturing behavior that has not yet been observable at low cost. This allows a more equitable distribution of risk. Examples of this are working and insurance contracts, product guarantees (e.g., "Has a customer carefully maintained his or her automobile?"), and emissions monitoring for harmful exhaust fumes. Based on economic analyses, an increase in the efficiency of economic trade can be foreseen. One of the main effects of computerized worlds in the context of value creation is the reduction of information asymmetries. In real markets, based on the asymmetric distribution of information, two effects may arise: adverse selection and moral hazard.

Adverse selection occurs because certain information is not observable by providers. For automobile insurance, this is information about whether a new policyholder is a good or a poor driver. The downside for good drivers that emerges from adverse selection is that they pay, in principle, just as high of an insurance premiums as poor ones (as long as the automobile insurance provider cannot distinguish a good driver from a poor one). Moral hazard causes a change in behavior, as the risk of discovery of especially bad behavior decreases. Thus, a driver can intentionally reduce the risk of accidents by driving at a reasonable speed, abstaining from alcohol, observing distances, and so on. As a result of the conclusion of an insurance policy, the incentive to avoid accidents is, at least theoretically, reduced. A stereotyped form of moral hazard is that drivers become even more risk-averse, as they feel that they are already well covered for financial risks of careless or risk-taking driving. Sensors are also able to observe behavior in an objective manner. Speed, travel times and distances, braking behavior, as well as attention and alcohol levels can be measured, in principle. An automobile insurance company can now introduce price differentiation according to the real behavior and abilities of the drivers. This has already occurred in 2004 in Great Britain in the insurance company Norwich Union, which offered the tariff "pay as you drive" to automobile drivers. As a part of the program, they installed a black box in the automobile, which collected the relevant data about the driving behavior and sent them to Norwich Union.

Not only technical but also socioeconomic networks will be much more abundant between companies, users, consumers, and even objects. From an economic perspective, this creates network effects on consumption and production. This means that the benefit of a technology will increase with increasing numbers of users in the market. In order to obtain market share for products or standards associated with network effects, low prices are to be expected at the very beginning in order to build up critical mass.

1.8.3 Increased Awareness for Information Spaces

For context-based services, information is often required, which is owned by different actors. Thus, such services may be based on information from the user, for example, his or her name and allergies, to information from the owner of an environment, for example, the position of a user in a supermarket, as well as products in his or her environment, and information from the service provider, for example, information on allergenic substances in a specific product. Therefore, context-based services cannot be offered if each actor would protect his or her information from external access. Rather, information spaces must be created in which different information systems are brought together by different actors. An information space thus includes all of the data and relevant information obtained in a smart environment to provide users with context-based services and applications. Access to an information space can be restricted to certain actors, but it can also be publicly accessible, so that third parties can utilize the information for innovative services. The main challenge is that third parties comprehend the information available in the information spaces. For this purpose, semantic technologies may constitute a useful approach. The management of information spaces can be viewed as a task of information management (Schoder, 2011). As indicated, the provision of smart, context-based services requires information spaces that encompass the information systems of different actors. This presents companies with the challenge of managing these information spaces to provide shared value with partners and for their own benefit. This management takes place in a relationship of tension between the potential for innovation induced by the opening up of information spaces and the desire to profit exclusively from closed information spaces with the presumed retention of full control and data integrity. On the one hand, opening up information spaces means that third parties can access the information and integrate it into new, innovative services. This is already currently apparent as an opening of information systems, such as Google Maps and Facebook, which has led to a huge number of mashups and externally developed, innovative applications. In a computerized world in which data about reality are available at a much higher level of quality, a dramatically greater potential for innovation is to be expected if the data are freely accessible. On the other hand, the question arises of how a company can benefit from the fact that third parties use their information to create innovations. Companies could therefore rely on keeping their information spaces closed in order to exclude competitors and to utilize access to the information space as a source of revenue. Such information spaces, however, would run counter the realization of a computerized world. For information space management, the question arises as to how far information spaces should be opened in order to increase the potential for innovation and, on the other hand, to profit as much as possible. In addition, this raises the question of who should own and control devices and their data? A simple use case illustrates the conflict (Cáceres and Friday, 2012), that is, augmented home thermostats (rendering them as smart objects) connected to a smart power grid. Who should own the data that are generated through the home thermostat at the user's home: the end user or the service provider? What happens when the user's (local) desires to be comfortable conflict with the provider's (global) goals to save energy?

1.8.4 Social, Ethical, Legal, and Risk Aspects

Informatized worlds exist in a constant tension between innovation (the technically feasible) and individual and social acceptance (the socially desirable). The difficulties with the above-mentioned pricing strategies are, above all, customer acceptance and related concerns regarding the violation of privacy (Shin, 2010). Obtaining fine-grained data on entities, and especially individuals, expose the core dilemma in a modern IoT. Specifically, any person-related information may do both: enrich context- and person-related, individual services, and constitute potential intrusion into privacy, leading to resistance (Garfield, 2005). Besides privacy, many other fundamental challenges exist with regard to (IT) security, trust, and so on. The lack of security across IoT in general and the Industrial Internet of Things in particular has come to light largely due to an experimental search engine called Shodan (Wright, 2017). Launched in 2009, the service crawls nearly four billion devices from which, at any given time, several hundred million devices are turned on (depending on network connectivity). As a threat analysis based on Shodan showed, more than 100,000 IoT devices can be easily attacked, among them being special-purpose industrial computers for regulating the flow of water, transportation systems, and even entire power grids (Leverett, 2011). Many of these systems were designed before the advent of IoT, and thus did not consider these types of security threats. IoT certainly is confronted with literally all security problems already known from other IT-based concepts and artifacts—and may add some more aspects if not just by the severity and importance.

Some examples of pressing research questions include the following[38]:

- How should we cope with privacy issues in Ubicomp scenarios focusing on system design considerations? (Langheinrich, 2001)
- Who is accountable for decisions made by autonomous systems? (Berman and Cerf, 2017)
- How do we promote the ethical use of IoT technologies? (Berman and Cerf, 2017)
- What role does trust management play in IoT scenarios? (Sicari et al., 2014; Yan et al., 2014)
- What can middleware do for security and privacy issues? (Atzori et al., 2010)
- What are the security requirements to deal with data confidentiality? (Miorandi et al., 2012)
- What are the relevant legislative challenges? (Weber, 2010)
- What are appropriate architectural options for security and privacy, in particular the advantages and disadvantages of centralized and distributed architectures? (Roman et al., 2013)
- What are the principal attack models and threats? (Hu, 2016)

In order to include some sensitivity in the scale and scope of social, ethical, legal, and risk aspects of IoT, there is a focus on "data" and related data privacy issues (Cáceres and Friday, 2012).[39] It is worth noting that this list of questions is not exhaustive.

---

38 An overview by Rose et al. (2015), published by the Internet Society, provides a broad spectrum of issues raised by IoT, including security, privacy, interoperability/standards, regulatory, legal, as well as emerging economy and development issues.

39 Cáceres and Friday (2012) created a list of relevant questions in the context of Ubicomp scenarios, which, by definition, have the "user" as a focal element and amplify privacy issues. On the other hand, IoT may encompass more generic scenarios, in which the user is not always the immediate focal element.

- When can we infer with certainty? We need to take into account that sensed data or interactions are imprecise observations of the world, often taken from multiple sensors and at varying points in time. IoT environments must consider this evidence and make a judgment of when and how to react. Full appreciation of the value and meaning of data is certainly application and context dependent. Do we have machines (or more precisely software) that are advanced enough to "understand" such context and data properly?
- Where are data located? It is easier to answer this question in the context of technical aspects of cloud and fog computing architectures. However, regarding ownership, control of data, and access to data, obtaining the answer is substantially more challenging. For many environments, such as rooms, homes, companies, and hospitals, the demand for security and privacy requires enforcing conventional, legal, and physical boundaries.
- How long should data persist? What does the environment know about us? What should it know and with what should we trust it? How long should data be retained? What is transient and what should persist? Can we delete data, and can they be forgotten? Who has (the right) to access data? Who is the owner of sensed data? Is it the sensor's owner, the owner of the environment where the sensor is working, or the collector of data (Foster, 2017)?

Most, if not all, of these questions have gained prominence recently with the advent of "Big Data" and its unprecedented scale of storing, computing, and executing data and gaining insight from them. In addition, questions exist that are related to who will pay for IoT infrastructures (less obviously, but eventually, the common citizen with his or her tax dollars). Regulations concerning who is responsible for managing and maintaining local, regional, national, and supernational infrastructures are needed, and to a large extent not yet defined. Due to technical issues and a general reluctance, it will not be the common user who will be manager of his or her own data, that is, it will be service providers. Managed services would certainly reduce complexity for the end user, and obscure complicated technological interfaces. However, managed services also introduce their own tension between manageability and cost for the provider versus flexibility and control for the end user (Cáceres and Friday, 2012). One example of an initiative that addresses policy and regulation issues is the Mauritius Declaration on the Internet of Things. Excerpted example statements include the following[40]:

- IoT sensor data are high in quantity, quality, and sensitivity, and as such should be regarded and treated as personal data.

40 Mauritius Declaration on the Internet of Things, Balaclava, https://icdppc.org/wp-content/uploads/2015/02/Mauritius-Declaration.pdf (October 14, 2014).

- Transparency for all stakeholders is key. Those who offer IoT devices should inform the user so that he/she becomes clear about what data are collected, for what purposes, and how long these data are retained.
- Privacy by design should be the default design principle.
- In order to cope with security challenges, one way to minimize the risk to individuals is to ensure that data can be processed on the device itself (local processing). Where this is not an option, companies should ensure end-to-end encryption to protect the data from unwarranted interference and/or tampering.
- The data protection and privacy authorities should ensure compliance with the data protection and privacy laws in their respective countries, as well as with internationally agreed privacy principles, including appropriate enforcement action, either unilaterally or through means of international cooperation.

Taking into account the enormous challenges faced by IoT developers, data protection authorities and individuals should engage in a strong, active, and constructive debate on the social, ethical, legal, and risk aspects of IoT.

1.9 Conclusion

While the concept of combining computers, sensors, and networks to monitor and control devices has existed for decades, the recent confluence of key technologies and market trends is catalyzing the idea of IoT.

In order to better structure the scale and scope of IoT, this chapter provided an introductory overview, and briefly sketched the conceptual core ideas as laid out prior to IoT with "ubiquitous computing." The chapter presented a four-layer "Internet of Things" framework that covers not only technical but also non-technical issues of IoT.

IoT promises to form the foundation of new products, processes, and business models, and may fundamentally affect both B2C and B2B markets, as well as the way that we produce goods as envisioned with derivatives, including the Industrial Internet of Things and Industry 4.0. While the ramifications are very likely significant, a number of potential challenges may obstruct this vision, particularly in the areas of security, privacy, interoperability, standards, as well as legal, regulatory, and rights issues, and the inclusion of emerging economies. IoT encompasses not only technological but also social and policy considerations. IoT is already rapidly becoming more and more of a reality, and a vast space currently exists for new designs and realizations of creators and developers.

References

Aarts, E., Harwig, R., and Schuurmans, M. (2001) Ambient intelligence, in Denning, P. (ed.), *The Invisible Future: The Seamless Integration of Technology in Everyday Life*, McGraw-Hill, New York, pp. 235–250.

Aggarwal, C.C., Ashish, N., and Sheth, A. (2013) The Internet of Things: a survey from the data-centric perspective, in Aggarwal, C.C. (ed.), *Managing and Mining Sensor Data*, Springer.

Ahlgren, B., Hidell, M., and Hgai, E. (2016) Internet of Things for Smart Cities: interoperability and open data. *IEEE Internet Computing*, 52–56.

Akyildiz, I.F., Su, W., Sankarasubramaniam, Y., and Cayirci, E. (2002) Wireless sensor networks: a survey. *Computer Networks*, **38**(2002), 393–422.

Allmendinger, G. and Lombreglia, R. (2005) Four strategies for the age of smart services. *Harvard Business Review*, **83**(10), S.131–S.145.

Anton, S. and Sodano, H. (2007) A review of power harvesting using piezoelectric materials (2003–2006). *Smart Materials and Structures*, **16**(3), R1–R21.

Armbrust, M., Fox, A., Griffith, R., Joseph A.D., Katz, R., Konwinski, A., Lee, G., Patterson, D., Rabkin, A., and Stoica, I. (2010) A view of cloud computing. *Communications of the ACM*, **53**, 50–58. doi: 10.1145/1721654.1721672

Ashton, K. (2009) That 'Internet of Things' thing: in the real world, things matter more than ideas. RFID Journal. C Available at http://www.rfidjournal.com/articles/view?4986 (retrieved December 2, 2016).

Atzori, L., Lera A., and Morabito, G. (2010) The Internet of Things: A Survey. *Computer Networks*, **54**(15), 2787–2805.

Atzori, L., Lera, A., and Morabito, G. (2014) From "smart objects" to "social objects": the next evolutionary step of the Internet of Things. *IEEE Communications Magazine*, **52**(1), 97–105.

Baras, K. and Brito, L. (2017) Introduction to the Internet of Things, in Hassan, Q.F. (ed.), *Internet of Things: Challenges, Advances, and Applications*, CRC Press.

Berman, F. and Cerf, V.G. (2017) Social and ethical behavior in the Internet of Things. *Communications of the ACM*, **60**(2), 6–7.

Bledowski, K. (2015) The Internet of Things: Industrie 4.0 vs. the Industrial Internet. Available at https://www.mapi.net/forecasts-data/internet-things-industrie-40-vs-industrial-internet.

Bonomi, F. (2011) Connected vehicles, the Internet of Things, and fog computing. The Eighth ACM International Workshop on VehiculAr Inter-NETworking (VANET 2011).

Bonomi, F., Milito, R., Zhu, J., and Addepalli, S. (2012) Fog computing and its role in the Internet of Things. Proceedings of MCC'12, August 17, 2012, Helsinki, Finland.

Borgia, E. (2014) The Internet of Things vision: key features, applications and open issues. *Computer Communications*, **54**, 1–31.

Buyya, R. and Dastjerdi, A. (2016) *Internet of Things: Principles and Paradigms*, Morgan Kaufmann.

Cáceres, R. and Friday, A. (2012) Ubicomp Systems at 20: progress, opportunities, and challenges. *IEEE Pervasive Computing,* **11**, 14–21.

Jain, F A.K., Hong, F L., Pankanti, S.. and CERP-IoT (2009) Internet of Things: Strategic Research Roadmap. Technical Report, Cluster of European Research Projects on the Internet of Things, September 2009. Available at http://www.internet-of-things-research.eu/pdf/IoT_Cluster_Strategic_Research_Agenda_2009.pdf.

Chan, H.C.Y. (2015) Internet of Things business models. *Journal of Service Science and Management,* **8**, 552–568. http://dx.doi.org/10.4236/jssm.2015.84056.

Cheshire, S. (2017) Multicast DNS. Available at http://www.multicastdns.org/ (accessed February 27, 2017).

Chui, M., Löffler, M., and Roberts, R. (2010) The Internet of Things. *McKinsey Quarterly,* **2**, 2010.

Conti, M., Das, S. K., Bisdikian, C., Kumar, M., Ni, L.M., Passarella, A., Roussos, G., Tröster, G., Tsudik, G., and Zambonelli, F. (2012) Looking ahead in pervasive computing: challenges and opportunities in the era of cyber–physical convergence. *Pervasive and Mobile Computing,* **8**(1), 2–21.

Coutaz, J., Crowley, J.L., and Dobson, S. (2005) Context is key. *Communications of the ACM,* **48**(3), S.49–S.53.

Dastjerdi, A.V. and Buyya, R. (2016) Fog computing: helping the Internet of Things realize its potential. *Computer,* **49**(8), 112–116.

Dey A.K. (2001) Understanding and using context. *Personal and Ubiquitous Computing,* **5**(1), S.4–S.7.

Dourish, P. (2004) What we talk about when we talk about context. *Personal and Ubiquitous Computing,* **8**(1), S.19–S.30.

Ducatel, K., Bogdanowicz, M., Scapolo, F., Leijten, J., and Burgelman, J. (2001) Scenarios for ambient intelligence in 2010. IST Advisory Group, Office for Official Publications of the European Communities.

Economist (2009) The connected car. The Economist Technology Quarterly, pp. 14–15 (June 6).

EPCglobal Inc. (2014) GS1 EPC Tag Data Standard 1.9, 2014. Available at http://www.gs1.org/epc/tag-data-standard.

Espada, J.P., Martínez, O.S., García-Bustelo, B.C., and Lovelle, J.M. (2011) Virtual objects on the Internet of Things. *International Journal of Interactive Multimedia and Artificial Intelligence,* **1**(4), 23–29.

ETSI (2010) TC M2M, ETSI TS 102 689 v1.1.1 (2010-08): Machine-to-Machine Communications (M2M); M2M Service Requirements. Available at http://www.etsi.org/deliver/etsi_ts/102600_102699/102689/01.01.01_60/ts_102689v010101p.pdf.

ETSI (2016) SmartM2M; IoT Standards Landscape and Future Evolutions, ETSI TR 103 375 V1.1.1 (2016-10); Available at http://www.aioti.org/wp-content/uploads/2016/05/tr_103375v010101p.pdf (assessed February 27, 2017).

Evans, D. (2011) The Internet of Things. How the Next Evolution of the Internet Is Changing Everything [White Paper], Cisco Internet Business Solutions Group (IBSG).

Fano, A. and Gershman, A. (2002) The future of business services in the age of ubiquitous computing. *Communications of the ACM*, **45**(12), S.83–S.87.

Ferguson, G.T. (2002) Have your objects call my objects. *Harvard Business Review*, **80**(6), S.138–S.144.

Fleisch, E. and Tellkamp, C. (2006) The business value of ubiquitous computing technologies, in Roussos, G. (ed.), *Ubiquitous and Pervasive Commerce*, Springer, pp. S.93–S.113.

Fortino, G. Ganzha, M., Palau, C., and Paprzycki, M. (2016) Interoperability in the Internet of Things. Guest Editors' Introduction, Computing now [IEEE]. Available at https://www.computer.org/web/computingnow/archive/interoperability-in-the-internet-of-things-december-2016-introduction.

Foster, T. (2017) Regulation of the Internet of Things. Available at http://www.scl.org/site.aspx?i=ed47967 (accessed February 27, 2017).

Garfield, M.J. (2005) Acceptance of ubiquitous computing. *Information Systems Management*, **22**(4), 24–31.

Gershenfeld, N. (1999) *When Things Start to Think*, Holt, New York.

Gubbi, J., Buyya, R., Marusic, S., and Marimuthu, P. (2013) Internet of Things (IoT): a vision, architectural elements, and future directions. *Future Generation Computer Systems*, **29**(7), S.1645–S.1660.

Hill, J., Horton, M., Kling, R., and Krishnamurty, L. (2004) The platforms enabling wireless sensor networks. *Communications of the ACM*, **47**(6), 41–46.

Hu, F. (2016) *Security and Privacy in Internet of Things (IoTs): Models, Algorithms, and Implementations*, CRC Press. ISBN 9781498723183.

Iansiti, M. and Lakhani, R. (2014) Digital ubiquity: how connections, sensors, and data are revolutionizing business. *Harvard Business Review*, November, 90–99.

IEC (2014) Internet of Things – Wireless Sensor Networks [White paper].

IEEE (2015) Towards a definition of the Internet of Things (IoT). Available at http://iot.ieee.org/images/files/pdf/IEEE_IoT_Towards_Definition_Internet_of_Things_Revision1_27MAY15.pdf.

IEEE (2017) Define IoT – IEEE Internet of Things. Available at http://iot.ieee.org/definition.html (accessed February 27, 2017).

ITU Internet Reports (2005) The Internet of Things.

Jeschke, S., Brecher, C., Song, H., and Rawat, D.B. (eds.) (2016) *Industrial Internet of Things*, Springer Series in Wireless Technology, Springer International Publishing, Cham.

Karzel, D., Marginean, H., and Tran, T. (2016) A Reference Architecture for the Internet of Things https://www.infoq.com/articles/internet-of-things-reference-architecture (Accessed March 14, 2017).

Kennedy, J.B. (1926) Woman is boss: an interview with Nikola Tesla by John B. Kennedy. Colliers Magazine. Available athttp://www.tfcbooks.com/tesla/1926-01-30.htm (accessed January 30, 1926; retrieved on January 29, 2017).

Kortuem, G., Kawsar, F., Sundramoorthy, V., and Fitton, D. (2010) Smart objects as building blocks for the Internet of Things. *IEEE Internet Computing,* **14**(1), 44–51.

Koyuncu, H. and Yang, S. H. (2010) A Survey of indoor positioning and object locating systems. *International Journal of Computer Science and Network Security,* **10**(5), 121–128.

Kranz, M., Holleis, P., and Schmidt, A. (2010) Embedded interaction: interacting with the Internet of Things. *IEEE Internet Computing,* **14**(2), 46–53.

Langheinrich, M. (2001) Privacy by design: principles of privacy-aware ubiquitous systems, in Abowd, G.D., Brumitt, B., and Shafer, S.A. (eds.), *Ubicomp,* LNCS 2201, Springer, pp. 273–291.

Laudon, K.C., Laudon, J.P., and Schoder, D. (2016) *Wirtschaftsinformatik: Eine Einführung,* 3rd completely revised edition (in German), Pearson, ISBN: 97838689-4269-9.

Leverett, E.P. (2011) Quantitatively assessing and visualizing industrial system attack surfaces. Ph.D. thesis, Computer Laboratory, Darwin College, University of Cambridge. Available at https://www.cl.cam.ac.uk/~fms27/papers/2011-Leverett-industrial.pdf.

Lyytinen, K. and Yoo, Y. (2002) Ubiquitous computing. *Communications of the ACM,* 45(12), S.62–S.65.

Manyika, J., Chui, M., Bisson, P., Woetzel, J., Dobbs, R., Bughin, J., and Aharon, D. (2015) *Unlocking the Potential of the Internet of Things,* McKinsey & Company.

March, J.G. (1991) Exploration and exploitation in organizational learning. *Organization Science,* **2**, 71–87.

Mattern, F. and Flörkemeier, C. (2010) Vom Internet der Computer zum Internet der Dinge. *Informatik-Spektrum,* **33**(2), S. 107–121.

Mell, D. and Grance, T. (2011) The NIST Definition of Cloud Computing. Special Publication (NIST SP) – 800-145.

Minoli, D. (2013) *Building the Internet of Things with IPv6 and MIPv6: the evolving world of M2M communications,* John Wiley & Sons, Inc., New York.

Miorandi, D., Sicari, S., De Pellegrini, F.D., and Chlamtac, I. (2012) Internet of Things: vision, applications and research challenges. *Ad Hoc Networks,* **10**(7), 1497–1516.

Opensensors (2017) How to choose the best connectivity network for your project. Available at https://publisher.opensensors.io/connectivity (accessed February 15, 2017).

Perera, C., Zaslavsky, A., Christen, P., and Georgakopoulos, D. (2015) Context aware computing for the Internet of Things: a survey. *IEEE Communications Surveys and Tutorials,* **16**(1), 414–454.

Polsonetti, C. (2014) Know the difference between IoT and M2M. Available at http://www.automationworld.com/cloud-computing/know-difference-between-iot-and-m2m (retrieved December 4, 2016).

Porter, M.E. and Heppelmann, J.E. (2015) How smart, connected products are transforming companies. *Harvard Business Review*, **October**, 96–112, 114.

Poslad, S. (2009) *Ubiquitous Computing: Smart Devices, Environments, and Interactions*, John Wiley & Sons, Inc. Chichester, UK.

Postscapes (2017) IoT standards and protocols. Available at http://www.postscapes.com/internet-of-things-protocols/ (accessed February 27, 2017).

Roman, R., Zhou, J., and Lopez, J. (2013) On the features and challenges of security and privacy in distributed Internet of Things. *Computer Networks*, **57**(10), 2266–2279.

Rose, K., Scott, E., and Lyman, C. (2015) *The Internet of Things: An Overview—Understanding the Issues and Challenges of a More Connected World*, The Internet Society.

Sánchez López, T., Ranasinghe, D. C., Patkai, B., and McFarlane, D. (2011) Taxonomy, technology and applications of smart objects. *Information Systems Frontiers*, **13**(2), 281–300.

Sangiovanni-Vincentelli, A. (2014) Let's get physical: adding physical dimensions to cyber systems. Internet of Everything Summit, Rome, July 2014.

Schoder, D. Informationsmanagement 2.0 – Nur der Wandel ist stetig. *Wirtschaftsinformatik & Management*, Ausgabe Nr. **2011-02**. S.54–S.59.

Shin, D.-H. (2010) Ubiquitous computing acceptance model: end user concern about security, privacy and risk. *International Journal of Mobile Communications*, **8**(2), 169–186.

Sicari, S., Rizzardi, A., Grieco, L., and Coen-Porisini, A. (2014) Security, privacy and trust in Internet of Things: the road ahead. *Computer Networks*, **76**, 146–164. doi: 10.1016/j.comnet.2014.11.008.

Sudevalayam, S. and Kulkarni, P. (2011) Energy harvesting sensor nodes: survey and implications. *IEEE Communications Surveys & Tutorials*, **13**(3), 443–461.

Sun, Y., Ringfang, B., Peter, T., and Xiuthen, C. (2016) [Editorial] New advances in data, information, and knowledge in the Internet of Things. *Personal and Ubiquitous Computing*, **20**, 653–655.

Sundmaeker, H., Guillemin, P., Friess, P., and Woelffle, S. (2010) Vision and Challenges for Realising the Internet of Things. Cluster of European Research Projects on the Internet of Things, European Commision.

Uckelmann, D., Harrison, M., and Michahelles, F. (eds.), (2011) *Architecting the Internet of Things*, 1st edn, Springer.

Unify-IoT (2016) Supporting Internet of Things Activities on Innovation Ecosystems [H2020 – UNIFY-IoT Project; Deliverable D03.01]. Report on IoT platform activities. Available at http://www.unify-iot.eu/wp-content/uploads/2016/10/D03_01_WP02_H2020_UNIFY-IoT_Final.pdf (retrieved January 29, 2017).

Vermesan, O. and Friess, P. (2016) *Digitizing the Industry: Internet of Things Connecting the Physical, Digital and Virtual Worlds*, River Publishers Series in Communications, River Publishers.

Vermesan, O., Friess, P., Guillemin, P., Gusmeroli, S., Sundmaeker, H., Bassi, A., Jubert, I.S., Mazura, M., Harrison, M., Eisenhauer, M., and Doody, P. (2011) Internet of Things strategic research roadmap. *Internet of Things: Global Technological and Societal Trends*, River Publishers, pp. 9–52.

Weber, R.H. (2010) Internet of Things: new security and privacy challenges. *Computer Law & Security Review*, **26**(1), 23–30. doi: 10.1016/j.clsr.20 09.11.0 08.

Weinhardt, C., Anandasivam, C., Blau, B., Borissov, N., Meinl, T., Michalk, W. and Stoesser, J. (2009) Cloud computing: a classification, business models and research directions. *Business & Information Systems Engineering*, **1**(5), 391–399.

Weiser, M. (1991) The computer for the 21st century. *Scientific American*, **265**(3), S.94–S.104.

Weiser, M. (1993) Some computer science issues in ubiquitous computing. *Communications of the ACM*, **36**(7), 75–84.

Weiser, M. and Brown, J.S. (1996) Designing calm technology. PowerGrid Journal, v 1.01.

Weiser, M., Gold, R., and Brown, J.S. (1999) The origins of ubiquitous computing research at PARC in the late 1980s. *IBM Systems Journal*, **38**(4), S.693–S.696.

Welbourne, E., Battle, L., Cole, G., Gould, K., Rector, K., Raymer, S., Balazinska, M., and Borriello, G. (2009) Building the Internet of Things using RFID: the RFID ecosystem experience. *IEEE Internet Computing*, **13**(3), 48–55.

Weyrich, M. and Ebert, C. (2016) Reference architectures for the Internet of Things. *IEEE Software*, **33**(1), 112–116.

World Economic Forum (2015) Industrial Internet of Things: unleashing the potential of connected products and services.

Wortmann, F. and Flüchter, K. (2015) Internet of Things: technology and value added. *Business & Information Systems Engineering*, **57**, 221. doi: 10.1007/s12599-015-0383-3.

Wright, A. (2017) Mapping the Internet of Things: researchers are discovering surprising new risks across the fast growing IoT. *Communications of the ACM*, **60**(1), 16–18.

Yan, Z., Zhang, P., and Vasilakos, A.V. (2014) A survey on trust management for Internet of Things. *Journal of Network and Computer Applications*, **42**, 120–134. doi: 10.1016/j.jnca.2014.01. 014. http://dx.doi.org/10.1016/j.jnca.2014.01.014.

Zhang, Q., Cheng, L., and Boutaba, R. (2010) Cloud computing: state-of-the-art and research challenges. *Journal of Internet Services and Applications*, **1**, 7–18. doi: 10.1007/s13174-010-0007-6.

2

Environment, People, and Time as Factors in the Internet of Things Technical Revolution

Jan Sliwa

Department of Engineering and Information Technology, Bern University of Applied Sciences, Bern, Switzerland

2.1 Introduction

Introducing the Internet of Things (IoT) causes an immense change in the human life. It is often presented in an idealized way: current state-of-the-art technology is used, software is bug free, devices never break, and the users are always excited by their new digital life. Real life is evidently different.

First, the main factors that introduce change and disorder to this ideal world are discussed: environment, time, and people. The environment may be aggressive. Cooperating systems live their own lives and the initial compatibility will degrade. Systems' support will fade and expire. Keeping them running requires an effort that after a certain time may not be available. Designing and supporting IoT systems need top specialists with a unique blend of skills. As they are rare, many projects will have to accept mediocre teams with limited abilities. Pushing the projects anyway is risky, as mediocrity in critical infrastructure or life saving devices is a road to disaster.

Cybersecurity is a prominent issue. IoT systems are often heterogeneous, evolve fast, and contain commercial commodity devices with security holes. Due to these challenging conditions, guaranteeing enduring security is virtually impossible.

As IoT devices and systems collect and treat large quantities of date, Big Data is a hot topic in this context. However, mixing any data in any way—as long as no division by zero is performed—gives some results. With a good visualization, they dangerously look like truth.

Developing and managing adaptive, learning systems resembles raising children. There is a trade-off between teaching good manners and allowing to freely adapt to the outside world. Digital children gain independence and may become difficult to control by their human parents.

Internet of Things A to Z: Technologies and Applications, First Edition. Edited by Qusay F. Hassan.
© 2018 by The Institute of Electrical and Electronics Engineers, Inc. Published 2018 by John Wiley & Sons, Inc.

IoT devices observe the world and act on it. This alone raises numerous ethical issues. Such an invasive, ubiquitous technology is never neutral. Apart from knowing *how* it is developed, it is important to ask *why* and *for whom*.

This chapter addresses several IoT aspects that are easy to omit in a purely technical discussion and nevertheless important.

2.2 Technical Revolutions

Several times in the human history life has changed in a revolutionary way (Sliwa and Benoist, 2011). Each time it appeared as something new, not comparable with the past. Sometimes the new events were seen as the final stage of the development, like the end of history or the end of science. But after every alleged termination, the next disruption arrives. It is always good to look in the past, where patterns similar to the present ones can be recognized, albeit in a different setting.

2.2.1 Past Experience

As a past technical revolution, typically the industrial revolution is considered—the invention of electricity or of computers. However, even much earlier inventions had far-reaching effects, like the harnessing fire or introduction of agriculture.

Characteristic properties of a technical revolution are as follows:

- Even if people see it coming, the extent of change is surprising
- The consequences are unpredictable
- It marks a point of no return

The individual changes act on an interconnected system and the reactions can be observed far from the initial cause. Let us consider the phases of the change from hunting/gathering to agriculture (the Neolithic revolution). All starts with a woman (the men are hunting) who is gathering spikes of wild grasses for eating.

- A woman preserves a part of the grains and sows them next spring
- The group owns the field and the seeds, property needs protection→professional warriors
- Surplus is exchanged according to rules→numbers, writing, law
- Greater units have to be managed→cities, kings, and administration
- Community needs ideology→religion, literature, monumental architecture
- Grain selection→more productive species
- Cooperation, not hunting as success factor→genetic adaptation
- Living in narrow space→intellectual exchange, but also spreading of infectious diseases

In this way, preserving a handful of grains—step-by-step—has changed the face of the Earth. Some are good, some bad, some neutral. In any case, civilization reached another point of no return. With agriculture, the same land could nourish a population 10 times bigger than before. Therefore, the hunters/gatherers will never win against the farmers. The price is that the population becomes dependent on the new technology and coming back is impossible. The consequences of a failure are tragic. In the nineteenth century, the food supply of Ireland was dependent on potatoes. In the Great Famine (1845–1852), a potato disease destroyed this basic crop. Mass emigration was a solution, what gave rise to Irish Boston.

A similar chain of events marked the progress of the industrial revolution (from the flying shuttle in the weaving loom to the British steamships in the bay of Hong Kong) or the computer revolution (from single ENIAC served by computer gurus to iPhone in everybody's pocket). Each of them has changed profoundly the lifestyle and the society.

In all cases, the change starts as a minor improvement of existing processes. An accumulation of small events, supported by favorable conditions, leads to a disruption. The computer revolution profited from the fact that due to the physical properties of matter and the substantial development effort the density of electronic components effectively increased according to the Moore's law, with a similar growth of the mass storage and communication links capacity. Ubiquitous presence of electrical power and Internet access is assumed and a safe fallback is rarely considered, except for such limited cases like fire doors that as default have to permit escape. A failure of a crucial IT system disables the non-IT part. Without the flight control system, the airplanes will not fly. A hospital attacked by a ransomware virus will not perform operations. In both cases, some limited service is still possible, but in a very reduced scale and with high risk.

2.2.2 Internet of Things as a Technical Revolution

What is new in the IoT? Evidently, the way has been paved by several developments. The proliferation of mobile smartphones has made ubiquitous wireless signal a standard. Cheap computing power and storage capacity has enabled cloud computing. New sensors are smaller and cheaper, therefore, can be produced massively and deployed everywhere—on small, movable objects and on/in people. This opens a way toward virtually unlimited application possibilities (Stankovic, 2014).

A few decades of incremental changes have transformed human lives. In the 1960s, as a computer required an air-conditioned room with strengthened floor, the idea to mark a handbag with a computer would be inconceivable. Now, it can be done with small, concealable devices. This shows that the technology used later by currently young people will relate to the current one like an iPhone to an

IBM 360 mainframe. And, exactly as in 1967, the technology of 2017 was inconceivable, the technology of 2077 (or even 2027) is inconceivable now. The same is true to the applications. In 1967, a computer was used to develop rockets (in FORTRAN) or to print out the payroll (in COBOL). Placing funny cat videos by individuals and watching them on a hand-held device was beyond imagination. Similarly, some development directions for the near future can be conceived, but the far-reaching consequences, especially in the social life, are unpredictable. When around 1995, spare bandwidth of the mobile phones were used for 160-character short message service (SMS), nobody could predict the impact on the social life, from dating to flashmobs, in which mobile messaging permits to gather rapidly and spontaneously a group of people.

There are several emerging application areas of the IoT: Industry 4.0 (Leitão et al., 2016), object tracking and logistics, smart cities (Zanella et al., 2014), telemedicine (Adibi, 2015), environmental sensors, and more. Their direct goal is clearly visible, it is however impossible to predict their long-term impact on the character and dynamics of the social interactions, man–machine symbiosis, uncontrolled evolution of learning machines, or subjugation of the individuals and society to automated decision algorithms. The experts always try to develop the visions of the future, but sooner or later the actual development diverges and new social structures and human behaviors emerge.

2.3 Cyber–Physical–Social Systems

The very goal of the IoT systems is an intense interaction with the physical world. This world is composed of the physical objects, other technical systems, and humans. Humans again have physical bodies, working according to natural laws, and minds, supporting mental processes. In order to build a system interacting with the real world, this world has to be understood. This will be discussed in this section.

Let us consider here the position of a typical IoT developer with the IT background. His/her mentality can be summarized in a following simplified (and exaggerated) way:

- Objects have binary states (0 or 1)
- The changes between states are immediate
- Hardware can be easily replaced
- Software can be easily upgraded and activated after a reset
- Last bugs will be found in action
- The users are like the programmer

Real-world objects are different. Objects have a continuous spectrum of multidimensional states. The answer to the question “does this medical device work?” is not a simple “yes” or “no.” It includes the control algorithm, weight and obtrusiveness, battery life, easy user interface, failure rate, and short and long-term efficacy.

Objects have dynamic properties. Heating a large object takes time and obeys to partial differential equations in three dimensions. If the process is measured by a grid of sensors distributed in the volume of the controlled object, the developer has to be aware of it. Those dynamic properties have proportional, integral, and derivative (PID) components, and if the control signal is not adequate (has a wrong waveform), the object may oscillate, become unstable and break.

A computer after an upgrade and reset restarts from zero. If a patient has an implanted defibrillator, it should be replaced after a failure. However, after a longer usage the wires leading to the heart are covered with tissue and a reoperation is risky. For an elderly person, it can be deadly. In the case of a fluid injecting device, if the substance, dosage, or timing were wrong, body needs some time to recover. Therefore, the reaction to the modified settings is not visible immediately.

Evidently, in the case of an industrial or medical application, the safety grade has to be high and well defined. The "best effort" attitude is not enough. A graceful degradation in case of a partial failure has to be defined.

The behavior, mentality, and preferences of the actual users have to be analyzed. This begins with the basic ergonomics, like the character size, for example, the diabetes patients under an insulin shock have an extremely reduced vision capacity. Operators of a nuclear power plant need no gamification in their user interface.

All those issues are not necessarily taught in a computer science course (Sliwa, 2014). In order to develop reliable IoT-based systems, a broader perspective is needed, especially because the consequences of the actions taken by the system and decided by its software can be quite severe. On every level of the development process this awareness is required. The programmer sees his/her code as follows:

```
if (condition)
then action1
else action2
```

The program acts on real people. The `action1` in the code may be evacuating an airport, a defibrillator impulse, calling for an ambulance, blocking the credit card or—in an extreme case—shooting by a drone. Therefore, the developer has to be aware of the importance of the code and specify and evaluate the `condition` accordingly.

Another trap is the unexpected error conditions after program actions (like `Exceptions` in Java). They should be handled properly, but as they disrupt the program flow, the programmers often just record them in a log, or make a pro forma handling. Especially when working under the deadline pressure, many are tempted to assume that this situation will never occur anyway. If the connect to a database is established or a file is opened, the programmer *knows* it will be there. But sometimes it is not, and the program crashes in a random way. In an IoT system, the consequence may be switching off the power to a part of a city, or else—depending on what is the goal of the system. For a programmer accustomed to just pressing the reset button, this requires a mental change.

2.4 Environment

The environment in which an IoT system operates can be very diversified. Depending on its usage and required safety level, the allowed conditions have to be defined and tested. A commercial airplane flies through turbulences in the test phase, lands with one engine in the sandstorm, and starts at −20 °C on an icy airfield. It is tested with various instrument landing systems, also in the presence of electromagnetic interference. Evidently, not all systems have to operate in an equally harsh environment—this has to be adapted to the specific case. Such procedure is the state of the art for high reliability equipment that has to be matched by IoT systems.

2.4.1 Physical Environment

An IoT system is initially developed in well controlled laboratory conditions, but the environment in which it will be deployed may be quite harsh. It will be exposed to varying temperature and humidity and vibrations. Indoor problems are interaction with other systems, magnetic fields, or radiation. Outdoor challenges include atmospheric conditions like rain, snow, frost, or overheating. In certain zones, hurricanes and typhoons are common. In countries with frequent seismic activity, the behavior during an earthquake has to be well defined.

If a device is not in a protected perimeter, it can be stolen or physically damaged. Even in a basically peaceful country, there is a risk of an extreme vandalism; for example, recently in Germany where during the G20 Summit in Hamburg (August 7–8, 2017) the antiglobalist militants converted the cite center into a war zone in three nights.

In a normal operation, material parts get worn out. Sensor surfaces can get dirty, which distorts the measurement. Injection tubes get clogged and the amount of effectively injected substance diminishes. They need cleaning and replacement.

Apparently, trivial cases also have to be taken into account. Let us consider infrastructure sensors placed on the top of bridge pylons. Fixing them with screws takes longer and makes the connection sensitive to corrosion. When fixed with glue, they can fall down and cause damage.

These examples show that the designer has to show a good knowledge of the physical world, of possible effects (present and future) and use imagination.

2.4.2 Other Technical Systems

The system often depends on the availability and correct functioning of other technical systems. In treating heart disease, vital signals can now be continuously monitored and the hospital can be automatically alarmed in the case of an emergency. On the other hand, a typical European restaurant toilet is in the basement, screened by a concrete floor, and no phone signal is available. Narrow staircase and bad air raise the risk of a heart failure. The local system basically works—it detects the event and generates the alarm—but the information flow is blocked, therefore, the action is useless. The designers should answer the questions: Should the system warn the patient about leaving the safety zone? Do not too many alarms pose themselves a health risk? And an important point: the designers have to be aware of the issue.

Another problem is a possible interaction. IoT Systems are new and relatively rare, but there are visions of space filled with communicating devices, with "smart dust" as an extreme form. If every designer considers his/her system as the sole owner of the physical space and of the bandwidth, this may lead to collisions (Sparber et al., 2017; Jeong and Shin, 2016). For instance, there can be a meeting of the users of heart monitoring systems—100 people in one room. Will there be enough bandwidth, and will the systems not interfere? Moreover, filling the space with IoT devices opens doors for malicious actions, as the eavesdropper/attacker device is small and difficult to detect. If many systems compete to offer a similar service, disturbing the function of the competitor is a tempting option.

Many applications rely on the GPS navigation. The GPS signal is fairly weak and easy to jam. The goal of an attack on GPS may be the wish to disable navigation in a certain region, as in the North Korean jamming of the South Korean space (Seo and Kim, 2013). For some places, like the English Channel or the Malacca Strait, essential for the world trade, the strategic importance is evident. Currently, in such places backup systems are deployed, such as eLORAN, an enhanced version of the older LORAN-C. Evidently, the passing ships have to be equipped with compatible devices. They may be different for different locations, as they depend on local regulations (Figure 2.1). This example shows how vulnerable is a composition of independent systems. "Navigable" is not a property of the ship alone, but of the ship together with the location, time, and current conditions. Similarly, a communicating medical device is not functional *per se* without the supporting environment it is useless.

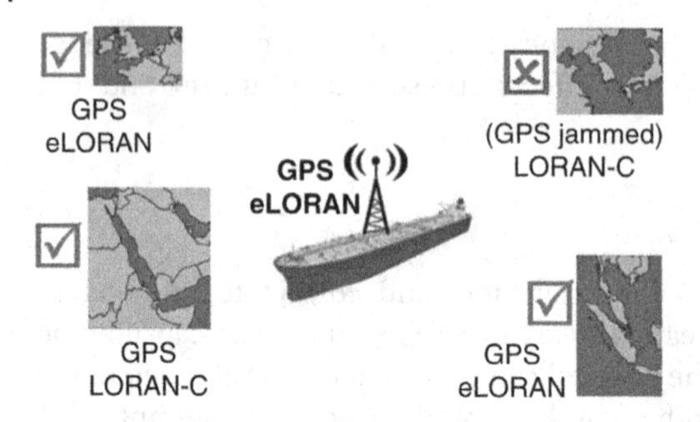

Figure 2.1 Navigable ship.

2.5 Time

IoT systems are often present in a timeless perspective. For quickly evolving systems, time, however, plays a major role (Rose et al., 2015).

2.5.1 Changing Goals and Values

Currently, developed systems are designed to solve current problems. With passing time, the problem may disappear or be reformulated. Let us start with a historical example. Around 1900 in Paris, the city administration was overwhelmed by the problem of garbage produced by horse carriages. Soon, the problem vanished together with the carriages. The problem of hay and manure management was replaced by the problem of providing car parking place and pollution by exhaust gases.

The evolution of the public phones is a more recent example. Long time ago, phones were available at the PTT (Post, Telegraph, and Telephone) offices. Then, public booths were deployed, first operated with single coins, then with complex mechanical system of payment and change, later replaced by coins and chip cards, and then by chip cards. The lifetime of each subsequent system was shorter. At a certain time, public phones were even installed in trains, in restaurant, or first-class cars. With the advent of the cellular phones, most of the public phones were removed.

At every stage, a system built to solve a certain problem was replaced by another one, at ever quicker pace. For a massively deployed system, every replacement requires an effort and may leave waste. It may be taken for granted, but smooth changes rely on healthy economy and technical capabilities. If these factors are missing, cities will degrade and with time will become interesting only as industrial archeology sites.

In the examples above, the main goal (transportation, communication) remained unchanged. However, due to perception change or disruptive events, those goals may also change. Traveling may become unattractive due to terrorism. Massive privacy breaches and data misuses may break up the social media and stop the quantified self-programs. Breaking public cryptography by quantum computing will disable all electronic payment systems. This again will destroy the entire economy, if in meantime physical cash is eliminated. Finally, typical habits may change—calculating the timing of Ramadan may become more important than quick dating. Such issues are mentioned just to stress that the assumptions about the common lifestyle and objects of desire are neither universal nor constant.

2.5.2 Interoperability Degradation

As already described, a system like heart monitoring with automatic alarm depends on many other systems, the existence of which is just assumed. They are controlled by autonomous parties, like hospitals with their emergency services or phone companies or power grid operators. With some there can be a cooperation agreement, with some not. After such a system is established, there is a risk that the initial interoperability will degrade with time. The community pushes for new standards that will replace the old ones. They will be implemented by the individual parties in their own chosen pace. Even minor modifications in communication protocols or data definitions (e.g., disease encoding) may have negative consequences.

A real-life problem will illustrate how apparently trivial causes can disable the interoperation. An IoT system consists of many communicating devices of various types. In the beginning, everything works correctly. The system integrator buys devices in large quantities, the device producers buy components from remote companies based on the current price. After a certain time, the components sold under the same name or as a direct replacement show slight differences in the implementation of the communication protocol. The problem is detected only in the customer's installation.

In a commodity market with low-profit margin some producers are tempted to use fake chips. It can happen that the producer of the original chips who also delivers the control software wants to fight the dishonest competition. The automatically updated control software stops to work when a fake chip is detected. The problem is detected as the malfunction of the final system and depending on the risk and safety level can have catastrophic consequences, if it is an airplane or a critical infrastructure with continuous operation. The problem is unexpected and its origin is at a very low level. The cost of the correction may be immense, even under the assumption that the component producer is available and ready for an immediate action. After a certain time, this company may be out of business, restructured, or not have enough capacities and skills to perform the correction.

2.5.3 Long-Term Support

Long-term operation of an IoT system faces the problem of incompatible life cycles of the related industrial system, the IoT system itself, and the involved people. The life of the industrial infrastructure (power stations, railways, bridges) is measured in decades. It will evolve—on the same rails, faster trains will go, with improved control systems, but with continuity of the basic structure. The control system, one day based on the IoT, could rapidly evolve, thanks to the progress of technology. However, the size and quantity (hundreds of stations with many track switches, thousands of trains) makes an upgrade costly and rare. Therefore, the equipment lives well beyond the time when it would be used for a new project.

Let us consider an industrial system (Figure 2.2) that lives for decades, like a power station. At the beginning, a computer-based control system using the technology A1 has been installed. It has been delivered by the producer A that in meantime has switched its product line many time and possibly has merged, has been overtaken, or went out of business. Getting support becomes over time more and more difficult. It is even worse with commercial off-the-shelf (COTS) products, coming from an anonymous source in the Far East, where documentation is scarce and support inexistent. When the existing technology finally has to be replaced by the current state of the art (B1), again a deep knowledge of the present state is necessary. In this moment, people and their skills play a role. The engineers who worked on the present system have retired or have been laidoff. As the recruitment system concentrates on a comprehensive list of modern skills, the engineers are incited to collect them. As an example, getting a young professional willing to upgrade a system of electromechanical switches with a control panel programmed in Visual Basic 4 is next to impossible. Moreover, if

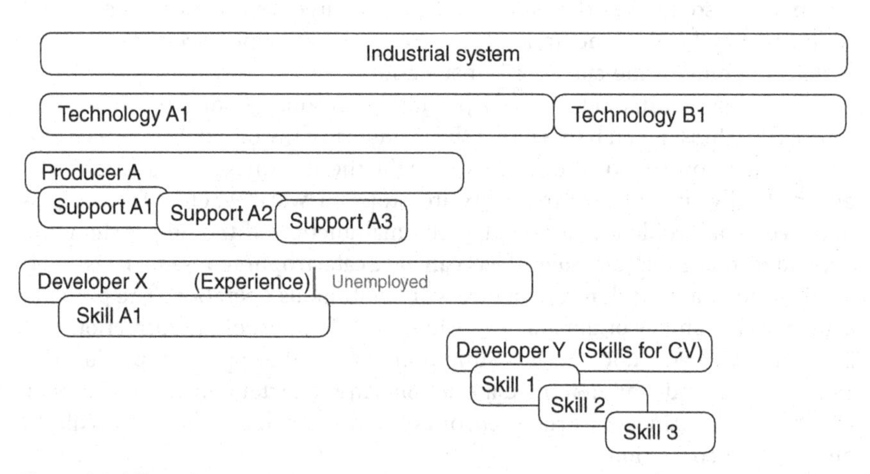

Figure 2.2 Life cycles.

the company is not perfectly organized, the project documentation will be incomplete, with texts stored in WordPerfect on 5 in. diskettes, which gets even worse if the development has been outsourced. This perfect example has to show the difference between the ideal and the real world. The fact that it describes the past should be no consolation. Only the time span is significant, and the future will look the same, especially with even faster progress pace, agile development, and programmers changing their jobs frequently.

2.5.4 Erosion and Economy

All technical systems (as well as all-natural formations) are subject to erosion. It is known from previous generations of technology. The infrastructure installed in a city, like power grid, gas pipes, fiber optic cables, or underground railway, does not last forever. It needs regular service and periodic upgrades, or otherwise it disintegrates, and finally leaves only useless relics. A recent case of such disintegration may be Detroit[1], once a burgeoning industrial city and now a place of spectacular ruins.

The problem of the discontinued support and of forgotten skills has been mentioned before. It is often tacitly assumed that providing adequate service will always be possible. With personal computers, the users got accustomed just to buy a new one if the capacity is too small or if more and more programs cease to work properly. In an industrial environment, it may not be a solution. The number of developed systems and their individual cost may be prohibitive. This causes certain inertia. Also, upgrading an industrial system when ensuring its continuous operation is much more difficult than copying the files to a new PC. If there are not enough capacities and skilled staff on the market, the system will lose its function.

The NASA moon flight program can be a warning. At Cape Canaveral, the tourist can marvel at the last Saturn V missile. In the 1970s, supported by a capable team, it could fly to the Moon. Now, it is just a museum object. This can happen to current systems, if people lose the interest, the will, and the resources. There is a risk of a digital Detroit. Not only the remains of the old systems will not work, but they will be a nuisance. This is bad enough for industrial installations but patients can be left with unsupported devices implanted in their bodies. A materially and functionally disintegrating deep brain stimulator is not an attractive option.

Also, more general issues have to be considered. For many decades, people have lived in prosperous times, at least in the Western world, and have forgotten what is a real crisis—a natural catastrophe, an economic depression, or a military conflict. Evidently, such conditions will have catastrophic consequences on any

1 Yves Marchand and Romain Meffre, The ruins of Detroit (2005–2010). Available at http://www.marchandmeffre.com/detroit (accessed August 23, 2017).

level of technical development. Such possibilities make contrast with the frequent visions of the *brave new world*, where the only problem will be the excess of free time—and no work for the 90% of the population.

2.5.5 Transferring Adaptable Objects

Intelligent objects adapt to the environment and learn from experience. A smart home detects the daily and weekly rhythms and habits of its residents. It also knows the local weather cycles. An intelligent car knows the driving style of the owner, his/her wishes and preferences. A driverless car gets accustomed to the typical routes, slows down in front of the school, and watches if big trucks are leaving the factory. Everything works perfectly if those rhythms are stable enough and every new event rather adds to the experience than destroys it. Each device and system behaves like an old butler, knowing the desires of its master, even before they are expressed.

What happens when the conditions change? Should the family give notice of departure to their home, so that police are not called when the rhythms change? If a temporary resident gets in, for example, a user of a home -sharing program like Airbnb, the home would be obliged to act – lock the doors and keep the intruder inside. The guest will be freed by the police patrol, provided they have electronic credentials to open the door. If the home has no option for temporary residents, the guest has to leave. In the same way, grandparents living in a home equipped with ambient assisted living cannot be visited by grandchildren without reprogramming the installation.

Personalized objects and environments also have to manage the ownership. The legal owner has all identity certificates and cryptography keys that permit to him/her (and to no one else) to communicate with the object, read the status, and to control it. If it is an object/environment composed of many elements, like a smart home, they have to know their neighbors, because they will exchange information only with known partners in a protected manner.

If a smart car is sold, it has to forget a large part of its experience. The new owner will have a different driving style and prefer different settings (temperature, seat position, radio stations, etc.). He/she will have a different voice and will express commands with other words. Possibly, he/she will take the car to a region or country with a different climate and where other drivers behave differently. However, basic driving skills—like avoiding traffic jams and obstacles—will still be useful, although the shape and color of the obstacle will change.

Moreover, the object has to be transferred completely, including all access rights. Otherwise the previous owner will be left with "spare keys." These subjects are known, although the practice is dominated by the wish to sell the car or home and leave the problems for later.

2.6 People

Technology is done by people for people. Behind a hardware circuit is a designer who orders chips from the catalog just on the basis of price. Behind a software system is a programmer who is working under time pressure and has no time to analyze all *what–if* questions. Evidently, most of the time they do a good job, but they are mere humans with strengths, weaknesses, and personal goals.

Demography is also rarely taken into account. For example, it can be required that all high-risk systems get an adequate protection against cyberattacks. From the number of the future systems and the required technical skills—assuming those values can be estimated—the number of necessary security specialists with certain skills can be derived. This simple, linear model is however wrong. The number of experts will not grow to infinity; it will saturate (nonlinearity). The recruitment method, favoring skills over experience, will influence the process. Finally, possible security experts, being human and looking for money and respect, may be attracted by other professions. Consequently, many future systems will not be sufficiently protected, whatever the requirements.

2.6.1 Users

There are several classes of users of the IoT devices and systems. The operators of industrial installations get professional training and using those systems belongs to their duties. More interesting are individual users. Their behavior is a real test of the design of the devices. If the portable devices are obtrusive or too heavy, they will be used rarely or put aside entirely. The user interface has to be adapted to the cognitive abilities of the users, like elderly patients. It has to leave no place for ambiguities. There were cases of programmable insulin pumps where missing data validation permitted to set the dosage to health impairing, possibly lethal values. A manual action is also a source of errors. If a patient has to place a sensor patch on his/her skin, the location can be wrong and the skin can be too dry or too wet.

Because of the unpredictable behavior of the users, care should be taken when analyzing data supplied by them. Already the division of the users who are willing to use the device and to share data produces a bias. Missing data will be a common problem. User-supplied data such as age, weight, education, and so on can be true or not.

2.6.2 Developers

What is the goal of a software developer? He/she wants to have an interesting, well-paid job for 40 years, from 25 to 65 years of age. During this time, the profession undergoes several disruptive changes. In the past, it was from

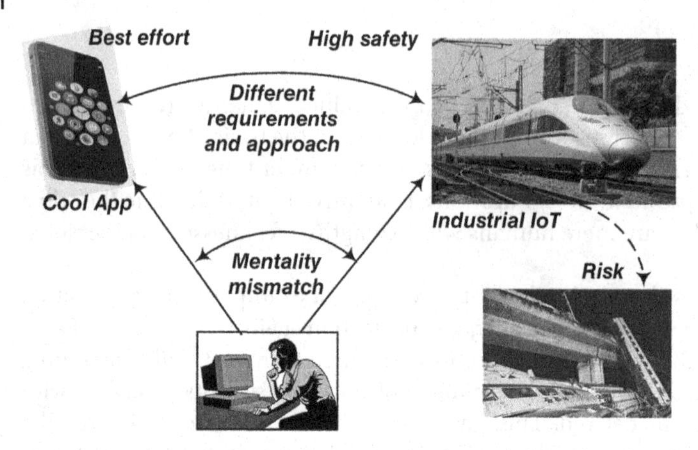

Figure 2.3 Mentality mismatch.

punched tape, teletype, and assembler to smartphones and Java. As the lifespan and the retirement age are expected to be similar, a similar—if not faster—sequence of disruptions can be expected. Human resources departments usually select candidates according to a skill list; some companies may automatize the process, especially in the initial phase. Therefore, an extensive collection of modern buzzwords gives advantage on the market. Persistence, meeting deadlines, readiness to do less spectacular tasks, or understanding other disciplines play a secondary role. Managing career in software development is a daunting task (Fowler, 2009), especially during major technological changes.

What is special in the development of the IoT systems? A new factor is that cheap elements permit to quickly implement devices and systems ready for a massive use. On the other hand, aside from toy-like systems, they can be used for responsible tasks where safety is crucial. On the high end, there are smart medical devices, possibly implanted into the body, such as cardiac defibrillators or deep brain stimulators. They not only measure vital signals, but they act directly on the body in essential functions. Evidently, such systems cannot be developed by young programmers with experience in lightweight applications for smartphones. Deep understanding of the physical, chemical, and biological processes is necessary, as well as a good imagination to think about possible risks. Such developers work for industrial companies, programming control systems for trains, power stations, and intensive care units. The highest goal is safety and reliability. Often such systems evolve for decades, domain knowledge is essential and all tools have to be well tested before use. Usability and clear dialog are crucial, visual appeal plays no role, may be even disturbing.

The other community is the "Cool Apps" programmers, knowing all the newest tools and eager to change them every month. The lifetime of their products is also much shorter; they work, therefore, in a faster time scale. Real-world programming is an addition to them and they often lack knowledge in natural sciences. This is stating of the facts, not a criticism. Such people bring the innovativeness necessary in a rapidly changing world.

Both worlds are fairly separated and there is a mentality mismatch between them (Figure 2.3). A balance between freshness and experience in the development team is needed. Still, capable workforce can be a limitation.

2.6.3 Supporters

All serious systems need an organized maintenance. In the case of a heterogeneous system, cooperating with other systems, the basic question is the responsibility. Who takes control, and where are the system's borders? Systems should not be divided by no man's land. What happens if one of the cooperating systems is upgraded, with extended or slightly changed functionality? Are all dependencies known and documented?

Evidently, such tasks require a dedicated, well-trained staff, having not only general skills but knowing the details of this specific application. Such knowledge can only be gained with time, and keeping good specialists is not easy.

A specific task is security management, essential for ensuring the continuity of operation. The demand for highly qualified specialists is constantly growing. Such skills seem to ensure optimal career prospects for the foreseeable future. However, the subject is demanding, and there is a difference between a formal certificate and real-life experience. The lack of security experts may be one of the major hindrances to the proliferation of safe IoT systems.

2.6.4 Project Managers

As IoT projects are typically multidisciplinary, one of the roles of a project manager is to enable the cooperation between people with various skills, backgrounds, mentalities, and habits. The case of smart medical devices is one of the hardest, as there are strict safety regulations and the market is difficult: The device has to be "prescribed" by doctor and reimbursed by insurances, based on the efficacy proof. Therefore, the following people have to cooperate: specialists in biological sensors and actuators (chemistry, micromechanics, nanotechnology, etc.); many classes of software developers (real-time software, smartphone apps, security); specialists in Big Data, stream processing, and data analytics; statisticians; experts in medicine and hospital processes, in approval procedures, and in legal and ethical procedures, in compliance with privacy rules; and maybe more.

This means that the project manager has to be able to communicate with so many different people and, evidently, to have an overview of the whole. It is useful to take advice from external experts from other domains who can bring another perspective and help to expose the omitted aspects.

Another important task of the project manager is to keep the team, especially in a long-term project, that develops a series of related products with guaranteed support. This leads to a game with the employees where the stakes are stable employment, sellable skills, and long-term job perspectives. In the software development, there are several levels of knowledge and abilities. The basic level is IT skills, like knowing languages and tools. The next level is knowing the domain, and then knowing the specific project. At the open job market, the IT skills are decisive and they are in constant motion. Domain knowledge is an asset for larger domains, like banking, but investing in narrow domains is risky, as they fluctuate strongly. Experience in a specific project is no asset at all at a job change. On the other hand, for a long-term project such expert is invaluable. He/she knows the structure of the software and the dependencies, therefore, when a modification request comes, he/she can react immediately. Getting this level of knowledge takes months. However, when the product support is phased out, such developer who worked for years on the same subject and was stuck in the same technology has very bad chances on the job market, especially if he/she is not so young.

These requirements define an ideal project manager, but evidently, such people are rare. They play a similar game as their employees, because constantly increasing expertise in an area that one day will fade away is a dead end. The positive side is that managing complex projects and working with people is a transferable skill.

2.6.5 Manufacturers

The manufacturers are interested in producing, selling, and making profit. They are innovation driven and want to be the first on the market. On the other hand, they are bound by rules and regulations defining the safety, personal data protection, and the necessary tests and certifications. For medical products, for example, formal clinical trials may be required. Therefore, many more devices (like fitness trackers) are sold as commercial products, for which the usability for user's health improvement is suggested but not formally proven. Evidently, quality is essential for the company's reputation and liability claims may be financially destructive.

The manufacturers also want to develop their products at a low cost. This relates to the questions how to keep an optimal team, how to balance freshness with experience, and whether to develop locally or to outsource.

2.6.6 Regulators

The regulators overview the technology and the market, and define and impose the regulations. Those tasks face many challenges. There are many markets with different issues. Driverless cars may cause accidents, monitoring cameras may breach privacy, and medical devices may harm the health. Those markets are not separate, because the technologies will be combined: A driverless car with a rooftop camera may document the trip but also monitor political demonstrations or spy on neighbors. Is it a car with a camera or a camera with four wheels? Such blurring of borders makes unclear which institution is actually responsible. Especially in the privacy regulations, the delimitations of competences are not sharp. The regulations are national or regional, like in the European Union, but the Internet is essentially borderless. It is possible to follow the rule that no data are explicitly transferred for processing to another country; however, guaranteeing that no data packet ever crosses the national border is virtually impossible. If a French company uses a Swiss server (presumed particularly secure) and stores data of worldwide customers, whose rules effectively apply?

Also, all rules forbidding the reuse of data have exceptions as those data are necessary to maintain and improve the service. Data are formally owned by the customers but managed by the company that is tempted to profit from them as much as possible.

Finally, the basic problem is the dynamics of the market: assessing the problem, considering all opinions, and making tests take time. Some effects are only visible much later, as in the medical treatments. In the meantime, the problem may not be relevant anymore. The answer is there, but new questions arise. In this way, the regulators are the natural adversaries of the manufacturers who see the regulations mainly as hindrances to innovation and initiative. Both positions can be defended, and in every case a balanced agreement is necessary.

2.7 Cybersecurity

A nonnetworked device, even if computer controlled, is subject to software bugs and has a limited range of possible faults. A traditional infrastructure, like a power station, can be protected by physically controlling the access to the perimeter, with sentinels, dogs, and searchlights. Networked devices and systems are accessible worldwide and need a different kind of protection.

An IoT system is especially vulnerable, as individual devices can be accessed without any physical contact. Many such devices are bought as a commodity. In consequence, their security depends entirely on the producer that cares more about sales and income. In networked devices, like cameras, many flaws have been detected, for example, hard-coded credentials, open Telnet ports, and classic errors, like SQL command injection or cross-site scripting. If misused by malicious agents, they can lead to such disasters like the DDoS (Distributed Denial of Service) attack by the Mirai botnet (Kolias et al., 2017).

As illustrated in Figure 2.4, such hidden vulnerabilities can propagate. If a malware takes control of a networked device that is externally visible, it can overtake the system servers. This evidently depends on the details of the

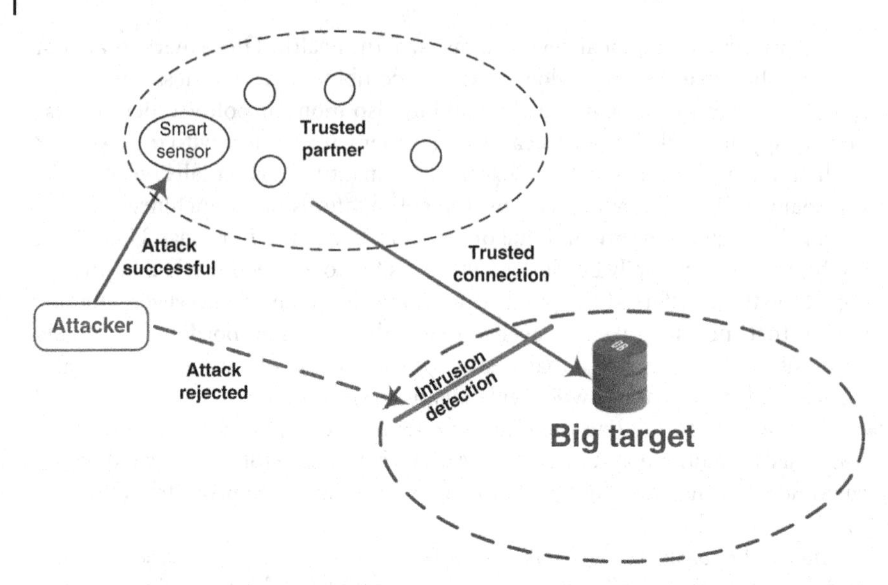

Figure 2.4 Indirect attack.

implementation—how good is the protection and how clever is the attacker. The owner of this system may underestimate the attack risk and save costs needed for a proper protection. This peripheral system may not be the target of the attack, but being a trusted partner of an important company, it may serve as a stepping stone and facilitate an indirect attack on the big target. Evidently, the attacked company can increase the sensitivity of its intrusion detection system, but there is a risk of blocking the necessary trusted connection on every spurious event.

In a networked system that does not control all elements and communicates with several partners, such margin of unsafety will exist. Evidently, security of a nuclear power station shall not rely on the settings of a camera in the partner's system. The control system of a strategic infrastructure has to be built in security circles and all weak points have to be identified. Currently, the attackers seem to have the upper hand, especially state-sponsored teams performing targeted attacks, as it (probably) was in the case of the Stuxnet worm (Kushner, 2013), the cyberattack on the Ukrainian power grid in 2015 (Lee et al., 2016), or the recent Petya worm that attacked many industrial companies, among them Maersk, the world's largest shipping company[2]. The ships could not be docked for a week and the financial losses are expected to reach $300 million.

2 Maersk, WPP and FedEx still struggling with cyber attack fallout. *Financial Times*, July 5 2017. Available at https://www.ft.com/content/b8432fc4-60c1-11e7-91a7-502f7ee26895 (accessed August 23, 2017).

A high-responsibility system needs a profound analysis of threats and mitigation options, for which structured methods exist (Tedeschi et al., 2017). There are several special cases, like medical device software (Fu and Blum, 2013), that manifest domain-specific issues.

2.8 Reasoning from Data

It is easy to generate some statistics from collected data and to present them in a visually attractive form. Statistics is, however, a difficult discipline and the road from collecting data to obtaining reliable knowledge is rough. First problem is the quality of data. In Paris, recently a system of microparticle pollution has been deployed. Sensors are placed on the top of cars belonging to an electric power supplier company. Being cheap, they have a lower sensitivity. Even worse, they get dirty and need recalibration (or cleaning). Therefore, they also show the pollution trend, but the absolute values are not reliable. This shows that reading a variable called "particle concentration" does not mean it really contains a credible value of actual particle concentration. It has to be checked how the value is measured, what is the precision, and do the conditions change in time. Are some values missing, how are they treated? How well are the positions of the sensors known?

Another problem is bias. In medical applications, consent for data reuse is required. Community projects, like quantified self, rely on voluntary data sharing. It can be expected that mostly young, computer savvy users will do it. Let us consider a data collection system used to get information about the progress of some medical condition. If conclusions are drawn from those data and applied to the entire population, elderly people will be treated like the young ones.

Typically, cause–effect dependencies are of interest, because they can be used to build predictive models. Many problems are described by a complex network of interdependent factors. If the search for results is too hasty, there is a risk to find spurious or incomplete correlations. In a medical case (Figure 2.5), many factors play a role. A rapid option is to extract only the correlation between directly measurable values (usage of a medical device—vital signals), not because it is the most important, but because it is the easiest to treat. As an anecdote says, some look for the lost key under a lamppost, not because the key has fallen there, but because the light is better. In this way, other important influences will be ignored and the resulting causation rule will be of limited value. However, identifying all parameters of a complex, evolving system is very difficult.

Finally, often long-term consequences of certain actions are interesting, like the effect of a medical treatment or the effect of security measures on criminality. On the other hand, the results are expected rapidly. These requirements are in

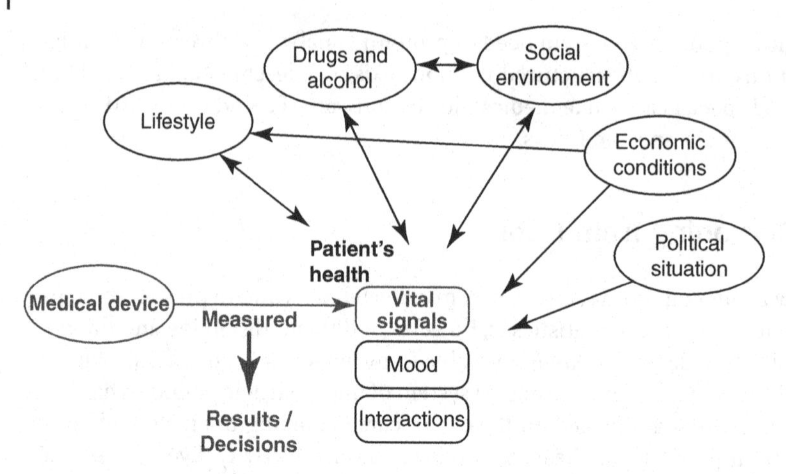

Figure 2.5 Data analysis/extracting knowledge.

conflict. Therefore, long-term predictions based on short-term observations have to be constantly verified.

Moreover, there is a risk that well-presented data will be taken for true, just because of their quantity and an appealing presentation. If they are uncritically used for decisions (actionable knowledge), the actions taken may be wrong.

2.9 Adaptable Self-Organizing Systems

In preceding technical revolutions, humans exploited new ways to use the properties of matter in order to reduce physical effort, typically without much regret. Nobody runs after cars, nobody competes against forklifts. Losing muscles was compensated with gymnastic exercises performed at will. With the computer revolution, people transferred simple mental skills to the machines. As effect, many have lost the ability to read the map, and wanting to fly to St. Petersburg, Florida, can now land in St. Petersburg, Russia. Many more complex tasks have been delegated, like air traffic control, power grid management, and so on. Delegation means reliance and consequently dependence, with no way back. In the meantime, even such quintessential artificial intelligence (AI) task, like playing chess, is better performed by a machine. This suggests that humanity is on the verge of a disruptive change, where humans get locked into a total dependence on intelligent technology and the development of this technology gets out of control. The first step has actually happened and essential services have been gradually delegated to machines. The future scenarios go from the stepwise automation of society (Helbing, 2015) to the takeover by AI surpassing the human one (Bostrom, 2014).

An ever-larger part of human lives is controlled by algorithms and data (Pasquale, 2015). A learning algorithm adapts to the observed reality and optimizes itself. The designer defines the initial algorithm and the method of learning, but the result of learning, that is, the effective algorithm depends on the data and changes dynamically. This evolution is similar to raising children: The parents prepare their children for independence, but they do not want them to cross certain borders and they want to intervene, if necessary. For instance, drug dealing may appear as an optimal solution to financial problems; the parents, however, know it is only a local, short-term optimum. The decision algorithm is not necessarily visible as verbalized rules. It may rather be hidden in the signal weights of a multilevel neural network. Therefore, at a certain complexity level, it is later impossible to summarize the actual rules or to modify them, especially if the control implemented by the algorithm is too fast for a human intervention.

Even if the designers intend to build a relatively simple control loop, the interactions with the external world—if strong enough—will make it a part of a complex, networked system. There are various approaches to the systems theory (Arnold, 2013). An important feature of such systems is self-organization and emergence of new properties on a higher complexity level (Nicolis and Nicolis, 2012; Damper, 2000). The higher level is based on the lower level but is not fully explained by it. Chemistry is based on physics and biology on chemistry. Although DNA is built of simpler molecules, those molecules are built of atoms and atoms of elementary particles; the replication of DNA or the production of proteins cannot be derived from the Schrödinger's equation. DNA and proteins are on a new organizational level, with own properties. In simple words, a complex system is more than the sum of its elements.

It can be expected that in complex learning systems, new behavior patterns will emerge, especially if many separately built systems will interact without human intervention. Humans will observe them but will have less and less influence on them, if they will be able to understand them at all. A special case of emerging properties is swarm behavior or swarm intelligence. A flock of birds or a shoal of fish consists of simple elements that interact with the neighbors according to simple rules, yet such an aggregation behaves like a higher level entity. This is a common property of such systems, independent of the nature of the components, therefore, it should be observed in large technical systems of adaptable elements. Observing flocks of birds dancing in the sky is fascinating, but are humans ready to enjoy self-organization and swarm behavior of their products?

Giving control in a limited field is not problematic. A pilot does not understand the details of how the flight control surfaces are steered. On the other hand, automatizing entire cities, their power supply, heating, and public and private traffic is risky. In this context, the use of IoT is very tempting. Environmental factors, car positions, human movement patterns, and many more can be measured. Intelligent algorithms will optimize predefined parameters. Various actuators will wield power over the city. Even without considering malicious action, the question arises—will it all work, forever? Smart cities in the brochures are populated by smiling Lego figures, happily enjoying the new possibilities. It is a strongly idealized picture.

Wallach and Allen (2008) presented a spectacular example of small causes leading to big consequences, with the fault propagating between not directly connected systems: On a hot day, spot price for oil rises; power stations switch to coal; an overloaded coal-burning generator explodes; power grid overloads, blackouts spread rapidly; blackouts are identified as a terrorist action; airports are closed, landing airplanes collide.

In general, many authors warn against endeavors too complex to be understood, designed, implemented, and maintained. Sloman and Fernbach (2017) show how overestimating own knowledge nearly leads to a catastrophe, like in the thermonuclear explosion, Castle Bravo in 1954, where a bomb designed for 6 Mt exploded with a force of 15 Mt. The calculation took into account several factors, but omitted the reaction of lithium-7. Taleb (2012) warns against *neomania*, the drive to buy or implement everything new, just because it is possible. He also warns of *black swans*, rare events that should not happen but sometimes happen anyway. If the system tries to control too much and has not enough flexibility, it can be severely damaged by such *black swans*, with unpredictable consequences for other technical systems and people relying on it. As a consequence, it has to be asked what is the measurable profit of a system, what are the risks (even unbelievable), what are the safety measures, how will it evolve and degrade, and if it has to be built at all.

Taleb (2012) in his highly recommendable, mind-opening book classifies systems with respect to their reaction to change: fragile (break), robust (resist), and antifragile (evolve). In the real world where only change is constant, antifragile systems flourish in the long term, for example, life or decentralized political organizations. Their evolution is naturally independent and unpredictable, and it may be asked if humans are ready to accept a parallel evolution of technology that gets out of human control.

2.10 Moral Things

An early vision of the relationship between humans and intelligent technology (in the form of humanoid robots or another) has been formulated by Isaac Asimov in his 1942 short story "Runaround" as *Three Laws of Robotics*. Asimov added later a fourth, or zeroth law, to precede the others, as more general:

0. A robot may not harm humanity or, by inaction, allow humanity to come to harm.

1. A robot may not injure a human being or, through inaction, allow a human being to come to harm.
2. A robot must obey the orders given to it by human beings except where such orders would conflict with the first law.
3. A robot must protect its own existence as long as such protection does not conflict with the first or second laws.

The major flaw is that humanity is taken as a homogeneous entity. In the real world, machines expand the capacities of various humans, having various, often diverging or conflicting goals. In a mild version, our stock trading bots compete against their stock trading bots. In an extreme case, our robot warriors fight their robot warriors.

People sometimes tend to idealize future technology (and sometimes tend to demonize it). Radio was already seen as a worldwide communication medium, facilitating understanding between the nations. Actually, it was soon used as a ruthless propaganda machine. Recently, many saw the Internet as an oasis of liberty, where anonymity will permit an unhindered free expression of opinions. A popular cartoon said: "On Internet, nobody knows you are a dog." Now, the users not only are not anonymous, their relationships, financial situation, political views, and sexual orientation are also known. As long as this information is only used to suggest books and videos, it is harmless, but the next step is near – one who is however anonymous and encrypts his/her communications often uses it for trading drugs, false passports, and arms or for preparing terrorist attacks. Free expression is dominated by insults by other humans and fake posts by artificial bots. Melvin Kranzberg said: "Technology is neither good nor bad; nor is it neutral" (Kranzberg, 1986). It can be used for good goals and for the bad ones, and its very existence transforms the world.

Wallach and Allen (2008) gave a broad coverage of the subject of machine ethics. They argue that as robots take on more and more responsibility, they must be programmed with moral decision-making abilities. The authors examine the challenge of building artificial moral agents that extend human decision making and ethics. From the technical point of view, the challenge is to encode the ethical rules, to implement them, and to manage them allowing for verification and modification.

IoT vastly expands the range of ethical challenges. It permits to monitor/invigilate at an unprecedented scale as well as act massively on objects and humans. In principle, all information is gathered for the benefit of people. Having detailed information about the city permits to optimize energy usage and public transport. As usual with large amount of information, there is a possibility of misuse. Things and people can be tracked electronically. Monitoring cameras can recognize faces and cars' license plates. This will be considered good if used to catch criminals and terrorists, but what is the guarantee it will never be used for oppression once the technology is in place? Another case: Observing elderly people at home makes possible to call for help in case of an emergency. But who has to be informed and under what conditions? Is the changed behavior pattern only a little personal freedom that should be respected?

The correct behavior of the citizens can be monitored and improved by the *nanny state*, for their own good. A new sensor measures the composition of blood and can detect traces of nicotine and other substances. Some nonsmokers agree to implant such devices in order to pay lower insurance premiums. As the usage of blood analyzers increases, they get cheaper. As smoking is considered harmful to own health, and consequently to the society, the pressure is growing for an obligatory use. Relevant law is passed. This scenario is realistic – in a similar way marking dogs with chips has become an obligation, and not performing prenatal tests is considered asocial. What can happen next? The technology is used in other cultures, to detect consumption of drugs and alcohol, with immediate notification and severe punishment. All those options are contained in the technology and the rest depends on human decisions.

2.11 Conclusion

IoT opens a vast area for creative applications that will change the world and human lives. Small sensing and acting devices, connected to the Internet, will make true pervasive computing possible. The support of large databases, Big Data analytics, and learning algorithms will enable building precise models of the world and a distributed AI environment, optimizing the life of the humans and preserving their ecosystem.

Every work is perfectly fine in the ideal world, as described already. In the real life, however, batteries get empty, deployed sensors are stolen or vandalized, and wireless signal is missing. Systems interoperable on paper have slight differences in implementation. Keeping systems running, especially if they depend on other systems of independent providers, requires constant support and the necessary staff may not be available.

There is also a basic conflict between the state-of-the-art, rapidly evolving technology and the longer lifetime, guaranteed support, safety and security of the critical infrastructure, and life-saving devices and systems. Two worlds with different tradition and mentality are coming together. Resolving this mismatch requires expanding multidisciplinary university curricula, training on real-life cases, and promoting cooperation between people of different skills, age and backgrounds.

As with the previous technical revolutions, a new technology like pervasive sensing or implemented artificial organs will change human life and the interactions between people. As before, long-term impact is unpredictable and is beyond imagination. Even the retrospective assessment of the impact – was it good or bad, for whom – is far from clarity.

In any case, this development will take place. Nothing can stop it. However, experts together with society should observe it with a critical eye and try to guide it according to the goals defined by humans and to their ethical rules.

References

Adibi, S. (ed.) (2015) *Mobile Health: A Technology Road Map*, vol. 5, Springer.

Arnold, D. (ed.) (2013) *Traditions of Systems Theory: Major Figures and Contemporary Developments*, vol. 11, Routledge.

Bostrom, N. (2014) *Superintelligence: Paths, Dangers, Strategies*, OUP, Oxford.

Damper, R. I. (2000) Editorial for the special issue on 'Emergent Properties of Complex Systems': emergence and levels of abstraction. *International Journal of Systems Science*, **31**, 811–818.

Fowler, C. (2009) *The Passionate Programmer: Creating a Remarkable Career in Software Development.* Pragmatic Bookshelf.

Fu, K. and Blum, J. (2013) Controlling for cybersecurity risks of medical device software. *Communications of the ACM*, **56**(10), 35–37.

Helbing, D. (2015) *The Automation of Society is Next*, CreateSpace Independent Publishing Platform.

Jeong, Y.-S. and Shin, S.-S. (2016) An IoT healthcare service model of a vehicle using implantable devices. *Cluster Computing*, 1–10. DOI https://doi.org/10.1007/s10586-016-0689-z.

Kolias, C., Kambourakis, G., Stavrou, A., and Voas, J. (2017) DDoS in the IoT: Mirai and other Botnets. *Computer*, **50**(7), 80–84.

Kranzberg, M. (1986) Technology and history:" Kranzberg's Laws". *Technology and Culture*, **27**(3), 544–560.

Kushner, D. (2013) The real story of stuxnet. *IEEE Spectrum*, **50**(3), 48–53.

Lee, R. M., Assante, M. J., and Conway, T. (2016) Analysis of the Cyber Attack on the Ukrainian Power Grid. E-ISAC. Available at https://ics.sans.org/media/E-ISAC_SANS_Ukraine_DUC_5.pdf (accessed August 7, 2017).

Leitão, P., Colombo, A. W., and Karnouskos, S. (2016) Industrial automation based on cyber-physical systems technologies: prototype implementations and challenges. *Computers in Industry*, **81**, 11–25.

Nicolis, G. and Nicolis, C. (2012) *Foundations of Complex Systems: Emergence, Information and Prediction*, World Scientific.

Pasquale, F. (2015) *The Black Box Society: The Secret Algorithms That Control Money and Information*, Harvard University Press.

Rose, K., Eldridge, S., and Chapin, L. (2015) *The Internet of Things: An Overview*, The Internet Society (ISOC), pp. 1–50.

Seo, J. and Kim, M. (2013) eLoran in Korea: current status and future plans. European Navigation Conference, pp. 23–25.

Sliwa, J. (2014) Ethics between the lines (of code). Proceedings of the IEEE 2014 International Symposium on Ethics in Engineering, Science, and Technology, p. 88.

Sliwa, J. and Benoist, E. (2011) Pervasive computing: the next technical revolution. Developments in E-systems Engineering (DeSE), pp. 621–626.

Sloman, S. and Fernbach, P. (2017) *The Knowledge Illusion: Why We Never Think Alone*, Penguin.

Sparber, T., Boano, C. A., Kanhere, S. S., and Römer, K. (2017) Mitigating radio interference in large IoT networks through dynamic CCA adjustment. *Open Journal of Internet Of Things (OJIOT)*, **3**(1), 103–113.

Stankovic, J. A. (2014) Research directions for the Internet of Things. *IEEE Internet of Things Journal*, **1**(1), 3–9.

Taleb, N. N. (2012) *Antifragile: Things that Gain From Disorder*, vol. 3, Random House.

Tedeschi, S., Mehnen, J., Tapoglou, N., and Roy, R. (2017) Secure IoT devices for the maintenance of machine tools. *Procedia CIRP*, **59**, 150–155.

Wallach, W. and Allen, C. (2008) *Moral Machines: Teaching Robots Right From Wrong*, Oxford University Press.

Zanella, A., Bui, N., Castellani, A., Vangelista, L., and Zorzi, M. (2014) Internet of Things for smart cities. *IEEE Internet of Things Journal*, **1**(1), 22–32.

Part II

Enablers

3

An Overview of Enabling Technologies for the Internet of Things

Faisal Alsubaei,[1,2] Abdullah Abuhussein,[3] and Sajjan Shiva[4]

[1]*Computer Science Department, University of Memphis, Memphis, TN, USA*
[2]*Faculty of Computing and Information Technology, University of Jeddah, Jeddah, Saudi Arabia*
[3]*Department of Information Systems, St. Cloud State University, St. Cloud, MN, USA*
[4]*Computer Science Department, The University of Memphis, Memphis, TN, USA*

3.1 Introduction

The Internet of Things (IoT) initially utilized current Internet infrastructure and existing technologies to transform stand-alone objects (i.e., devices) into interconnected smart objects. Before the development of IoT-specific technologies, computers and networking technologies were applied, with major tweaks, to support IoT applications. For example, IPv4 that was used in wireless sensor networks (WSNs) was used extensively in the beginning of IoT to connect nodes to the Internet using one IP address (i.e., Network Address Translation (NAT)), which requires extra effort to setup. However, many issues emerged from the use of technologies that were not designed with IoT in mind (e.g., the lack of autoconfiguration in IPv4). As such, there was a need to develop new technologies to overcome various issues, such as efficient routing, scalability, and mobility, to make developing IoT applications easier and more efficient. These advancements in IoT technologies have resulted in extensive real-world deployment of IoT applications in many domains, such as healthcare, industry, and smart buildings.

Considering the development of IoT enabling technologies during the past decade, it is clear that the most distinct trends are related to efficiency, which requires the optimal use of resources (e.g., battery and memory). Low-power consumption is an important factor for designing new IoT applications. Devices save energy while they are in "sleep" mode and consume energy when they are active (or awake). Devices are designed to wake up based on an optimum schedule to perform some operations (i.e., collecting and sending data). The complexity of these operations must be optimized to increase the lifespan of

Internet of Things A to Z: Technologies and Applications, First Edition. Edited by Qusay F. Hassan.
© 2018 by The Institute of Electrical and Electronics Engineers, Inc. Published 2018 by John Wiley & Sons, Inc.

devices. Designers strive to achieve low device complexity, low energy consumption (i.e., long network lifetime), and an appropriate balance between signal/data processing capabilities and communication. An important issue in efficiency is the lack of interoperability in IoT (i.e., devices from different brands and models cannot work well together without a translation layer). The interoperability issue remains a major challenge in current IoT technologies because of the lack of unified standards for the IoT. Hopefully, as the IoT matures, these problems will be minimized.

This chapter details the key concepts of the enabling technologies based on their positions in the well-known five-layer model, which consists of the perception layer, network layer, middleware layer, application layer, and business layer.

3.2 Overview of IoT Architecture

Managing millions of heterogeneous connected devices via the Internet requires a flexible, layered architecture. As introduced in Chapter 1, there are various IoT architectural models, namely, Internet of Things—Architecture (IoT-A), Industrial Internet Reference Architecture (IIRA), Reference Architecture Model Industrie 4.0 (RAMI 4.0), and Cisco's Internet of Things Reference Model. However, no standard reference architecture has been adopted (Al-Fuqaha et al., 2015; Al-Qaseemi et al., 2016).

There are also ongoing projects that aim at designing new architectures for IoT such as P2413 - Standard for an Architectural Framework for the IoT[1], European Research Cluster on the Internet of Things (IREC)[2], Internet of Things Reference Architecture (IoT RA)[3], and Arrowhead Framework[4]. Additionally, there are other architectures that are related to the IoT but only focus on subsets of the IoT. For example, the European Telecommunications Standards Institute Technical Committee (ETSI TC) for M2M[5] is focused only on the communication standards while the International Telecommunication Union Telecommunication Standardization Sector (ITU-T)[6] focuses only on the identification systems (Weyrich and Ebert, 2016).

In order to systematically discuss the IoT enabling technologies, they will be discussed in this chapter based on their appearance on an architecture model. For the purpose of this chapter (i.e., to provide an overview of IoT generic technologies), none of the above reference architectures are suitable for the

1 http://standards.ieee.org/develop/project/2413.html
2 http://www.internet-of-things-research.eu/
3 https://www.iso.org/standard/65695.html
4 http://www.arrowhead.eu/
5 http://www.etsi.org/technologies-clusters/technologies/internet-of-things
6 http://www.itu.int/en/ITU-T/Pages/default.aspx

following reasons. First, IoT-A is relatively old and does not define IoT key concepts like cloud and fog computing, which are left to be identified in the implementation (Weyrich and Ebert, 2016). Second, IoT-A focuses only on the WSNs and the application activities (Gubbi et al., 2013; Li et al., 2015) and does not include other technologies that are important in modern IoT applications (such as processing Big Data for analytics). Third, RAMI 4.0 and IIRA are mostly for Industrial IoT (IIoT) and do not expand to other IoT domains such as Internet of Medical Things (IoMT), Internet of Vehicles (IoV), and so on. Moreover, Cisco's reference model cannot be used because it has not been finalized yet. Finally, because technologies will be presented without a specific IoT scenario in mind, a pseudophysical viewpoint is appropriate; however, none of the above architectures provides this viewpoint.

Therefore, this chapter categorizes the IoT enabling technologies based on a comprehensive, generic, and simple architecture, namely, the IoT five-layer model as depicted in Figure 3.1. This model captures many of the essences of other models (and simplifies and systematizes them). Hence, this architecture has been used extensively in recent publications (Al-Fuqaha et al., 2015; Al-Qaseemi et al., 2016). In Figure 3.1, each layer includes examples of technologies that are categorized by their functionality. These layers are conceptually similar to the Transmission Control (TCP)/Internet Protocol (IP) layers, as will be discussed in detail in the next section. Although many IoT technologies can fit into the TCP/IP (or Open Systems Interconnection (OSI)) model, categorizing all IoT technologies based on the TCP/IP model does not encompass all technologies. Thus, their classification based on the five-layer model is more appropriate.

3.3 Enabling Technologies

IoT enabling technologies vary depending on the domain and scenario. For example, smart transportation requires flexible technologies that ensure the connectivity of a vast number of mobile nodes. In contrast, the focus in healthcare is reliability and integrity. Therefore, generic technologies are briefly presented here based on their functionality in the IoT five-layer model.

3.3.1 Perception Layer Technologies

The perception layer (also known as the objects layer) is the first layer (from the bottom, as shown in Figure 3.1) in the IoT model. It includes various types of physical devices that are responsible for collecting data and acting accordingly, such as object identifiers, temperatures, locations, and humidity measurements. The power consumption and communication ability (i.e., unidirectional or

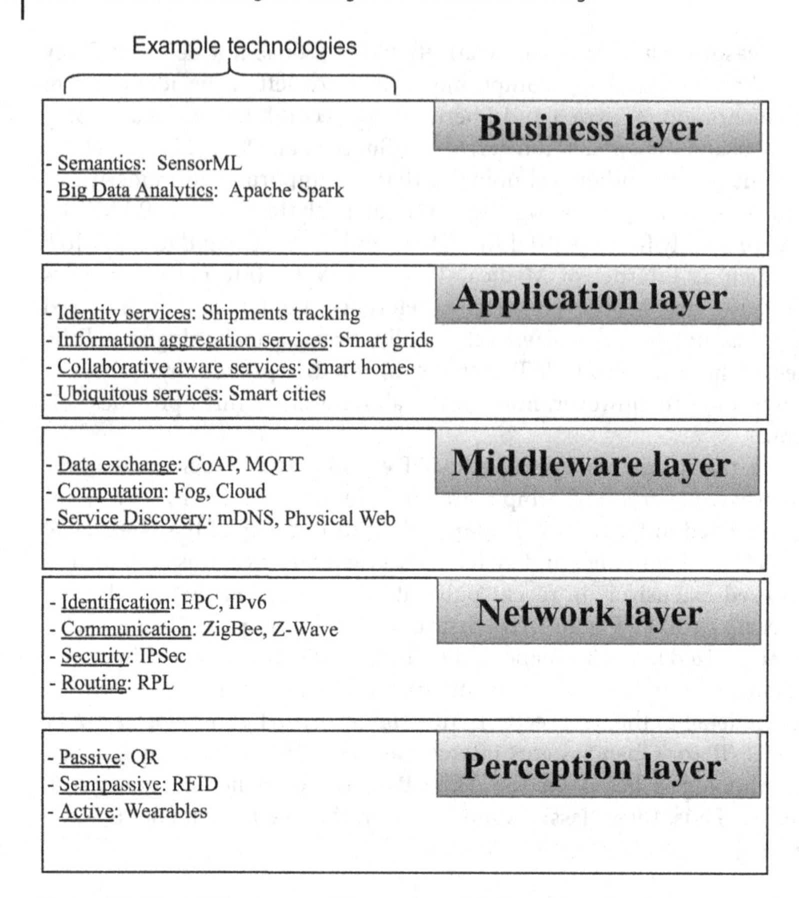

Figure 3.1 The IoT five-layer architectural model. (Reproduced with permission from IEEE Press.)

bidirectional) are important aspects of this layer. The classification of the perception layer technologies is as follows:

3.3.1.1 Passive

Passive devices have the lowest radio coverage because they do not have onboard power supplies and rely on readers in close proximity for their energy supply. The limitation of passive enabling technologies is that communication with them is unidirectional. Data can be read from nodes but cannot be written back to the nodes. However, for many applications, unidirectional communication is sufficient. Examples of passive perception layer technologies are as follows:

- *Quick Response (QR) Codes.* These codes function as a method of storing information on an IoT node and can be used to recognize nodes. QR codes are probably the easiest and cheapest technology in this category. They can potentially be applied to almost anything. By printing a 1 sq. in. code on nodes, they become readily recognizable using various readers, such as cameras or light sensors, although they must appear within the line of sight of the reader. QR codes can be employed in many scenarios, such as shipments or manufacturing tracking.

- *Passive Radio Frequency Identification (RFID) Tags.* These tags are small electronic chips that—similar to QR codes—can be attached to any node enabling the nodes to be recognized for tracking or automatic identification purposes, which enable applications such as inventory management. Readers recognize these chips when they are within a short range by sending a signal that provides sufficient power for the tag to respond; the response contains the identification data.

3.3.1.2 Semipassive

Semipassive devices have batteries that power tags while receiving signals from a reader. Unlike passive technologies, the extra available power enables them to read and write data. Examples of semipassive technologies are as follows:

- *Semipassive RFID Tags.* These tags are similar to passive tags, but their batteries enable them to have a longer range, which supports a variety of additional uses. One example is smart transportation, for example, blocking railway crossings when a train is approaching.
- *Infrared (IR) Readers.* These readers are devices that sense electromagnetic radiation that is invisible to humans. Modern uses of this technology include automatic and remote body temperature sensing and remote controls. In the context of IoT, connected IR readers can trigger some actions (e.g., alarm) when detecting abnormalities (e.g., high body temperature). Since the IR is cheap, simple, and consumes low energy, it might be preferable in some IoT scenarios where IR readers might be stolen (e.g., monitoring body temperature at airports). However, it is limited to the line of sight and is subject to interference or failure due to objects blocking the light signals.

3.3.1.3 Active

Active devices often have the longest range because they have larger batteries or are attached to power supplies that enable them to receive and transmit data. Examples of these devices are as follows:

- *Active RFID Tags.* Similar to semipassive RFID tags but have a longer range due to their larger onboard battery. These tags are mostly used in real-time locating systems (RTLS)[7], for example, to monitor valuable assets (e.g., to

7 Real-time location system (RTLS) is a set of technologies to track targets (e.g., vehicles, items in a factory, or people) based on their recent geolocation.

track lost or stolen expensive medical devices). In such IoT scenarios, connected nodes equipped with active RFID can trigger an alarm when they leave a particular area.
- *Smart Actuators.* Mechanical devices that can be automatically controlled, without human intervention, to apply a specific motion to a product or process based on sensed data or based on the defined workflow. Connected actuators can help industries or smart cities to be more efficient and save money because actuators can react automatically based on sensed data, which requires less staff.
- *Wearable, Embedded, or Stand-Alone Smart Sensors.* Devices that include embedded sensors, such as gyroscopes, Global Positioning System (GPS) radios, and heart rate monitors used in modern smartphones or smart watches to monitor users' activities and health. Smart thermostats also use connected sensors to efficiently manage energy. In modern IoT applications, such connected devices (e.g., smartwatch, smartphone, etc.) often collaborate to achieve a common goal (e.g., pervasive health monitoring, alert care givers, etc.).

3.3.2 Network Layer Technologies

The network or object abstraction layer is the second layer. It is considered the infrastructure layer because its technologies transform the conventional sensors described in the perception layer into smart and connected nodes. Network layer technologies enable nodes to be identifiable over the Internet or any local area network, which allows them to securely communicate with each other. Many technologies in this layer are also found in the first three layers in the IP suite (TCP/IP) (i.e., Link, Internet, and Transport layers). Due to the restricted capabilities (i.e., power and computing) of most IoT nodes, scalable and efficient routing techniques are required to ensure interoperability among IoT devices. Note that many of IoT technologies are actually utilized in WSNs or machine-to-machine (M2M) communications but were improved to meet IoT requirements (i.e., Internet connectivity). Many modern devices use more than one technology; as in smart watches, which often have Wi-Fi, Bluetooth, and NFC. In this section, we categorize the enabling technologies based on their functionality in the network layer.

3.3.2.1 Identification

When a node connects to the network, it is assigned an identifier that enables it to communicate with other nodes. Identification is important to control the excessive use of bandwidth in large networks (e.g., flooding) by ensuring that only identified nodes communicate. To easily locate devices in the massive network, names and addresses are assigned to the devices. The naming conventions assign structured names to nodes; these names allow the nodes to be readily identified and may even specifically describe the functions. Naming solutions are as follows:

- *Uniform Resource Identifier (URI)*[8]. A unique sequence of characters in the form of a web link that refer to an abstract or physical resource (Masinter et al., 2016). These identifiers are used to locate and interact with resources (including IoT nodes) on networks using human-friendly structured names that describe their purpose (e.g., http://www.example.com/temperature/actual).
- *Electronic Product Code (EPC)*[9]. An EPC global tag standard that assigns a unique identity to each physical object anytime worldwide (e.g., EPC = urn:epc:id:sgtin:0614141.012345.62852). EPCs are encoded on passive RFID tags, which can be used to track all types of objects, such as vehicles on toll roads.
- *Ubiquitous Code (uCode)*[10]. A numeric identification system that uniquely identifies physical objects and places in the real world. Using uCode, information can be associated with objects and locations and can be retrieved from www.uidcenter.org. An example uCode is 00001C00000000000001000285E7A6E3. Unlike EPC, uCode offers more flexibility because the tags can take many forms, such as print (i.e., QR and barcodes), passive RFID, active RFID, infrared, and acoustic tags.

Addressing techniques enable the assignment of unique network addresses to nodes for better network management. Addressing protocols are as follows:

- *IPv4*. A popular networking protocol that uses 32-bit addresses to assign nodes to unique addresses and aggregate them into smaller networks. Although IPv4 is extensively applied, it is not recommended for IoT due to the small number of addresses (i.e., 2^{32} addresses) that can be employed.
- *IPv6*. A networking protocol that aims to overcome the limitations of IPv4 by supporting 2^{128} addresses, which allows every device to have a unique address. Unlike IPv4, IPv6 provides autoconfiguration, which enables devices to be assigned IPs without the need for a Dynamic Host Configuration Protocol (DHCP) server. Autoconfiguration is an important feature for IoT because it simplifies the large-scale deployment of WSNs. Due to the large header size (i.e., 320 bits) in IPv6, *IPv6 over Low Power Wireless Personal Area Networks* (6LoWPAN) is used to compress the address to make it compatible with traditional WSN protocols (i.e., IEEE 802.15.4), as will be subsequently described.

8 https://tools.ietf.org/html/rfc3986
9 www.gs1.org/epc-rfid
10 www.uidcenter.org

3.3.2.2 Communication

After a node has been identified (i.e., addressed and named), it can start communicating with other nodes or with backend servers. However, this communication requires the selection of a suitable communication medium, which depends on the capabilities of the node (e.g., power and coverage). Nodes can communicate horizontally (i.e., ad hoc with another node) or vertically (i.e., with servers in the middleware layer). Communication technologies are discussed in this section with emphasis on their use in the IoT. These commonly employed technologies are listed in ascending order based on their wireless range (i.e., wired and short, medium, and long range):

- *Power-Line Communication (PLC).* This comprises a set of communication protocols that use power-line wiring to simultaneously transmit both data and alternating current. Using this approach, a person can power devices and control/retrieve data using only the standard power cables that run to the device. PLC can be used to connect the nodes in an IoT environment to share data with the backend that then performs some actions. It is mainly preferable in stationary nodes because they rely on power lines anyway as well as in IoT environments that have high wireless interference in smart vehicles, smart grids, and smart homes.
- *X10.* Similar to PLC, X10 is an industry standard that uses electrical cabling for signaling and controlling devices, in which the signals contain short Radio Frequency (RF) bursts that can include data. Despite its lower data rate and range compared with PLC, it remains a popular choice due to its low cost. X10 is primarily employed to connect nodes in smart homes where outlet modules (e.g., light switches, AC switches, etc.) can be controlled from a mobile application or a web portal.
- *Near-Field Communication (NFC)*[11]. A set of protocols that enable communication between two devices over very short distances. NFC is unlike RFID, which is used primarily for identification, it offers a simple way to authenticate, access, and share data between devices and users in IoT systems. NFC is employed in many IoT applications such as smart payments and access control systems.
- *Ultra-Wide Bandwidth (UWB).* An older wireless technology that uses low-energy transmissions to provide high-bandwidth communications over a wide radio spectrum (Fontana, 2004). This technology is suitable for IoT due to its resistance to interference, which makes it highly scalable since adding more nodes will not cause interference. This technology can help wireless networks to be more adaptive and efficient because each node can be tracked using the

---

11 www.nfc-forum.org

UWB. UWB is usually employed in location-based services (e.g., asset tracking) (Anonymous, 2017).

- *Wi-Fi.* Wi-Fi is particularly useful for ad hoc configurations, such as Wi-Fi Direct, which does not require a wireless access point. The main limitation of Wi-Fi is its power consumption. However, in some IoT applications (e.g., smart homes), power is not an important issue. The Wi-Fi Alliance is launching a new energy-efficient Wi-Fi technology named Wi-Fi HaLow[12] that is specifically designed for IoT nodes.
- *IEEE 802.15.4*[13]. A standard for low-rate wireless personal area networks (LR-WPANs) that specifies the physical layer and media access control. Because IEEE 802.15.4 was designed for IPv4, the larger sized datagrams of IPv6 are not a natural fit for IEEE 802.15.4 networks. The solution to this problem is to use an adaptation layer (i.e., 6LoWPAN), which encapsulates and compresses headers to allow IPv6 packets to be carried over IEEE 802.15.4 networks. ZigBee[14], ISA100.11a[15], WirelessHART[16], MiWi[17], and Thread[18] are based on IEEE 802.15.4 and only differ in their upper layers. ZigBee and Thread are commonly employed in smart homes and smart meters, whereas ISA100.11a, MiWi, and WirelessHART are commonly used in industrial applications.
- *Bluetooth Low Energy (BLE)*[19]. A wireless standard is intended to exchange data over short distances and build *personal area networks* (PANs). For Bluetooth v4.0 and later versions, the energy consumption is improved, which renders Bluetooth well suited for sensors and other small devices that require low power. Internet-enabled nodes often use BLE to act with local nodes and send collected data to the backend for more actions. BLE can be used in an extensive range of applications, such as smart buildings, smart transportation, and wearables.
- *ANT+*[20]. A proprietary (but open access) wireless communications protocol stack enables nodes to communicate by creating rules for coexistence, signaling, data representation, authentication, and error detection. It is primarily employed in sports, wellness management, and home health monitoring IoT applications where each IoT node use or send sensed data to the cloud, for more actions.

---

12 www.wi-fi.org/discover-wi-fi/wi-fi-halow

13 www.standards.ieee.org/about/get/802/802.15.html

14 www.zigbee.org/

15 www.isa.org/isa100/

16 www.fieldcommgroup.org/technologies/hart/hart-technology

17 www.microchip.com/design-centers/wireless-connectivity/embedded-wireless/802-15-4/miwi-protocol

18 www.threadgroup.org

19 www.bluetooth.com/what-is-bluetooth-technology/how-it-works/low-energy

20 www.thisisant.com/developer/ant-plus/

- *Z-Wave*[21]. A technology extensively applied in smart homes. Z-Wave devices can be attached to home appliances, which enable them to be controlled over the Internet.
- *Weightless*[22]. A set of Low-Power Wide-Area Network (LPWAN)[23] open wireless technology standards for exchanging data between a base station and thousands of surrounding nodes. There are three different standards for Weightless, namely, Weightless-W, Weightless-P, and Weightless-N; each of which has different use cases due to its features. Weightless-W offers two-way communication that is ideal for use in the smart oil and gas sector, where TV white space (TVWS) is likely to be available. Although weightless-P also provides two-way communication, it is perfect for private networks and situations that require both uplink and downlink capabilities. Weightless-N offers an uplink that makes it ideal for sensor-based networks, such as temperature readings, tank level monitoring, and metering.
- *Cellular Networks.* These networks exist in various generations (e.g., 3G and 4G) of cellular network standards that are often employed in smartphones but are also suitable for IoT due to their high mobility and higher speed. Substantial progress has been made in recent generations in terms of power consumption and speed. For example, a recent specification of 4G (called LTE-Advanced) offers a power saving mode that extends the device battery life to a decade with a maximum speed of 10 Mbps. For comparison, other versions of 4G (less energy efficient) can have a speed of up to 1 Gbps (Nokia, 2017).
- *SigFox*[24]. A subscription-based service that offers connectivity solutions (in some countries) over dedicated LPWAN networks. SigFox is ideal for IoT applications that need to send infrequent and small bursts of data, such as alarm systems or smart meters.
- *DASH7 Alliance Protocol (D7A)*[25]. An open-source wireless sensor and actuator network protocol that provides multiyear battery life and low latency to connect moving nodes. Its IoT applications include smart buildings, location-based services, logistics, and smart vehicles.
- *Long-Range Wide-Area Network (LoRaWAN)*[26]. A technology from LoRa Alliance that offers low-cost, mobile, and secure bidirectional communication. It is optimized for low power consumption and is designed to support large networks with millions of nodes, which renders it ideal for IoT applications such as smart cities and industrial applications.

21 www.z-wave.com
22 www.weightless.org
23 LPWAN is a form of wireless network that allows connected objects (things) to communicate over a wide coverage area at a low speed.
24 www.sigfox.com
25 www.dash7-alliance.org
26 www.lora-alliance.org/For-Developers/LoRaWANDevelopers

Some additional technologies that are expected to be adopted in IoT applications are currently under development and will likely be used in the next few years because they will address some of the communication limitations in current technologies. This includes, for example, the following:

- *5G.* A medium-range communication candidate that should be available by 2020 (Prasad, 2014). It offers many improvements over 4G, such as increased speed (almost equivalent to Ethernet) and better coverage, and supports a large number of users. Thus, it will enable a larger number of nodes to be connected with extra mobility.
- *Light Fidelity (Li-Fi).* A new wireless technology that offers high-speed data transmission using visible light communication (VLC) (Ackerman, 2015). Li-Fi can also utilize energy-efficient light-emitting diode (LED) lights to reduce energy costs. However, the Li-Fi scheme is limited to the line of sight and is subject to interference or failure due to objects blocking the light signals. Although Li-Fi has not been extensively adopted in IoT, its high speed indicates that it has substantial potential for use in many applications, such as smart theaters and smart classrooms.
- *Software Defined Network (SDN).* A networking architecture that is dynamic, manageable, adaptable, and cost effective. These characteristics make SDNs very effective, given the high-bandwidth and dynamic nature of the IoT (Jacquenet and Boucadair, 2016). Underutilized network resources and intelligent route trafficking will facilitate planning for the data flood that is expected as IoT usage increases. SDNs minimize bottlenecks and introduce network efficiencies (Duffy, 2014).

Each of these current communication technologies has special features that enable different scenarios to be implemented efficiently. Table 3.1 compares the important characteristics of most IoT communication technologies and lists them in ascending wireless range order. The data rate aids in the selection of the best technology based on the data flow requirements of applications. Similarly, range is an important consideration for IoT nodes in terms of mobility. Security is also a vital feature because large-scale sensors can cause catastrophic damage when compromised. Therefore, technologies that do not offer any built-in security should only be employed with additional security protocols provided by other layers, as discussed in the next section. Ad hoc communications are critical in applications where nodes communicate with each other without a controller. Technologies that are considered native TCP/IP protocols can be easily integrated into current systems. Header size may affect the choice of technologies with low data rates. Latency can be important in some IoT applications (e.g., remote surgery (West, 2016)). The total latency often increases when a large number of sensors are deployed. Frequency is important to avoid signal interference or restrictions in some places. Scalability is also an important feature of the IoT. Technologies that are scalable often allow for addition of new

Table 3.1 Comparison of communication technologies.

Technology	Data rate	Range	Mobility	Security	Ad hoc	Native TCP/IP	Header size	Latency	Frequency	Scalable
X10	20 bps	~20 m	No	No	Yes	No	8 bytes	Low	120 KHz	No
PLC	10 Mbps	~ 9 km	No	No	Yes	Yes	133 bytes	Varies	Narrowband (3–500 KHz) Broadband (1.8–250 MHz)	Yes
NFC	106, 212, 424, and 848 kbps	~20 cm	Yes	LPI/D[a)]	Yes	No	4 bytes	Low	13.56 MHz	No
UWB	480 Mbps–1.6 Gbps	~10 m	Yes	LPI/D	Yes	Yes	40 bits	Low	3.1–10.6 GHz	Yes
Z-Wave	9.6, 40, and 100 kbps	~30 m	Yes	No	Yes	No	Varies	Low	Regional sub-GHz bands	Yes
Thread	40–250 kbps	~30 m	Yes	TLS 1.2	No	No	40 bytes	Low	Global 2.4 GHz	Yes
MiWi	250 kbps	~20–50 m	Yes	TLS 1.2	Yes	No	11 bytes	Low	Regional sub-GHz and global 2.4 GHz	Yes
BLE	1 Mbps	~50 m	Yes	AES 128-bit	Yes	No	2 bytes	Low	Global 2.4 GHz	No
ANT+	1 Mbps	~50 m	Yes	AES 128-bit	Yes	No	14 bytes	Low	Global 2.4 GHz	Yes
Wi-Fi	250 Mbps	~60 m	Yes	WPA2 with AES	Yes	Yes	2 bytes	Low	Global 2.4, 5.8 GHz	Yes
ZigBee	20, 40, and 250 kbps	~100 m	Yes	128-bit Encryption	Yes	No	15 bytes	Low	Regional sub-GHz and global 2.4 GHz	Yes
ISA100.11a	250 kbps	~200 m	Yes	AES 128-bit	Yes	No	Varies	Low	Global 2.4 GHz	Yes
WirelessHART	250 kbps	~200	Yes	AES-128 bit	Yes	No	21 bytes	Low	Global 2.4 GHz	Yes

Wi-Fi	11 Mbps, 54 Mbps, 600 Mbps, 1300 Mbps, and 6.9 Gbps	~200 m	Yes	WPA2 with AES	Yes	Yes	2,346 bytes	High	Global 2.4, 5.8 GHz	Yes
D7A	167 kbps	1–5 km	Yes	AES 128-bit	Yes	No	3–38 bytes	Low	433 MHz, 868 MHz, 915 MHz	Yes
Sigfox	100 bps up, 600 bps down	~15 km	Yes	No	No	No	Varies	High	Regional sub-GHz bands	Yes
LoRa	0.3–50 kbps	~13 km	Yes	128-bit encryption	Yes	No	Varies	Low	Regional sub-GHz bands	Yes
3G	144–400 kbps (while moving)	Vary	Yes	KASUMI[b)]	Yes	No	Varies	Medium	UMTS 850 MHz–2100 MHz	Yes
4G	Up to 1 Gbps	Vary	Yes	Enhanced SNOW 3G[c)]	Yes	No	Varies	Low	LTE bands	Yes
Weightless-W	1–10 Mbps	~5 km	Yes	128-bit encryption	Yes	No	10 bytes or more	High	TVWS	Yes

a) Low Probability of Intercept/Detect (LPI/D) is a set of wireless security techniques that allow devices to see but not to be seen by modern and capable intercept receivers.
b) KASUMI is a block cipher with a 128-bit key and 64-bit input and output.
c) SNOW 3G is a word-based synchronous stream cipher.

large number of nodes without affecting the performance, whereas nonscalable technologies are generally limited to fewer than 100 nodes per network.

3.3.2.3 Security

Due to a large number of nodes with limited capabilities, security is an important and challenging task because a successful attack is likely to cause excessive damage (e.g., Distributed Denial-of-Service (DDoS) attacks) (Stephen, 2016). As discussed in the communication technologies section, security may not be built in to the various communication technologies (e.g., Z-wave and Sigfox). Therefore, additional security mechanisms must be provided by different layers to reduce the likelihood of attacks. Attacks can occur in this layer due to insecure communication among nodes. Therefore, lightweight security mechanisms are required for secure communication. Some suggested security techniques are as follows:

- *Internet Protocol Security (IPsec)*[27]. A well-known network layer protocol that is employed with IPv6 for authentication and end-to-end encryption among nodes. Because IPsec is implemented in the network layer (in TCP/IP), it also serves the upper layers. A lightweight implementation of IPsec is possible, which renders it ideal for IoT nodes (Raza et al., 2011).
- *Transport Layer Security (TLS)*[28] *and Datagram TLS (DTLS)*[29]. These are well-known cryptographic protocols that are also employed with TCP protocols and user datagram protocols (UDPs), respectively, to provide secure communications.
- *IEEE 1888*[30]. This protocol is part of a family of standards for the Ubiquitous Green Community Control Network. IEEE 1888.3 prevents unauthorized access to resources and accidental data leaks while enhancing the confidentiality and integrity of transferred data.

3.3.2.4 Routing

Once a node knows a destination address, it must have an efficient method for routing data to that destination node. Efficient routing is critical in IoT environments because of the massive number of nodes that can be connected in an ad hoc manner. Due to the limited capabilities of IoT nodes, each node must efficiently learn about the optimal route. Therefore, a special routing protocol was introduced for these environments. *Routing Protocol for Low Power and Lossy Networks* (RPL)[31] is an IPv6 routing standard protocol for resource-limited devices. This protocol was proposed to support different types of links,

27 https://tools.ietf.org/html/rfc4301
28 https://tools.ietf.org/html/rfc5246
29 https://tools.ietf.org/html/rfc6347
30 https://standards.ieee.org/findstds/standard/1888-2014.html
31 https://tools.ietf.org/html/rfc6550

such as IEEE 802.15.4, and common traffic types, including one-to-one, many-to-one, and one-to-many. The protocol can be compactly represented as a set of graphs in which each graph is a Destination-Oriented Directed Acyclic Graph (DODAG). In this type of graph, each node knows at least one path to its root node (i.e., border router), and each node is aware of its parent. Nodes exchange RPL special messages to maintain the graph and ensure that a valid route(s) to the root is always available. As shown in Figure 3.2, routers can work in two operational modes: nonstoring mode (as in DODAG 1) and storing mode (as in DODAG 2). In nonstoring mode, nodes forward each message to the root, which could be useful in scenarios where nodes are programmed to send sensed data (e.g., temperature readings) periodically. In contrast, in storing mode, communications are bidirectional, and each node stores the routes to all nodes under it. This can be helpful in scenarios where nodes only send sensed data when they are requested; hence, downward routes are needed to deliver the request messages.

3.3.3 Middleware Technologies

The middleware layer (also known as the service management layer) is the core of the IoT environment. It can be mapped to the application layer in the IP suite (TCP/IP). Technologies in this layer are often supported by IoT platforms. This layer enables services to be identified and requested based on names and addresses and enables programmers to interact with heterogeneous objects, regardless of the specific hardware setup. This layer also processes received data, makes decisions, and delivers required services (Al-Fuqaha et al., 2015). The

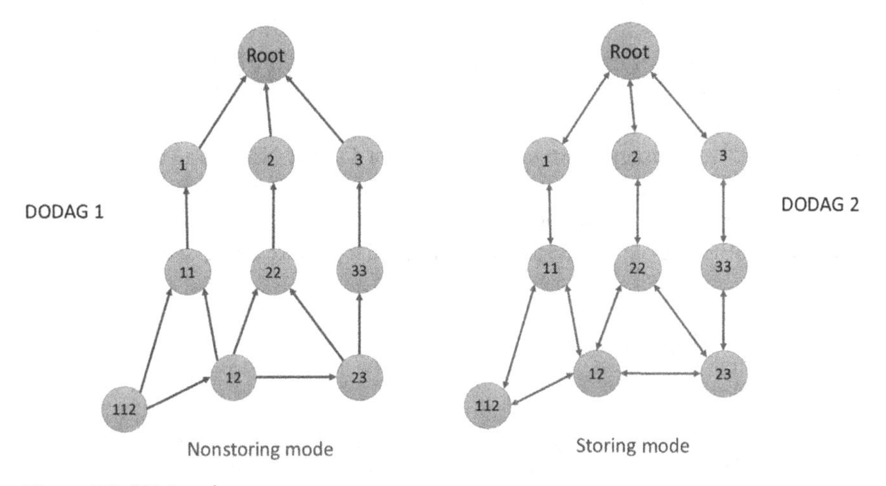

Figure 3.2 RPL topology.

middleware layer performs those tasks using the following technologies, which are divided into three groups based on their functionalities.

3.3.3.1 Service Discovery

IoT applications and users need to be able to use names or addresses to request services without knowing the underlying infrastructure details. Therefore, services need to be discoverable and registered. Scalable and heterogeneous environments, service registration, and discovery should be autonomous, dynamic, and efficient. The protocols that are most commonly employed to provide service discovery are as follows:

- *Multicast DNS (mDNS)*[32]. A zero-configuration and infrastructure-independent protocol that can continue to function even during an infrastructure failure because they can use their local caches. It binds host names to IP addresses inside small networks by multicasting an IP request message asking a device with a particular host name to send its IP address. This host replies with a multicast message that contains its IP address. All other devices in the subnet use this response to update their mDNS local caches. This protocol, when combined with IPv6 (with autoconfiguration), can entirely eliminate the need for a local server to manage local WSNs.
- *DNS Service Discovery (DNS-SD)*[33]. Another zero-configuration protocol that complements mDNS because it pairs services to hosts within a domain. IoT clients can find a service by sending a multicast request for a service type (e.g., temperature readings), and all machines that provide this service will respond with their host names. Names are important because the nodes names, unlike IP addresses, rarely change. Then, mDNS is used to pair the names of the nodes with their IP addresses. Avahi[34] and Bonjour[35] are two popular implementations of both mDNS and DNS-SD.

  Another approach to service discovery is to publish the uniform resource identifiers (URIs) of nodes that provide services over the Web. This approach is known as the Web of Things (WoT). Technologies that adopt this approach are as follows:
- *HyperCat*[36]. An open, lightweight JavaScript Object Notation (JSON)-based hypermedia catalog format is intended to be used for publishing collections of URIs. Each HyperCat catalog can include any number of URIs, each with any number of Resource Description Format (RDF)[37]-like triple statements that

---

32 http://www.multicastdns.org/
33 http://www.dns-sd.org/
34 http://www.avahi.org/
35 https://support.apple.com/bonjour
36 www.hypercat.io
37 RDF is a World Wide Web Consortium (W3C) standard model for data interchange on the Web.

provide information about this URI. HyperCat enables developers to publish linked data descriptions of IoT nodes that can be fetched by HTTP GET requests, which requires IoT clients to know the URI address.

- *Physical Web*[38]. A technology that enables nodes to broadcast their Uniform Resource Locators (URLs) or some local data inside web pages. Unlike HyperCat, IoT clients do not need to know the URLs because they can fetch them via Bluetooth beacons (e.g., Eddystone[39]), mDNS, or Wi-Fi Direct (Anonymous, 2016a). The published URLs enable users to interact with the nodes (e.g., sensors) from their smartphones or laptops. A proxy service is used to prevent malicious URLs and sort lists of URLs based on users' preferences.
- *Universal Plug and Play (UPnP)*[40]. A group of networking protocols that enable networked IoT nodes to seamlessly discover each other's presence in the network and supply functional network services (i.e., data sharing, communications, and entertainment). However, UPnP is known to be vulnerable to security problems (Anonymous, 2016b). UPnP+ is an emerging initiative that aims to address the issues associated with UPnP (Anonymous, 2016c).

3.3.3.2 Data Exchange

To manage the vast amount of data created by the large number of nodes, a manager must transmit data to receivers in a secure and efficient manner. This management occurs on the TCP/IP application layer. Thus, the technologies that reside here are known as application protocols. In IoT environments, each node can capture data, analyze data, and control other nodes. However, before a node can begin communicating and exchanging data in the IoT network, it must be identified and registered as a part of the service network. This operation is referred to as node subscription. A subscribing node must make a request to join a service network (i.e., subscribe) before it can obtain or publish service-related data (e.g., temperature) from or to the network. Once the node is subscribed to publishers, it will receive service-related data when they are published. This model is known as the publish/subscribe model. An alternative approach is called request/response in which a joined node always requests a service-related data explicitly and receives a response that contains the requested data. Once the node joins the network, it is assigned network resources and can start exchanging data and operating as a component of the IoT environment. The publish/subscribe model is generally preferable for IoT scenarios that require the large deployment of nodes that frequently exchange data. Indeed, in such scenarios, using the request/response model leads to high network overhead, which can

38 https://google.github.io/physical-web/

39 https://developers.google.com/beacons/

40 https://openconnectivity.org/resources/specifications/upnp/specifications

cause increased power consumption. In contrast, the request/response model can be easily integrated into client/server applications, as will be discussed for the Constrained Application Protocol (CoAP). Some of the commonly used protocols are as follows:

- *CoAP*[41]. A request/response protocol was created by the *Constrained RESTful Environments* (CoRE) group for constrained devices. To provide resource-oriented interactions in a request/response architecture, it uses the HTTP commands (i.e., GET, POST, PUT, and DELETE) over UDP. Unlike TCP, UDP is suitable for lightweight IoT implementations. HTTP-CoAP proxies can be used to effortlessly convert interactions between HTTP and CoAP. To provide the reliability that UDP lacks, the header of CoAP messages has two bits that determine the required Quality of Service (QoS) level. CoAP includes four types of messages: *confirmable* (i.e., must be acknowledged by the receiver with an acknowledgment [ACK] packet), *nonconfirmable* (i.e., "fire and forget" messages that do not require acknowledgment), *acknowledgment* (i.e., an acknowledgment of a confirmable message), and *reset* (i.e., rejecting a message or removing an observer). CoAP has a simple back-off retransmission mechanism to detect and prevent duplicates for confirmable messages via a unique message ID, which is a 16-bit header field. DTLS can be used to provide security to CoAP applications; however, DTLS has problems because (1) its handshakes cause resource-limited nodes to consume additional resources, leading to reduced battery life; (2) it increases the conversion complexity between HTTP and CoAP; and (3) it prevents CoAP's multicast support.
- *Extensible Messaging and Presence Protocol (XMPP)*[42]. An IETF protocol for communicating over the Internet. It is appropriate for IoT because it provides near real-time, low-latency, and platform-independent communications in a decentralized manner. It runs over TCP and works in both publish/subscribe and request/response models. In XMPP, a client connects to a server using eXtensible Markup Language (XML) messages. The XML parsing creates high network overhead; however, methods are available to reduce this overhead (Waher and DOI, 2015). XMPP has built-in core TLS/Secure Sockets Layer (SSL) security features. QoS is not provided in XMPP; it inherits the TCP mechanisms for reliability, which are not suitable for the IoT (Karagiannis et al., 2015). To overcome this issue, a draft standard (i.e., XEP-0184: Message Delivery Receipts) has been proposed (Saint-Andre and Hildebrand, 2011).
- *Message Queue Telemetry Transport (MQTT)*[43]. A lightweight M2M communication protocol that utilizes the publish/subscribe model. It is suitable

41 https://tools.ietf.org/html/rfc7252
42 https://xmpp.org/
43 http://mqtt.org/

for slow, unreliable connections and computation-limited devices. MQTT includes three types of modules: *broker, publishers,* and *subscribers.* A subscriber can register for a specific service; then, the broker notifies interested subscribers when publishers start a service that interests them. MQTT runs on top of TCP and provides three levels of QoS: *at most once* (i.e., a message will be delivered once with no confirmation), *at least once* (i.e., a message will be delivered at least once, with confirmation required), and *exactly once* (i.e., a message will be delivered exactly once using a four-step handshake) (Karagiannis et al., 2015). A slightly different version of MQTT for sensor networks (i.e., MQTT-SN) is aimed at embedded devices on non-TCP/IP networks, such as ZigBee. Security is implemented by the MQTT broker using TLS/SSL, which requires subscriber authentication.

- *Advanced Message Queuing Protocol (AMQP)*[44]. An open standard for the IoT that provides message-oriented publish/subscribe communications capabilities. Similar to MQTT, message reliability is maintained using message delivery guarantees, such as *at most once, at least once,* and *exactly once.* TLS/SSL and/or Simple Authentication and Security Layer (SASL)[45] are used for security over TCP. Communications in AMQP are managed by two components: exchanges and message queues. Exchanges route messages to or among appropriate queues based on predefined rules. Message queues store messages in queues before they are sent to receivers. Although it lacks protocol support for Last-Value Queues (LVQs)[46], AMQP provides a richer group of messaging scenarios (e.g., one-to-one, broadcast, topic filtered, and request-reply) (StormMQ, 2017).
- *Data Distribution Service (DDS)*[47]. A M2M standard that was created by the Object Management Group (OMG). It is a data-centric publish/subscribe protocol that does not rely on a broker. Due to its multicast support, it is a very reliable protocol because it offers 23 different QoS policies. The DDS architecture consists of two interface levels. The first level—Data-Centric Publish-Subscribe (DCPS)—is responsible for information delivery to subscribers. The second level—the Data-Local Reconstruction Layer (DLRL)—is an optional layer that breaks data objects into separate elements to enable the simple integration of DDS into the application layer (Anonymous, 2016d). In addition to publishers and subscribers, a typical DDS application has the following elements: (1) *a domain* (e.g., medical devices), (2) *a topic* (i.e., a group of data from similar devices, such as electrocardiogram (ECG) readings from many devices), (3) an *instance,* (i.e., a target where data is changing, such

44 https://www.amqp.org/

45 https://tools.ietf.org/html/rfc4422

46 LVQs are special queues that only retain the last value of the messages (i.e., messages are discarded when a newer message with the same value is received).

47 http://www.omg.org/spec/DDS/

as “patient #13 ECG readings”), (4) *a sample* (i.e., a snapshot of an instance at a point in time), (5) *a Data-Writer* (i.e., a source of information about the topic), and (6) a *DataReader* (i.e., an observer of the topic).

Table 3.2 compares the most important features of data exchange protocols. The first two columns list the messaging model type (i.e., request/response or publish/subscribe). The Brokerless column indicates whether a broker between subscribers and publishers is required. Brokerless protocols do not rely on a broker to deliver messages, thereby eliminating the possibility of a single point of failure. The RESTful column indicates whether services are accessible via stateless operations, which is relevant in situations where communication is required upon request (as in HTTP). The Transport column shows the type of transport protocol that is employed in the technology. TCP is connection oriented, which offers high QoS but can be slow for large applications. The Multicast column specifies whether the technology enables multicast, which is preferable to reduce unwanted communications and add extra QoS. The Security column lists the security features supported by each technology. The QoS policies column specifies the number of QoS policies: Typically, a high number of policies is better. The Header Size column indicates the minimum size of a packet, which can be a consideration when the data rate is low, and communications are unreliable. The Discovery column specifies whether the protocol has discovery abilities in terms of services and node properties.

3.3.3.3 Computation

Computation is required to process the collected data and manage sensors. Computations in IoT environments can range from lightweight to complex operations. Lightweight computations are related to managing the platform, such as service access lists that involve encryption/decryption. Complex computations address the actual data (e.g., logic). Computations can be performed based on IoT application requirements in the following locations:

- *Local.* Local computations are performed using a processing unit or *system on a chip* (SoC). Applications may include networking and one-board sensing technologies, such as Arduino[48], UDOO[49], Intel’s IoT products[50], and Raspberry PI[51]. This hardware is typically operated by real-time operating systems (RTOSs), such as Contiki OS[52] and RIOT OS[53]. This type can be used in small

48 https://www.arduino.cc/
49 http://www.udoo.org/
50 http://www.intel.com/content/www/us/en/internet-of-things/products-and-solutions.html
51 https://www.raspberrypi.org/
52 http://www.contiki-os.org/
53 https://www.riot-os.org/

Table 3.2 Comparison of IoT data exchange protocols.

Protocol	Request/ response	Publish/ subscribe	Brokerless	RESTful	Transport	Multicast	Security	QoS policies	Header size (Byte)	Discovery
CoAP	Yes	Yes	No	Yes	UDP	No	DTLS	4	4	Yes
MQTT MQTT-SN	No	Yes	No	No	TCP	Yes	SSL	3	2	No
XMPP	Yes	Yes	No	No	TCP	Yes	SSL	0	Varies	Yes
AMQP	No	Yes	No	No	TCP	No	SSL SASL	3	8	No
DDS	No	Yes	Yes	No	TCP UDP	Yes	SSL DTLS	23	Varies	Yes

IoT project (e.g., do-it-yourself (DIY) security systems (Anonymous, 2016e)) or in border routers for lightweight operations, such as filtering and compressing data, to reduce network overhead and load on servers in the cloud or fog.

- *Cloud.* Cloud computing is preferable for applications in which sensors that reside in different places send generated data to the cloud for centralized processing. This approach is flexible because cloud services can scale-up or scale-down on demand (Hassan et al., 2012). Cloud-based IoT platforms are the most common architecture in current real IoT deployments. Kaa[54] and DeviceHive[55] are examples of cloud-based IoT *Platform as a Service* (PaaS) implementations. Cloud *Infrastructure as a Service* (IaaS) is used by IoT service providers to allow their customers to use IoT services (i.e., *Software as a Service* (SaaS)).
- *Fog.* A fog computing layer is ideally employed with the cloud to improve performance because it can be deployed near end users or nodes. This is important because cloud servers might be far away from the nodes; thus, fog computing servers can perform some operations to reduce the latency of transferring data all the way to the cloud. Also, fog is preferred over cloud because only a small amount of the data transferred is relevant for actionable insights. Hence, cloud is not always efficient. Thousands of fog computing units are provided by mobile network operators but do not have the same computing capabilities as the cloud. Similar to cloud computing, fog computing improves scalability. Low latency and location awareness, widespread graphical distribution, device density, mobility, and real-time capabilities are advantages of fog computing (Saint-Andre and Hildebrand, 2011). For more details about fog computing in the IoT, refer to Chapter 4.

3.3.4 Application Layer Technologies

This layer is responsible for providing requested services to IoT users via a simple interface without knowing how service requests are processed in the underlying layers. IoT users can access a service (e.g., reading or setting temperature conditions remotely or tracking and managing vehicles) using many platforms (e.g., laptops, smartphones, and smartwatches) via web portals or applications. Services vary based on the IoT scenario but can be categorized into four main classes as described in the following sections (Gigli and Koo, 2011).

3.3.4.1 Identity-Related Services

Identity-related services require an identifier that is embedded in a node and a reader device, such as an RFID device. Identity-related services can be active or

54 https://www.kaaproject.org/
55 http://devicehive.com/

passive (as discussed in the RFID subsections). These services are extremely important in IoT applications because they keep track of devices in large deployments (e.g., recording device maintenance and maintaining inventories of used parts and stock levels.) A package-tracking application is an example of this type of service.

3.3.4.2 Information Aggregation Services

Aggregation services summarize the collected raw sensory measurements from various types of sensors and networks that must be processed and reported to the IoT application. Load distribution among smart grids is an example of this type of service.

3.3.4.3 Collaborative Aware Services

Collaborative aware services are built on top of information aggregation services and are used to make decisions about the acquired data. This type of service can be found in smart homes, smart agriculture, and smart manufacturing, among other applications. For example, in a smart home, a security system and smart thermostats are employed together to improve security and energy efficiency based on sharing data between them (Anonymous, 2016f).

3.3.4.4 Ubiquitous Services

These are the most advantageous services of the IoT because they advance collaborative aware services to the next level by offering complete access to everything at (almost) any time and from anywhere. Access and control can be achieved using a computer, smartphone, or any smart device. Smart cities are examples of these services.

3.3.5 Business Layer Technologies

In this layer, unlike the application layer, service data and IoT environmental data, such as business models, flowcharts, and graphs, can be accessed. This access helps administrators in the design, analysis, implementation, evaluation, monitoring, and development of IoT systems because the output of each of the previously mentioned layers is analyzed in this layer to improve services and protect user privacy (Al-Fuqaha et al., 2015). The technologies in this layer can be divided into two categories based on their functions (i.e., semantics and Big Data analytics), as described in the subsequent sections.

3.3.5.1 Semantics

After data are captured by sensors, they can be analyzed to extract knowledge. Knowledge is crucial for enhancing services and obtaining useful results that cannot be discovered without examining all IoT environment data. In order to analyze data efficiently, it has to be well-formed using some semantics technologies. Many semantic XML-based technologies can be used for knowledge extraction, resource discovery and usage, and modeling. An Efficient XML Interchange (EXI) is often used to resolve the performance issue that arises from the use of XML (e.g., XML parsing) to make it suitable for IoT applications (Al-Fuqaha et al., 2015). Many nodes from different brands are likely to generate data that has different form in each type of nodes. Such problem requires more work to unify the data so that it can be processed efficiently. Therefore, dynamic semantics is an important factor in improving the interoperability of the IoT. IoT technologies for semantics include the following:

- *Sensor Model Language (SensorML)*[56]. A standard model based on an XML schema to describe sensors and measurement procedures. SensorML is useful in IoT systems for creating electronic description sheets for sensor modules and collecting metadata to be used to discover sensor systems and observe processes. It also enables sensor networks to be autonomous because of the self-describing features of SensorML-supported sensors.
- *Media Types for Sensor Markup Language (SenML)*[57]. It is a new, simple model for acquiring sensed data and to control actuators. It provides semantics for the data and allows for additional metadata with links and extensions. This simple model can be used in many IoT applications (Keränen and Jennings, 2017). For example, a sensor, such as a humidity sensor, could use this media type in CoAP to transport the sensors' measurements.
- *IoT Database (IOTDB)*[58]. A new technology with unlimited expandability that supports semantics for providing formal definitions of all necessary items. Unlike the aforementioned technologies, IOTDB uses JSON dictionaries to manipulate and monitor nodes, which makes it relatively fast since JSON parsing is always more efficient than XML parsing. IOTDB is compatible with protocols, such as CoAP, and MQTT.
- *RESTful API Modeling Language (RAML)*[59]. This language is used to define HTTP-based APIs that represent most of the principles of Representational State Transfer (REST)[60]. Since it is RESTful, it is more likely to be used in IoT scenarios that are suitable for CoAP (as discussed in Section 3.3.3.2), where the network overhead is negligible.
- *Wolfram Data Drop*[61]. An open service that allows accumulating data of any type from anywhere (including IoT nodes) to prepare it semantically for instant computation, querying, analysis, visualization, or other operations.

56 http://www.opengeospatial.org/standards/sensorml
57 https://tools.ietf.org/html/draft-jennings-senml-10
58 https://iotdb.org
59 http://raml.org
60 Fielding, R.T. (2000) Representational State Transfer (REST). PhD dissertation. Available at https://www.ics.uci.edu/~fielding/pubs/dissertation/rest_arch_style.htm.
61 https://datadrop.wolframcloud.com

Computable data (i.e., collections and time series) are saved in named data bins in the Wolfram Cloud and are immediately accessible from all other systems/applications.

3.3.5.2 Big Data Analytics

IoT environments involve an immense number of sensors that collect vast amounts of data, resulting in extremely large datasets (i.e., Big Data) that may be computationally analyzed to extract knowledge, such as patterns, trends, and associations. The generated Big Data continues to increase as the IoT collects additional data. Therefore, special technologies that can address the ever-increasing amount of data are required. The most efficient strategy for addressing continuously increasing data is to process them in real time (streaming). Real-time processing enables up-to-date analytics that reflect recent changes in data and conserves storage. Machine learning can also utilize Big Data to make accurate predictions. Additionally, parallel computing can be employed using some IoT nodes (or fog computing units) to process data in parallel with back-end servers for better performance and load balancing. The following technologies aid in the generation of real-time (streaming) analytics from Big Data:

- *Apache Spark*[62]. A well-known open-source distributed processing technology that utilizes in-memory caching and features enhanced execution for better performance. Apache Spark supports streaming analytics, batch processing, graph databases, ad hoc queries, and machine learning. Although Spark has been employed in numerous IoT systems, it requires some extra effort to overcome some limitations, such as time delays in real-time processing (Freedman, 2016). Spark is a very flexible technology since it supports static and streaming data and many programming languages (as presented in Table 3.3).
- *Apache Apex*[63]. An open-source platform, similar to Spark, unifies stream and batch processing in a distributed, scalable, fault-tolerant, performant, stateful, and secure manner. Unlike Spark, it has a very low latency and processes big-data-in-motion (i.e., processes streaming data, including in-flight modification). Its ability for processing big-data-in-motion, enables it to be integrated easily with fog computing units. However, it lacks out-of-order processing and only supports Java as a programming language. Although it is a relatively new platform, it is widely adopted in many large-scale IoT projects, such as smart grids (Anonymous, 2016g; DataTorrent, 2016) and industrial applications (Apache Apex, 2016).
- *Apache Flink*[64]. An open-source platform that is conceptually very similar to Apex. However, it is older than Apex, it has not been used frequently in

62 http://spark.apache.org
63 https://apex.apache.org
64 https://flink.apache.org

Table 3.3 Comparison of Big Data streaming processing technologies.

Technology	API languages	QoS policies	Autoscaling	Latency	In-flight modifications
Spark	Scala, Java, R, Python	Exactly once At least once	Yes	Medium	No
Apex	Java	Exactly once At least once At most once	Yes	Very low	Yes
Flink	Scala, Java, Python	Exactly once	No	Low	No
Kafka	Java	At least once	Yes	Very low	Yes

real-life IoT projects due to its lack of autoscaling and limited message delivery guarantees (QoS).

- *Apache Kafka*[65]. A client library used to process and analyze the data stored in Kafka and either forward the output to an external system or write it back to Kafka. The first scenario allows Kafka to serve as a middle layer in which it collects the data generated by IoT sensors, then applies some operations to it (e.g., filtering and sorting), and finally streams it to other components in other frameworks (e.g., Flink and Spark).

Table 3.3 highlights the main differences between these technologies. The first column describes the languages that these Big Data frameworks support. This helps IoT adopters to choose Big Data frameworks based on their employee's expertise. The next column describes the delivery policies, which might be an important element in critical IoT scenarios in which "at least once" policy is appropriate. In contrast, in less critical scenarios (or for more efficiency), at most once would be sufficient. In this feature, Apex has an advantage over the others because it has the highest number of policies. Autoscaling is a very important feature for IoT because it ensures the efficiency of computation as extra resources are used only when needed. Latency can be relevant in some IoT scenarios (e.g., traffic control) to ensure the accuracy of real-time analytics. Finally, the last column indicates the ability of the framework to perform some computations on data before they reach the back-end servers, as in fog computing.

65 http://kafka.apache.org

3.4 IoT Platforms and Operating Systems

There are a vast number of IoT platforms and operating systems that can integrate many of the abovementioned technologies to provide IoT services. In the context of IoT, both terms (i.e., platforms and operating systems) are used interchangeably. However, there is a slight difference between them, that is, operating systems (i.e., embedded operating systems or RTOS) are focused only on some communication-related functions such as connecting sensors to the Internet and allowing data to be collected from sensors. Other functions, such as analytics, are often occurring, if needed, in partner systems. This makes operating systems suitable for small IoT applications. On the other hand, IoT platforms cover almost all functionalities from the five mentioned layers (e.g., communication, data exchange, analytics, etc.). Another advantage of IoT platforms is ease of setup and deployment. IoT platforms, in contrast to operating systems which require extra effort to integrate with other systems, enable IoT adopters to easily choose suitable technologies based on their needs and give them better compatibility and support. This section discusses only the popular generic IoT operating systems and platforms.

The following are two well-known C-based IoT operating systems:

- *RIOT*[66]. A lightweight operating system for memory-constrained, wireless networked systems with a focus on low-power IoT devices. RIOT supports a wide range of devices and offers higher complexity that allows for larger and time sensitive IoT applications.
- *Contiki*[67]. An open-source operating system for connecting memory-constrained, low-power wireless networked systems to the Internet. It is similar to RIOT but less advanced in terms of latency and performance (i.e., multithreading) capabilities. However, one of the main advantages of Contiki is its simulator (i.e., Cooja). Cooja helps researchers and developers to test their systems before purchasing devices. Since Contiki is relatively older than RIOT, it is widely used and hence, has an active community.

Other operating systems can also be used, such as Linux, but due to its requirement for larger ROM and RAM (i.e., ~1 MB), it is not suitable for tiny IoT devices. Table 3.4 compares main features in RIOT and Contiki operating systems. RIOT OS is more advanced because it supports multithreading as well as for its very low latency (almost real-time).Most IoT platforms are cloud based and provide IoT technologies for most functions mentioned in the previous section. The following are some platforms among the vast number of current platforms:

---

66 https://riot-os.org/

67 http://www.contiki-os.org/

Table 3.4 Comparison of IoT operating systems.

OS	Programming languages	Simulator	Required RAM, ROM	Multithreading	Real-time capabilities
RIOT	C, C++	None	1.5 KB, 5 KB	Yes	Yes
Contiki	Partial C	Cooja	2 KB, 30 KB	No	No

- *AWS IoT*[68]. A commercial IoT platform that exploits the popular AWS cloud services to provide everything that an enterprise needs (i.e., device management, visualization). Data ingestion, storage, processing, and visualization are handled by services such as Amazon Redshift, Amazon EMR, AWS Lambda, Amazon DynamoDB, Amazon Kinesis, and Amazon QuickSight.
- *IBM Watson*[69]. A commercial IoT platform that is strongly integrated with Bluemix to bring the power of cognitive computing and machine learning to IoT. This platform can be deployed on the cloud or on-premises in which onboarding of devices to the platform is automated using the SDKs and APIs. It also allows for blockchaining in which IoT integration with the evolving distributed ledger technology is based on HyperLedger.
- *ThingWorx*[70]. A commercial platform for the development of connected smart devices using an integrated IoT development tool that supports connectivity, production, analysis, and other areas of IoT. ThingWorx allows users to connect devices, establish a data source, establish device behaviors, and build an interface without any coding.
- *Bosch IoT Suite*[71]. A commercial flexible cloud-based IoT that allow developers to test the applications before implementing them, deploying them, and operating them under normal conditions. Its device management capabilities (i.e., executing software roll-out processes, connecting third-party systems and services, and analyzing data) can also be used stand-alone and on-premise.
- *Xively*[72]. A commercial cloud-based platform that makes the development of connected smart products for business easy due to its strong partnerships with hardware partners and one-click integrations with business tools. Xively enables visualization of data graphically in real time as well as updating devices remotely.
- *EVRYTHNG*[73]. A cloud-based platform to manage IoT identities for products in which all products are given a persistent, addressable web presence. It

68 https://aws.amazon.com/iot-platform/
69 https://www.ibm.com/internet-of-things/
70 https://www.thingworx.com/
71 https://www.bosch-si.com/iot-platform/bosch-iot-suite/homepage-bosch-iot-suite.html
72 https://www.xively.com/
73 https://evrythng.com/

enables elastic semantic data store to customize these dynamic data profiles for any product so that authorized applications can interact with them and exchange data in real time during their lifecycle.

- *Kaa*[74]. An open-source cloud-based platform for managing IoT devices and analyzing generated data to provide complete end-to-end IoT solutions, connected applications, and smart products. It is compatible with virtually any connected devices and gateways. It enables devices to be used almost as plug and play units with minimal code.

Table 3.5 gives an overview of the main features of the popular IoT platforms. Data exchange protocols column indicates the protocols used to collect and send data among devices. The security column indicates security protocols used for authentication and confidentiality. The integration column shows the way

Table 3.5 Comparison of IoT platforms.

Platform	Data exchange	Security	Integration	Device management
AWS IoT	MQTT, HTTP	TLS, SigV4[a], X.509[b]	REST API	Yes
IBM Watson	MQTT, HTTPS	TLS, IBM Cloud SSO[c], LDAP[d]	REST and Real-time APIs	Yes
ThingWorx	MQTT, AMQP, XMPP, CoAP, DDS, WebSockets[e]	ISO 27001[f], LDAP	REST API	Yes
Bosch IoT Suite	MQTT, CoAP, AMQP, STOMP[g]	Unknown	REST API	Yes
Xively	HTTP, HTTPS, WebSocket, MQTT	SSL/TSL	REST API	No
EVRYTHNG	MQTT, CoAP, WebSockets	SSL	REST API	No
Kaa	MQTT, HTTP	RSA and AES	REST API	Yes

a) http://docs.aws.amazon.com/general/latest/gr/signature-version-4.html
b) https://www.ietf.org/rfc/rfc5280.txt
c) https://www.ibm.com/security/cloud/cloud-identity-service/
d) https://tools.ietf.org/rfc/rfc4511.txt
e) https://tools.ietf.org/html/rfc6455
f) https://www.iso.org/isoiec-27001-information-security.html
g) https://stomp.github.io/

74 https://www.kaaproject.org/

that a platform can be integrated with other systems. Finally, device management indicates whether the platform enables devices to be remotely monitored, updated, disabled, and so on.

3.5 Conclusion

This chapter presented current and commonly employed technologies and protocols and their functionalities in a logical manner based on the IoT five-layer model. The perception layer technologies discussed are generic physical devices that can be used to send sensed data. The networking technologies mentioned in the second layer enable nodes to be identified and reached over a secure medium using special routing protocols, such as RPL. The technologies discussed in the middleware layer utilize physical components to provide services that are easily discoverable and organized for use by IoT users in the upper layers. The application layer provides the ultimate goal of IoT systems (i.e., services) to allow end users to access and use them on their favorite platform. Finally, the technologies in the business layer help system administrators to monitor and enhance IoT activities through analytics performed on collected data.

The technologies were discussed in a generic manner without suggesting any preferences because the most suitable IoT technologies cannot be recommended without knowing the requirements of the specific scenario in which they will be applied. For example, the IoMT probably requires low latency and secure technologies; hence, choosing technologies that have such features would be appropriate. The informative comparisons in this chapter highlighted the main differences among the various technologies.

Several challenges remain associated with the currently available IoT technologies. First, better horizontal integration among protocols is needed in each layer to enable existing IoT applications to collaborate and share resources. For example, in the middleware layer, a CoAP system should be able to integrate with DDS. This integration will enable collaboration among organizations and allow them to effectively share resources (e.g., sensors) or data. Second, service continuity for mobile Internet remains challenging because all current technologies rely on fixed points (e.g., cellular towers and border routers) for connectivity (Elsaleh et al., 2011). Third, IoT scalability and interoperability remain challenging as compatibility issues are inevitable in its current state because many different technologies are involved (Sarkar et al., 2014). Finally, security in the IoT is a distinct problem affecting currently available technologies due to the large-scale potential impact of successful attacks.

IoT enabling technologies are the key to building effective and successful IoT applications. As the number of technologies continues to expand, understanding the characteristics and limitations of currently available technologies is important because they provide the foundation for new technologies. Hopefully, current challenges in the IoT will be minimized by the rapid advancement of IoT technologies driven by many organizations worldwide.

References

Ackerman, E. (2015) Disney seeks to make visible light communication practical. IEEE Spectrum: Technology, Engineering, and Science News, Available at http://spectrum.ieee.org/tech-talk/computing/networks/disney-seeks-to-make-visible-light-communication-practical (accessed Mar 30, 2017).

Al-Fuqaha, A., Guizani, M., Mohammadi, M., Aledhari, M., and Ayyash, M. (2015) Internet of Things: a survey on enabling technologies, protocols, and applications. *IEEE Communications Surveys and Tutorials*, **17**(4), 2347–2376.

Al-Qaseemi, S.A., Almulhim, H.A., Almulhim, M.F., and Chaudhry, S.R. (2016) IoT architecture challenges and issues: lack of standardization. 2016 Future Technologies Conference (FTC), pp. 731–738.

Anonymous (2016a) GitHub – google/physical-web: The Physical Web: walk up and use anything. Available at https://github.com/google/physical-web (accessed Sep 23, 2016).

Anonymous. (2016b) Vulnerability Note VU#357851 - UPnP requests accepted over router WAN interfaces. Available at http://www.kb.cert.org/vuls/id/357851 (accessed Sep 24, 2016).

Anonymous. (2016c) OCF - UPnP+ Initiative. Open Connectivity Foundation (OCF).

Anonymous. (2016d) DDS-DLRL 1.4. Available at http://www.omg.org/spec/DDS-DLRL/1.4/ (accessed Sep 30, 2016).

Anonymous. (2016e) Raspberry Pi security system with motion detection/Camera. *Hackster.io*. Available at https://www.hackster.io/FutureSharks/raspberry-pi-security-system-with-motion-detection-camera-bed172 (accessed May 6, 2017).

Anonymous. (2016f) SimpliSafe | Home Security Systems. Available at http://simplisafe.com/simplisafe-now-works-with-nest (accessed Dec 2, 2016).

Anonymous. (2016g) IOT Big Data Ingestion and Processing in Hadoop by Silver Spring Network. Available at http://www.slideshare.net/ApacheApex/iot-big-data-ingestion-and-processing-in-hadoop-by-silver-spring-networks (accessed Oct 17, 2016).

Anonymous. (2017) UWB: Back from the dead, bound for IoT location-based services | FierceWireless. Available at http://www.fiercewireless.com/tech/uwb-back-from-dead-bound-for-iot-location-based-services (accessed May 31, 2017).

Apache Apex, (2016) GE IOT Predix Time Series & Data Ingestion Service using Apache Apex. Available athttps://www.slideshare.net/ApacheApex/ge-iot-predix-time-series-data-ingestion-service-using-apache-apex-hadoop (accessed Mar 31, 2017).

Bonomi, F., Milito, R., Zhu, J., and Addepalli, S. (2012) Fog computing and its role in the Internet of Things, in Proceedings of the First Edition of the MCC Workshop on Mobile Cloud Computing, New York, NY, USA, pp. 13–16.

DataTorrent, *Meetup + Slides: GE IOT Predix Time Series & Data Ingestion Service Using Apache Apex (Hadoop)*, 2016.

Duffy, J. (2014) SDN vital to IoT. Network World, Available at http://www.networkworld.com/article/2601926/sdn/sdn-vital-to-iot.html (accessed Nov 27, 2016).

Elsaleh, T., Gluhak, A., and Moessner, K. (2011) service continuity for subscribers of the mobile real world Internet. 2011 IEEE International Conference on Communications Workshops (ICC), pp. 1–5.

Fontana, R.J. (2004) Recent system applications of short-pulse ultra-wideband (UWB) technology. *IEEE Transactions on Microwave Theory and Techniques*, **52**(9), 2087–2104.

Gigli, M. and Koo, S. (2011) Internet of Things: services and applications categorization. *Advances in Internet of Things*, **1**(2), 27–31.

Gubbi, J., Buyya, R., Marusic, S., and Palaniswami, M. (2013) Internet of Things (IoT): A vision, architectural elements, and future directions. *Future Generation Computer Systems*, **29**(7), 1645–1660.

Hassan, Q.F., Riad, A.M., and Hassan, A.E. (2012) Understanding cloud computing, in Yang, H. and Liu, X. (eds), *Software Reuse in the Emerging Cloud Computing Era*, IGI Global.

Jacquenet, C. and Boucadair, M. (2016) A Software: Defined Approach to IoT Networking. September.

Karagiannis, V., Chatzimisios, P., Vazquez-Gallego, F., and Alonso-Zarate, J. (2015) A survey on application layer protocols for the internet of things. *Transaction on IoT and Cloud Computing*, **3**(1), 11–17.

Keränen, A. and Jennings, C. (2017) SenML: simple building block for IoT semantic interoperability. Available at https://www.iab.org/wp-content/IAB-uploads/2016/03/IAB_IOTSI_Keranen_Jennings_SenML.pdf (accessed May 4, 2017).

Li, S., Xu, L.D., and Zhao, S. (2015) The Internet of Things: a survey. *Information Systems Frontiers*, **17**(2), 243–259.

Masinter, L., Berners-Lee, T., and Fielding, R.T. (2016) Uniform Resource Identifier (URI): Generic Syntax. Available at https://tools.ietf.org/html/rfc3986 (accessed Sep. 1, 2016).

Freedman, M. (2016) Spark streaming and IoT. Available at https://spark-summit.org/east-2016/events/spark-streaming-and-iot/ (accessed Oct 17, 2016).

Nokia (2017) LTE evolution for IoT connectivity white paper. Available at http://resources.alcatel-lucent.com/asset/200178 (accessed May 3, 2017).

Prasad, R. (2014) *5G: 2020 and Beyond*, River Publishers.

Raza, S., Chung, T., Duquennoy, S., Dogan, Y., Voigt, T., and Roedig, U. (2011) Securing Internet of Things with Lightweight IPsec.

Saint-Andre, P. and Hildebrand, J. (2011) Message Delivery Receipts. Available at https://xmpp.org/extensions/xep-0184.html (accessed Mar 30, 2017).

Sarkar, C., Nambi, S.N.A.U., Prasad, R.V., and Rahim, A. (2014) A scalable distributed architecture towards unifying IoT applications. 2014 IEEE World Forum on Internet of Things (WF-IoT), pp. 508–513.

Stephen C. (2016) 10 things to know about the October 21 IoT DDoS attacks. *We Live Security*, (posted on October 24, 2016 at 07:16 pm). Available at http://www.welivesecurity.com/2016/10/24/10-things-know-october-21-iot-ddos-attacks/ (accessed Dec 01, 2016)

StormMQ, (2017) A Comparison of AMQP and MQTT white paper. Available at https://lists.oasis-open.org/archives/amqp/201202/msg00086/StormMQ_WhitePaper_-_A_Comparison_of_AMQP_and_MQTT.pdf (accessed May 4, 2017).

Waher, P. and DOI, Y. (2015) Efficient XML Interchange (EXI) Format. Available at http://xmpp.org/extensions/xep-0322.html (accessed Sep 27, 2016).

West, D.M. (2016) How 5G technology enables the health internet of things. Center for Technology Innovation Brookings, vol. 3.

Weyrich, M. and Ebert, C. (2016) Reference architectures for the Internet of Things. *Software IEEE*, **33**(1), 112–116.

4

Cloud and Fog Computing in the Internet of Things

Daniel Happ

Telecommunication Networks Group (TKN), Technische Universität Berlin, Berlin, Germany

4.1 Introduction

In contrast to classical wireless sensor networks (WSN) that usually only serve a single application, one of the core benefits of the shift to the IoT lies in the common usage of sensor hardware by heterogeneous applications (Tschofenig et al., 2015). Additionally, the revolution of the IoT does not stem from the number of connected things alone, but from the solutions and services offered on top of the data. The basic requirements of such value-added services can be briefly summarized into nonvolatile storage of historical sensor data, sensor data processing, and efficient near-real-time distribution of sensor data.

However, although everyday objects are increasingly connected to the Internet and becoming more and more powerful, they are usually not capable enough to fulfill all those requirements themselves. One of the major challenges is that commonly used devices, such as sensor nodes, smartphones, and wearable techs, usually run on battery power, making storage or complex processing of a large amount of data unfeasible. Mains-operated connected objects may also often be too constrained to perform those tasks as reliably and quickly as required.

The recent advances in cloud computing have led to an increasing usage of this model to meet the aforementioned requirements and enable value-added services in the IoT context. Cloud computing offers convenient, ubiquitous, and on-demand access to a shared pool of configurable computing resources, which are accessible over the Internet, and usually reside in third-party datacenters (Hassan et al., 2012; Mell and Grance, 2011). Along with these resources, cloud providers offer fast and configurable networking for data distribution and reliable, nonvolatile, replicated storage. Thanks to its flexibility, reliability, and

Internet of Things A to Z: Technologies and Applications, First Edition. Edited by Qusay F. Hassan.
© 2018 by The Institute of Electrical and Electronics Engineers, Inc. Published 2018 by John Wiley & Sons, Inc.

usage-based cost model, cloud computing is well positioned to meet the specific requirements of value-added services in the IoT context. Constraint devices can save energy by transferring their data to a cloud-based platform where it will be distributed to multiple relevant applications and services, which will process the data accordingly.

Though this architecture works well today, it is not suitable for latency-sensitive applications since cloud datacenters are colocated neither with connected objects nor with consumers of value-added services (Zhang et al., 2015). From a network topology view, cloud datacenters are located several hops away from IoT data producers and consumers, and are most often separated by a constrained last-mile link. The physical distance alone causes additional latencies that may not be acceptable for latency-sensitive applications such as control loops. Thus, instead of forcing all IoT communications through a cloud intermediary, there has been a push to move storage, processing, and distribution of data closer to the edge of the network, toward data producers and consumers.

One of those concepts is fog computing, which is seen as an extension of cloud computing, but enable services closer to the edge of the network (Bonomi et al., 2012). While fog and cloud computing paradigms share many mechanisms, fog computing mainly addresses applications and services that would not be feasible in the cloud. One reason would be to eliminate bandwidth bottlenecks and improve on latency for local control loops commonly found in the IoT context. To achieve this, fog computing uses processing power that is available locally today, such as on network hardware or local gateway nodes, mobile phones, or additional hardware that would have to be deployed in the future between the user and the cloud.

This chapter gives an introduction to the concepts of cloud- and fog-based computing in the IoT context. For both approaches, advantages and challenges are presented. Additionally, specific IoT use cases are outlined that benefit from cloud and fog computing, respectively. Since fog computing is a rather new model, the overview also includes potential future use cases that are not yet fully realized.

4.2 IoT System Requirements

A central concern when analyzing cloud and fog computing is the actual role these technologies are expected to play in IoT systems. As discussed in Chapter 1, the basic architectural reference model has three layers: device layer, connectivity layer, and application layer. In general, cloud and fog technology will be used to realize the connectivity layer by relaying device data and enabling value-added services in the application layer. In particular, this chapter identifies the following basic functions as what the cloud or fog backend must provide for IoT systems:

1) *Data Distribution.* IoT environments will generate enormous amounts of data that can be useful in many ways. The first requirement is to provide ubiquitous, cross-vendor real-time access to the data by distributing it from data sources (e.g., sensors, wearables, and smartphones) to consumers (e.g., actuators, value-added services, and applications). The great diversity in hardware and software highlights the need for a unified messaging middleware interconnecting data sources to consumers by providing uniform and standardized APIs.
2) *Scalable Storage.* The value of the vast amount of data available from things might not be directly evident on collection, but the data might provide very valuable insights in the future. Hence, for many use cases, scalable storage of and ubiquitous access to historic data will be a valuable function.
3) *Processing Services.* As mentioned in Chapter 1, analytics and big data processing have become increasingly relevant to the IoT ecosystem. Raw IoT data alone are not of much value, but value-added services can use analytics as well as temporal and spatial aggregation and correlation to further provide important insights.

Apart from those basic services, there are additional high-level qualities imposed by the specific IoT use case:

1) *Flexible Self-Organization.* As stated in Chapter 1, the IoT has to be self-organizing. While things are to be uniquely identifiable, new data producers may join or leave the system at any time, so it must enable the discovery of relevant data sources and services out of a heterogeneous and constantly changing blend. On the other side, the system has to be able to automatically adapt to the changing needs of data consumers.
2) *Reliability.* Depending on the particular use case, stringent reliability and quality of service requirements must be respected, such as low latency and reliable end-to-end data delivery.
3) *Scalability.* Due to the large number of connected things as well as data consumers that are expected, the system has to be scalable, that is, the system should provide adequate basic services independent of the number of connected devices, services, and consumers.
4) *Data Confidentiality and Security.* Device data or derived insights might be sensitive, possibly also not allowed to be shared with certain entities or not allowed to be transferred or stored in other juridical areas. The IoT system has to support those requirements and ensure data are not accessed by restricted entities.

The aforementioned requirements motivate the use of cloud and fog computing to enable the IoT vision. The chapter continues by giving a brief introduction to cloud computing and how it is used in IoT systems; then fog computing and its integration in modern IoT systems is also addressed.

4.3 Cloud Computing in IoT

Cloud computing refers to both a subset of applications delivered as services over the Internet and the underlying hardware and software systems in the datacenters that provide those services. Cloud computing is usually divided into Infrastructure as a Service (IaaS), Platform as a Service (PaaS), and Software as a Service (SaaS), as shown in Figure 4.1 (Hassan et al., 2012; Mell and Grance, 2011).

IaaS describes a business model in which no complete solution is offered, but only the hardware necessary to implement specific applications. PaaS, on the other hand, allows developers to develop and run their applications on the managed infrastructure using well-defined interfaces. For this purpose, PaaS providers usually maintain development environments in the form of frameworks. SaaS is a model in which the software application is no longer sold to the customer for a one-time fee, but is provided as a service for a recurring fee. In this model, infrastructure and platform are also fully managed by the service provider.

Cloud computing can help realize IoT systems that meet most of the requirements mentioned in the previous section. Indeed, the approach to send device data to a cloud-based service for messaging and processing is widely adopted by developers (Gubbi et al., 2013; Menzel et al., 2014). Additionally, solutions for many common tasks are already offered as hosted services from cloud providers, including storage, messaging, and processing. For instance, there are PaaS providers specifically for IoT systems, such as Amazon IoT, that offer a so-called device gateway for IoT messaging needs, a rule engine

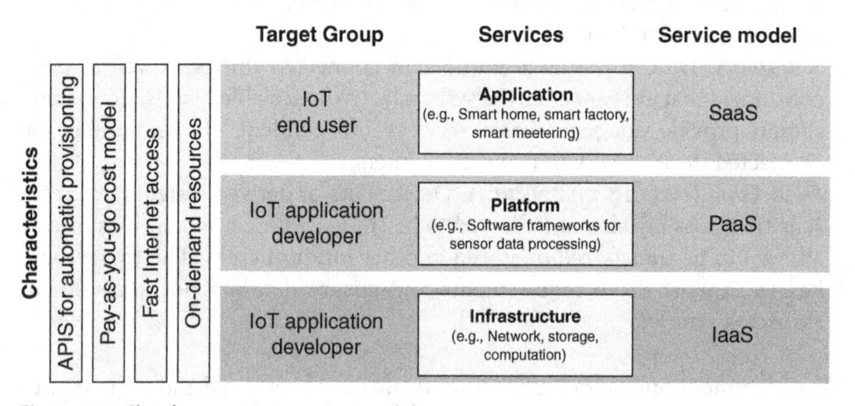

Figure 4.1 Cloud computing service models.

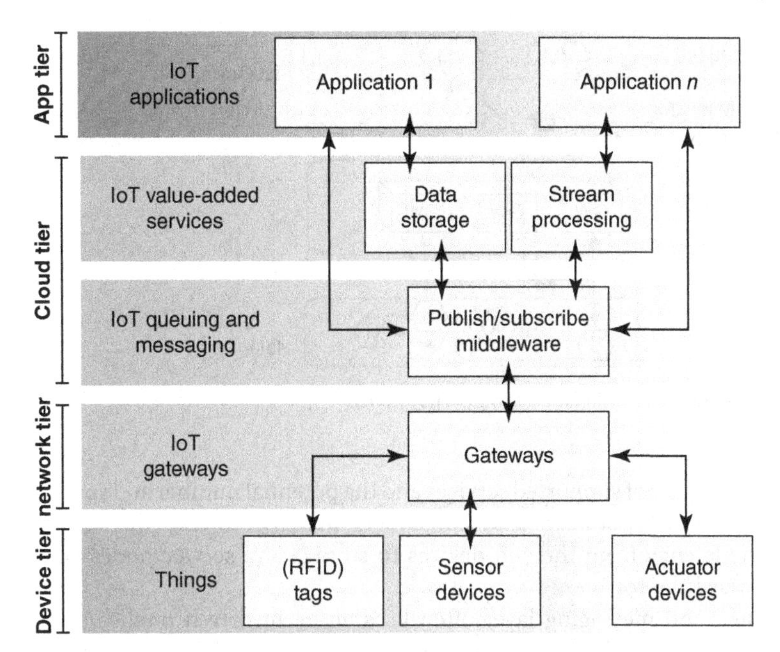

Figure 4.2 Common cloud-based IoT architecture.

for data processing, and a database backend (Dynamo DB) for storage and querying. Other vendors, such as Microsoft, IBM, and Xively, also offer similar services. There are also SaaS providers in the IoT ecosystem that offer ready-made services for specific purposes, such as smart heating or building automation. Examples include Nest[1] and Tado,[2] which market smart thermostats for home users, aiming to reduce the users' heating cost.

Most IoT service providers use a three-layered architecture similar to the one depicted in Figure 4.2. The typical cloud-based IoT solution is characterized by many data-producing end devices that usually use gateway nodes to communicate to a wide area network, usually the public Internet. The storage, messaging, and processing that are crucial for IoT applications are offered as services in the cloud.

An overview of the general placement of things, gateways, and cloud is shown in Figure 4.3. Usually connected things themselves are severely constrained and connect to a gateway, which is connected to the Internet. However, most often this link uses a relatively slow last-mile technology, such as DSL (digital subscriber line) 3G/4G, making it the bottleneck between devices and the cloud in terms of delay and throughput. Regarding scale, because more resources are

1 https://nest.com
2 https://www.tado.com

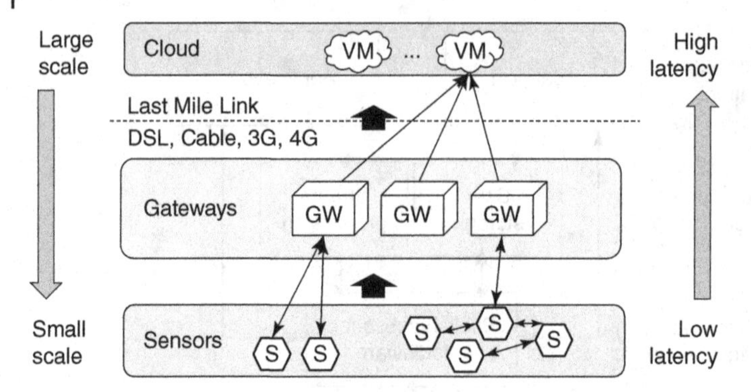

Figure 4.3 Scale and latency in cloud-based IoT systems.

available, the number of supported services and the potential number and spatial distribution of data producers and consumers increase toward the cloud. However, the latency from the end-devices to storage and service nodes also increases toward the top.

The queuing and messaging layer often uses an event-driven publish/subscribe system in the cloud-based IoT approach (Happ et al., 2015; Antonic et al., 2014; Hunkeler et al., 2008; Banks and Gupta, 2014). As a result, data producers and consumers are loosely coupled in space (do not have to know each other), time (do not have to be available at the same time), and in synchronization (asynchronous, nonblocking messaging, that is, calling applications do not have to wait for system to complete call).

4.3.1 Advantages of Using the Cloud for IoT

Cloud computing offers the means to meet some of the IoT systems' requirements outlined above.

1) *Flexibility.* Cloud instances are usually based on virtualized physical hardware in datacenters. Resources are pooled and can be used in an on-demand fashion, often with a pay-as-you-go pricing model, where only resources actually used are billed. This enables on-demand resource provisioning matching the constantly changing requirements of IoT applications.
2) *Reliability.* Intra- and inter-datacenter redundancy can be used to create scalable and reliable services, such as distributed object storage. The basic services of distribution, storage, and processing can be easily realized using a cloud-based infrastructure.
3) *Fast Ubiquitous Access.* Cloud datacenters are well connected using both private peering and multiple uplink transit providers. This enables ubiquitous and fast access to services from anywhere over the Internet.

4) *Fast Time to Market.* Cloud providers usually offer readily available software components to use for the development of customized solutions, which can be accessed through well-defined APIs. Some providers specifically target IoT use cases. This reduces both the time to market for applications and the development cost.

4.3.2 Examples of Cloud-Based IoT

To further emphasize the impact of cloud computing on the current IoT ecosystem, the chapter continues with an overview of recent use cases where the use of the cloud-based IoT paradigm was proposed. The following examples are based on the taxonomy of IoT domains that was introduced in Chapter 1, namely, industrial, smart city, and health care domains.

4.3.2.1 Industrial Domain

The combination of IoT and cloud computing makes several innovative use cases in the industrial domain possible (Tao et al., 2014). Smart industrial IoT systems allow monitoring of industrial plants to not only improve productivity but also safety, especially for high environmental risks. For instance, the approach has been used for a monitoring and prealarm system for tracking tailings dam failures in the mining industry (Sun et al., 2012). The system offers real-time monitoring of the saturated line, water level, and dam deformation. Data are gathered by a local sensor network and processed in the cloud. The results are disseminated to Personal Digital Assistants (PDAs) and other client devices. The system has since been applied and proven useful in several mines.

The combination of IoT and cloud computing can enable advanced solutions in the transportation industry as well. Sensors in vehicles can monitor various aspects related to the efficiency of the transport: tire pressure, fuel consumption, location, and speed (Borgia, 2014).

4.3.2.2 Smart Cities

IoT together with cloud-based analytics has the potential to increase the quality of everyday urban life. Particular focus lies in more efficient and environment-friendly use of energy, transportation, environmental protection, and urban planning.

Efficient management of renewable energy sources is a core building block for achieving sustainability in urban environments. The future smart grid will transform the electrical distribution infrastructure into an intelligent multi-directional system, which will enable the decentralized generation of energy and real-time transparency for producers and consumers. Application areas include demand response and forecasting, dynamic pricing, micro-grid management, and real-time monitoring (Bera et al., 2015).

One application of cloud computing in this context is the use of readily available cloud storage for data from smart meters and other related sensors. Cloud storage allows not only the necessary data storage but also easy access and means for additional analytics. It further provides high fault tolerance through redundancy and can use versioned copies for transparency and roll-back (Rusitschka et al., 2010; Fang et al., 2012).

The combination of cloud computing, IoT, and crowd sensing can further be used to protect the environment and raise awareness of inherent problems, such as the air quality. For example, a crowd-sensing application can be used for monitoring urban air quality, using sensors that are carried by citizens across the city (Antonić et al., 2014a, 2014b). The data gathered are sent to cloud servers, which will filter out the redundant or unneeded data. This particular system additionally provides personalized real-time notifications about air quality to mobile users using Google Cloud Messaging.

SmartSantander (Sanchez et al., 2014) provides an experimental research facility in a real urban deployment in the smart city context. It focuses on the sensor network part of smart city research and provides a large-scale sensor testbed that help develop solutions for connecting and managing smart city sensor devices. To support applications on top of the data, a private cloud infrastructure is provided.

4.3.2.3 Health/Well-Being

IoT will enable applications in the smart health care and well-being domain that will improve the experience for patients and physicians alike. The spectrum of useful applications ranges from remote health monitoring to drug inventory management. For various use cases, offloading heavy processing to cloud-based backends has been proposed.

A key challenge for health-related use cases is the remote monitoring of patients (Delmastro 2012; Hossain et al., 2016; Kan et al., 2015). There have been multiple proposals to bridge the gap between physical world and software using cloud-based approaches. Medical parameters and vital functions (e.g., blood pressure and heart rate) can be monitored in real-time and data gathered using Body Area Networks (BANs), which pass the data on to gateways and ultimately to a cloud backend. For instance, data from an electroencephalogram (EEG) can be collected using connected devices and sent to cloud-based virtual machines, which will further analyze them (Peddi et al., 2017).

Another important aspect in this area is the well-being of users. Smart applications, such as fitness trackers, can help motivate its users to exercise on a regular basis by giving positive feedbacks, such as the number of steps walked during a day. Many applications can make use of knowledge about the patients' mood or feelings. Multiple readings from sensors already deployed can be combined for emotion detection. For example, Hossain et al. (2016) use audio and visual data from a camera that might be available in future smart homes, a cloud-based Hadoop cluster, and 5G networking to achieve over 80% accuracy in detecting emotions. They also show the reduction of processing time when using multiple cluster nodes in contrast to running the same workload on a single server. The presented solution thus is an example of how the on-demand resources available in the cloud can be used to speed up IoT-related processing services, especially for data-intensive workloads such as video. Similar work has studied a feature extraction method for emotion recognition using EEG-based human brain signals (Mehmood and Lee 2016).

4.3.3 Key Challenges of Cloud-Based IoT

There are also several specific use cases where cloud computing is not prepared for IoT applications. The limitations of cloud computing in the IoT context can be summarized as follows:

1) *High or Unpredictable Latency.* Low and predictable latency are fundamental requirements for many control loops, for instance, those commonly found in industrial automation, but also other IoT use cases, such as home automation. As cloud datacenters are mostly located where energy is inexpensive, they are not located close to potential IoT users, adding unavoidable latency to control loops, since the physical distance between users and the cloud dictates the minimum latency that can be achieved.
2) *High Uplink Bandwidth Requirements.* Gateways that do not have the bandwidth capacity to upload certain types of sensor data to the cloud will not be able to use the cloud-based storage and processing approach. For instance, a video analysis value-added service may not be feasible as a remote cloud-based service, since the data that have to be transferred are too large for the link available. This is especially true for rural areas or mobile settings, such as a 3G uplink.
3) *No In-Network Filtering or Aggregation.* Some applications cover a large geographical space, whereas only an aggregate value of the sensors is actually important. For instance, if the maximum temperature across various locations is important, sending every sensor reading to the cloud will not be the most efficient solution. Instead, in-network processing could greatly reduce the amount of data that actually need to be sent to, processed on, and stored in the cloud.
4) *Uninterrupted Internet Connection Required.* Some applications, such as smart connected vehicles, may not have a connection to the Internet and, thus, to the cloud at all times. During network outages, the cloud approach would stop working, and in such cases, local actuation loops would not run. More local processing might be used instead to at least provide a fallback provisional service until network connectivity is restored.

5) *Privacy and Security Concerns.* Cloud providers are usually expected to respect the privacy of the user and provide sufficiently secure services. However, there is no easy way to measure or monitor the security of cloud services, so ultimately those requirements cannot be easily verified and the provider has to be trusted to some extent. Even if the provider is trusted, since resources are virtualized, containers of different tenants are located on the same physical machine, where software bugs could potentially leak private data to third parties. Moreover, laws might demand that certain data must not be stored outside certain juridical areas, which are not obviously verifiable with most cloud providers.

These limitations imposed by the use of cloud computing motivate the need for an additional technology that can mitigate these issues especially in critical IoT applications. Fog computing, as will be discussed in the following sections, seems like a viable option in such an environment.

4.4 Fog Computing in IoT

Although the benefits of cloud computing in the context of IoT are widely recognized by industry and research communities, there has been criticism about offloading all data to and processing it on the cloud (Happ and Wolisz 2017; Zhang et al., 2015). In many use cases, the consumer of data or the user of related services is in the vicinity of the data producer, but still the data take the route via the cloud. One problem with this is the unnecessary load on the ISPs (Internet service providers) and the additional delay introduced. The unnecessary traffic could make the participation in the network expensive or even prohibitive where no broadband connection is available and alternative technologies, for example, cellular data links, have to be used. This concerns both developing countries and rural areas where sensors have to be deployed.

Fog computing is a term coined by Cisco for a concept similar to cloud computing that offers a highly virtualized resource pool at the edge of the network (Bonomi et al., 2012). It provides computation, storage, and networking services to nearby end users, opposed to the cloud that is typically located at the edge of network. Hence, the fog will have a special widely distributed deployment, in addition to a large heterogeneity of devices. The fog concept can be illustrated as a cloud near the ground or the user. Traditional content delivery networks (CDNs) share a similar concept of bringing data closer to the user, which helps in meeting the increasing demand of streaming media traffic by placing data at the edge of the network. Akamai's Edge Computing, Cisco's IOx, and Intel's Intelligent Edge extend this concept by bringing cloud-like services—in particular, the execution of customized value-added services—closer to the user.

Microsoft has promoted a similar concept with their micro-datacenter approach as a smaller version of the cloud (Brown 2015). The envisioned micro-datacenters are self-contained computing environments with computing and storage resources, and they are connected to the Internet using high-speed connections. They basically follow the typical cloud approach, but bring the hardware to the premise on a smaller scale to avoid unnecessary delays. A customer may host tens of servers on the premise that have single-digit terabytes of storage when combined.

Fog computing has substantial overlap with edge computing, which is a similar technology that enables storage and processing at the edge of the network. The research community has not yet come to a precise definition of fog computing and the specific differences to edge computing. However, one work (Garcia Lopez et al., 2015) mentions human-driven applications, such as video streaming or web browsing, which are usually triggered by human interaction, as an indicator for more traditional edge computing, while machine-to-machine communication as seen in the IoT would be an indicator for fog computing.

The overall architecture including cloud and fog instances is depicted in Figure 4.4. Instead of a central cloud layer, several hierarchical layers of fog nodes are introduced, which are increasingly closer to the edge of the network. Gateways are now seen as part of an abstract fog layer. Since those devices are

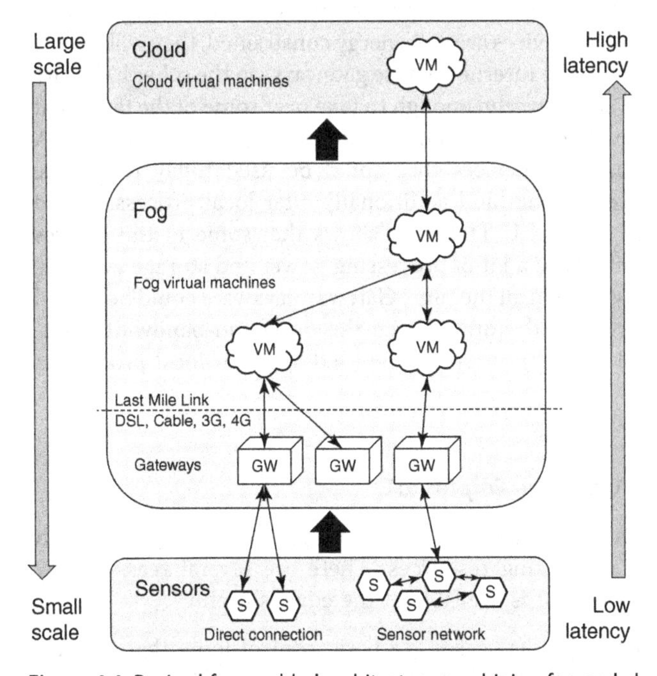

Figure 4.4 Revised fog-enabled architecture combining fog and cloud-based IoT.

Table 4.1 Overview of some gateway devices suitable for local computation, processing, and storage.

Device	CPU clock	Memory	Storage	Price
Intel NUC	1.3 GHz	16 GB	SATA	~$300
BeagleBone Black	1 GHz	512 MB	4 GB EMMC	$55
Raspberry Pi 3	1.2 GHz	1 GB	SD	$35
TP-Link TL-WR841N	650 MHz	32 MB	4 MB	420

still constrained in terms of processing power available, the remote cloud provides storage and processing capabilities when they are not sufficiently available in the fog. Generally, fog instances will also be available on other network hardware, such as the routers at the ISPs, providing the means to analyze and process data within the network closer to the end user than in a centralized remote cloud. More powerful fog nodes may also offer the provisioning of virtual machines (Yi et al., 2015; Osanaiye et al., 2017). However, while available resources and thus scalability increase toward the top, the physical distance, latency, and cost of communication also increase.

Local processing on gateway devices is made possible by advanced IoT hardware, which includes powerful smartphones and embedded single-board computers. While most IoT devices are still energy constrained, they will mostly use a gateway to connect to the Internet. Those gateways, on the other hand, are usually mains-powered and powerful enough to take over some of the tasks that cloud services are providing today, such as data processing and storage. A selection of possible gateway devices that could be used today to do the interfacing to low-power sensors and additionally offer local processing and storage is compiled in Table 4.1. The table shows that some of the devices available today potentially have a lot of processing power and storage available that is probably not fully utilized all the time. Gateway hardware could be a good option for data processing and storage since the delay-and-bandwidth-constrained Internet uplink would not have to be used at all for local processing loops.

4.4.1 Advantages of Using the Fog for IoT

This section highlights the advantages of extending cloud computing with closer-to-the-edge fog computing resources. There are several reasons why utilizing processing power that is available at the edge is useful:

1) *Minimizing Latency.* Some use cases are basic control loop; that is, on a certain condition, a certain action is triggered. These control loops, however,

are often quite time-sensitive or users expect a certain action almost instantly. Since cloud datacenters are usually deployed where it makes sense economically, the physical distance between the data source and the utilized cloud service results in a concerning latency. Analyzing raw data and making decisions locally or close to the data source can greatly reduce the latency of those control loops, since the distant cloud service will not be contacted.

2) *Improving Reliability.* The vision of IoT includes using sensor data for public safety or critical infrastructure. While most enterprise cloud providers offer very reliable datacenters with redundant network connections, storage, and processing infrastructure, the uplink to the distant cloud could turn out to be easily breakable in those scenarios. It would be very valuable to be able to at least do some of the processing locally as a fallback option, or exclusively use local processing altogether.
3) *Addressing Privacy Concerns.* Some IoT data will be sensitive or required by law to not be stored outside specific geographical boundaries. While cloud service providers are usually considered trusted, the user has no control over where the data are actually stored and who can access it. It also cannot be fully ruled out that due to bugs, third parties could potentially unlawfully access sensitive data. Local gateway or other edge nodes, on the other hand, can be trusted since they are under the control of the local operator. Of course, this is assuming that no security vulnerabilities in the local systems.
4) *Conserving Bandwidth.* The uplink bandwidth of IoT gateways is often severely limited, such as DSL or a 3G connection. It is not always feasible to transport vast amounts of data from edge devices to the cloud. Performing data processing locally and only sending aggregates and filtered data to the cloud can significantly reduce the uplink bandwidth needed, making it possible to implement IoT systems that are connected to the cloud using limited or intermittent connections.

4.4.2 Potential Future Fog Use Cases in the IoT

There are various possible use cases where IoT applications can take advantage of fog computing. The following sections provide some examples of such applications.

4.4.2.1 Smart Grid

Smart grid is the next-generation electricity network that aims to provide more effective load balancing and higher reliability, as well as reduced electricity cost by automating the metering process. Smart meters will collect and relay information on electricity consumption, enabling more fine-grained pricing models based on actual supply and demand. Fog computing can be used for storing metering information locally in such scenarios, where local gateways would act as the lowest tier of semipermanent storage. Fog nodes between end user and the cloud would only be updated in batches, reducing bandwidth requirement for smart meters. Instead of relying on higher layers, such as the cloud tier, for advanced analytics, fog nodes close to the user could use such information to either help customers change their behavior regarding power usage or to better predict future demand and supply. Fog or edge computing could also be used to improve local energy load balancing applications. Based on current price and local energy needs, the system can switch on or off appliances that need a lot of power, or it could switch them automatically to different power sources, such as solar and wind.

4.4.2.2 Connected Vehicles

Connected vehicles refer to techniques providing wireless connectivity for vehicles enabling direct communication between vehicles themselves, and between vehicles and the environment. The application that would benefit the most from fog-based approaches in that context would be the automatic intelligent reaction to sensor readings found in today's cars. For example, a smart traffic light could stop or slow down approaching traffic to avoid traffic jams or accidents when multiple braking vehicles are detected. Those use cases need inter-vehicle communication as well as local processing, as sending all sensor readings to a distant cloud is expensive and introduces a delay that is not acceptable for timely reaction to traffic conditions. Local storage and sensor data processing can help overcome those issues and enable those use cases.

4.4.2.3 Education

Students are increasingly using devices such as desktop computers, laptops, or tablets for their studies. This enables students to study at their own pace and to have the same information available at their homes as in the classroom. By tracking students' progress on these devices, teachers can gather real-time performance data on individual students and gain actionable insights into which students need further assistance. Analyzing performance data and storing additional course materials fundamentally pose the same challenges to the infrastructure as IoT data, which also has to be stored and processed. Fog techniques can also be used to enhance the privacy of student data by not uploading sensitive data about their performance to cloud-based systems.

4.4.2.4 Health Care

Sensors can collect health information, such as electrocardiograms, temperature, or blood glucose level. Additional environmental sensors could detect if elderly persons are following their daily routine, for example, they could monitor if the person gets up in the morning or is eating regularly. Most of these applications have high reliability and strict latency requirements. On top of that, the privacy of and regulations on patient data must be considered. Patient data could be directly transferred to physicians for diagnosis. Fog-based storage and processing could enable the better monitoring of patients and elderly people without sacrificing privacy or reliability.

4.4.2.5 Smart Buildings

Smart buildings can optimize the energy consumption of buildings and increase comfort of living as well as safety. These smart environments are made possible by the fusion of multiple technologies, such as temperature and humidity sensors among other sensors, ubiquitous connectivity, and data analytics. Smart buildings could, for instance, detect when nobody is home and turn down heating. Gas or air quality sensors could improve safety by giving a warning signal for bad air quality or health-threatening concentrations of certain gases in the home or even take actions on their own by opening windows to let fresh air in. Of course, these actions can be triggered over a cloud-based service, but since data producers and consumers are mostly local, the increase in latency and traffic is unnecessary. A local fog-based system could use external data from cloud-based services, but analyze local sensors readings without sending them to distant clouds. As system reliability is crucial in some use cases, it is important to note that fog-based services would keep on working in case Internet connectivity is intermittent.

4.4.2.6 Surveillance

Surveillance, including intelligent video surveillance, relies on cloud-based systems for complex video analytics (VSaaS). Instead of a human operator, a cloud-based service employs computer vision algorithms and pattern recognition to evaluate the scene. Other sensor readings can help with the recognition of potential threats. A fog-based system can help improve the detection and response time of such a system by avoiding the delay introduced by communicating with a distant cloud service. Multimedia data in particular are large in size and it is not always feasible to send high-resolution high-frame-rate video to the cloud. Instead, the video can be analyzed locally, potentially at a higher resolution and frame rate than would be possible to send over the Internet uplink. While gateway hardware is constantly becoming more suitable for those computing tasks, they are still limited in their capabilities in comparison with the vast resources available in the cloud.

4.4.2.7 Wearables

The rise of wearable sensors, such as fitness trackers and smart watches, has led to an increasing demand in processing sensor data, such as activity recognition based on accelerometer and gyroscope readings. The data have to be sampled at relatively high frequencies to derive meaningful activities. The activities themselves might be highly private and need to be protected from unauthorized access. Finally, the result of the recognition algorithm may be needed with low latency in settings where no Internet connection is available. In these settings, local processing on fog nodes might be a good compromise between offloading heavy computation from battery-powered devices and keeping data and processing close to the user.

4.4.2.8 Virtual Reality

In recent years, there have been several virtual reality solutions, such as Oculus Rift, HTC Vive, and Google Cardboard, that made it to the market and were inexpensive enough to see moderate adoption. In the future, some of these devices may become wireless and portable. The amount of data that has to be processed to make a convincing virtual reality experience possible could potentially be offloaded to fog nodes, saving energy and hence battery power, while still having a latency small enough to go unnoticed.

4.4.3 Examples of Fog-Based IoT

Recent research has highlighted the applicability of fog computing in the IoT context. A summary of the recent findings on the industrial domain, smart cities, and health is provided in the following sections.

4.4.3.1 Industrial Domain

In the industrial domain, there have been several proposals to process and store IoT data closer to the edge in order to overcome some of the challenges specific to the use case. For instance, a platform aiming to improve the effectiveness of applying equipment failure models was proposed (Gazis et al., 2015). The prototypical implementation uses existing sensors to monitor the deployed industrial machinery to detect potential anomalies. The platform builds failure models using machine learning algorithms and makes them available to the manufacturer and the administrator of the equipment. For moving this data processing closer to the edge, a Cisco ISR 819 router supporting the data in motion (DMo) framework is used. The study finds that fog computing in two realistic use cases reduces network traffic and increases resource utilization.

To support the messaging requirements of the industrial use case, an extension to the MQTT protocol was proposed (Peralta et al., 2017). A modified MQTT broker is placed at the fog layer, which supports predicting future measurements and the capability to offload computationally expensive data processing jobs from the cloud to the fog. The study shows that the energy efficiency is improved compared to using the standardized MQTT in those industrial settings using fog computing.

4.4.3.2 Smart Cities

Combining IoT with fog computing has the potential to move smart city analytics tasks closer to the user. Going in this direction, Stack4Things (Bruneo et al., 2016) was proposed, a model for fog-based IoT for smart cities based on

OpenStack, a software toolkit for controlling computing, storage, and networking resources in cloud datacenters. A smart mobility scenario in which a smart car interacts with smart city objects to achieve a variety of geolocalized services, such as signaling presence to traffic lights, is considered. A quantitative analysis shows that the framework allows managing of small areas within an urban region, locally on fog nodes.

Other works have focused on the analytics involved in the smart city domain. One work proposes a fog-based context-aware real-time data analytics platform (Jayaraman et al., 2014). The example use case of monitoring citizen activity in a smart city is given. The approach of combining local analytics, optionally combined with smart data reduction and on-demand sensing, is shown to significantly reduce the energy footprint of smart city analytics.

A recent study considers the use case of a smart pipeline monitoring system based on fiber optic sensors (Tang et al., 2015). It uses a hidden Markov model to sequentially learn and detect hazardous events threatening pipeline safety. Using a prototype and experimentation, it is shown that this is feasible even on a city-wide scale. Also, using the fog computing approach, the amount of data that has to be transmitted over the Internet, as well as the response time, is decreased substantially.

4.4.3.3 Health/Well-Being

Especially in the health domain, there have been privacy concerns when using cloud-based services. Fog computing could be one option to address these concerns by storing and processing data partially on devices the user trusts or even owns. A recent example is Health Fog (Mahmood et al., 2016), a framework that enables sharing and processing of health-related sensor data. The proposed system introduces an intermediate fog layer between the devices and the cloud to give the patient control over the flow of their health data and thus enable for better control over data privacy and security while still enabling processing and sharing of their data.

A design of a fog-based system to identify the chikungunya virus based on the patient's symptoms and surrounding environment conditions was also proposed (Sood and Mahajan 2017). It introduces different layers for different tasks and users. The task of the fog layer is to diagnose the virus and generate emergency alerts for patients and doctors. These alerts are relayed to a cloud-based service, which calculates the probability of the disease spreading and generates alerts for government agencies to counter the outbreak.

In a similar fashion, one work (Gia et al., 2015) studies ECG feature extraction on smart gateways. The study considers features that are important for the diagnosis of many cardiac diseases, including heart rate, P wave, and T wave. It shows that the approach is feasible by providing a proof-of-concept implementation on real hardware and show a reduction in traffic of more than 90% and a significant reduction in latency when using fog computing.

4.4.4 Key Challenges of Fog-Based IoT

Despite the advantages of fog computing, several issues need to be addressed toward the realization of the fog-based IoT approach as well as the integration with the current cloud-based model. Some of the key challenges are discussed in the following.

1) *Technological Interoperability.* The seamless interaction between devices and systems from different vendors is a major challenge for IoT as well as for the fog-based extensions to IoT. As discussed in several chapters of this book, there is still a lack of common standards for communication protocols, both locally and for the uplink to the cloud. For example, for connecting sensors to gateways, technologies used include Bluetooth Low Energy (BLE), 802.15.4 or ZigBee, or Wi-Fi (802.11). Also, long-range wireless technologies, such as LoRaWAN or Sigfox, are being used. Uplink protocols include proprietary technologies, in addition to open pub/sub protocols such as MQTT, XMPP (Extensible Messaging and Presence Protocol) or AMQP (Advanced Message Queuing Protocol), and request/response protocols such as COAP (Constrained Application Protocol). Protocols have to be standardized and agreed upon for the success of IoT. Additionally, techniques for storage and processing migration have to be developed.
2) *Semantic Interoperability.* For interoperability, it is also necessary that the devices and software involved interpret the information gathered, processed, and shared in the same way and act upon commands accordingly. A semantic model has to be available for every aspect of the fog/cloud approach; this should not only be for understanding the data but also for expressing the requirements and constraints when processing or moving these data, including quality of service or privacy requirements. There is still a lack of insight into which ontologies may be suitable for these tasks. Semantic web technologies could also help enhance device discovery or implement automatic reasoning.
3) *Programmability.* Data processing is of exceptional importance to the IoT ecosystem. Due to the variability of the requirements of event-based processing tasks, it is further crucial to automatically relocate processing tasks between the fog and cloud nodes. However, it is still unclear how processing jobs should be defined in the first place. There is still a lack of insight into which programming language and which interfaces are necessary for IoT data processing and for enabling seamless offloading of tasks between different systems, potentially using different hardware architectures and different formats. In particular, there has not been any consensus yet about whether to use an interface in the form of discrete functions or in the form of containers or virtual machine images.
4) *Scalability.* In the near future, the IoT will be composed of billions (or even trillions) of devices. The number of connected nodes will outnumber by

several orders of magnitude the number of hosts in today's Internet. Despite the problems that fog-based processing and storage may solve, it is still unclear how a seamless interaction between devices behind different gateways should work. Questions such as how devices should be discovered on a large scale, where the registry in a cloud–fog system should be located, and where data should sit to minimize communication latency and increase throughput are still not fully answered.

5) *Resilience and Reliability*. The fog-based approach is appealing to developers for implementing applications where relying on the permanent availability of the link to the cloud is not acceptable due to the possibility of temporary outages. Examples include industrial control loops or emergency response systems.

4.5 Conclusion

Cloud-based services are increasingly used to assist constrained IoT devices with storage and processing. The cloud offers various advantages in the IoT domain, including the flexible on-demand availability of resources, fast and reliable networking, and a multitude of hosted services ready to be used. While the benefits of cloud computing in the IoT context are evident by the vast adoption seen in existing systems, the main shortcoming is the position of cloud datacenters, which are often several hops away from data producers and consumers. Data usually have to pass a constraint link between data producers and the cloud and between the cloud and data consumers. This link adds additional delay and may limit the amount of data that can be transmitted. Over the next few years, fog computing will increasingly help to tackle those shortcomings by introducing cloud-like resources closer to the user.

This chapter gave an overview of the advantages, potential use cases, recent examples, and open challenges of cloud and fog computing in the IoT context. Both approaches have distinct benefits and will see wider adoption. Most likely, they will be combined to achieve a seamless integration of fog and cloud resources into a common IoT resource pool. Future developments will include mechanisms for fog–cloud interaction, such as automatic resource provisioning, replication, and migration, which are essential for meeting the IoT's requirements for resilience and reliability. Value-added service will have to be defined only once and will be automatically provisioned and relocated on demand to a suitable processing node. Existing services could be combined to create new innovative services. Advances in security and privacy will help keep sensitive data confidential and secure across processing nodes from different providers. Despite the remaining challenges, cloud as well as fog computing will prove to be indispensable tools in realizing this IoT vision.

References

Antonić, A. et al. (2014a) Urban crowd sensing demonstrator: sense the Zagreb air. 22nd International Conference on Software, Telecommunications and Computer Networks, pp. 423–424.

Antonic, A., Roankovic, K., Marjanovic, M., Pripuic, K., and Zarko, I.P. (2014b) A mobile crowdsensing ecosystem enabled by a cloud-based publish/subscribe middleware. 2014 International Conference on Future Internet of Things and Cloud (FiCloud), IEEE, pp. 107–114.

Banks, A. and Gupta, R. (2014) MQTT Version 3.1.1, OASIS Standard.

Bera, S., Misra, S., and Rodrigues, J.J.P.C. (2015) Cloud computing applications for smart grid: a survey. *IEEE Transactions on Parallel and Distributed Systems*, **26**(5), 1477–1494.

Bonomi, F., Milito, R., Zhu, J., and Addepalli, S. (2012) Fog computing and its role in the Internet of Things. Proceedings of the First Edition of the MCC Workshop on Mobile Cloud Computing, ACM, pp. 13–16.

Borgia, E. (2014) The Internet of Things vision: key features, applications and open issues. *Computer Communications*, **54**, 1–31.

Brown, B. (2015) Why Micro Datacenters really matter to mobile's future.

Bruneo, D. et al. (2016) Stack4Things as a fog computing platform for Smart City applications. IEEE Conference on Computer Communications Workshops, C IEEE, San Francisco, CA.

Delmastro, F. (2012) Pervasive communications in healthcare. *Computer Communications*, **35**(11), 1284–1295.

Fang, X., Misra, S., Xue, G., and Yang, D. (2012) Managing smart grid information in the cloud: opportunities, model, and applications. *IEEE Network*, **26**(4). doi: 10.1109/MNET.2012.6246750

Garcia Lopez, P. et al. (2015) Edge-centric computing: vision and challenges. *SIGCOMM Computer Communication Review*, **45**(5), 37–42.

Gazis, V., Leonardi, A., Mathioudakis, K., Sasloglou, K., Kikiras, P., and Sudhaakar R. (2015) Components of fog computing in an industrial Internet of Things context. 12th Annual IEEE International Conference on Sensing, Communication, and Networking: Workshops (SECON Workshops), IEEE.

Gia, T.N., Jiang, M., Rahmani, A.-M., Westerlund, T., Liljeberg, P., and Tenhunen, H. (2015) Fog computing in healthcare Internet of Things: a case study on ECG feature extraction. IEEE International Conference on Computer and Information Technology; Ubiquitous Computing and Communications; Dependable, Autonomic and Secure Computing; Pervasive Intelligence and Computing (CIT/IUCC/DASC/PICOM), IEEE.

Gubbi, J., Buyya, R., Marusic, S., and Palaniswami, M. (2013) Internet of Things (IoT): a vision, architectural elements, and future directions. *Future Generation Computer Systems*, 7(29), 1645–1660.

Happ, D. and Wolisz, A. (2017) Towards gateway to cloud offloading in IoT publish/subscribe systems. 2nd International Conference on Fog & Mobile Edge Computing, IEEE, Valencia, Spain.

Happ, D., Karowski, N., Thomas, H., Menzel, V., and Wolisz, A. (2015) Meeting IoT platform requirements with open pub/sub solutions. 1st International Conference on Cloudification of the Internet of Things (CIoT'15), Paris, France.

Hassan, Q.F., Riad, A.M., and Hassan, A.E. (2012) Understanding cloud computing. *Software Reuse in the Emerging Cloud Computing Era*, IGI Global, pp. 204–227.

Hossain, M.S., Muhammad, G., Alhamid, M.F., Song, B., and Al-Mutib, K. (2016) Audio-visual emotion recognition using big data towards 5G. *Mobile Networks and Applications*, **21**(5), 753–763.

Hunkeler, U., Truong, H.L., and Stanford-Clark, A. (2008) MQTT-S: a publish/subscribe protocol for Wireless Sensor Networks. 3rd International Conference on Communication Systems Software and Middleware and Workshops (COMSWARE'08), Bangalore, India, pp. 791–798.

Jayaraman, P.P., Gomes, J.B., Nguyen, H.L., Abdallah, Z.S., Krishnaswamy, S., and Zaslavsky, A. et al. (2014) East European Conference on Advances in Databases and Information Systems, Springer, Cham.

Kan, C., Chen, Y., Leonelli, F., and Yang, H. (2015) Mobile sensing and network analytics for realizing smart automated systems towards health Internet of Things. IEEE International Conference on Automation Science and Engineering (CASE), IEEE, pp. 1072–1077.

Mahmood, A., Amin, M.B., Hussain, S., Ho Kang, B., Cheong, T., and Lee, S. (2016) Health fog: a novel framework for health and wellness applications. *The Journal of Supercomputing*, **72**(10), 3677–3695.

Mehmood, R.M. and Lee, H.J. (2016) A novel feature extraction method based on late positive potential for emotion recognition in human brain signal patterns. *Computers & Electrical Engineering*, **53**, 444–457.

Mell, P. and Grance, T. (2011) The NIST Definition of Cloud Computing, Computer Security Division, Information Technology Laboratory, National Institute of Standards and Technology, Gaithersburg.

Menzel, T., Karowski, N., Happ, D., Handziski, V., and Wolisz, A. (2014) Social sensor cloud: an architecture meeting cloud-centric IoT platform requirements. 9th KuVS NGSDP Expert Talk on Next Generation Service Delivery Platforms, Berlin, Germany.

Osanaiye, O., Chen, S., Yan, Z., Lu, R., Choo, K.K.R., and Dlodlo, M. (2017) From cloud to fog computing: a review and a conceptual live VM migration framework. *IEEE Access*, **5**, 8284–8300.

Peddi, V.B., Kuhad, P., Yassine, A., Pouladzadeh, P., Shirmohammadi, S., and Shirehjini, A.A.N. (2017) An intelligent cloud-based data processing broker for mobile e-health multimedia applications. *Future Generation Computer Systems*, **66**, 71–86.

Peralta, G., Iglesias-Urkia, M., Barcelo, M., Gomez, R., Moran, A., and Bilbao, J. (2017) Fog computing based efficient IoT scheme for the Industry 4.0. IEEE International Workshop of Electronics, Control, Measurement, Signals and Their Application to Mechatronics, IEEE.

Rusitschka, S., Eger, K., and Gerdes, C. (2010) Smart grid data cloud: a model for utilizing cloud computing in the smart grid domain. *First IEEE International Conference on Smart Grid Communications*, doi: 10.1109/SMARTGRID.2010.5622089.

Sanchez, L. et al. (2014) SmartSantander: IoT experimentation over a smart city testbed. *Computer Networks*, **61**, 217–238.

Sood, S.K. and Mahajan, I. (2017) Wearable IoT sensor based healthcare system for identifying and controlling chikungunya virus. *Computers in Industry*, **91**, 33–44.

Sun, E., Zhang, X., and Li, Z. (2012) The Internet of Things (IOT) and cloud computing (CC) based tailings dam monitoring and pre-alarm system in mines. *Safety Science* **50**(4), 811–815.

Tang, B., Chen, Z., Hefferman, G., Wei, T., He, H., and Yang, Q. (2015) A hierarchical distributed fog computing architecture for big data analysis in smart cities. Proceedings of the ASE Big Data & SocialInformatics, ACM.

Tao, F., Cheng, Y., Xu, L.D., Zhang, L., and Li, B.H. (2014) CCIoT-CMfg: cloud computing and Internet of Things-based cloud manufacturing service system. *IEEE Transactions on Industrial Informatics*, **10**(2), 1435–1442.

Tschofenig, H., Arkko, J., McPherson, D., Thaler, D., and McPherson, D. (2015) Architectural Considerations in Smart Object Networking.

Yi, S., Hao, Z., Qin, Z., and Li, Q. (2015) Fog computing: platform and applications. Third IEEE Workshop on Hot Topics in Web Systems and Technologies (HotWeb), pp. 73–78.

Zhang, B. et al. (2015) The cloud is not enough: saving IoT from the cloud. 7th USENIX Workshop on Hot Topics in Cloud Computing (HotCloud '15), Santa Clara, CA, USENIX Association.

5

RFID in the Internet of Things

Akaa Agbaeze Eteng,[1] *Sharul Kamal Abdul Rahim,*[2] *and Chee Yen Leow*[2]

[1]*Department of Electronic and Computer Engineering, University of Port Harcourt, Port Harcourt, Nigeria*
[2]*Wireless Communication Centre, Universiti Teknologi Malaysia, Johor, Malaysia*

5.1 Introduction

The expansion of the current "internet of computers" to the Internet of Things (IoT) requires an ecosystem characterized by interactions between everyday objects with embedded online intelligence. By implication, such smart objects would be capable of data-driven real-time interventions without human mediation. One of the requirements for this pervasive information acquisition and sharing environment is that objects should be characterized by an innate ability to gather contextual data about their internal or external environments (Roselli et al., 2015). This suggests a need to embed sensors in everyday objects. Furthermore, smart objects should be able to communicate acquired data to other entities. Historically, radio frequency identification (RFID) is one of the earliest technologies whose utility required embedding or placing electronics on an object. Given the fact the RFID is also a communications technology, it is currently touted as a promising technology to be incorporated in smart objects for the IoT (Vermesan and Friess, 2013).

This chapter aims to provide a background to the use of RFID in the IoT. It begins with a brief historical perspective of RFID technology, showing how it is associated with the birth and development of the IoT.

5.2 Historical Perspective

The 1906 demonstration of a 2 kW, 100 Hz alternator by Ernst Alexanderson was an important milestone in the development of wireless communication

Internet of Things A to Z: Technologies and Applications, First Edition. Edited by Qusay F. Hassan.
© 2018 by The Institute of Electrical and Electronics Engineers, Inc. Published 2018 by John Wiley & Sons, Inc.

through continuous wave (CW) radio generation and transmission (Ernst, 1906). Radar systems were one of the early technologies to harness the potentials of CW radio wave transmission, using the reflection of such waves to detect the presence and locations of objects. In 1948, the seminal work by Stockman (Stockman, 1948) was published, which explored the potentials of communication through reflected radio waves. Not long after this, the invention of the transistor and the development of the integrated circuit enabled further explorations into ideas inspired by Stockman's work, thus, birthing RFID.

RFID is a technology that employs transmitted and received radio frequency RF) energy for the automatic identification of objects. In a basic sense, RFID systems comprise of two components, namely, the RFID transponder or tag and the RFID interrogator or reader. The RFID tag is a miniaturized electronic circuit containing data, and is attached to or embedded in the object to be identified. The RFID reader is a relatively larger electronic circuit that reads data wirelessly communicated to it by RFID tags.

One of the earliest RFID applications was in electronic article surveillance (EAS) systems, developed in the 1960s. EAS systems provided an inexpensive and effective antitheft measure. Basically, these systems employed inductive coupling between a resonant reader circuit and a tag. Single-bit tags were placed on pieces of merchandise, and EAS systems could detect either their presence or absence. The 1970s were marked by notable advances in RFID technology as a consequence of research at institutions such as Los Alamos Scientific Laboratory in the United States and the Microwave Institute Foundation in Sweden (Landt, 2005). Backscatter RFID systems were developed for operation in the ultrahigh frequency (UHF) spectrum, enabling the deployment of RFID systems at greater ranges than were hitherto possible (Koelle et al., 1975). Also, development efforts were directed at miniaturization of RFID tags and circuitry, and improvements in functionality.

While the 1970s were characterized by RFID research and development, the 1980s were the decade of commercialization of the technology. Short-range RFID technologies were adopted in Europe for livestock tracking, as well as in business and industrial applications. In the United States, however, access control and transportation were the major RFID applications (Landt, 2005). Electronic toll collection was also fast becoming an important niche application for RFID. It is important to note the role played by the personal computers in the development of RFID within this era. The availability of personal computers meant that the data about the presence and location of objects could be transmitted to computing systems for storage and further processing. Consequently, a third component of RFID systems—the middleware—became the focus of considerable research attention.

The decade of the 1990s saw a further consolidation in the use of RFID in transportation management, inventory management, supply chain management, and access control. New applications in health care management were also being developed. Many companies all over the world became actively involved in the development of RFID solutions. This led to the need for the development of standards to ensure interoperability between RFID hardware developed by the different vendors. The US-based Uniform Code Council (UCC) commenced discussions with a similar standards organization, the European Article Number (EAN) International, on global standardization. The Auto-ID Centre, founded in 1999, developed the Electronic Product Code (EPC), with support from the UCC, to serve as a universal RFID system. In 2003, the Auto-ID Centre morphed into two separate entities: the Auto-ID Labs and EPCglobal. While the Auto-ID labs are involved in research activities, with laboratories in seven countries, the EPCglobal is involved in EPC standardization activities. In 2005, EAN International was reorganized and its name changed to GS1 (GS1, 2013). A direct consequence of these standardization efforts was the emergence of multiuse RFID tags, which could be used across different business segments. For instance, a single tag could be used to gain access to an office building, and still serve as an e-wallet for electronic toll collection.

The late 1990s were also characterized by a rapid growth in internet usage. Personal computers were becoming cheaper and wireless connectivity was more readily available. The global surge in internet use, which has continued till present, meant that enormous amounts of data were being continuously generated and communicated through computer networks. Initially, the bulk of data input to computer systems, and hence the Internet, came through human interventions. Especially in industry and business applications, such data often provided qualitative or quantitative descriptions of physical processes and variables. Developers in the RFID community, consequently, began toying with the idea of data input to the Internet coming directly from the objects exposed to the physical processes and variables themselves, without human intermediaries. The idea was that RFID could be adopted to provide a means for computers to gather information about the physical world for themselves. This led to the coinage of the term, "Internet of Things" (IoT), an ecosystem in which data input to the Internet came directly from "things," and not human beings (Ashton, 2009).Thus, RFID has been associated with the IoT, right from the latter's conceptual origins.

5.3 RFID and the Internet of Things

The specific role of RFID in the IoT can be observed from noting some fundamental components of the IoT, namely, sensing, communication, services, semantics, computation, and identification (Zhu et al., 2015). While sensing deals with data collection by objects, communication technologies enable connections between these heterogeneous objects for specific services. Such services may include information aggregation, real-time decision-making, or ensuring seamless on-demand access to data. The semantic component of the IoT deals with the intelligent extraction of knowledge from collected data to provide required services. Computation, on the other hand, refers to the hardware and software components that process collected data in real time. Finally, the ability to uniquely identify objects enables services to be matched with demand in the physical world. Automatic identification provided by RFID is currently the most widely used functionality to provide a wide range of services. RFID inventory and supply chain management solutions are extensively deployed in many organizations globally.

As a means of data acquisition, RFID technology can provide the link between the physical world and the virtual elements of the IoT (see Figure 5.1). This potential has been demonstrated by researchers at the University of Washington in a building-scale microcosm of the IoT, which they called the "RFID Ecosystem" (Welbourne et al., 2009). Multiple RFID readers were positioned in an 8000 m^2 building, while tracking tags were attached to persons and personal objects. Using various tools, low-level RFID data was transformed into higher level contextual information about the tracked persons and objects.

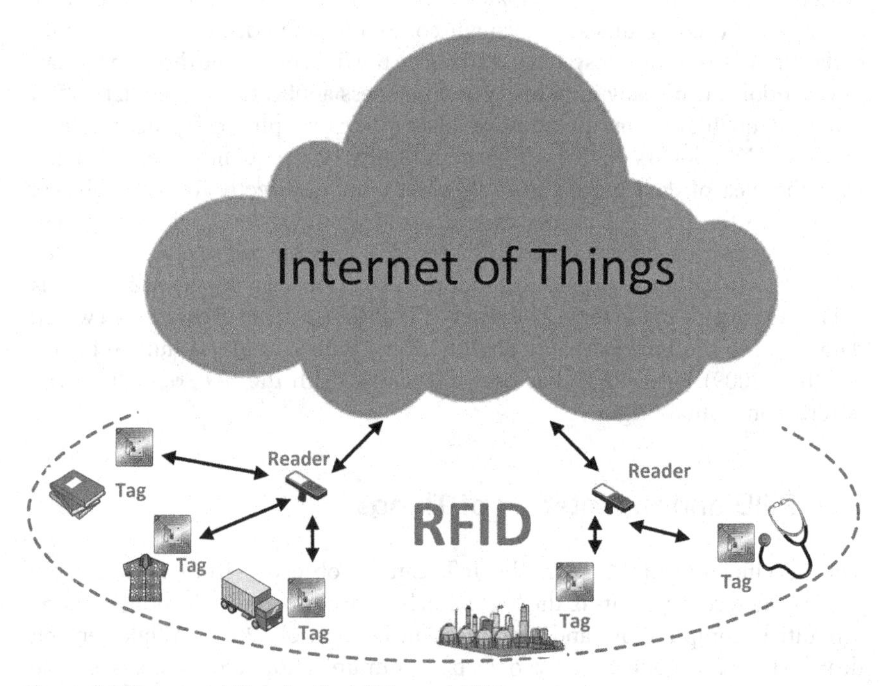

Figure 5.1 Data acquisition for the IoT through RFID.

5.3.1 Object Identification using RFID

Readers and tags are the basic hardware used in RFID. Readers and tags are equipped with antennas for communication. In addition, a typical tag contains a chip that stores a unique identification number (ID) (Liu et al., 2008). However, chipless tags are also available, where the tag ID is encoded onto the physical structure of the tag. Normally, the RFID tag is embedded in or attached to an object to be tracked, so that the tag ID is associated with the object. RFID tags are classified as active, semipassive, or passive, depending on how they are powered. Generally speaking, active and semipassive tags are battery powered, while passive RFID tags do not have independent power sources.

In a conventional RFID interaction, a reader interrogates the tag to obtain its unique ID, which in effect identifies the associated object. Active tags are able to transmit their unique IDs through electromagnetic fields in response to reader electromagnetic transmissions. Semipassive and passive tags, however, are unable to generate their own electromagnetic fields for transmitting data. These tags rather communicate with readers by modulating the magnetic or electromagnetic fields transmitted by readers. Magnetic field-based tag-reader interactions are facilitated by load modulation. Here, a tag changes its impedance in step with the ID data to be transmitted. This variation in impedance is sensed by the reader, and interpreted accordingly. For electromagnetic systems, passive and semipassive tags communicate with the reader using backscatter modulation. In this case, the reader's electromagnetic transmission is reflected back from the tag. This is achieved by intentionally mismatching the tag antennas' input impedance in step with the tag ID data. Thus, tag ID data is encoded in the amplitude and phase of the reflected electromagnetic field and is interpreted accordingly by the reader.

RFID object identification enables multiple applications spanning from access control to RFID-enabled passports (Bogari et al., 2012). This functionality also has practical industrial relevance. Production processes are usually monitored by registering manufactured items at certain checkpoints in the production chain. With RFID technology, this registration process can be automated by equipping such items with tags. Additional scan points can be introduced in the production process, even in hostile environments, thereby enabling closer monitoring of production processes (Evdokimov et al., 2010). Extended further, these same tags can enable the identification and tracking of each manufactured item, right through the supply chain, and possibly throughout their operational life with the final consumer.

The benefits of incorporating RFID technology into the operation of an organization are well known. However, the potentials of the technology are significantly magnified when multiple parties can access and use data acquired through RFID. For example, the efficiency and transparency of supply chains is greatly improved with the IoT serving as a global information service architecture for RFID-tagged items (Evdokimov et al., 2010). For instance, RFID has been employed by an industrial laundry service in the United Kingdom as part of its IoT infrastructure to manage its supply chain (Palmer, 2017). Inconspicuous RFID tags sewn into laundry are connected to an internet-based network, which enables effective management of shipment cycles, consisting of up to a million items daily, across multiple locations. Also, IoT systems based on RFID item tagging are currently being developed to minimize baggage mishandling incidents at airports in compliance with International Air Transport Association (IATA) resolutions (Strukhoff and Yaroshko, 2017). It should be noted that extending the basic identification functionality of RFID tags to include data about the environments in which they are deployed significantly enriches the content of RFID-acquired data, and enhances the utility of RFID item tagging. Environmental sensing, therefore, is potentially a key application of RFID in the IoT (Want, 2004).

5.3.2 RFID Sensors

RFID tags can be enhanced to sense physical or chemical variables in an environment. Some examples of common physical parameters RFID sensors have been developed to sense include temperature (Bhattacharyya et al., 2010; Brenk et al., 2011; Trevisan and Costanzo, 2016), humidity (Nair et al., 2013), pressure (Beriain et al., 2014), light (Colella et al., 2015a), and movement. There are various real-world applications for these RFID sensor enhancements. For example, RFID-based temperature sensors can be used in a cold supply chain to monitor the temperature of frozen goods, while still providing basic article surveillance. Also, RFID-based strain sensors find ready application in structural health monitoring. RFID chemical sensors, on the other hand, can be used in applications such as determining the presence of a potentially harmful gas (Occhiuzzi et al., 2011; Shrestha et al., 2009), and measuring the pH level of bioprocess components (Potyrailo and Wortley, 2011). In health care, RFID chemical sensors can be used to measure glucose levels (Tankiewicz et al., 2013). Motion sensors depend on sensing the acceleration of a sensor from its rest position. Wearable RFID motion sensors have been developed, which are able to track the movement of the human wearer and find medical applications in patient management (Hong et al., 2008; Occhiuzzi et al., 2010; Occhiuzzi and Marrocco, 2010). A key advantage of RFID-based sensing over other wireless sensor approaches is that, with passive tags, wireless sensing can be achieved using battery-free implementations, making them ideal for embedded IoT applications where sensors cannot be periodically retrieved for battery replacement.

Sensing functionality may be incorporated in an RFID tag by exploiting some properties of the tag antenna or chip, such as antenna directivity or impedance. This is possible because any variation in the environment within the near-field region of a tag will alter the tag's electrical properties (Tedjini et al., 2016). On the other hand, it is possible for external sensors to be connected to the RFID tag circuitry. For instance, a passive surface acoustic wave (SAW)-based RFID tag typically comprises of a coding area and an interdigital transducer (IDT) coupled to an antenna. The IDT converts received reader signals into a surface acoustic wave, which is coded according to the configuration of the coding area, and reflected back after reconversion to an electrical signal by the IDT. Directly connecting a sensor as an external load to the IDT alters the tag reflection characteristics according to the measuring signal of the sensor, so that the signal reflected back to the reader contains sensor data. With CMOS RFID tag implementations, however, it is necessary to design interface circuitry in order to connect the external sensor to tag electronics (Ussmueller et al., 2014). The inclusion of external sensor modules invariably increases the complexity and energy demand of RFID tag implementations. Innovative schemes need to be introduced to simplify the tag-sensor operation, as well as minimize energy consumption. For example, the read-out of a resistive temperature sensor can be achieved by tracking the nonlinear characteristic of the tag rectifier as it is loaded by the sensor (Trevisan and Costanzo, 2014). This approach avoids the need for embedding additional active electronics on the sensor for its read-out.

Multisensor RFID tag platforms with additional computational abilities have also been developed. In a recent project, near-field communication (NFC) has been combined with the wireless identification and sensing platform (WISP) to provide a near-field RFID sensing platform (Zhao et al., 2015). This programmable sensing and computing platform can be interrogated by RFID readers and NFC-enabled smartphones. Another example of a multisensor RFID tag platform is the RFID augmented module for smart environmental sensing (RAMSES) (De Donno et al., 2014). RAMSES harvests radio frequency energy to energize circuits that perform environmental sensing, computation, and data communication. RAMSES presents an improvement over the WISP (Sample et al., 2008), which was one of the earliest UHF-RFID multisensor platforms. More recently, the self-powered augmented RFID Tag for autonomous computing and ubiquitous sensing (SPARTACUS) has been developed (Colella et al., 2015b), which combines sensing, computation, and communication functionalities.

5.3.3 RFID Sensor Localization

Information mined about objects and their environments using RFID-based sensors is only useful in the IoT if the locations of sensed processes and events are known. Consequently, the real-time determination of the position of the IoT object is of utmost importance. In supply chain management, location is the primary information required about the tracked assets. Generally, RFID localization technologies can be categorized as tag based, reader based, transceiver-free, or hybrid approaches (Ni et al., 2011). In a tag-based approach, the tracked item is required to carry a tag that periodically transmits beacon signals. In reader-based schemes, readers are attached to tracked objects to gather location information from nearby tags. In transceiver-free approaches, object tracking is achieved without the need for reader or tag in the tracked item. They are rather based on observations of disturbances to wireless signals by movement of the tracked object. One of the hybrid RFID localization techniques involves utilizing the richer bidirectional connectivity provided by traditional wireless sensor networks (WSN) to augment the RFID localization implementation. An example of this approach is COCKTAIL (Zhang et al., 2010), which is implemented by combining a very sparse deployment of WSN motes with a denser deployment of RFID tags over an area. To obtain the spatial position of an intended tag, the WSN motes first segregate a cluster of reference RFID tags closest to the target tag. The final localization of the target tag is then performed by this cluster of reference tags.

5.3.4 Connecting RFID Sensors to the Internet

The incorporation of sensor functionalities into RFID implementations opens up opportunities for pervasive low-cost data gathering and processing. RFID sensors could be interconnected in networks to ensure that sensed information can be extracted over long distances. This has significant implications on the availability of social services and quality of life in general. For example, in virtual health care, patients can be under medical observation irrespective of their physical locations. Ultimately, RFID-acquired data should be available anywhere and anytime. This vision implies IoT implementation on a global scale, as opposed to localized realizations. One way this goal can be achieved is through the integration of RFID sensors with WSN platforms. The argument for this integration is predicated on appropriating the benefits of WSNs, such as increased range through the use of multihop links, non-line-of-sight operability, reliable communications even with node failures, and so on.

Some network topologies for integrating RFID and WSN platforms include the agent network topology, reader-as-sensor topology, and the tag-as-sensor topology (Yang and Yang, 2007). As illustrated in Figure 5.2, the agent network topology is one in which the RFID links and WSN networks logically operate at the same system layer, but are not directly connected to each other. An agent network rather operates as a backbone, connecting both networks to central servers. In the reader-as-sensor topology, the RFID reader is at the same system layer as WSN motes. Consequently, the sensor gateway device linking the WSN motes to a central server also acts as the gateway device connecting the readers to the central server. All RFID tag information sent to the RFID reader is sent to the central server through the sensor gateway. Conceptually, the sensor gateway device treats all incoming data as sensor inputs, irrespective of whether they

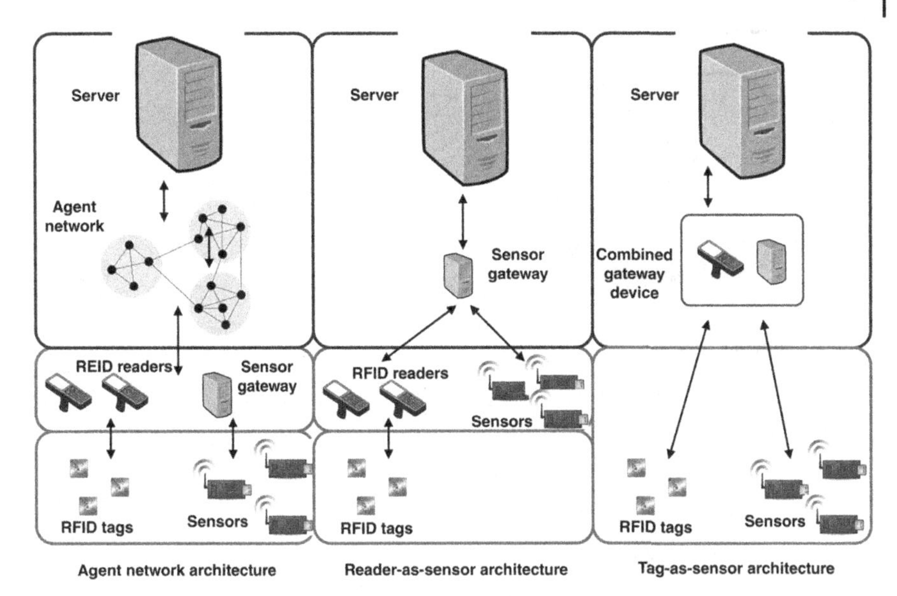

Figure 5.2 Network topologies for linking RFID sensors to the Internet.

were sent by WSN motes or RFID readers. On the other hand, the tag-as-sensor topology considers RFID tags and WSN sensors as belonging to the same system layer. Above this layer is the layer in which RFID readers and sensor gateways operate as combined gateway devices, linking the tags and motes to central servers.

Alternatively, WSN nodes can be utilized as the link between RFID sensors and internet gateways. In this role, WSN nodes act as data routers that relay sensed information emanating from RFID sensors through multihop links to gateways. In this scheme, RFID readers are not required, since RFID tags communicate directly with WSN sensors. Consequently, RFID sensors can be viewed as the lowest-level network devices in the hierarchy, as shown in Figure 5.3. Reasons for the elimination of RFID readers from the scheme include the fact that they are comparatively more expensive and bulkier than WSN sensors (Lakafosis et al., 2014).

Having RFID tags communicate directly with WSN sensors requires a protocol to handle communications between the devices. This implies that signal transmissions from tags to sensors have to occur at an agreed frequency and modulation scheme, and with proper data encapsulation. Using a simple proprietary protocol, this concept has been demonstrated in a prototype solar-powered UHF tag, which communicates with WSN nodes in a localization application (Lakafosis et al., 2010). Connecting such RFID–WSN IoT realizations to the internet, however, requires a two-way translation of addresses and commands at the internet gateway.

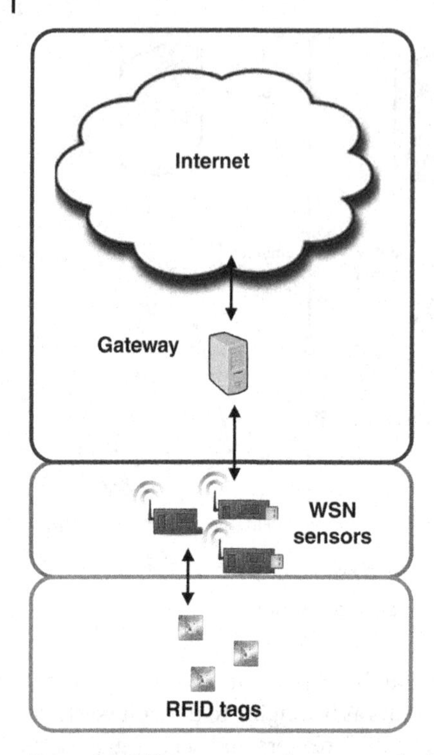

Figure 5.3 WSN sensors as data routers for RFID tags.

In order to realize full intrinsic internet capabilities, smart devices need to provide some form of support for internet integration. The 6LowPAN protocol enables the compression of IPv6 addresses into a few bytes, thereby enabling native support of the IPv6 protocol on a resource-constrained device. An example of this approach is realized in an IPv6-enabled shoe-mounted sensor platform, which is developed for health monitoring (Mariotti et al., 2013). By enabling IPv6 support, the shoe-mounted sensor is able to connect to any existing IP network infrastructure without translation gateways or proxies, and still allow for mesh routing within Wireless Personal Area Network (WPAN) domains. (Doraswamy and Harkins, 2003).

5.4 Emergent Issues

The incorporation of RFID in the IoT no doubt provides numerous novel opportunities. However, these prospects bring along with them several technical challenges, especially with respect to new security vulnerabilities. Each connected IoT end node – RFID devices, WSN motes, sensor gateways, and so on – is a potential security loophole in an IoT infrastructure. The core

IoT security issues border on data confidentiality, privacy, and trust (Li et al., 2016).

In the development of standards for RFID, security features have been introduced to address data confidentiality and reader-tag authentication. Methods to ensure the data confidentiality in RFID schemes include the use of signal interference, antenna energy analysis, encryption, among others (Jing et al., 2014). Related to data confidentiality is the issue of location confidentiality, where it is necessary to prevent unauthorized tracking of the location of a tag. Reader-tag authentication, on the other hand, is an issue of trust management. Given the limited resources of RFID tags, it is necessary that lightweight trust management algorithms are developed to ensure data and location privacy (Jing et al., 2014).

WSN implementations are also susceptible to security threats bordering on privacy, authentication, and availability. Unlike RFID systems, the threats to WSN privacy may be internal, that is, an entity acquiring more information than it should (Shah et al., 2016). Furthermore, it is necessary to guard against fake network nodes, fake routes, and fake locations. Jamming signals are one of the most prevalent attacks on the availability of a WSN network.

The integration of RFID and WSNs for the realization of the IoT extends the security challenges of both technologies. This is more especially so given the fact that such integrated networks deal with heterogeneous data. Privacy, data confidentiality, and trust must be ensured throughout the integrated network. End-to-end communication between the multiprotocol devices in such integrated networks must be secure, while the authenticity and integrity of these devices must be assured.

The low cost of RFID sensor-tag implementation provides an impetus for the massive deployment of these devices in various sensing environments. However, the question arises as to the means for preventing these devices from becoming environmental pollutants. In order to address this issue, deployed RFID devices can be categorized as being either locatable or nonlocatable (Roselli et al., 2015). Locatable devices should be recyclable at the end of their operational lives. Nonlocatable devices, which cannot be recovered once dispersed in the environment, have to be implemented using materials that biodegrade soon after the expiration of the operational life of the devices. Consequently, constraints of eco-friendliness must apply to RFID devices in the IoT, namely, recyclability of locatable device and biodegradability of nonlocatable devices. Currently, there is considerable interest in the use of organic materials for the implementation of RFID hardware. Some examples include RFID tags built using organic transistors (Cantatore et al., 2007), organic frequency doublers for chipless RFID tags (Virili et al., 2014), the use of microfluidics in organic UHF-RFID antennas (López et al., 2013), textile tags (Manzari et al., 2012), and inkjet printed antennas on paper substrates (Jankowski-Mihułowicz, 2015).

Sensors will be more increasingly deployed in scenarios for which alternative powering solutions must be sought. Indeed, with sensor deployments running into billions of devices, it is virtually impossible to connect all these devices to the grid. Especially for nonlocatable sensor deployments, battery-powered solutions may not be the best option, as these devices cannot be retrieved when the batteries run down. This is one major advantage passive RFID has over other proposals for IoT sensor implementations, since the power for tag operation is derived from the reader's transmission. However, the incorporation of sensor electronics increases the energy demand of such tags. UHF-RFID readers are based on RF broadcast methods, where the power density of the radiation from the reader antenna obeys an inverse-square law. The power available at the tag is, therefore, considerably lower than what was transmitted. Although inductive RFID links can be perceived as wireless power transfer links, conventionally, these links trade power efficiency for operating distance (Sample et al., 2011). Consequently, it is imperative for RFID links to find an optimal trade-off between operational range and power efficiency to support the expanded functionality.

5.5 Conclusion

One of the key features of the IoT is the embedding of intelligence within objects, enabling them to acquire contextual data about their environments, and transmit such data as required without human intervention. The low-cost, low-complexity, and automatic identification features of RFID technology make it attractive as a means of acquiring data in the IoT. This chapter has presented a brief overview of RFID technology, and highlighted its data acquisition role in the IoT. Although, presently its most widespread application is in its capacity as a replacement to barcodes, environmental sensing will likely be the most important application of RFID in the IoT. The integration of RFID with WSN is an important step toward the realization of ubiquitous sensing, as it provides a means to connect RFID infrastructure to the Internet. Full internet integration can be provided through a linkage between RFID tags and IPv6 addressing. However, these extensions of RFID functionality present new challenges in security, eco-friendliness, and energy consumption, which must be tackled in order to realize the vision of the IoT on a global scale.

References

Ashton, K. (2009) That "Internet of Things" thing. RFID Journal, Available at http://www.rfidjournal.com/articles/view?4986

Beriain, A., Berenguer, R., Jimenez-Irastorza, A., Montiel-Nelson, J., Sosa, J., and Pulido, R. (2014) Full passive RFID pressure sensor with a low power and low voltage time to digital interface. Design of Circuits and Integrated Systems Conference, pp. 1–6.

Bhattacharyya, R., Leo, C. D., Floerkemeier, C., Sarma, S., and Anand, L. (2010) RFID tag antenna based temperature sensing using shape memory polymer actuation. Sensors, 2363–2368.

Bogari, E. A., Zavarsky, P., Lindskog, D., and Ruhl, R. (2012) An investigative analysis of the security weaknesses in the evolution of RFID enabled passport. *International Journal of Internet Technology and Secured Transactions*, **4**(4), 290. Available at http://doi.org/10.1504/IJITST.2012.054060

Brenk, D., Essel, J., Heidrich, J., Agethen, R., Kissinger, D., Hofer, G., Holweg, G., Fischer, G., and Weigel, R. (2011) Energy-efficient wireless sensing using a generic ADC sensor interface within a passive multi-standard RFID transponder. *Sensors Journal, IEEE*, **11**(11), 2698–2710.

Cantatore, E., Geuns, T. C. T., Gelinck, G. H., Van Veenendaal, E., Gruijthuijsen, A. F. A., Schrijnemakers, L., et al. (2007) A 13. 56-MHz RFID System Based on Organic Transponders. *IEEE Journal of Solid-State Circuits*, **42**(1), 84–92.

Colella, R., Catarinucci, L., and Tarricone, L. (2015) EM design of a passive RFID-based device with sensing and reasoning capabilities. 2015 IEEE 15th Mediterranean Microwave Symposium (MMS) pp. 1–4. Available at http://doi.org/10.1109/MMS.2015.7375387

Colella, R., Tarricone, L., and Catarinucci, L. (2015) SPARTACUS: self-powered augmented RFID tag for autonomous computing and ubiquitous sensing. *IEEE Transactions on Antennas and Propagation*, **63**(5), 2272–2281. Available at http://doi.org/10.1109/TAP.2015.2407908

De Donno, D., Catarinucci, L., and Tarricone, L. (2014) RAMSES: RFID augmented module for smart environmental sensing. *IEEE Transactions on Instrumentation and Measurement*, **63**(7), 1701–1708. Available at http://doi.org/10.1109/TIM.2014.2298692

Doraswamy, N. and Harkins, D. (2003) *IPSec: The New Security Standard for the Internet, Intranets, and Virtual Private Networks*, Prentice Hall.

Ernst, A. (1906) Alternating Current Generator. U.S. Patent and Trademark Office.

Evdokimov, S., Fabian, B., Gunther, O., Ivantysynova, L., and Ziekow, H. (2010) RFID and the Internet of Things: technology, applications, and security challenges. *Foundations and Trends® in Technology, Information and Operations Management*, **4**(2), 105–185. Available at http://doi.org/10.1561/0200000020

GS1. (2013) Celebrating our 40th anniversary: a history of standards. Available at http://40.gs1.org/historic-timeline.php

Hong, Y.-J., Kim, I.-J., Ahn, S. C., and Kim, H.-G. (2008) Activity recognition using wearable sensors for elder care. 2008 Second International Conference on

Future Generation Communication and Networking pp. 302–305. IEEE. Available at http://doi.org/10.1109/FGCN.2008.165

Jankowski-Mihułowicz, P. (2015) Flexible antenna design for semi-passive HF RFID transponder in ink-jet technology. *Przegląd Elektrotechniczny*, **1**(4), 3–7. Available at http://doi.org/10.15199/48.2015.04.01

Jing, Q., Vasilakos, A. V., Wan, J., Lu, J., and Qiu, D. (2014) Security of the Internet of Things: perspectives and challenges. *Wireless Networks*, **20**(8), 2481–2501. Available at http://doi.org/10.1007/s11276-014-0761-7

Koelle, A. R., Depp, S. W., and Freyman, R. W. (1975) Short-range radio-telemetry for electronic identification, using modulated RF backscatter. *Proceedings of the IEEE*, **63**(8), 1260–1261. Available at http://doi.org/10.1109/PROC.1975.9928

Lakafosis, V., Vyas, R., Mariotti, C., Le, T., and Tentzeris, M. M. (2014) Integrating tiny RFID- and NFC-based sensors with the Internet, in Roselli, L. (ed.), *Green RFID Systems*, Cambridge University Press, pp. 152–175.

Lakafosis, V., Vyas, R., and Tentzeris, M. M. (2010) A localization and position tracking solution utilizing solar-powered RFID tags. Proceedings of the Fourth European Conference on Antennas and Propagation (EuCAP) pp. 1–4.

Landt, J. (2005) The history of RFID. *IEEE Potentials*, **24**(4), 8–11. Available at http://doi.org/10.1109/MP.2005.1549751

Li, S., Tryfonas, T., and Li, H. (2016) The Internet of Things: a security point of view. *Internet Research*, **26**(2), 337–359. Available at http://doi.org/10.1108/IntR-07-2014-0173

Liu, H., Bolic, M., Nayak, A., and Stojmenović, I. (2008) Taxonomy and challenges of the integration of RFID and wireless sensor networks. *IEEE Network*, **22**(6), 26–32. Available at http://doi.org/10.1109/MNET.2008.4694171

López, A. L. V., Giles, D. B., Khan, W. T., Chlieh, O., and Ponchak, G. E. (2013) Microfluidic channel on organic substrates as size reducing technique for 915 MHz antenna designs, Microwave Conference Proceedings (APMC), 2013 Asia-Pacific, pp. 3–6.

Manzari, S., Occhiuzzi, C., and Marrocco, G. (2012) Feasibility of body-centric systems using passive textile RFID tags. *IEEE Antennas and Propagation Magazine*, **54**(4), 49–62. Available at http://doi.org/10.1109/MAP.2012.6309156

Mariotti, C., Lakafosis, V., Tentzeris, M. M., and Roselli, L. (2013) An IPv6-enabled wireless shoe-mounted platform for health-monitoring. WiSNet 2013 - Proceedings: 2013 IEEE Topical Conference on Wireless Sensors and Sensor Networks - 2013 IEEE Radio and Wireless Week, RWW 2013, 46–48. Available at http://doi.org/10.1109/WiSNet.2013.6488629

Nair, R., Perret, E., Tedjini, S., and Baron, T. (2013) A group delay-based chipless RFID humidity tag sensor using silicon nanowires. *Antennas and Wireless Propagation Letters, IEEE*, **12**, 729–732.

Ni, L. M., Zhang, D., and Souryal, M. R. (2011) RFID-based localization and tracking technologies. *IEEE Wireless Communications,* **18**(2), 45–51. Available at http://doi.org/10.1109/MWC.2011.5751295

Occhiuzzi, C., Cippitelli, S., and Marrocco, G. (2010) Modeling, design and experimentation of wearable RFID sensor tag. *IEEE Transactions on Antennas and Propagation,* **58**(8), 2490–2498. Available at http://doi.org/10.1109/TAP.2010.2050435

Occhiuzzi, C. and Marrocco, G. (2010) The RFID technology for neurosciences: feasibility of limbs' monitoring in sleep diseases. *IEEE Transactions on Information Technology in Biomedicine,* **14**(1), 37–43. Available at http://doi.org/10.1109/TITB.2009.2028081

Occhiuzzi, C., Rida, A., Marrocco, G., and Tentzeris, M. (2011) RFID passive gas sensor integrating carbon nanotubes. *IEEE Transactions on Microwave Theory and Techniques,* **59**(10), 2674–2684.

Palmer, D. (2017) The Internet of Laundry: How RFID tags in your sheets help you get a better night's sleep IT director Duncan Macmillan explains how RFID chips and scanners are helping to manage over a million pieces of linen a day for Berendsen. Available at http://www.zdnet.com/article/the-internet-of-laundry-how-rfid-tags-in-your-sheets-help-you-get-a-better-nights-sleep/

Potyrailo, R. A. and Wortley, T. (2011) Passive multivariable RFID pH sensors. 2011 IEEE International Conference on RFID-Technologies and Applications, pp. 533–536. Available at http://doi.org/10.1109/RFID-TA.2011.6068596

Roselli, L., Mariotti, C., Mezzanotte, P., Alimenti, F., Orecchini, G., Virili, M., and Carvalho, N. B. (2015) Review of the present technologies concurrently contributing to the implementation of the Internet of Things (IoT) paradigm: RFID, Green Electronics, WPT and Energy Harvesting. 2015 IEEE Topical Conference on Wireless Sensors and Sensor Networks, WiSNet 2015, 1–3. Available at http://doi.org/10.1109/WISNET.2015.7127402

Sample, A. P., Meyer, D. A., and Smith, J. R. (2011) Analysis, experimental results, and range adaptation of magnetically coupled resonators for wireless power transfer. *IEEE Transactions on Industrial Electronics,* **58**(2), 544–554. Available at http://doi.org/10.1109/TIE.2010.2046002

Sample, A. P., Yeager, D. J., Powledge, P. S., Mamishev, A. V., and Smith, J. R. (2008) Design of an RFID-based battery-free programmable sensing platform. *IEEE Transactions on Instrumentation and Measurement,* **57**(11), 2608–2615.

Shah, A., Pal, A., and Acharya, H. B. (2016) The Internet of Things: Perspectives on Security from RFID and WSN. Available at http://arxiv.org/abs/1604.00389

Shrestha, S., Balachandran, M., Agarwal, M., Phoha, V. V., and Varahramyan, K. (2009) A chipless RFID sensor system for cyber centric monitoring applications. *IEEE Transactions on Microwave Theory and Techniques,* **57**(5), 1303–1309. Available at http://doi.org/10.1109/TMTT.2009.2017298

Stockman, H. (1948) Communication by Means of Reflected Power. *Proceedings of the IRE*, **36**(10), 1196–1204. Available at http://doi.org/10.1109/JRPROC.1948.226245

Strukhoff, R. and Yaroshko, S. (2017) IoT for Airlines: Smart Baggage Tracking with RFID Tags and Cloud Foundry. Available at https://www.altoros.com/blog/iot-for-airlines-smart-baggage-tracking-with-rfid-tags-and-cloud-foundry/. (accessed June 13, 2017)

Tankiewicz, S., Schaefer, J., and DeHennis, A. (2013) A co-planar, near field communication telemetry link for a fully-implantable glucose sensor using high permeability ferrites. 2013 IEEE Sensors pp. 1–4 Available at http://doi.org/10.1109/ICSENS.2013.6688255

Tedjini, S., Zurita, M., Freire, R. C. S., Duroc, Y., Federal, U., Grande, C., and Lyon, U. C. (2016) Augmented RFID Tags, 67–70.

Trevisan, R. and Costanzo, A. (2014) Exploitation of passive RFID technology for wireless read-out of temperature sensors. 2014 IEEE RFID Technology and Applications Conference (RFID-TA) pp. 150–154 Available at http://doi.org/10.1109/RFID-TA.2014.6934218

Trevisan, R. and Costanzo, A. (2016) A UHF near-field link for passive sensing in industrial wireless power transfer systems. *IEEE Transactions on Microwave Theory and Techniques*, **64**(5), 1634–1643. Available at http://doi.org/10.1109/TMTT.2016.2544317

Ussmueller, T., Hagelauer, A., and Weigel, R. (2014) Technologies for RFID sensors and sensor Tags, in Roselli, L. (ed.), *Green RFID Systems*, Cambridge University Press, pp. 76–115.

Vermesan, O. and Friess, P. (eds.) (2013) *Internet of Things: Converging Technologies for Smart Environments and Integrated Ecosystems*, River Publishers.

Virili, M., Casula, G., Mariotti, C., Orecchini, G., Alimenti, F., Cosseddu, P., et al. (2014) 7.5–15 MHz organic frequency doubler made with pentacene-based diode and paper substrate. Available at http://doi.org/10.1109/MWSYM.2014.6848395

Want, R. (2004) Enabling ubiquitous sensing with RFID. *Computer*, **37**(4), 84–86. Available at http://doi.org/10.1109/MC.2004.1297315

Welbourne, E., Battle, L., Cole, G., Gould, K., Rector, K., Raymer, S., and Balazinska, M. (2009) Building the Internet of Things using RFID. *Internet Computing, IEEE*, **13**(3), 48–55. Available at http://doi.org/10.1109/MIC.2009.52

Yang, H. Y. and Yang, S.-H. (2007) RFID sensor network – network architectures to integrate RFID, sensor and WSN. *Measurement and Control*, **40**(2), 56–59.

Zhang, D., Yang, Y., Cheng, D., Liu, S., and Ni, L. M. (2010) COCKTAIL: An RF-based hybrid approach for indoor localization. 2010 IEEE International

Conference on Communications, pp. 1–5. Available at http://doi.org/10.1109/ICC.2010.5502137
Zhao, Y., Smith, J. R., and Sample, A. (2015) NFC-WISP: A sensing and computationally enhanced near-field RFID platform. 2015 IEEE International Conference on RFID (RFID), 174–181. Available at http://doi.org/10.1109/RFID.2015.7113089
Zhu, C., Leung, V. C. M., Shu, L., and Ngai, E. C. H. (2015) Green Internet of Things for smart world. *IEEE Access*, **3**(January), 2151–2162. Available at http://doi.org/10.1109/ACCESS.2015.2497312

6

A Tutorial Introduction to IoT Design and Prototyping with Examples

Manuel Meruje,[1] Musa Gwani Samaila,[1,2] Virginia N. L. Franqueira,[3] Mário Marques Freire,[1] and Pedro Ricardo Morais Inácio[1]

[1]*Instituto de Telecomunicações and Department of Computer Science, Universidade da Beira Interior, Covilhã, Portugal*
[2]*Centre for Geodesy and Geodynamics, National Space Research and Development Agency, Toro, Bauchi State, Nigeria*
[3]*Department of Electronics, Computing & Mathematics, University of Derby, Derby, UK*

6.1 Introduction

The evolution of the Internet of Things (IoT) has brought about tremendous digital transformations in many different sectors of human endeavor, including, but not limited to, healthcare, transportation, power grid, manufacturing, and logistics (Haase et al., 2016). This has resulted partly from the fact that communication technologies, sensor technologies, and computer memory and processing power have become readily available and inexpensive (Javed et al., 2016) such that manufacturers of consumer products can afford to put them on a wide range of products (McEwen and Cassimally, 2014). Over the years, the ubiquity of the aforementioned technologies has transformed the semiconductor industry by creating the enabling environments that facilitate chip development and manufacturing. Today, despite the twilight of Moore's law, chipmakers are still racing to develop smaller, smarter, less expensive, and more energy-efficient chips (DeBenedictis, 2017; James, 2016; Lentine and DeRose, 2016), and this has in turn significantly enhanced the development of different IoT hardware development platforms.

IoT hardware development platforms are basically kits or prebuilt development boards that combine microcontrollers and processors with wireless communication chips and other components in a ready-to-build and ready-to-program bundle. They come in different configurations and include a variety of peripherals for connecting sensors and interfacing with other hardware components or devices for designing and prototyping IoT devices. IoT hardware development platforms are generally divided into two categories: microcontroller-based boards

Internet of Things A to Z: Technologies and Applications, First Edition. Edited by Qusay F. Hassan.
© 2018 by The Institute of Electrical and Electronics Engineers, Inc. Published 2018 by John Wiley & Sons, Inc.

(e.g., Arduino boards, Particle Photon, and Adafruit Feather Bluefruit) and Single-Board Computers (SBCs) (e.g., Raspberry Pi, BeagleBone Black, and C.H.I.P).

The emergence of the IoT along with the creation of several innovative consumer-centric applications and services in various domains have led to an explosion of a wide variety of powerful but small IoT hardware development platforms (Musa et al., 2015). A fascinating trend today is the fact that many vendors and companies are competing to make their hardware platforms look a lot more attractive and appealing to IoT designers, developers, and electronics enthusiasts. This quest for competitive advantage is rapidly transforming the IoT hardware development platform landscape. As a result, many new IoT hardware development platforms are now less expensive, require minimal system configurations, have more memory and processing capacity, support a number of programming languages, and come embedded with other functionalities that could include hardware-embedded security and out of the box communication features like wireless connectivity options (Postscapes, 2017; Johnston et al., 2016). Consequently, IoT hardware development platforms are now becoming more user-friendly to the extent that even people with less technical skills can use them. However, lack of fundamental knowledge about proper design and development can potentially lead to poor design and implementation of IoT devices, which could have serious performance and security repercussions.

This chapter seeks to proffer a solution to the problem highlighted above. The aim is to develop a comprehensive design and prototyping guidelines for the IoT that will be easy to understand as well as effective for the learner. Thus, the chapter is presented in a tutorial format that can serve as a guide for beginners who are interested in learning how to develop IoT prototypes. Given the challenges the design and development of embedded systems like IoT devices present, both on the software and hardware parts (Chen et al., 2016; Ruiz-de-Garibay et al., 2014), the chapter could also serve to broaden the experience and knowledge of practicing designers and developers.

6.2 Main Features of IoT Hardware Development Platforms

IoT hardware development platforms have a number of features that make them suitable for designing and prototyping IoT devices. However, this section only considers some of the most important features of these platforms, namely, processing and memory/storage capacity; power consumption, size, and cost; Operating Systems (OSes) and programming languages; connectivity and flexibility/customizability of peripherals of hardware platforms; and onboard sensors and hardware security features. The main focus here is on Arduino boards and Raspberry Pi SBCs.

6.2.1 Key Features of Arduino Hardware Development Platforms

The Arduino is basically an open-source electronics prototyping platform that is made up of two essential parts: the hardware, which is the Arduino board, and the software, the Integrated Development Environment (IDE). While there are different types of Arduino boards, this section does not intend to cover all the Arduino boards that have been created since 2007. It rather provides a comprehensive description of the aforementioned features for a few Arduino boards, namely, the *Uno* (released in 2010), the *Mega 2560* (released in 2010), the *Due* (released in 2012), the *Yún* (released in 2013), the *Arduino/Genuino 101* (release in 2015), and the *Arduino/Genuino MKR1000* (released in 2016). The boards are selected based on their popularity, functionality, newness to the market, and applicability in IoT projects. Figure 6.1a–f shows the pictures of the selected Arduino boards.[1]

6.2.1.1 Processing and Memory/Storage Capacity

At the heart of every Arduino board is a Microcontroller Unit (MCU), a type of Integrated Circuit (IC) with a processor, embedded memory, and programmable I/O peripheral devices (Alaraje et al., 2016). Whether 8, 16, or 32-bit MCU, the processor clock rate or clock speed, measured in Megahertz (MHz) or Gigahertz (GHz), determines how many instructions per second the MCU can execute. The 8, 16, or 32-bit refer to the size of the registers, which are high-speed memory areas on the processor that hold instructions and data currently being processed. Larger number of registers enables the processor to simultaneously handle larger memory areas (Aditya et al., 2012). Thus, the bigger the size of the registers, the faster the MCU can process a set of data. The embedded memory on an MCU consists of a Random Access Memory (RAM) for volatile data storage, a flash memory for storing program code and constants, and an Electrically Erasable Programmable Read-Only Memory (EEPROM) for storing persistent data, such as system configurations or other data that ensures normal operations upon reboot.

The *Uno* is based on the ATmega328, which is a microcontroller chip produced by Atmel. The chip has an 8-bit Central Processing Unit (CPU) with a clock speed of 16 MHz, 2 KB of Static RAM (SRAM), 1 KB of EEPROM, and 32 KB of flash memory (of which 0.5 KB is used by the boot loader). The *Mega* 2560 is based on the ATmega2560 MCU. The 8-bit CPU on the ATmega2560 chip is clocked at 16 MHz. The *Mega 2560* board has an SRAM of 8 KB, 4 KB of EEPROM, and 250 KB of flash memory (of which 8 KB is used by the boot loader). The *Due* is an Arduino board that is based on the AT91SAM3X8E MCU, and the CPU is a 32-bit SAM3X8E ARM Cortex-M3 that has a clock speed of 84 MHz. It has 96 KB of SRAM that is divided into two

1 https://www.arduino.cc/en/Main/Products.

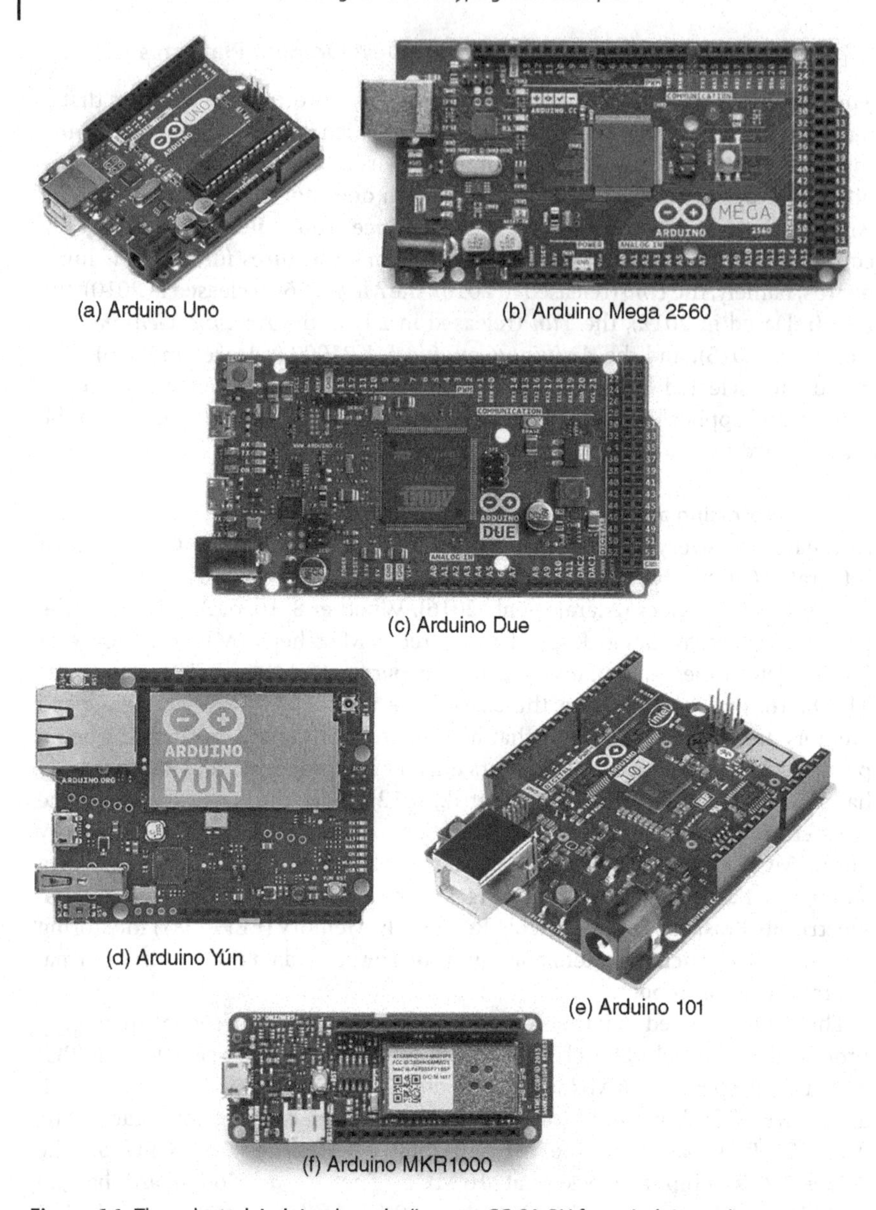

(a) Arduino Uno
(b) Arduino Mega 2560
(c) Arduino Due
(d) Arduino Yún
(e) Arduino 101
(f) Arduino MKR1000

Figure 6.1 The selected Arduino boards. (Images CC-SA-BY from Arduino.cc.)

banks: 64 KB and 32 KB. The 512 KB flash memory on the board is completely available for the user.

The *Yún* is a unique Arduino board because it has two different 8-bit processors, the ATmega32U4 MCU and the Atheros AR9331 microprocessor, all clocked at 16 MHz. The ATmega32U4 MCU has 2.5 KB of SRAM, 1 KB of EEPROM, and 32 KB of flash memory (of which 4 KB is used by the boot loader). The Atheros AR9331 processor has 64 MB of Double Data Rate 2 (DDR2) Synchronous Dynamic RAM (SDRAM), 2.5 KB of SRAM, 1 KB of EEPROM, and 16 KB of flash memory. The *Yún* also has a micro Secure Digital (SD) card reader that is linked to the Atheros AR9331 processor.

The *Arduino/Genuino 101* uses an Intel Curie module that is based on Intel Quark SE System on Chip (SoC). The module has two tiny cores, x86 (Intel Quark) and 32-bit ARC EM core architecture that are clocked at 32 MHz. The tiny cores operate simultaneously and share the same memory: 24 KB of SRAM and 196 KB of flash memory. The *Arduino/Genuino MKR1000* board is based on the Atmel ATSAMW25 SoC that is made up of three main blocks. While the first block is the SAMD21 Cortex-M0+ 32-bit low-power ARM MCU with a clock speed of 48 MHz, the other two main blocks will be explained subsequently. The board also has an onboard Real-Time Clock (RTC) used for coordinating auto-wakeup from stop/standby mode, which is clocked at 32.768 kHz. The *MKR1000* has 32 KB of SRAM and 256 KB of flash memory. Table 6.1 presents the summary of the processing and storage capacity of the Arduino boards.

6.2.1.2 Power Consumption, Size, and Cost of Arduino Boards

A key requirement for IoT devices is the ability to consume very low power while maintaining an acceptable level of performance, which could enable them to run

Table 6.1 Summary of processing and storage capacity of Arduino Boards.

	Microcontroller			Memory/storage		
Board name	**MCU chip**	**Processor bit**	**Clock speed**	**RAM**	**EEPROM (kb)**	**Flash memory (kb)**
Uno	ATmega328	8	16 MHz	2 KB	1	32
Mega 2560	ATmega2560	8	16 MHZ	8 KB	4	250
Due	AT91SAM3X8E	32	84 MHz	96 KB	–	512
Yún	ATmega32U4	8	16 MHz	2.5 KB	1	32
	AR9331	8	16 MHZ	64 MB, 2.5 KB	1	16
101	Intel Quark SE SoC	32	32 MHz	24 KB	–	196
MKR1000	ATSAMW25 SoC	32	48 MHz, 32.768 KHz	32 KB	–	256

on batteries for long periods of time. Strategies for achieving ultralow-power operation in IoT devices include, among others, employing low-power communication technologies and implementing low-duty-cycle operations (Maksimović et al., 2014). Nonetheless, power consumption or battery life of IoT devices still remains a challenging research topic (Ghasempour, 2016; Nishimura and Sugita, 2015; Patil et al., 2015). The battery life (in hours) of an IoT device can be calculated by dividing the capacity of the battery (in milliamp-hour (mAh)) by the average current consumption (in milliampere (mA)) of the device. The power consumption or battery life of a hardware development board will largely depend on the operating voltage and operating current of the board as well as on the components that are connected to it. This section covers operating voltage, active current consumption, and power consumption of bare-Arduino boards under consideration. It also discusses size and cost of the hardware development boards. Note that the cost refers to the prices of the boards at the time of writing this chapter.

Coincidentally, the voltage and current specifications of the *Uno* and the *Mega 2560* are the same: Their operating voltage is 5 V, DC current per I/O pin is 20 mA, and DC current for 3.3 V pin is 50 mA. Power consumption can be calculated using Equation 6.1.

$$P_{(W)} = I_{(A)} \times V_{(V)} \tag{6.1}$$

where P is the power in Watts, I is the operating current in Amperes, and V is the operating voltage in Volts. Hence, using 50 mA (0.05 A) and 5 V as the operating current and voltage, respectively, and substituting the values into Equation 6.1 yields

$$P = 0.05 \times 5 = 0.25 \quad \mathrm{W} = 250\ \mathrm{mW} \tag{6.2}$$

Therefore, the average power consumption of both the *Uno* and the *Mega 2560* is 250 mW. The *Due* has different specifications, its operating voltage is 3.3 V, DC current for 3.3 V pin is 800 mA, and DC current for 5 V pin is 800 mA, hence its average power consumption is 2.64 W. The voltage and current specifications of the ATmega32U4 MCU on the *Yún* are 5 V, DC current per I/O pin is 40 mA, and DC current for 3.3 V pin is 50 mA, and the operating voltage on the Atheros AR9331 processor is 3.3 V; since the *Yún* has built-in Ethernet and Wi-Fi support, the operating current can be more than 200 mA, therefore the average power consumption can be $\geq$ 1W.

Although the *Arduino/Genuino 101* uses a low-power Intel Curie SoC with a 20 mA DC current per I/O pin, the DC current for 3.3 V pin (i.e., operating current) is not specified. Considering that the board has a Bluetooth LE functionality, the current consumption of the board cannot be 20 mA, hence its average power consumption cannot be estimated correctly from the given specifications. The low-power capability of the *Arduino/Genuino MKR1000* MCU makes the DC current per I/O pin to be 7 mA at 3.3 V operating voltage.

Table 6.2 Summary of power consumption, size, and cost of Arduino boards.

Board name	Average power consumption	Size (mm)	Cost ($)
Uno	250 mW	68.8 × 53.4	22.19
Mega2560	250 mW	101.5 × 53.3	38.83
Due	2.64 W	101.5 × 53.3	39.93
Yún	≥1 W	73 × 53	64.90
101	–	68.6 × 53.4	30.00
MKR1000	≥ 396 mW	65 × 25 × 6	34.38

While the current consumption of the MCU is about 20 mA, the Wi-Fi alone can consume about 100 mA or more when operational.[2] Thus, for IoT applications, the total *MKR1000* average current consumption can be 120 mA or more, which implies an average power consumption larger than 396 mW.

The length and width of the *Uno* are, respectively, 68.6 and 53.4 mm, and the price of the *Uno*-R3 is $22.19. The dimensions of the *Mega 2560* in terms of length and width are 101.52 and 53.3 mm, respectively, and *Mega 2560*-R3 is currently selling for $38.83. The *Due* has the same dimensions as *Mega 2560*, but its price is $39.93. The length and width of the *Yún* are 73 and 53 mm, respectively, and it is selling for $64.90. The length and width of the *Arduino/Genuino 101* are 68.6 and 53.4 mm, respectively, and its price is $30.00. The dimensions of the *Arduino/Genuino MKR1000* in terms of length, width, and height are, respectively, 65, 25, and 6 mm, and it is selling for $34.38. Table 6.2 summarizes the power consumption, size, and cost of the Arduino boards.

6.2.1.3 Operating Systems and Programming Languages for Arduino Boards

Compared to standard computers, tablets, and smartphones, many IoT devices are resource constrained, especially in terms of memory footprint, and hence they cannot run High-Level Operating System (HLOS) such as Windows and Linux. In addition, the diversity of MCU families and architectures (e.g., there are 8-bit, 16-bit, and 32-bit processors) is one of the greatest bottlenecks to the development of generic OSes for these devices. The abovementioned diversity is also aggravating the restrictions on providing a more generic OS support for IoT heterogeneous hardware (Hahm et al., 2016). In this section, OSes and programming languages for Arduino hardware platforms are discussed.

Owing to stringent resource restrictions, particularly low RAM, the *Yún* is the only Arduino board under consideration that supports an OS. The onboard Atheros AR9331 processor supports the OpenWrt Linux distribution.

2 https://www.arduino.cc/en/Tutorial/MKR1000BatteryLife.

Processes/threads in the other models of Arduino are managed in the program. However, a number developers/people in the Arduino community have been exploring possibilities for developing portable OSes for many of the Arduino boards.

The Arduino programming language or Arduino Language (AL) is based on C/C++. AL is composed of a set of C/C++ functions that users can easily call from their code (also known as sketch). While virtually all support libraries are subset of C standard library and not C++ standard library, the Arduino IDE basically uses a simplified version of C++ language. The IDE, which can be downloaded from the Arduino website, is used for writing and uploading the programs to the Arduino board. Another alternative is to use the online IDE (Arduino Web Editor), which allows users to save their sketches in the Cloud. The sketch automatically generates function prototypes after which it is directly passed to a C/C++ compiler (avr-g++).[3]

Although Arduino boards cannot be programmed directly in JavaScript (JS), it is possible to leverage the web client scripting functionalities of JS using both Firmata and Johnny-Five libraries (Cvjetkovic and Matijevic, 2016). This is particularly important for IoT applications, especially considering how JS is recently gaining more and more popularity among IoT developers and designers as a result of the appearance of Node.JS (Poulter et al., 2015; Singh et al., 2015).

6.2.1.4 Connectivity and Flexibility/Customizability of Peripherals of the Arduino Boards

The ability to communicate with other devices, especially through the Internet is a very essential feature of an IoT hardware development platform. Similarly, the availability and accessibility of both low-level (I/O pins) and high-level (hardware communication interfaces) peripherals are important factors that determine the types of projects that can be built with a particular hardware platform.

Among the Arduino boards being considered, only the *Yún* and *MKR1000* can directly connect to the Internet without using a *shield* (an Arduino *shield* is a compatible board that can be plugged on top of an Arduino board for the purpose of extending its capabilities). The *Yún* has both Ethernet and Wi-Fi support, while the *MKR1000* can only connect to the Internet using the WINC1500, a low power 2.4 GHz IEEE 802.11b/g/n Wi-Fi, which is the second block of the ATSAMW25 SoC on the *MKR1000*. The other boards can connect to the Internet using Ethernet, Wi-Fi, or GSM *shields*. The *Arduino/Genuino 101* has Bluetooth Low Energy (BLE) capabilities.

Despite their versatility and flexibility, Arduino boards cannot be used for general purpose computing. The flexibility of an Arduino board lies in the availability of the peripherals on the MCU as well as their accessibility by users via the program. Virtually all Arduino boards have serial communication

---

3 https://www.arduino.cc/en/Main/FAQ.

interfaces like Universal Serial Bus (USB) port. The USB port can be used to power an Arduino board from a computer; it is also used for uploading programs from the IDE to the board. All the Arduino boards feature a number of digital I/O pins that are used as low-level peripherals, which are also known as General Purpose I/O (GPIO) pins. The analog pins on the Arduino boards also have all the functionality of GPIO pins.

In addition to the digital I/O pins, the Arduino boards support other hardware communication interfaces, such as Serial Peripheral Interface (SPI) communication using the SPI library as well as Universal Asynchronous Receiver/Transmitter (UART) communication. They also support Inter-Integrated Circuit/Two Wire Interface (I2C/TWI) communication using the wire library,[4] and apart from the *MKR1000*, all the other boards have the In-Circuit Serial Programming (ICSP) header.

The *Uno* features a USB port, 14 GPIO digital pins (of which 6 are Pulse Width Modulation (PWM) output), and 6 analog input pins. The Uno also has the ICSP header, and it supports SPI, UART, and I2C/TWI communications. The *Mega 2560* also has a USB port, 54 GPIO digital pins (of which 15 are PWM output), and 16 analog input pins. It also features 4 UART hardware serial ports and an ICSP header, as well as supports SPI and I2C/TWI communications. The *Due* features 2 USB ports, 54 GPIO digital pins (of which 12 are PWM output), 12 analog input pins, and 2 Digital to Analog Converter (DAC) analog output pins. In addition, the *Due* has 4 UART ports, 1 ICSP header, 1 SPI header, 1 I2C, and 2 TWI headers. The *Yún* has 2 USB ports, an Ethernet port, 20 GPIO digital pins (of which 7 are PWM output), and 12 analog input pins. Additionally, it features 1 UART port, 1 ICSP header, and supports SPI and I2C/TWI communications. The *Arduino/Genuino 101* features a USB port, 14 GPIO digital pins (of which 4 are PWM output), and 6 analog input pins. It also comes with an ICSP header and supports SPI, UART, and I2C/TWI communications. The *MKR1000* has a USB port, 8 GPIO digital pins, 12 PWM output pins, 1 UART, 1 SPI, and 1 I2C peripheral. The other pins include 7 analog input pins (Analog to Digital Converter (ADC) 8/10/12 bit) and 1 analog output pin (DAC 10 bit). A summary of the connectivity and flexibility/customizability of peripherals of the Arduino boards is presented in Table 6.3.

6.2.1.5 Onboard Sensors and Hardware Security Features of Arduino Boards

While a sensor is meant to perceive and measure some type of input (e.g., temperature, pressure, motion, light, heat, or other environmental phenomena) from the physical environment, an onboard sensor is a device that detects and measures not only what is happening in its surrounding, but also within the board. The *Arduino/Genuino 101* is the only board among the Arduino boards under consideration that features some onboard sensors. The board comes with

4 https://www.arduino.cc/en/Reference/Wire.

Table 6.3 Summary of connectivity and flexibility/customizability of peripherals of the Arduino boards.

Board name	Onboard connectivity	GPIO pins	Analog input	USB ports	ICSP header	Other hardware interfaces
Uno	–	14	6	1	1	SPI, UART, I2C/TWI
Mega2560	–	54	16	1	1	SPI, 4 UART, I2C/TWI
Due	–	54	12	2	1	SPI, 4 UART, I2C, 2 TWI
Yún	Ethernet, Wi-Fi	20	12	2	1	SPI, UART, I2C/TWI
101	BLE	14	6	1	1	SPI, UART, I2C/TWI
MKR1000	IEEE 802.11b/g/n	8	7	1	–	SPI, UART, I2C

a six-axis accelerometer and a gyroscope. The two sensors can be used together to form an Inertial Monitoring Unit (IMU) that can be employed to identify with accuracy the orientation of the Arduino board.[5]

The computational and memory limitations of most of the Arduino boards have imposed restrictions on their security capabilities, and hence it will be very difficult for them to run mature technology stacks (Bonetto et al., 2012). The stacks used for securing Hypertext Transfer Protocol (HTTP), Constrained Application Protocol (CoAP), or Message Queuing Telemetry Transport (MQTT) communications are: Internet Protocol Security (IPsec), Secure Sockets Layer/Transport Layer Security (SSL/TLS) or Datagram Transport Layer Security (DTLS). Apart from the *Arduino/Genuino MKR1000*, all the other Arduino boards covered in this chapter do not have crypto engine or hardware cryptographic module. A hardware crypto engine is a self-contained cryptographic module with a dedicated processor, making it difficult for hackers to access valuable data during cryptographic operations. It is designed to be integrated into devices as an alternative to software-based security implementations. The module performs encryption and decryption computations much faster than the software implementation of the same operations. The *MKR1000* has an onboard crypto chip; essentially, the third block in the ATSAMW25 is the ECC508 that provides cryptoauthentication. Additionally, the built-in Wi-Fi module supports SHA-256 certificates for ensuring secure communication. For more details about security, please refer to Chapter 8.

5 https://www.arduino.cc/en/Tutorial/Genuino101CurieIMUOrientationVisualiser

6.2.2 Major Features of the Raspberry Pi Hardware Platforms

The Raspberry Pi is a credit card-sized SBC that provides almost full capabilities of a desktop or a laptop computer while remaining small, lightweight, and inexpensive. The Raspberry Pi minicomputer can also be used for prototyping electronics projects.[6] First created in the United Kingdom in February 2012 by the Raspberry Pi Foundation, the SBC was developed for the purpose of promoting the teaching of programming skills and computer hardware in schools and in the developing countries (Lyons, 2015). With reference to the abovementioned features of IoT hardware development platforms, this section describes the following Raspberry Pi models: Raspberry Pi Zero (released in November 2015), Raspberry Pi Zero Wireless (W) (released in February 2017), Raspberry Pi 1 model B+ (released in July 2014), Raspberry Pi 2 model B (released in February 2015), and Raspberry Pi 3 model B (released in February 2016). For the purpose of describing current features of the Raspberry Pi SBCs, only the latest models are selected. Figure 6.2a–e shows the images of the selected models.[7]

6.2.2.1 Processing and Memory/Storage Capacity

In comparison to the Arduino hardware development platforms, the Raspberry Pi SBCs have faster processors, bigger memory, and larger storage capacity. All Raspberry Pi models are based on the Broadcom SoC that includes ARM CPU and on-chip Graphics Processing Unit (GPU), a VideoCore IV low-power mobile multimedia processor that supports up to 1920 × 1200 resolution. For instance, both the Raspberry Pi Zero and the Pi Zero W feature a Broadcom BCM2835 SoC with a 1 GHz ARM1176JZF-S CPU core, a member of the ARM11 family (which are 40% faster than Raspberry Pi 1) and have a 512 MB Low-Power DDR2 (LPDDR2) SDRAM. The Raspberry Pi 1 B+ is based on the same Broadcom BCM2835 SoC, but with a 700 MHz ARM1176JZF-S Core processor and a 512 MB RAM (Bell, 2014). The Raspberry Pi 2 B uses a Broadcom BCM2836 SoC with a 900 MHz 32-bit quad-core ARM Cortex-A7 CPU and a 1 GB RAM. Similarly, the Raspberry Pi 3 B uses a Broadcom BCM2837 SoC with a 1.2 GHz 64-bit quad-core ARM Cortex-A53 CPU and 1 GB RAM.[8] All Raspberry Pi SBCs feature a micro SD card slot. The summary of processing speed and memory/storage capacity of the Raspberry Pi hardware development platforms is presented in Table 6.4.

---

6 https://www.raspberrypi.org/magpi-issues/Projects_Book_v1.pdf.

7 https://www.raspberrypi.org/products/.

8 https://www.raspberrypi.org/products/

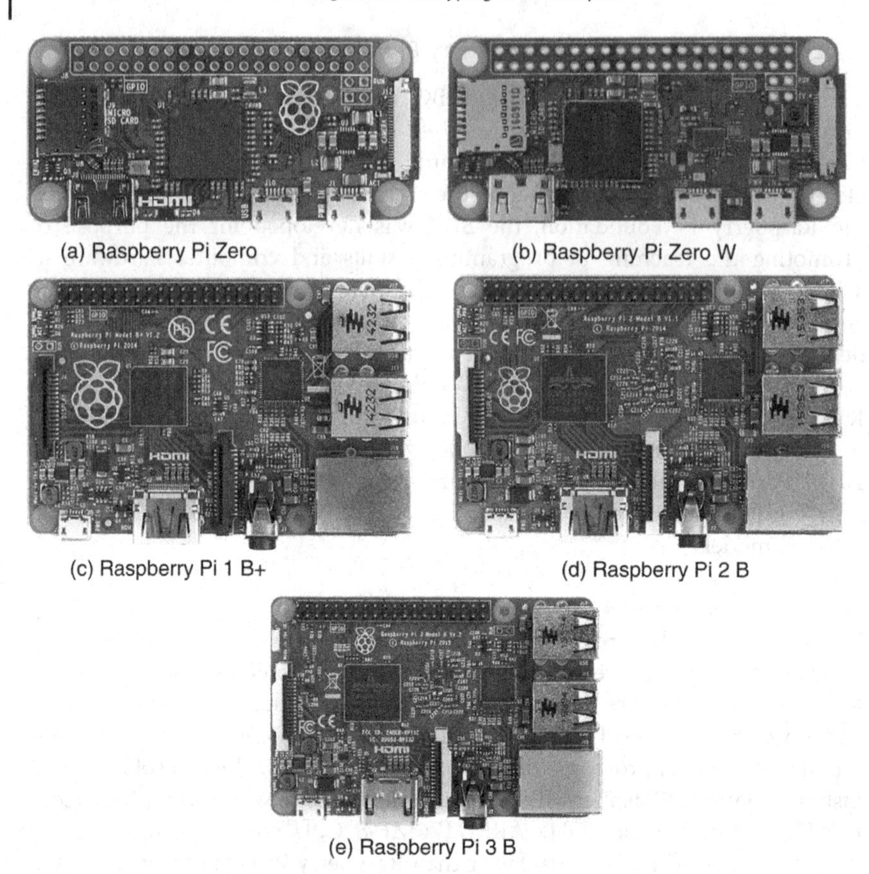

(a) Raspberry Pi Zero (b) Raspberry Pi Zero W

(c) Raspberry Pi 1 B+ (d) Raspberry Pi 2 B

(e) Raspberry Pi 3 B

Figure 6.2 The selected Raspberry Pi models. (*Source:* Raspberrypi.org2017.)

Table 6.4 Summary of processing speed and memory/storage capacity of Raspberry Pi hardware platforms.

	Broadcom SoC				Memory/storage	
Raspberry Pi	**SoC**	**CPU core**	**Processor architecture**	**Clock speed**	**RAM**	**Storage**
Zero	BCM2835	ARM1176JZF-S	–	1 GHz	512 MB	Micro SD
Zero W	BCM2835	ARM1176JZF-S	–	1 GHz	512 MB	Micro SD
1 B+	BCM2835	ARM1176JZF-S	–	700 MHz	512 MB	Micro SD
2 B	BCM2836	Quad-core ARM Cortex-A7	32	900 MHz	1 GB	Micro SD
3 B	BCM2837	Quad-core ARM Cortex-A53	64	1.2 GHz	1 GB	Micro SD

6.2.2.2 Power Consumption, Size, and Cost of Raspberry Pi Hardware Platforms

Being a low-power minicomputer, the Raspberry Pi operates on a 5 V DC power supply at 1–2.5 A, powered through a micro-USB port. Note that the cost in this section refers to the price of the Raspberry Pi at the time of writing this chapter.

Although the Raspberry Pi consumes different amounts of power in its four distinct power modes (Maksimović et al., 2014), namely, run, standby, shutdown, and dormant modes, the current draw of each model presented in the official Raspberry Pi magazine are in two modes: idle (standby) and load (run) modes.[9] The current consumptions in the two power modes for each model are: 0.1 and 0.25 A for Raspberry Pi Zero; 0.25 and 0.31 A for the Raspberry Pi 1 B+; 0.26 and 0.42 A for the Raspberry Pi 2; and 0.31 and 0.58 A for the Raspberry Pi 3. The conditions under which those measurements were carried out are not explicitly stated in the magazine.

However, other sources, including the Raspberry Pi Foundation and the element14 community show more realistic current consumption. For example, the average current consumption of the Raspberry Pi Zero when running is approximately 160 mA, which shows that the average power consumption is about 0.8 W.[10] According to Klosowski (2017) and RasPi.TV (2017), the new Raspberry Pi Zero W requires about 20 mA more than the Pi Zero. This implies that its average current consumption when running is about 180 mA, and therefore, the average power consumption is about 0.9 W. While the Raspberry Pi 1 B+ consumes about 600 mA, implying that average power consumption is about 3.0 W, the average current consumption of the Raspberry Pi 2 B is about 800 mA, and hence the average power consumption is about 4.0 W (Brown, 2015). According to Shabaz (2016), the current draw of Raspberry Pi 3 that runs the Rasbian but doing nothing is 266 mA and the current draw when running *cpuburn-a53* (i.e., a stress test that maxes out the CPU of a Pi completely) is 1.45 A. The tests results show that the power consumption of the Raspberry Pi 3 ranges from 1.33 to 7.25 W.

While the power consumption of the Raspberry Pi actually depends on the usage as well as the model of the device, a few best practice techniques that could be applied during operation to reduce power consumption include the following (Bekaroo and Santokhee, 2016):

1) Disconnect every peripheral that is not in use
2) Switch-off Internet connectivity when not needed
3) Shutdown the device or put it in dormant mode when not in use
4) When using the device in headless mode (i.e., accessing it via network connections without a keyboard or display), the High-Definition Multimedia Interface (HDMI) could be switched off

9 https://www.raspberrypi.org/magpi/raspberry-pi-3-specs-benchmarks/.
10 https://www.raspberrypi.org/blog/raspberry-pi-zero/.

Table 6.5 Summary of power consumption, size, and cost of Raspberry Pi hardware platforms.

Raspberry Pi	Average power consumption (W)	Size (mm)	Cost ($)
Zero	0.8	65 × 30 × 5.4	5
Zero W	0.9	65 × 30 × 5.4	10
1 B+	3.0	85.6 × 56 × 21	25
2 B	4.0	85.6 × 56 × 21	35
3 B	1.33–7.25	85.6 × 56 × 21	35

5) Avoid running several daemons at a time, and run only power efficient applications

The dimensions of the Raspberry Pi Zero and Pi Zero W, which are the smallest of them all, are 65 mm × 30 mm × 5.4 mm. But the Raspberry Pi 1 B+, the Raspberry Pi 2 B, and the Raspberry Pi 3 B have the same dimensions: 85.60 mm × 56 mm × 21 mm. While the price of Raspberry Pi Zero is just $5, the Zero W costs $10. The Raspberry Pi 1 B+ is selling for $25. However, both the Raspberry Pi 2 B and the Raspberry Pi 3 B are selling for $35. Table 6.5 summarizes the power consumption, size, and cost of the Raspberry Pi hardware platforms.

6.2.2.3 Operating Systems and Programming Languages for Raspberry Pi

The Raspberry Pi does not come with an OS. Hence, based on their projects, users can choose the type of OS that best suits their needs. This section considers the Raspberry Pi OSes and programming languages.

Raspbian is a Debian-based OS freely provided by the Raspberry Pi Foundation for the Raspberry Pi hardware. Over the years, the Raspbian OS has remarkably gained popularity among Raspberry Pi users for its high performance (Biglesp, 2015). Like a full-fledged OS for traditional computers, it comes with all the basic program utilities and more than 35,000 packages.[11] Despite the fact that the Raspbian OS is officially supported by the Raspberry Pi Foundation, the Raspberry Pi SBC is capable of supporting a wide variety of OSes (Keeler and Wolfer, 2016). The Linux-based OSes supported by the Raspberry Pi include Ubuntu MATE, Snappy Ubuntu, Pidora (a Fedora OS for Raspberry Pi), Linutop, SARPi (i.e., Slackware ARM on a Raspberry Pi), Arch Linux ARM, Gentoo Linux, FreeBSD, and Kali Linux. The Raspberry Pi 2 and 3 can also run Windows based OSes like Windows 10 IoT Core (Windows Dev Center, 2017; Klosowski, 2016).

11 https://www.raspbian.org/.

The easiest way to install Raspbian on a Raspberry Pi is to use the New Out Of Box Software (NOOBS), which is an easy-to-use Raspberry Pi operating system installation manager. Users can purchase SD cards with NOOBS preinstalled from Raspberry Pi distributors. An alternative way of obtaining NOOBS is to download the zip file for free from the Raspberry Pi website. An 8 GB SD card is needed to load the latest version, Jessie.[12] NOOBS comes with quite a number of OSes that users can choose from.

The two programming languages that normally come with the Raspberry Pi by default are Scratch and Python,[13] even though Python is more popular (Astudillo-Salinas et al., 2016; Patil et al., 2016). However, over the course of years, several programming languages have been adapted for the Raspberry Pi. Additionally, skilled individuals in the user community who desired to see their favorite languages on the Raspberry Pi have played a significant role in ensuring that their languages of choice have been adapted for the Raspberry Pi. Some of the programming languages now available for users to program on the Raspberry Pi using different IDEs include Java, C, C++, Objective C, JS, and Ruby (Richards, 2017).

6.2.2.4 Connectivity and Flexibility/Customizability of Peripherals of the Raspberry Pi

Like standard desktop and laptop computers, almost all the Raspberry Pi SBCs under consideration can be connected to the Internet directly. But unlike standard computers, the I/O pins and the hardware communication interfaces on all the Raspberry Pis are accessible to the user. In this section, connectivity, flexibility, customizability of I/O pins, and hardware communication interfaces of the Raspberry Pi will be discussed.

The Raspberry Pi Zero does not have an onboard Ethernet port and Wi-Fi capability. It can nonetheless be connected to the Internet in different ways. For example, it can be connected to the Internet by using USB On The Go (OTG) cable along with RJ45 USB converter, or by using USB OTG and Wi-Fi dongle. On the other end of the spectrum, the Pi Zero W has connectivity functionality that includes IEEE 802.11 b/g/n wireless LAN, Bluetooth 4.1, and BLE.[14] Both the Raspberry Pi 1 B+ and Pi 2 B have Ethernet port but no embedded Wi-Fi chip. The Raspberry Pi 3 B, however, is equipped with both Ethernet and Wi-Fi (IEEE 802.11n Wireless LAN) connection capabilities. The Raspberry Pi 3 B also features both Bluetooth 4.1 and BLE.[15]

---

12 https://www.raspberrypi.org/downloads/raspbian/.

13 https://www.sos.sk/productdata/16/72/87/167287/Raspberry_Pi_for_Beginners-Second_Revised_Edition_2014.pdf.

14 https://www.raspberrypi.org/products/pi-zero-wireless/.

15 https://www.raspberrypi.org/products/.

Table 6.6 Summary of connectivity and flexibility/customizability of the peripherals of the Raspberry Pi.

Raspberry Pi	Onboard connectivity	GPIO pins	USB ports	Display ports/ interfaces	Camera port	Other hardware interfaces
Zero	–	40	1 mini	Mini-HDMI	CSI	UART, SPI, I2C
Zero W	Wi-Fi, Bluetooth 4.1, BLE	40	1 mini	Mini-HDMI	CSI	UART, SPI, I2C
1 B+	Ethernet	40	4	HDMI, DSI, 3.5 mm Video Jack	CSI	UART, SPI, I2C
2 B	Ethernet	40	4	HDMI, DSI, 3.5 mm Video Jack	CSI	UART, SPI, I2C
3 B	Ethernet, Wi-Fi, Bluetooth 4.1, BLE	40	4	HDMI, DSI, 3.5 mm Video Jack	CSI	UART, SPI, I2C

The flexibility of the Raspberry Pi allows users to use the device for general purpose computing as well as for electronics projects (Maksimović et al., 2014). All the four Raspberry Pis feature the 40 GPIO pins, Camera Serial Interface (CSI) and HDMI, although the HDMI on the Raspberry Pi Zero and the Zero W is a mini version. Except for the Raspberry Pi Zero and the Pi Zero W, all the other Pis have Display Serial Interface (DSI). The Pi Zero and the Pi Zero W do not have the combined 3.5 mm audio and composite video jack (i.e., the 4-pole socket that carries both audio and video signal), which is on the other Raspberry Pis. Apart from the Pi Zero and Zero W that feature a mini USB OTG port, all the other Pis have 4 USB 2.0 ports. Aside from the 40 GPIO pins, other hardware interfaces that can be found on Raspberry Pi are SPI, UART, and I2C. For example, the BCM2835 ARM (i.e., the SoC on the Pi Zero, Pi Zero W, and Pi 1 B+) has three auxiliary peripherals, a mini UART, two SPI masters, and I2C.[16] Similarly, the GPIO pins on the Pi 2 B and the Pi 3 B can be configured as I2C, SPI, and UART.[17] Table 6.6 provides a summary of the connectivity and flexibility/customizability of the peripherals of the Raspberry Pi minicomputers.

6.2.2.5 Onboard Sensors and Hardware Security Features of Raspberry Pi

Additional features like onboard sensors and hardware security features on IoT hardware development platforms increase user acceptability, especially now that

16 https://www.raspberrypi.org/wp-content/uploads/2012/02/BCM2835-ARM-Peripherals.pdf.
17 https://www.raspberrypi.org/documentation/usage/gpio-plus-and-raspi2/

there are serious concerns about the security of IoT device. This section discusses onboard sensors and hardware security features of the Raspberry Pi SBCs.

Essentially, no member of the Raspberry Pi family features any onboard sensors; however, a typical Linux computer would likely come with onboard monitoring sensors (Suehle and Callaway, 2014). Thus, the Broadcom SoC includes on-chip temperature and voltage sensors that can be queried using the *vcgencmd* utility in the firmware package of the Raspberry Pi. The on-chip temperature sensor can be used to measure the temperature of the CPU and GPU. Similarly, the on-chip voltage sensor can be used to measure different voltages, including the core voltage, the SDRAM I/O voltage, and the SDRAM physical memory voltage.[18] The temperature or voltage values can be viewed by typing the appropriate commands in the terminal window of the Raspberry Pi (LDighera, 2015). For instance, the `vcgencmd measure_volts` command returns the voltage values for some vital components of the Raspberry Pi. To prevent the system from returning the voltage value for the core each time the command is executed, each of the following components must be passed to the *vcgencmd measure_volts* command as an option (Suehle and Callaway, 2014):

a) *sdram_c* (returns voltage value for the SDRAM controller)
b) *sdram_i* (returns the SDRAM I/O voltage)
c) *sdram_p* (returns the SDRAM physical memory voltage)
d) *core* (returns the GPU processor core voltage)

Despite the fact that Raspberry Pi is a versatile Linux-based hardware development platform that provides unlimited possibilities for different applications, it lacks hardware-based security engines. Additionally, securing private keys for public key cryptography or shared keys for symmetric key cryptography is a very big issue. This is because both data and user code reside on the same SD card, which is usually exposed (Miller, 2017).

6.3 Design and Prototyping of IoT Applications

This section focuses on the design and prototyping of IoT-enabled applications using both Arduino and Raspberry Pi development platforms.

6.3.1 IoT Design and Prototyping Using Arduino Boards

When designing IoT applications for Arduino development boards, the developer usually starts by doing a survey of the necessary components for the project,

18 https://wiki.archlinux.org/index.php/Raspberry_Pi#Temperature.

which may include hardware modules (and respective datasheets), diagrams, Arduino shields, programming libraries, and/or additional software.

Arduino shields, as explained in Section 6.2.1.4, are add-on circuit boards or additional hardware that are attached on top of the Arduino devices to provide enhanced capabilities, or supplementary functionalities for users. Shields are especially intended for beginner designers and developers to overcome the complexity of soldering components and attaching additional hardware resources to their projects.

In case the designer chooses to use alternative electronic modules instead of shields, datasheets from these components enable the designer to understand how they internally work by describing data and technical characteristics, such as power or current specifications. This information also helps the designer to know how the Arduino development platform will respond to the signals received from sensors, or other hardware modules. Additionally, it provides information on whether a voltage regulator is needed or not, given an existing power source.

Usually, in order to ease the work of the designer, hardware companies or third-party programmers develop software libraries for a high-level use of the modules for different programming languages and platforms, such as Adafruit modules and libraries (Earl, 2013). There are several tools that can facilitate the development of IoT projects. Electronic design and prototyping software like Fritzing and Oscad can be used to draw different types of schematic diagrams that can help in the development of IoT projects.

Fritzing (Knörig et al., 2009) is an open-source software application for electronic prototyping that is written in C++ and available for several OSes. It is developed by the Potsdam University of Applied Sciences. The software allows users to design projects in several different views, such as the breadboard view (a graphic representation of the prototyping board and respective circuits), electronic circuit view (a schematic representation of the electronic connections), and printed circuit board schematic view (a schematic representation of the printed circuit). This means designers are able to design a prototype in breadboard view mode and then evolve it into a final project by using the printed circuit board view. Another alternative software application is Oscad (Save et al., 2013), an open-source tool for circuit and Printed Circuit Board (PCB) design, simulation, and analysis. The software helps designers to plan, test, and examine their circuits. It supports Python, KiCad, Ngspice, and Scilab. One of the main features of the Oscad tool is its capability for simulation of circuits developed in the tool.

Arduino IDE (Earl, 2013), as previously mentioned, is the official Arduino open-source software developed in Java language that enables the development and uploading of code to the development board. This software is available on the official Arduino webpage.[19] The IDE is bundled with built-in libraries that facilitate I/O operations (Nayyar and Puri, 2016). A typical Arduino program is

19 https://www.arduino.cc/en/main/software

structured with, at least, two main functions, namely, `setup()` and `loop()`. The `setup()` function is used for setting initialization, which may include initialization of variables, setup of library variables, and serial communications. The `loop()` function is the main function that will run iteratively until the power is turned off. Data operations and I/O manipulations are usually made within these functions.

IoT devices are typically designed to communicate with other devices within a network using specific messaging protocols, which include CoAP, MQTT, and Extensible Messaging and Presence Protocol (XMPP) (Al-Fuqaha et al., 2015). Although there are several possible protocols that can be used, as previously mentioned in Chapter 3, MQTT will be briefly described here, since it will be employed in the Arduino project presented in Section 6.4.

MQTT is a lightweight messaging protocol designed for devices with constrained hardware, which are usually connected to unreliable or lossy networks with limited and unpredictable bandwidth as well as high latency. At the time of writing, the most recent version of the protocol is 3.1.1 and it was last reviewed in October 2014 (Banks and Gupta, 2014). MQTT is built on top of the Transmission Control Protocol (TCP). The protocol aims to connect embedded devices and networks through applications and middleware using the publish/subscribe pattern, which allows the implementation to be simple. MQTT consists of three components: (1) the broker, (2) the subscriber, and (3) the publisher. Both the subscriber and the publisher are clients to the broker, as explained:

1) *The Broker.* The broker acts as a gateway; it receives messages from a publisher (a client) and delivers the messages to a subscriber (another client). Brokers are sometimes referred to as *servers*.
2) *The Subscriber.* The subscriber declares its topics of interest to the broker, and the broker sends messages published to those topics.
3) *The Publisher.* The publisher sends messages to the broker using a namespace or a topic name, and the broker forwards the messages to the respective subscribers.

In MQTT, clients can subscribe to multiple topics, and a subscriber can unsubscribe to any topic. Additionally, MQTT has one-to-many message delivery functionality. The protocol also has levels of Quality of Service (QoS), which ensures that the message is delivered or transmitted from the sender to the receiver. There are three QoS levels of delivery in MQTT: at most once, at least once, and exactly once.

An example that illustrates IoT project design and prototyping using the tools and the MQTT protocol described above is the temperature logger, in which the current temperature is published via an MQTT publisher to an MQTT broker. This example, which will be further described in Section 6.4.1, is implemented on an Arduino platform.

6.3.2 IoT Design and Prototyping using Raspberry Pi Platforms

As mentioned in Section 6.2, one of the main differences between the Raspberry Pi and the Arduino board is that the former is a computer based on a System on a Chip processor, while the latter is a programmable microprocessor. This implies that an OS is necessary in order to use the Raspberry Pi for designing and prototyping IoT projects. The choice of OS is usually motivated by its features, the type, and requirements of the application the developer is aiming to design. A list of officially compatible OSes for this type of device is available on the website of the Raspberry Pi Foundation.[20]

Design and prototyping of IoT projects using both platforms are, in a way, similar. Usually, the designer starts by doing a survey of what hardware and software is necessary for the desired IoT project. This may include the use of hardware datasheets, modules and modules libraries, Raspberry Pi shields, and/or additional software. Sensing and actuating modules can also be connected to the platform by using the GPIO header of the device. Raspbian includes a GPIO controller application that permits users to read and write the state of the GPIO pins. To facilitate the integration of the module in the project that is being designed and prototyped, there are module libraries for different programming languages that are usually written by the community or by the module manufacturer available to the designer. Shields for Raspberry Pi are expansion boards that enable the integration of modules with extra functionalities, such as LCD modules, GPS receiver modules, and Internet connection (GPRS/HSDPA) modules.

The main difference between the development of IoT applications for the Raspberry Pi and for Arduino boards lies in the programming languages. While C/C++ is used for programming the Arduino boards, Python is often used for programming the Raspberry Pi due to its simplicity and readability, which allow beginner designers to catch up fast. One method of designing the hardware part of an IoT project is by using Computer-Aided Design (CAD) software applications. CAD can help in drawing the physical and schematic diagrams of the project, as discussed in the Section 6.3.1. The IoT application prototyping for a Raspberry Pi platform can be developed in the device itself using text editors, such as vi, nano, gedit, and leafpad. Other options include employing IDEs such as NINJA-IDE, Adafruit, and WebIDE, as well as using compilers for the respective programming language being used.

An IoT project usually follows communication protocol for compatibility and compliance purposes. This allows, for example, the project to connect with other devices that operate following the same protocol within a network. Application communication protocols such as MQTT, CoAP, XMPP, and Web Application

---

20 https://www.raspberrypi.org/downloads/raspbian/

Programming Interfaces (APIs) are usually implemented when designing and prototyping IoT solutions.

A simple example of the design and prototyping of Raspberry Pi IoT application project that can be used to illustrate prototyping using the tools mentioned above is the web-controlled Light-Emitting Diode (LED) project. In this project, an LED connected to the Internet is switched on and off by accessing a web server. While an LED is used in this example, any actuator like electric motor can be used to replace the LED. This example will be further described in Section 6.4.2.

6.4 Projects on IoT Applications

In order to understand how Arduino and Raspberry Pi platforms can be used, this section describes the basic usage of the Arduino IDE and how to write code in C language that can manipulate digital pins, and explains how to upload the code to the Arduino board. It also covers the basic usage of the Raspberry Pi platform, including how to configure the GPIO pins using Bash Shell scripts and Python programming language.

6.4.1 An Arduino Project for IoT Application

The purpose of this project is to transmit temperature readings collected by a sensor to an MQTT broker. This example makes use of a router with Ethernet and Wi-Fi connection, a 4.7 kΩ resistor, a Maxim Integrated DS18B20 temperature sensor, an ENC28J60 Ethernet module, and two applications: Arduino IDE,[21] where the microprocessor program will be developed, and Mosquitto (Eclipse, 2009), an open-source MQTT broker. It is assumed that the computer on which the project is going to be developed is connected to the router via Ethernet or Wi-Fi. For the purpose of demonstration, the computer is assigned the IP address 192.168.1.1.

6.4.1.1 Installing the Arduino IDE and Mosquitto Software

On Linux OS, the Arduino IDE (arduino and arduino-core) and the Mosquitto packages (mosquitto and mosquitto-clients) as well as their respective dependencies can be installed using the Linux terminal. Please note that required packages will only be installed with the following command if one is using the Debian software package manager:

```
$> sudo apt-get update
$> sudo apt-get install arduino arduino-core mosquitto
mosquitto-clients
```

21 https://www.arduino.cc/en/main/software.

On Windows OS, both applications can be downloaded from their official websites. After the download is completed, the software can be installed by simply executing the downloaded files. Please note that the Windows version of Mosquitto requires OpenSSL and *pthreads* library for Windows. Instructions and both download links are given to the user while installing Mosquitto.

6.4.1.2 Using Arduino IDE and Downloading Libraries

Arduino IDE needs to be launched with administrator rights, which can be done by opening the Linux terminal and executing the following command:

```
$> sudo arduino &
```

After opening the IDE, the steps for both OSes (Linux and Windows) are similar. At this point, the development board needs to be connected to the computer so that the IDE can be configured. This is done by selecting a development board and determining a serial port to be used to upload the code by `Board` and `Port` menu in `Tools`. After opening the `Board` menu, in which the list of compatible development boards with the IDE will appear, the user can select the desired board, which in this example is Arduino/Genuino *Uno,* as depicted in Figure 6.3.

The port that will be used can be configured by accessing the `Tools` menu, and then selecting the `Port` menu. In this menu, assuming that there are no more serial devices connected to the computer, the user must select the only available item, e.g.,/`dev/ttyUSB`* or `COM`*, where * indicates a number that corresponds to the development board serial port, as shown in Figure 6.4.

The next step is to download the necessary libraries for the development of the Arduino application that will be uploaded to the development board. From

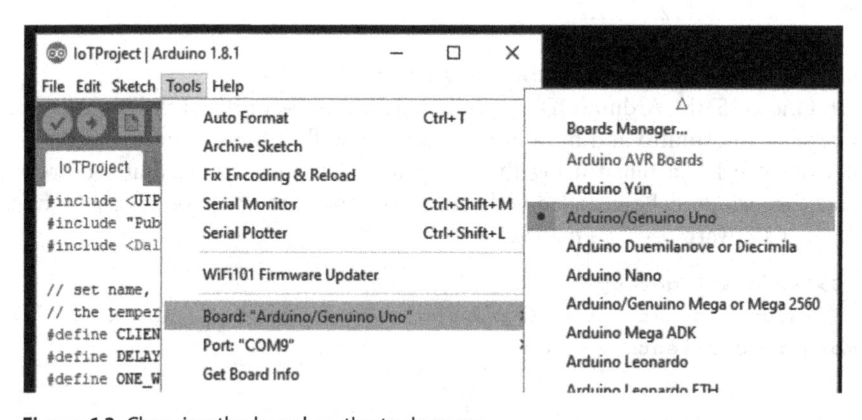

Figure 6.3 Choosing the board on the tools menu.

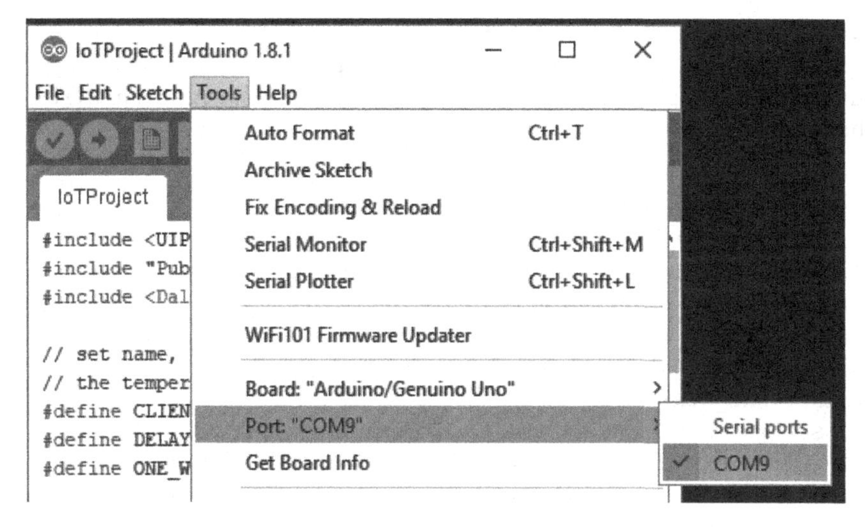

Figure 6.4 Choosing the serial port.

version 1.6.2, Arduino IDE includes a functionality that allows the developer to manage libraries without downloading them using a browser. This functionality may be accessed through the `Sketch` menu, `Include Library`, and selecting `Manage Libraries` as illustrated in Figure 6.5. Using the Search bar, the user can download the required libraries which are the following: `UIPEthernet`, `OneWire`, `DallasTemperature`, and `PubSubClient`.

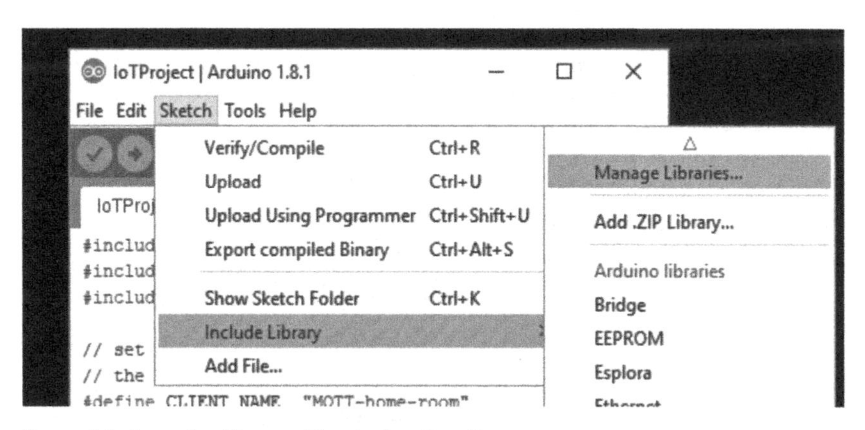

Figure 6.5 Accessing *Manage Libraries* functionality.

6.4.1.3 Project Source Code and Arduino Wiring Diagram

Listing 1 shows a typical implementation of an application (using C language) that uses MQTT protocol to publish temperature measurements to an MQTT broker (please note the inline comments):

```
#include <UIPEthernet.h>
#include "PubSubClient.h"
#include <DallasTemperature.h>
// set name, interval between readings,
// and the temperature module PIN
#define CLIENT_NAME  "MQTT-home-room"
#define DELAY        5000
#define ONE_WIRE_PIN 7
// Set MAC address and IP. Just make sure you don't have
// the same MAC / IP addresses set in different devices
uint8_t mac[6] = {0x00, 0x54, 0x45, 0x4d, 0x50, 0x01};
IPAddress ip(192, 168, 1, 2);
// Declare the modules variables
EthernetClient ecEth;
PubSubClient pscMQTTClient;
OneWire owOneWire(ONE_WIRE_PIN);
DallasTemperature dtSensor(&owOneWire);
// Declare variables
long lPrevMillis;
float fTemp;
void setup(){
    // Setup serial connection and modules
    Serial.begin(9600);
    dtSensor.begin();
    Ethernet.begin(mac, ip);
    // Set the ethernet module variable to PubSub library
    // and set the gateway / broker IP and Port.
    // Assume the broker is running on
    // ip-address:  192.168.1.1
    // port-number: 1883
(this is the default port of the MQTT Broker)
    pscMQTTClient.setClient(ecEth);
    pscMQTTClient.setServer("<ip&hyphen;address>",
<port-number>);
}
void loop() {
    // If the defined delay is exceeded.
    if(millis() - lPrevMillis >= DELAY) {
```

```
        dtSensor.requestTemperatures(); //
 Request temperature to the module
        fTemp = dtSensor.getTempCByIndex(0); //
 and set it to a variable again
        publishMessage(fTemp);
        lPrevMillis = millis();
    }
    pscMQTTClient.loop(); //
 Maintain the connection with the broker
}
void publishMessage(float fTemp) {
    char caMsgBuf[10];
    if(pscMQTTClient.connect(CLIENT_NAME)) {
        pscMQTTClient.publish("home/room/temp", dtostrf
(fTemp, 6, 2, caMsgBuf));
// Note the MQTT topic where the values will be published
(home/room/temp)
// The topic may describe, with few words, what the
code will publish: in this example, we are
publishing temperature data from a room within a home.
    }
}
```

Listing 1: Arduino code for the IoT temperature logger project.

After writing the code, it can be compiled and uploaded to the Arduino development board. This can be done by accessing the menu `Sketch` and selecting the `Upload`, as illustrated in Figure 6.6. A completion bar will be presented to the user. Since the components are not connected to the board, the code can be executed but nothing will happen.

At this point, the board can be disconnected from the computer in order to safely connect the components to the Arduino board. Figure 6.7 shows how the different components are connected on the breadboard using jumper wires.

With the Arduino board and the other components fully connected as shown in Figure 6.7, the user can connect an Ethernet cable to the ENC28J60 module and to the router. After connecting the Ethernet cable, both MQTT broker and MQTT subscriber can be executed using default settings in order to test the Arduino application.

6.4.1.4 Executing Mosquitto and Testing the Application

For both OSes, it is suggested that an instance of a Mosquitto broker be opened in one window and an instance of the Mosquitto subscriber in another. On Linux, both applications can be executed by opening two different terminal

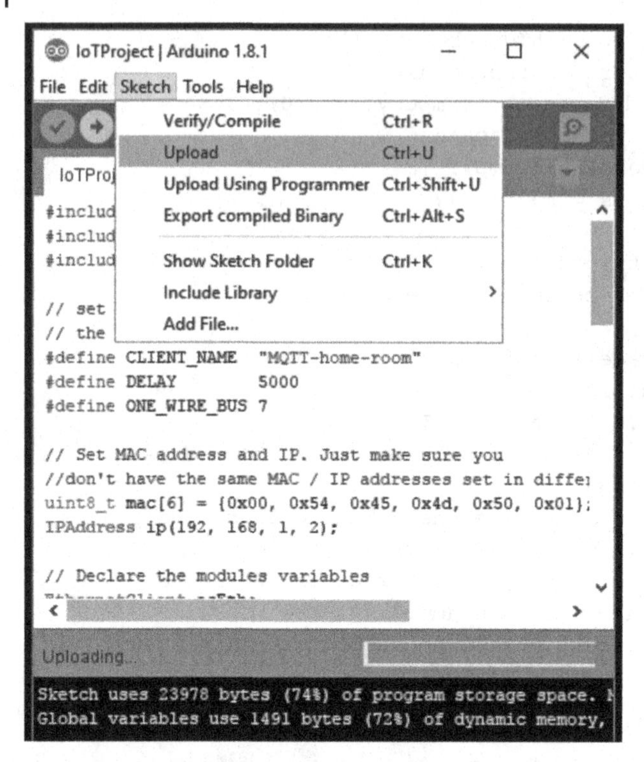

Figure 6.6 Uploading the code to the device.

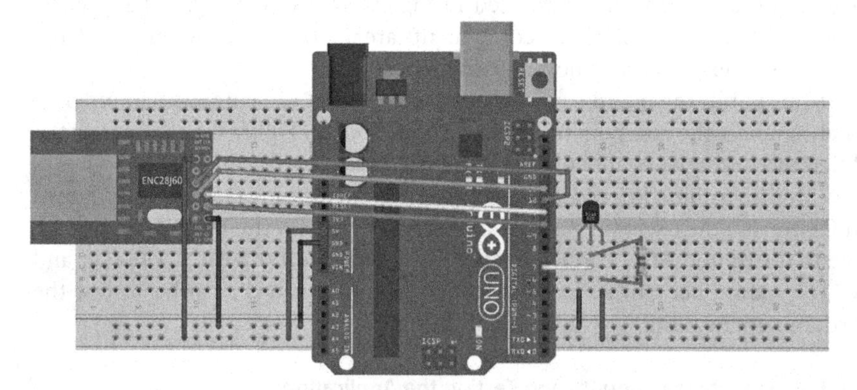

Figure 6.7 A breadboard diagram of the Arduino temperature logger project using the Fritzing software.

windows. The following commands can be executed in the same order, one in the first window, and the other in the second window:

```
$> sudo mosquito
$> sudo mosquito_sub -h localhost -p 1883 -t "/home/room/
temp"
```

For Windows, the user can use the `Run` functionality of the system by pressing CTRL + R, followed by typing the absolute path to the Mosquitto broker, which in this example will be assumed to be `C:/Program Files (x86)/mosquitto/mosquitto.exe`. The Mosquitto subscriber application should be installed in the same path as the Mosquitto broker, which can be executed by opening the `Run` functionality and typing:

```
"C:/Program Files (x86)/mosquitto/mosquitto_sub.exe"
-h localhost -p 1883 -t "/home/room/temp"
```

The flag `-t` is used to specify which MQTT topic will be subscribed. Note that both the Mosquitto broker and the Mosquitto subscriber will not present any messages until the Arduino application publishes them to the broker. With the computer prepared to receive data from Arduino, one is able to power it on and see the incoming messages on the Mosquitto subscriber. Figure 6.8 shows the Mosquitto broker running on a Linux computer (accessed via SSH from the Windows OS).

The Mosquitto broker presents log messages on which the subscriber and the publisher connections are announced, as seen in Figure 6.8. While the log is presented in the Mosquitto broker, messages are starting to be shown in the Mosquitto subscriber, as can be seen in the screenshot of PuTTY (a free telnet, SSH server, and serial ports terminal software that allows users to remotely access devices over the Internet) in Figure 6.9.

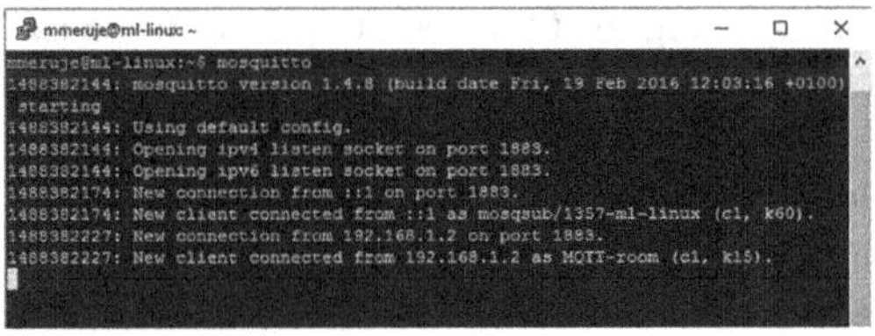

Figure 6.8 Mosquitto broker running on a Linux computer (accessed via SSH on Windows).

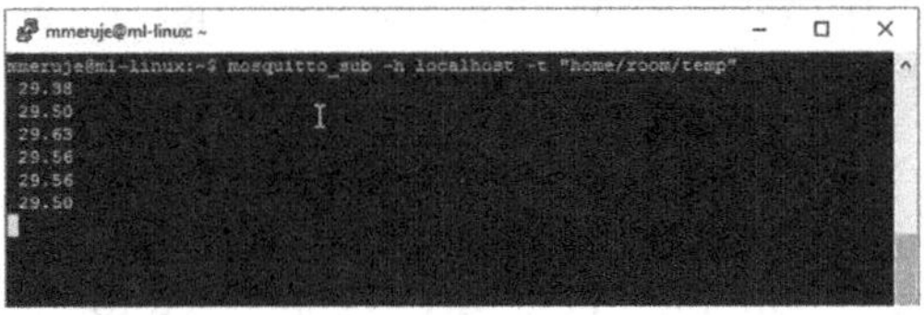

Figure 6.9 Mosquitto subscriber running on a Linux computer (accessed via SSH on Windows).

This is a simple IoT application that works as a temperature publisher within an MQTT environment. Another interesting application that could be developed to complement the one described herein is an MQTT subscriber that can be configured to send an alarm email when the temperature has risen above a certain threshold.

6.4.2 A Raspberry Pi Project for IoT Application

Like the older versions, the newer versions of the Raspberry Pi SBCs also include the 40-pin GPIO headers or a provision to solder them. This means that the project described here can be replicated on all Raspberry Pi versions available in the market. The project is built on the Raspberry Pi Model B Rev. 2, and the Raspbian OS is used (introduced in Section 6.2.2.3). The OS comes preinstalled with some of the necessary software for the project, such as the Python interpreter and the GPIO manipulation tool.

6.4.2.1 Installing the Operating System

For the Linux user, the Raspbian image can be copied to a clean micro SD card by using the `dd` tool or any appropriate tool. Before writing the image into the memory card, one has to connect it to a computer, and identify the device letter and make sure that it is unmounted. After unmounting the volume, the following command will allow the user to write the OS image into the memory card, where * is the identifier letter of the micro SD card:

```
$> dd if=~/2017-01-11-raspbian-jessie.img of=/dev/sd* bs=4M
```

For the Windows user, the Raspbian image can be copied to the micro SD card by using the `Win32DiskImager` utility that is available on the Sourceforge Project web page. Before writing the image into the memory card, one has to run the utility and choose the image file by using the file selector option. When the file is selected, the user can select the `Write` option, and the image will be written on the micro SD card.

After writing the image on the micro SD card, the user can insert the card into the micro SD card slot on the Raspberry Pi device. The user can now connect all the peripherals needed to use the device (mouse, keyboard, Ethernet cable, and monitor). Finally, the device can be turned on by plugging the USB cable to the power. In this example, Ethernet will be used for connection purposes. After booting up, Raspberry Pi will run a Bash shell script asking the user to login. The default login of the Raspbian distribution is the username `Pi` and password `raspberry`.

6.4.2.2 Downloading and Installing the Required Packages

The first step after login is to make sure that the device is connected to the Internet by pinging a website. The ping command is used to test connectivity between two devices by measuring the time between a request packet and the corresponding response packet. The following command will make the device send exactly four ping packets (`-c 4`) to one of the Google servers:

```
$> ping www.google.com -c 4
```

Although some software is preinstalled, some additional Linux packages are needed for the project, such as the *wiringPi* (`wiringpi`) package, the *Python Package Manager* (`python-pip`), and the header files for Python (`python-dev`):

```
$> sudo apt-get install python-dev python-pip wiringpi
```

After the *Python Package Manager* is successfully installed, one will need two Python modules for this project. The first is the *wiringPi* module, which allows the user to manipulate the GPIO pins through Python. The second one is Flask, a micro web-development framework for Python. Both modules can be installed using the following command:

```
$> sudo pip install flask wiringPi
```

Since all the dependencies are installed in the system, one should be able to start the project development. Before starting to code, it is suggested to run the command `gpio readall`. This will give the user an overview of the states of GPIO pins, as well as the pin numbering, which will be displayed in a table on the screen. Different numbering exists due to varying implementations of GPIO libraries. The command is as follows:

```
pi@raspberrypi:~ $ gpio readall

+-----+-----+---------+------+---+-Model B2-+---+------+---------+-----+-----+
| BCM | wPi |   Name  | Mode | V | Physical | V | Mode | Name    | wPi | BCM |
+-----+-----+---------+------+---+----++----+---+------+---------+-----+-----+
|     |     |    3.3v |      |   |  1 || 2  |   |      | 5v      |     |     |
|   2 |   8 |   SDA.1 |   IN | 1 |  3 || 4  |   |      | 5V      |     |     |
|   3 |   9 |   SCL.1 |   IN | 1 |  5 || 6  |   |      | 0v      |     |     |
|   4 |   7 | GPIO. 7 |   IN | 1 |  7 || 8  | 1 | ALT0 | TxD     | 15  | 14  |
|     |     |      0v |      |   |  9 || 10 | 1 | ALT0 | RxD     | 16  | 15  |
|  17 |   0 | GPIO. 0 |   IN | 0 | 11 || 12 | 0 | IN   | GPIO. 1 | 1   | 18  |
|  27 |   2 | GPIO. 2 |   IN | 0 | 13 || 14 |   |      | 0v      |     |     |
|  22 |   3 | GPIO. 3 |   IN | 0 | 15 || 16 | 0 | IN   | GPIO. 4 | 4   | 23  |
|     |     |    3.3v |      |   | 17 || 18 | 0 | IN   | GPIO. 5 | 5   | 24  |
|  10 |  12 |    MOSI |   IN | 0 | 19 || 20 |   |      | 0v      |     |     |
|   9 |  13 |    MISO |   IN | 0 | 21 || 22 | 0 | IN   | GPIO. 6 | 6   | 25  |
|  11 |  14 |    SCLK |   IN | 0 | 23 || 24 | 1 | IN   | CE0     | 10  | 8   |
|     |     |      0v |      |   | 25 || 26 | 1 | IN   | CE1     | 11  | 7   |
+-----+-----+---------+------+---+----++----+---+------+---------+-----+-----+
|  28 |  17 | GPIO.17 |   IN | 0 | 51 || 52 | 0 | IN   | GPIO.18 | 18  | 29  |
|  30 |  19 | GPIO.19 |   IN | 0 | 53 || 54 | 0 | IN   | GPIO.20 | 20  | 31  |
+-----+-----+---------+------+---+----++----+---+------+---------+-----+-----+
| BCM | wPi |   Name  | Mode | V | Physical | V | Mode | Name    | wPi | BCM |
+-----+-----+---------+------+---+-Model B2-+---+------+---------+-----+-----+
```

Listing 2: Output of the *gpio readall* command.

Listing 2 shows the different numbering, including the physical, the Broadcom (BCM), and the wiringPi, as well as the respective modes (e.g., In, Out, . . .) and states. The `gpio` command allows users to control the GPIO pins on the device (i.e., setting the mode, changing their values, etc.). The user can now construct the electronic circuit on the breadboard. At this point, it is recommended to shut down the device by typing the following command in the terminal:

```
$> sudo shutdown now
```

6.4.2.3 Constructing and Testing the Circuit

When the device is turned off, the LED and the 4.7 kΩ resistor can be connected on the breadboard using two jumper wires, as shown in the breadboard diagram of Figure 6.10. The positive pin (pole) of the LED is usually longer than the negative one. Thus, it must be ensured that the positive pin of the LED is connected to the GPIO 15 pin of the GPIO header on the Raspberry Pi, while the negative pin of the LED is connected to any of the resistor pins. The other pin of the resistor should be connected to the GND (Ground) pin of the Raspberry Pi GPIO header.

After connecting the components as shown in Figure 6.10, the Raspberry Pi can be powered on. When the device asks for the login input, the Raspbian OS default username and password can be entered. Right after the login step, the manipulation of the GPIO pins can be done via `gpio` command. The pin

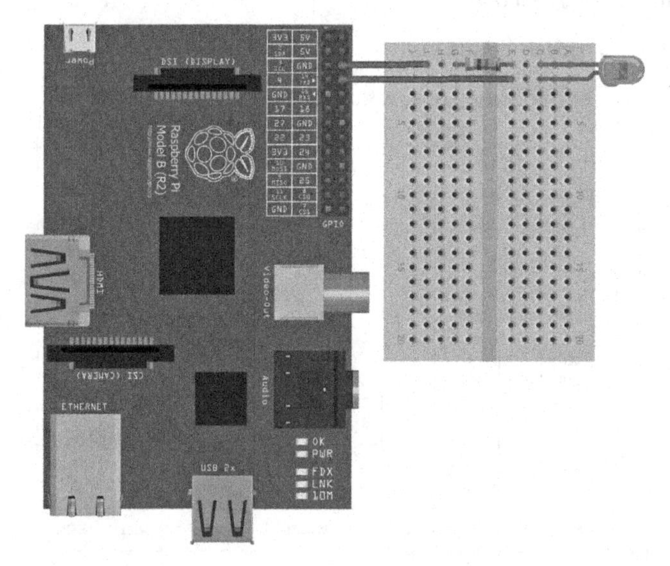

Figure 6.10 A breadboard diagram of the Raspberry Pi web-controlled LED project using the Fritzing software.

(GPIO 15) that is being used needs to be configured to output mode before its state can be manipulated, which can be done by using the following command:

```
$> sudo gpio mode 15 out
```

After the pin is correctly configured to output mode, it can have two distinct states, HIGH and LOW, which are represented by 1 and 0, respectively. The GPIO pins will send 3.3 V when they are in HIGH state (with the exception of the Ground and 5 V pins, as can be observed on the Listing 2) and 0 V when the GPIO pins are in LOW state. The following command allows the change of state of a specified GPIO pin to HIGH:

```
$> sudo gpio write 15 1
```

As soon as this command is executed in the device, one can notice the LED turning on. Alternatively, if the number 1 in the command is switched to 0, the pin will be set to LOW and the LED will turn off. Two Bash script files will be used to turn the LED on and off. The following commands will create two Bash script files named `turnOn.sh` and `turnOff.sh`. Each of these files will allow one to turn on and off the LED using the previous tested commands:

```
$> echo -e '#!\\bin\\bash\nsudo gpio mode 15 out\nsudo
gpio write 1' > turnOn.sh
$> cp turnOn.sh turnOff.sh && truncate -s-2 turnOff.sh
$> echo "0" >> turnOff.sh && chmod +x turn*.sh
```

These script files will allow the user to manipulate the GPIO pin via Bash shell, and they can be executed by simply typing `./turnOn.sh` or `./turnOff.sh`. As such, control can only be made when the user has direct access to the Raspberry Pi (considering SSH and Telnet connections are disabled).

6.4.2.4 Development and Testing of a Python Web Application

The next step consists of the development of a small web application that will act as a switch, allowing the user to control the state of the LED through a simple browser. The following command can be used to create the file that will host the web application script and the file permission change:

```
$> touch ledServer.py && chmod +x ledServer.py
```

Listing 3 includes the code one needs to write in the file, as well as the explanation of each line of code in inline comments. The file can be edited and saved by using the preferred text editor of the user (e.g., nano, vi, or gedit).

```
#!/usr/bin/python
import wiringpi # Import the wiringPi module for Python
from flask import Flask # Import the Flask
app = Flask(__name__)
```

```
pin, state = 15, 0
@app.route("/switch") # When someone access the http server endpoint
def switchLED():      # the switchLED() function is executed.
    global state      # It starts by using the global variable state,
    if state == 0 :   # and if the variable is LOW (0)
        state = 1     # change it to HIGH (1)
    else:             # otherwise
        state = 0     # change it to LOW
    wiringpi.digitalWrite(pin, state)
 # Write the state to the specified pin
    return "{'pin':"+str(pin)+", 'state':"+str
(state)+"}" # Return a response with the pin and state
if __name__ == "__main__":
    wiringpi.wiringPiSetup() # Initialization of the module
    wiringpi.pinMode(pin, 1) # Input = 0; output = 1;
    app.run(host='0.0.0.0', port=8080)
 # Run the application on all interfaces using port 8080
# Giving the variable host the value 0.0.0.0, will allow
to run the application in all interfaces the device has available
```

Listing 3: Python code for a Raspberry Pi IoT application example of an online LED switch.

Before running the server in the device, it is necessary to know the IP address of the Ethernet interface (`eth0`) of the Raspberry Pi so that when the server is running, one can access it via the browser. The IP address can be known by running the command `ipconfig`. When the script is prepared, the user can execute it, enabling the control of the LED via the browser by executing the following command:

```
$> ./ledServer.py
```

Now, with another device connected to the same network, one may open a browser and go to the server by using the IP address of the Raspberry Pi. Since the server was configured to use the port 8080 of the device, one may go to `http://<ip-address>:8080/switch` and verify if the LED is turning on or off according to the HIGH or LOW states.

6.5 Conclusion

IoT is fast becoming ubiquitous in everyday life and giving rise to a myriad of applications. Arguably, the advancements and convergence of hardware and software in a single package, also known as development kit, will undoubtedly remain the central backbone providing a major boost and support for IoT deployments.

This chapter has presented a basic, yet comprehensive introduction to IoT design and prototyping. It has described in detail the main features of different IoT hardware development platforms, focusing on a number of Arduino boards and some models of the Raspberry Pi family. The chapter has also discussed design and prototyping in the context of IoT, where some basic examples of IoT design and prototyping that showed step-by-step procedures for IoT project implementations in both software and hardware have been presented. These projects are meant to teach beginners some basic principles of design and prototyping as well as to show how easy it is to get up and running with IoT design and implementation.

Acknowledgments

The authors wish to thank the Centre for Geodesy and Geodynamics, National Space Research and Development Agency, Toro, Bauchi State, Nigeria for supporting this work. This work was supported by National Funding from the FCT—Fundação para a Ciência e a Tecnologia, through the UID/EEA/50008/2013 Project.

References

Aditya, A. L. G. N., Chowdary, G. R., and Meenakshi, J. (2012) Delay and power optimized register blocks for the low power microcontrollers. IEEE International Conference on Devices, Circuits and Systems (ICDCS), 408–412. doi: 10.1109/ICDCSyst.2012.6188793.

Alaraje, N., Sergeyev, A., Reutter, J., Kief, C., Matar, B., and Hata, D. (2016) Technology driven university and community college collaboration: faculty training and ARM microcontrollers. IEEE Portland International Conference on Management of Engineering and Technology (PICMET), 1681–1687. doi: 10.1109/PICMET.2016.7806819.

Al-Fuqaha, A., Guizani, M., Mohammadi, M., Aledhari, M., and Ayyash, M. (2015) Internet of Things: a survey on enabling technologies, protocols, and applications. *IEEE Communications Surveys & Tutorials*, **17**(4), 2347–2376.

Astudillo-Salinas, F., Barrera-Salamea, D., Vázquez-Rodas, A., and Solano-Quinde, L. (2016) Minimizing the power consumption in Raspberry Pi to use as a remote WSN gateway. IEEE 8th Latin-American Conference on Communications (LATINCOM), 1–5. doi: 10.1109/LATINCOM.2016.7811590.

Banks, A. Gupta, R. (eds), (2014) MQTT Version 3.1.1. OASIS Standard. Available at http://docs.oasis-open.org/mqtt/mqtt/v3.1.1/os/mqtt-v3.1.1-os.html. Latest version available at http://docs.oasis-open.org/mqtt/mqtt/v3.1.1/mqtt-v3.1.1.html.

Bekaroo, G. and Santokhee, A. (2016) Power consumption of the Raspberry Pi: a comparative analysis. IEEE International Conference on Emerging Technologies and Innovative Business Practices for the Transformation of Societies (EmergiTech), 361–366. doi: 10.1109/EmergiTech.2016.7737367.

Bell, L. (2014) Raspberry Pi Model B+ Review. The Inquirer, Available at http://www.theinquirer.net/inquirer/review/2357623/raspberry-pi-model-b-review (accessed February 17, 2017).

Biglesp, (2015) Raspberry 2: Which OS is Best? element14 Community. Available at https://www.element14.com/community/polls/2103 (accessed February 19, 2017).

Bonetto, R., Bui, N., Lakkundi, V., Olivereau, A., Serbanati, A., and Rossi, M. (2012) Secure communication for smart IoT objects: protocol stacks, use cases and practical examples. IEEE International Symposium on a World of Wireless, Mobile and Multimedia Networks (WoWMoM), 1–7. doi: 10.1109/WoWMoM.2012.6263790.

Brown, E. (2015) Raspberry Pi 2 has quad-core SoC, Keeps $35 Price. LinuxGizmos. Available at http://linuxgizmos.com/raspberry-pi-gets-quad-core-soc-keeps-same-price/ (accessed February 18, 2017).

Chen, D., Cong, J., Gurumani, S., Hwu, W.-m., Rupnow, K., and Zhang, Z. (2016) Platform choices and design demands for IoT platforms: cost, power, and performance tradeoffs. *IET Cyber-Physical Systems: Theory Applications*, **1**, 70–77. doi: 10.1049/iet-cps.2016.0020.

Cvjetkovic, V. and Matijevic, M. (2016) Overview of architectures with Arduino boards as building blocks for data acquisition and control systems. 13th IEEE International Conference on Remote Engineering and Virtual Instrumentation (REV), 56–63. doi: 10.1109/REV.2016.7444440.

DeBenedictis, E.P. (2017) It's time to redefine Moore's law again. *IEEE Computer Journal*, **50**, 72–75. doi: 10.1109/MC.2017.34.

Earl, B. (2013) Adafruit Arduino Libraries. Adafruit Industries, Available at https://learn.adafruit.com/adafruit-all-about-arduino-libraries-install-use/arduino-libraries (accessed February 28, 2017).

Eclipse (2009) Mosquitto: an open-source MQTT broker. The Eclipse Foundation. Available at https://mosquitto.org (accessed February 28, 2017).

Ghasempour, A. (2016) Optimum number of aggregators based on power consumption, cost, and network lifetime in advanced metering infrastructure architecture for smart grid Internet of Things. 13th IEEE Annual Consumer Communications & Networking Conference (CCNC), 295–296. doi: 10.1109/CCNC.2016.7444787.

Haase, J., Alahmad, M., Nishi, H., Ploennigs, J., and Tsang, K.F. (2016) The IOT mediated built environment: a brief survey. IEEE 14th International Conference on Industrial Informatics (INDIN), 1065–1068. doi: 10.1109/INDIN.2016.7819322.

Hahm, O., Baccelli, E., Petersen, H., and Tsiftes, N. (2016) Operating systems for low-end devices in the Internet of Things: a survey. *IEEE Internet of Things Journal*, **3**, 720–734. doi: 10.1109/JIOT.2015.2505901.

James, D. (2016) Moore's Law continues into the 1x-nm era. IEEE 27th Annual SEMI Advanced Semiconductor Manufacturing Conference (ASMC), 324–329, doi: 10.1109/ASMC.2016.7491159.

Javed, A., Larijani, H., Ahmadinia, A., and Gibson, D. (2016) Smart random neural network controller for HVAC using cloud computing technology. *IEEE Transactions on Industrial Informatics*, doi: 10.1109/TII.2016.2597746 (accessed February 3, 2017).

Johnston, S. J., Apetroaie-Cristea, M., Scott, M., and Cox, S. J. (2016) Applicability of commodity, low cost, single board computers for Internet of Things devices. IEEE 3rd World Forum on Internet of Things (WF-IoT), 141–146. doi: 10.1109/WF-IoT.2016.7845414.

Keeler, W. J. and Wolfer, J. (2016) A Raspberry PI Cluster and Geiger Counter supporting random number acquisition in the CS operating systems class. IEEE 13th International Conference on Remote Engineering and Virtual Instrumentation (REV), 353–354. doi: 10.1109/REV.2016.7444500.

Klosowski, T. (2016) Best operating systems for your Raspberry Pi projects. lifehacker. Available at http://lifehacker.com/the-best-operating-systems-for-your-raspberry-pi-projec-1774669829 (accessed February 19, 2017).

Klosowski, T. (2017) How much power the Raspberry Pi Zero W uses compared to other models. lifehacker. Available at http://lifehacker.com/how-much-power-the-raspberry-pi-zero-w-uses-compared-to-1792854782 (accessed March 5, 2017).

Knörig, A., Wettach, R., and Cohen, J. (2009) Fritzing: a tool for advancing electronic prototyping for designers. Proceedings of the 3rd International Conference on Tangible and Embedded Interaction (TEI '09). ACM, New York, NY, USA, pp. 351–358.

LDighera, (2015) Shell Script to Display Temperature and Voltage Sensor Output. Raspberry Pi, Available at https://www.raspberrypi.org/forums/viewtopic.php?t=118641&p=805362 (accessed February 22, 2017).

Lentine, A. L. and DeRose, C. T. (2016) Challenges for optical interconnect for beyond Moore's Law computing. IEEE International Conference on Rebooting Computing (ICRC), 1–5, doi: 10.1109/ICRC.2016.7738696.

Lyons, C. (2015) A History of the Raspberry Pi Nova, Available at http://novadigitalmedia.com/history-raspberry-pi/ (accessed February 16, 2017).

Maksimović, M., Vujović, V., Davidović, N., Milošević, V., and Perišić, B. (2014) Raspberry Pi as an Internet of Things hardware: performance and constraints. Proceedings of 1st International Conference on Electrical, Electronic and Computing Engineering IcETRAN, 1–6.

McEwen, A. and Cassimally, H. (2014) *Designing the Internet of Things*, John Wiley & Sons, Ltd, Chichester, 1.

Miller, S. (2017) Enhancing Raspberry Pi security. Zymbit, Available at https://zymbit.com/securing-your-iot-devices-2/ (accessed February 22).

Musa, A., Minotani, T., Matsunaga, K., Kondo, T., and Morimura, H. (2015) An 8-mode reconfigurable sensor-independent readout circuit for trillion sensors era. IEEE Tenth International Conference on Intelligent Sensors, Sensor Networks and Information Processing (ISSNIP), 1–6. doi: 10.1109/ISSNIP.2015.7106913.

Nayyar, A. and Puri, V. (2016) A review of Arduino board's, Lilypad's & Arduino shields. 2016 3rd International Conference on Computing for Sustainable Global Development (INDIACom), New Delhi, pp. 1485–1492.

Nishimura, K. and Sugita, K. (2015) Evaluation of power consumption for smart devices in web applications. 9th IEEE International Conference on Innovative Mobile and Internet Services in Ubiquitous Computing, 47–52. doi: 10.1109/IMIS.2015.12.

Patil, P. S., Doshi, J., and Ambawade, D. (2015) Reducing power consumption of smart device by proper management of wakelocks. IEEE International Advance Computing Conference (IACC), 883–887. doi: 10.1109/IADCC.2015.7154832.

Patil, S. S., Katwe, S., Mudengudi, U., Shettar, R. B., and Kumar, P. (2016) Open ended approach to empirical learning of IOT with Raspberry Pi in modeling and simulation lab. IEEE 8th International Conference on Technology for Education, 258–259. doi: 10.1109/T4E.2016.67.

Postscapes, IoT Development Kits, (2017) Available at http://www.postscapes.com/cellular-internet-of-things-development-kits/ (accessed February 3, 2017).

Poulter, A. J., Johnston, S. J., and Cox, S. J. (2015) Using the MEAN stack to implement a RESTful service for an Internet of Things application. IEEE 2nd World Forum on Internet of Things (WF-IoT), 280–285. doi: 10.1109/WF-IoT.2015.7389066.

RasPi.TV, (2017) How much power does Pi Zero W use? Available at http://raspi.tv/2017/how-much-power-does-pi-zero-w-use (accessed March 5, 2017).

Richards, M. (2017) Young persons guide to BCPL programming on the Raspberry Pi: part 1. Computer Laboratory University of Cambridge. Available at http://www.cl.cam.ac.uk/~mr10/bcpl4raspi.pdf (accessed February 19, 2017).

Richardson, M. and Wallace, S. (2012) *Getting Started with Raspberry Pi*, O'Reilly Media, Sebastopol, pp. 33–57.

Ruiz-de-Garibay, J., Almeida, A., Kados, S. A., García-Corcuera, A., and López-de-Ipiña, D. (2014) Codesign-oriented platform for agile internet of things prototype development. IEEE Eighth International Conference on Innovative Mobile and Internet Services in Ubiquitous Computing, 387–392. doi: 10.1109/IMIS.2014.52.

Save, Y. D., Rakhi, R., Shambhulingayya, N. D., Srivastava, A., Das, M. R., Choudhary, S., and Moudgalya, K. M. (2013) Oscad: An open source EDA tool for circuit design, simulation, analysis and PCB design. 2013 IEEE 20th International Conference on Electronics, Circuits, and Systems (ICECS), Abu Dhabi, pp. 851–854.

Shabaz, (2016) Raspberry Pi 3 dynamic current consumption, power and temperature tests. element14 Community, Available at https://www.element14.com/community/community/raspberry-pi/blog/2016/05/11/raspberry-pi-3-dynamic-current-consumption-power-and-temperature-tests (accessed February 18, 2017).

Singh, H.V.P., Rizvi, S. R., and Mahmoud, Q. H. (2015) Two architectures for real-time sensor data streaming for cloud applications. IEEE International Symposium on Signal Processing and Information Technology (ISSPIT), 133–138. doi: 10.1109/ISSPIT.2015.7394315.

Suehle, R. and Callaway, T. (2014) *Raspberry Pi Hacks: Tips & Tools for Making Things with Inexpensive Linux Computer*. O'Reilly Media, Sebastopol, 8.

Windows Dev Center. (2017) Windows 10 IoT Core: The operating system built for the Internet of Things. Microsoft. Available at https://developer.microsoft.com/en-us/windows/iot (accessed February 19, 2017).

7

On Standardizing the Internet of Things and Its Applications

Kai Jakobs

Computer Science Department, RWTH Aachen University, Aachen, Germany

7.1 Introduction

"Standards are not only a technical question. They determine the technology that will implement the Information Society, and consequently the way in which industry, users, consumers, and administrations will benefit from it." (EC, 1996). This quote conveys two important insights that are overlooked all too often. The first one is that Information and Communication Technology (ICT) systems simply would not work without underlying standards. The second one is that today's ICT standards are tomorrow's technology. That is, those who lead the standardization initiatives today are likely to be in the driving seat when it comes to the actual technology development and implementation.

Scores of standards are implemented in every ICT system, from the most complex international infrastructure down to the humble PC on the desk back home. There are standards for operating systems, programming languages, user interfaces, communication protocols, disk drives, cables and connectors, and so on. Biddle et al. (2010) found that at the very least 251 interoperability standards are implemented in a laptop computer; they reckon that the total number of standards relevant to such a device is much higher.

These days the economic importance of standards is no longer questioned. Swann (2010) provides a very thorough review of the relevant literature. The reported findings include, among many others, that standards contribute at least as much as patents to economic growth (DIN, 2000). Studies from different parts of the world show that the contribution of standards to the growth rate of the gross domestic product (GDP) is about 0.9% in Germany, 0.8% in Australia, 0.3% in the United Kingdom, and 0.2% in Canada. In absolute terms, this means that the economic benefit of the current body of standards for, for example, Germany amounts to almost €17 billion a year (Blind et al., 2011). In addition, numerous

Internet of Things A to Z: Technologies and Applications, First Edition. Edited by Qusay F. Hassan.
© 2018 by The Institute of Electrical and Electronics Engineers, Inc. Published 2018 by John Wiley & Sons, Inc.

case studies exist that highlight the economic benefits of standards for both nation-states and firms[1].

Specifically, standards frequently (albeit not necessarily) have a positive impact on innovation. For example, Blind (2009) argues that there are several ways in which standards can promote innovation. This holds particularly for the ICT sector, where compatibility standards are the one major basis for innovations. Indeed, the GSM (Global System for Mobile Communications) platform standards, for example, have been the basis for the numerous mobile services we are offered today.

Finally, standards are also of considerable interest to policy makers. In Europe, for instance, harmonized standards contributed a great deal to the creation of the European single market—they help to remove technical barriers to trade and to enable people, services, goods, and capital to move more freely.

The above holds for standards in general and for ICT and the Internet of Things (IoT) in particular. However, the standardization of the IoT offers some additional challenges. By definition, the IoT is based on standard communication protocols to interconnect uniquely addressable objects. However, "object" is a very broad term; it may be a sensor or a mainframe computer. The most prominent characteristic of the majority of IoT nodes will be "power constrained." The communication infrastructure needs to take this restriction into account through, for example, appropriate modifications of existing protocols and/or through dedicated new ones. In their current form, most standards are too complex for the constrained devices in the IoT. To make things worse, many of these devices run proprietary protocols, thus creating isolated data silos. This increased variety also implies that interoperability will be even harder to achieve than in other areas of the ICT sector. Indeed, Jari Arkko, the Chair of the Internet Engineering Task Force (IETF), cannot think of a better example of where interoperability is important than the Internet of Things[2]. Interoperability standards are a *sine qua non* for both the IoT and the various "smart" applications that will be based on it.

In addition, standardizing the IoT will frequently require multidisciplinary cooperation, at least between the following:

- *Electrical Engineers*, who build the—frequently power constrained—nodes and specify associated requirements on, for example, the operating system and the communication software and protocols.
- *Computer Scientists/Telecommunication Engineers*, who design, for instance, the communication protocols.

---

1 A collection of such studies may be found at http://www.iso.org/iso/home/standards/benefitsofstandards/benefits_repository.htm?type=EBS-MS (accessed October 31, 2016).

2 https://www.ietf.org/blog/2016/01/an-interoperable-internet-of-things/ (accessed December 2, 2016).

- *Software Engineers*, who develop, for example, the operating system and take care of its integration with the communication software.

This need for multidisciplinary cooperation holds even more for cyber-physical systems[3] (CPSs) and, particularly, for the various IoT application areas (see, for example, Ho and O'Sullivan (2015) and Section 7.4).

This chapter looks at the development over the past 20 years of the standardization entities for the IoT and for Intelligent Transport Systems (ITS), Smart Manufacturing (SM), the Smart Grid (SG), and Smart Cities (SC) and at the links between them. In doing so, it aims at taking a glimpse into the future of standardization in these sectors, informed by developments of the past. This represents a first step toward an answer to the question: How should the standardization environment for (e)merging applications look like in the future? This, in turn, should help industry, policy makers, and Standards Setting Organizations[4] (SSOs) to optimally position themselves when dedicated standardization activities for (e)merging applications will eventually truly get off the ground.

7.2 Current Status

A number of SSOs have become active in IoT standardization and have already developed a considerable number of standards. These organizations mostly develop what not just Sherif (2001) calls "anticipatory standards"; they are typically specified at the introduction of a technology and are crucial for interoperable communication systems. Anticipatory standards stand in contrast to "participatory" and "responsive" standards. The former proceed in parallel with implementations, thus enabling testing of the specifications prior to their adoption. The latter basically rubber-stamp existing successful specifications.

According to the International Organization for Standardization (ISO), there are currently (late 2016) over 900 IoT-related standards. Of those, around 140 come from the Institute of Electrical and Electronics Engineers (IEEE), 200 from the International Telecommunication Union (ITU), and 300 from the joint committee for ICT standardization of the ISO and the International Electrotechnical Committee (IEC; ISO/IEC JTC1). Although most of these are rather

---

3 CPSs are closely linked to, and extend, the idea of the Internet of Things (IoT). CPSs are a new generation of systems with integrated computational and physical capabilities that can interact with humans. The ability to interact with, and expand the capabilities of, the physical world through computation, communication, and control is a key enabler for future technology developments (Baheti and Gill, 2011).

4 This term is used to denote both, "formal" Standards Developing Organisations (SDOs), such as the International Organization for Standardization (ISO) and the International Telecommunication Union (ITU), and private standards consortia like the World Wide Consortium (W3C).

more generic standards in the field of wireless communication systems that were not necessarily developed specifically for the IoT they may well be deployed by it as well.

Cybersecurity and privacy are other important fields of standards setting in which a vast array of SSOs are active (including ITU, ISO, IEC, CEN, ETSI, W3C, OASIS, and the IETF). Again, these standards are not necessarily unique to the IoT.

7.2.1 IoT Standardization

Specific IoT-related standards for the communication infrastructure have mostly been developed for the field of (power-)constrained devices. Relevant activities in this field are ongoing in, for example, oneM2M, where standards for a common M2M (Machine-to-Machine) Service Layer are being developed. Within the European Telecommunications Standards Institute (ETSI), the "Smart M2M Communications" committee works on the interface between the service layer and the application layer. Other ETSI Technical Committees (TCs) as well as groups in IEEE, ITU, and several other SSOs work on wireless applications. Within the IETF, several Working Groups (WGs) focus on constrained devices.

The ITU has identified a "List of Internet of things (IoT) relevant organizations and forums."[5] Updating and adapting this list to include only organizations and entities that develop native IoT standards (as opposed to those that develop more generic standards that may also be deployed by the IoT) yields the list of SSOs shown in Table 7.1; without any claim for completeness.

The links that exist between these (and other) SSOs will be discussed in Section 7.3.

The number of entities that are devoted to the standardization of a particular technology may be seen as an indicator of the increasing perceived (market) relevance of this technology. For the IoT, this number has skyrocketed over the past 8 years (see Section 7.5). These entities focus primarily on the two upper layers depicted in Figure 7.1.

The different application areas will be discussed in Section 7.2.2. The IoT layer provides the application layer with IoT-specific functions (e.g., data management, security, privacy). The standardization focus here is still largely on the architectural level. Existing communication standards may be reused for the deployment in the network layer. IoT-specific efforts in this layer primarily address problems related to the power constrainedness of many IoT devices (e.g., sensors). Dedicated requirements will emerge in some application areas and will require new or adapted communication protocols ((Fettweis et al., 2014), think, for example, autonomous driving).

5 http://www.itu.int/en/ITU-T/jca/iot/Pages/sdo.aspx.

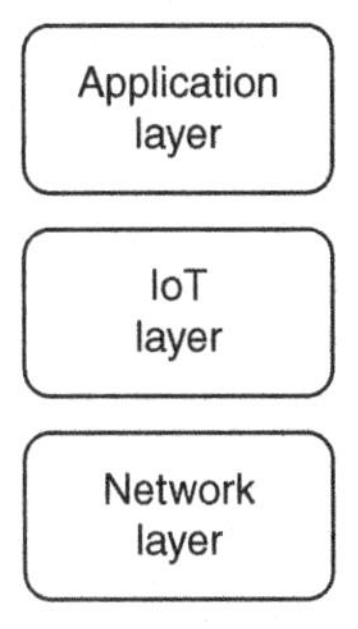

Figure 7.1 The generic three-tier IoT architecture. (Adapted from AIOTI (2015c).)

Table 7.1 Major SSOs developing dedicated IoT-specific standards.

CEN: European Committee for Standardization CEN provides a platform for the development of European Standards for all sectors excluding electrotechnology and telecommunications. • TC 225/WG 6: Internet of Things—Identification, data capture, and edge technologies.
ETSI: European Telecommunications Standards Institute ETSI is a regional (European) telecommunication SDO. It contributes to M2M (Machine-to-Machine) standardization through a dedicated Technical Committee (TC). • TC SmartM2M
IEEE The Institute of Electrical and Electronics Engineers is a professional association that is also active in standards setting. The two major IoT-related "projects" are • P2413: Standard for an Architectural Framework for the Internet of Things • P1451: Smart Transducer Interface Standards
IETF The IETF is the Internet's standardization body. It comprises a number of Working Groups (WGs). A WG covers a comparably narrow aspect and is dissolved once it has achieved its goal. • 6LoWPAN: IPv6 over Low power WPAN – dissolved • 6Lo: IPv6 over Networks of Resource-constrained Nodes • CoRE: Constrained RESTful Environments • ACE: Authentication and Authorization for Constrained Environments • DICE: DTLS In Constrained Environments – dissolved
Industrial Internet Consortium (IIC) The IIC aims to improve the integration of the physical and digital worlds. Strictly speaking, it is not an SSO; it defines common architectures to connect smart devices, machines, people, processes, and data. • "Architecture" Task Group
ISO/IEC JTC1 JTC1 is ISOs and IEC's "ICT-arm." It develops standards for the IoT primarily through two dedicated Working Groups. • WG 7: Sensor Networks

(*continued*)

Table 7.1 *(continued)*

- WG 10: Internet of Things
- Sub-Committee SC 41: Internet of Things and related technologies (to be established)

ITU-T

The ITU-T, the ITU's standardization arm, is subdivided into a number of Study Groups (SGs), each of which deals with a number of "questions," covering different technical aspects.

- SG 11: Protocols and test specifications
  - Question 12/11: Internet of things test specifications
  - Question 7/11: Signaling and control requirements and protocols for network attachment supporting multiscreen service, future networks, and M2M
  - Question 3/11: Signaling requirements and protocol for emergency telecommunications
- SG 13: Future networks (and cloud).
- SG 16: Multimedia
  - Question 27/16: Vehicle gateway platform for telecommunication/ITS services/applications.
- SG 20: IoT and its applications including smart cities and communities
- Internet of Things Global Standards Initiative (IoT-GSI) – *completed*
- Joint Coordination Activity on Internet of Things and Smart Cities & Communities (JJCA-IoT and SC&C)

oneM2M

oneM2M is an alliance of eight regional telecommunication SDOs that develops standards for Machine-to-Machine (M2M) communication and for the IoT.

Open Geospatial Consortium (OGC)

Through its Domain Working Groups, the OGC develops interfaces that enable real-time integration of heterogeneous sensors into the IoT information infrastructure.

- Sensor Web Enablement DWG

TIA: Telecommunications Industry Association

TIA is a regional (US) telecommunication SDO. It contributes to M2M standardization through a dedicated Engineering Committee.

- TR-50 M2M: Smart Device Communications

TSDSI: Telecommunications Standards Development Society

TSDSI is a regional (Indian) telecommunication SDO. It contributes to M2M standardization through a dedicated Working Group.

- Working Group on M2M

TTA: Telecommunications Technology Association

TTA is a regional (Korean) telecommunication SDO. It contributes to IoT standardization through a dedicated Special Technical Committee.

- IoT Special Technical Committee (STC1)

W3C

The World Wide Web Consortium develops most of the standards related to the WWW. Its WoT interest group is a fairly new player. Ultimately, it aims to bridge incompatible IoT platforms through the Web.

- Web of Things Interest Group

7.2.2 IoT-Based Applications

The merger of formerly separate technologies has been an ongoing trend over the past couple of years. The—almost completed—integration of (tele)communication and information technology led to ICT. More recent examples of (e)merging technologies include, among others, intelligent transport systems (ITS; comprising transport telematics, traffic engineering, power engineering, automotive, ICT), smart manufacturing (SM; production engineering, robotics, control engineering, ICT), smart grid (SG; power engineering, ICT). Smart cities are pretty much a superset of the other three, plus a number of others (see also Section 7.4.4).

One common characteristic of these technologies is the prominent role of ICT. In fact, the integration of ICT into "traditional" technologies (transportation, manufacturing, power distribution) is decisive; it enables the "smart" bit. And while compatibility and interoperability are important aspects in many technologies, they are the *sine qua non* for the IoT. This, in turn, implies that (compatibility/interoperability) standards play a pivotal role. Without them, smart technologies simply will not materialize.

A considerable number of (e)merging "smart" application areas may be identified. Table 7.2 summarizes the most prominent ones as identified in reports by the US National Institute of Standards and Technology (NIST) and two major European research and innovation initiatives, respectively. This is complemented by the findings of a survey from the academic literature. They all identify smart manufacturing, e-health, ITS, and smart (power) infrastructure, with other application areas identified by a subset of the reports and papers considered.

Each of these application areas is inherently multidisciplinary. Table 7.3 shows the most important disciplines involved in the five most frequently named areas. One of their common characteristics is the reliance on Telecommunication Engineering and Computer Science.

7.2.3 Security and Privacy

Security encompasses a set of services, including authentication, authorization, integrity, and confidentiality. In addition, privacy needs to be guaranteed and, ideally, mechanisms to support the development of a certain level of trust between parties should be provided.

In fact, a widely perceived lack of security and privacy may well be a potential showstopper for the IoT and its applications. It is very likely that people will resist them if there is no confidence that they will not cause serious threats to privacy (Atzori et al., 2010). Likewise, virtually all application areas have strong security requirements; accordingly, industry concerns also very much focus on this aspect (Li et al., 2015). These concerns as well need to be addressed by standardization.

Since such concerns are not unique to the IoT, a large number of protocols to ensure security and privacy already exist. Accordingly, one might be tempted to

Table 7.2 (E)merging application areas as identified in the literature.

(Gunes et al., 2014)	(NIST, 2013)	(NIST, 2015)	(AIOTI, 2015a)	(IoT-A, 2013)
Smart Manufacturing	Smart Manufacturing and Production	Manufacturing	Manufacturing/Industry Automation	Productive Business Environment
Health care	Health care	Health care	Health care	Health care
Intelligent Transport Systems	Transportation and Mobility	Transportation	Vehicular/Transportation	Smart Transport Logistics
Critical Infrastructure	Civil Infrastructure & Energy	Infrastructure (communication, power, water)	Energy	Smart Energy
Smart City		Cities	Cities	Smart Cities
	Buildings and Structures	Buildings	Home/Building	Smart Homes
Emergency Response	Emergency response	Emergency response		
			Living Environment	Ambient Assisted Living
		Environmental monitoring	Environment	
		Agriculture	Farming/Agri-food	
		Supply chain/retail		Retail
	Defense	Defense		
			Wearables	

Table 7.3 Disciplines involved in different application areas (excerpt).

Intelligent transport systems	Smart manufacturing	Smart grid	Health care	Smart buildings
Transport Telematics Traffic Engineering Power Engineering Automotive Computer Science Tele- communication Engineering	Production Engineering Tele- communication Engineering Computer Science Robotics Control Engineering	Power Engineering Computer Science Tele- communication Engineering	Medicine Tele- communication Engineering Computer Science Mechanical Engineering	Architecture Civil Engineering Computer Science Tele- communication Engineering

think there is no real need to design new, dedicated security protocols for the IoT (Keoh et al., 2014). However, the major issue that stands in the way of a one-to-one adoption of existing protocols is—again—the power constrainedness of embedded "smart" devices (like sensors and actuators), which will represent the vast majority of IoT nodes. As a consequence, additional efforts will need to go into the adaptation of existing protocols to the limited capabilities of these devices. This may well amount to the development of dedicated, IoT-specific security mechanisms.

Despite their crucial importance, security and privacy standardization will not be considered in this chapter. For one, their inclusion would go beyond its scope; thorough discussions of the associated technical aspects and of ongoing standardization activities are provided in, for example, (Keoh et al., 2014; Sicari et al., 2015; Li et al., 2016). Moreover, the socioeconomic and policy ramifications of the topic are actually too important to be hidden in a general chapter on IoT standardization.

7.3 The Standardization Environment

Most industry sectors have a rather simple standardization environment. A number of National Standards Organizations (NSOs) contribute to the work of ISO and IEC at the international level. An additional regional level in between has been established in Europe through the European Standards Organizations (ESOs).

The situation is different in the ICT sector (specifically in telecommunications). This sector is characterized by a number of national/regional bodies and, particularly, by a huge number (more than 200) of private standards setting consortia. The proliferation of these consortia began in the late 1980s and was primarily triggered by the fast-paced development of ICT technologies and a

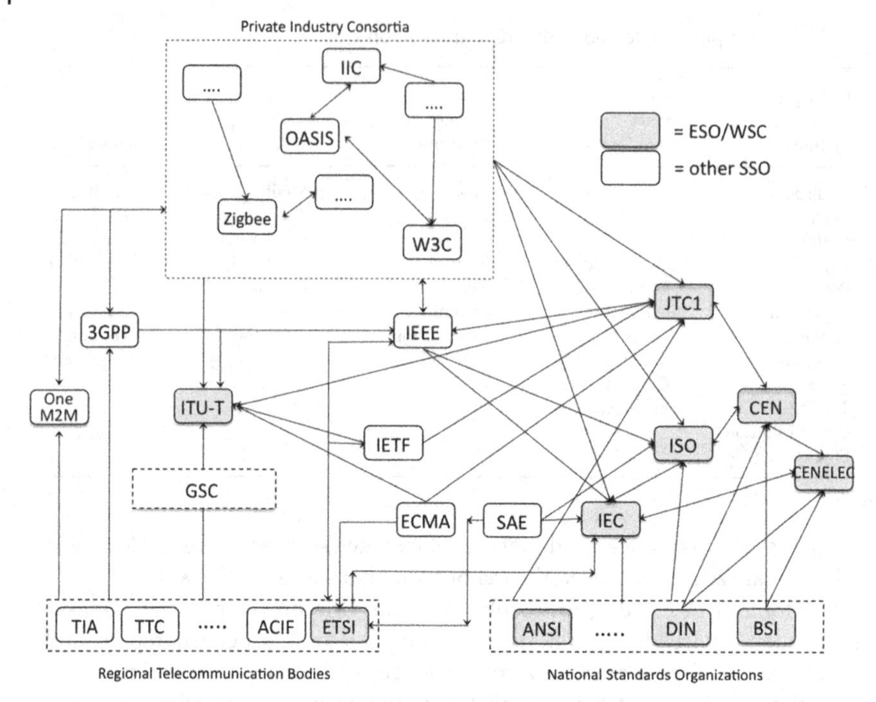

Figure 7.2 The web of relevant SSOs.[6] (Adapted from Jakobs (2017).)

widely perceived slowness (Lehr, 1992) and nonresponsiveness to users' needs on the side of the SDOs (Cargill, 1995).

The number of consortia and the complex links that exist between them and the SDOs yield an almost impenetrable web of SSOs. Figure 7.2 gives a rough idea of this complexity.

The figure shows the major SSOs that are active in the standardization of the IoT and/or its applications and some of the links that exist between them. Generally speaking, such a link represents some level of formal cooperation. Such cooperation may take the form of exchanging information about planned new work items, the joint development of common standards, or anything in between. In the absence of a central coordinating entity, these links currently represent the most important (distributed) coordination mechanism in standards setting (see also further).

The rectangle at the top shows some private standards consortia. The links between them are not normally particularly well developed, but some do exist.

6 It should be noted that Figure 7.4 by no means claims to be complete in the sense that all (potentially) relevant standardization entities and links between them are listed. In a dynamic environment like ICT/IoT standardization, any such attempt would be futile—the findings would be outdated by the time of publishing.

There are more links between consortia and "formal" SDOs; for instance, consortia may submit their specifications for an SDO for (international) standardization ("reactive standardization"; see Section 7.2.1). SSOs active in telecommunication standardization are shown at the bottom left. Their number is comparably limited, with relatively strong links between them. For instance, 3GPP and oneM2M are joint initiatives by several regional SSOs. Specifically, oneM2M develops standards for Machine-to-Machine (M2M) communication and for the IoT. The Global Standards Collaboration (GSC) is a less formal form of cooperation between regional SSOs and the ITU. The IT standardization landscape (bottom right) is the most densely populated and most complex part of the ICT standardization environment. Here as well, comparably strong links exist between individual players.

Coordination of the different standardization activities remains an important issue. A lack of coordination, eventually resulting in the development of functionally equivalent (and thus competing) standards, will reduce market transparency, decrease interoperability and ease of use, fragment the market, and increase transaction costs (Egyedi, 2014). Indeed, various coordination mechanisms have been developed over time, ranging from highly formalized high-level regulatory documents to very informal coordination through individuals who contribute to the work of multiple SSOs. Figure 7.3 shows some of the existing formal coordination mechanisms between some major global players.

In any case, all forms of coordination will be of particular importance in the complex world of IoT standardization.

7.4 Standardization in Selected Application Areas

This section will look at the environment of four typical IoT application areas. These include intelligent transport systems (ITS), smart manufacturing (SM), the smart grid (SG), and smart cities (SC). These areas represent one comparably mature field (ITS), two more recent developments including a trend in the private manufacturing sector (SM) and a typically private utility that is nonetheless a crucial part of the public infrastructure (SC) and, finally, the overarching and perhaps even slightly futuristic concept of smart cities (SC).

7.4.1 Intelligent Transport Systems (Automotive Sector)

ITSs result from the integration of ICT with vehicles and transport infrastructure to improve economic performance, safety, mobility, and environmental sustainability[7].

---

7 http://ec.europa.eu/research/transport/multimodal/intelligent_transport_systems/index_en.htm.

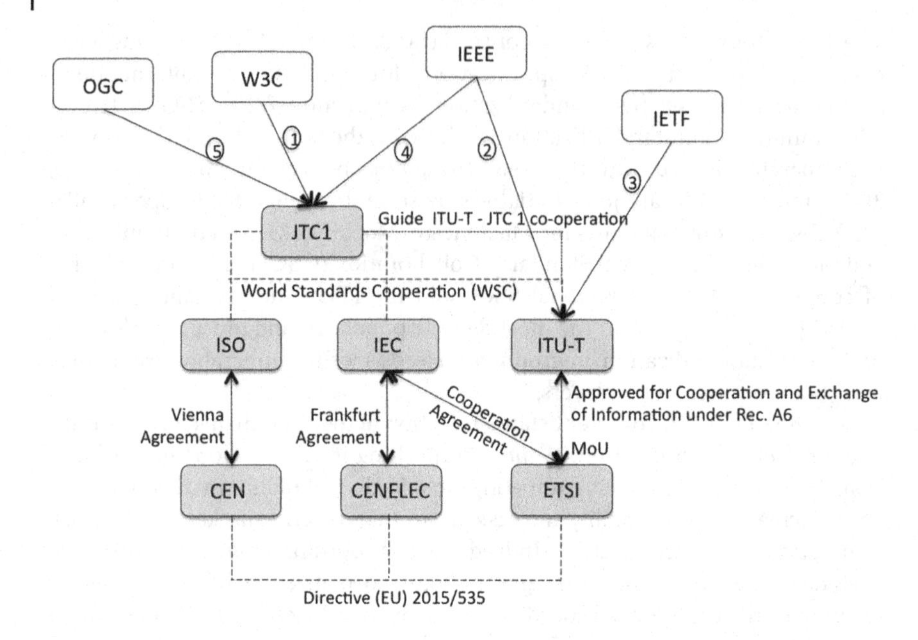

Figure 7.3 Formal coordination between SSOs. (Adapted from Jakobs et al. (2010).)

The ITS standardization landscape is populated by three different types of entities. For one, there are those from the "traditional" automotive sector. Their major common characteristic is their comparably old age (certainly by ICT standards). They may be further subdivided into specialized entities that deal exclusively with automotive/transport issues and into more general SSOs that have covered transport-related aspects for quite some time. Typically, they have added ITS-related topics to their portfolio only more recently, either through dedicated groups or by expanding the coverage of existing groups (e.g., IEC). Private consortia that focus exclusively on ITS topics form the third group. Most members of this group were founded only in this millennium, for example, CAR2CAR, the Car Connectivity Consortium (CCC) and the Open Automotive Alliance (OAA).

Figure 7.4 shows the very thinly populated ITS standardization environment in 1996, with hardly any links between the individual entities. This is not too surprising since ITSs became popular only in the late 1980s–early 1990s.

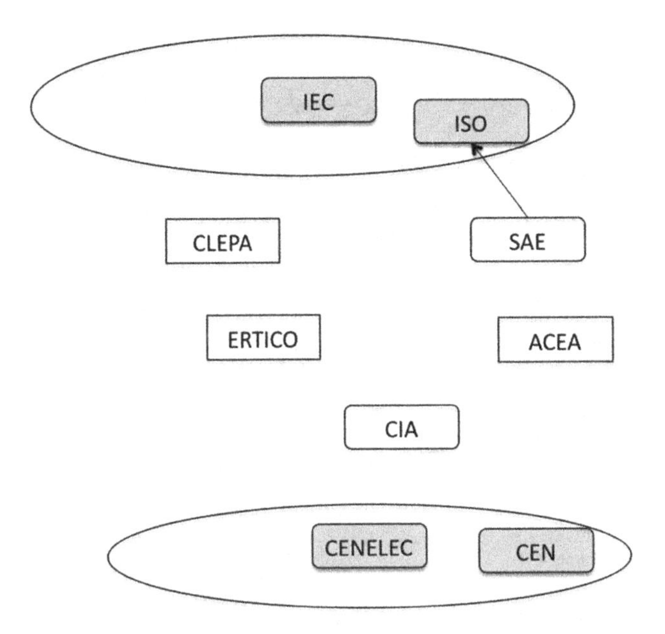

Figure 7.4 Entities and links between them in ITS (automotive) standardization 20 years ago (well-defined links exist within and between the ⬭).

Today, the environment has become much more complex, with a number of additional players forming a much more elaborate web with different types of links between them (see Figure 7.5). On the one hand, this is in line with the general trend in standards setting toward closer cooperation. On the other hand, the figure also shows that links between private consortia are still very limited; for quite a few, no links to other SSOs active in the same sector could be identified[8]. This situation may also be observed in other parts of the standardization universe.

The time line depicted in Figure 7.6 also offers some interesting insights. As noted above, the idea of ITS emerged in the late 1980s. In the early 1990s, the first standardization-related entities that were founded (in 1991) were not SSOs, but rather more policy-related entities. The associated Technical Committees (TCs) of the major international and European SDOs were established afterward.

A second "wave" of standardization-related bodies started almost 10 years later, again led by a non-SSO (EasyWay, a program run by European road operators and authorities and the European Commission). Between 2002 and 2010, major specialized consortia (Car2Car, Autosar) emerged, as well as

8 Informal links, for example, through joint membership, may well exist, though.

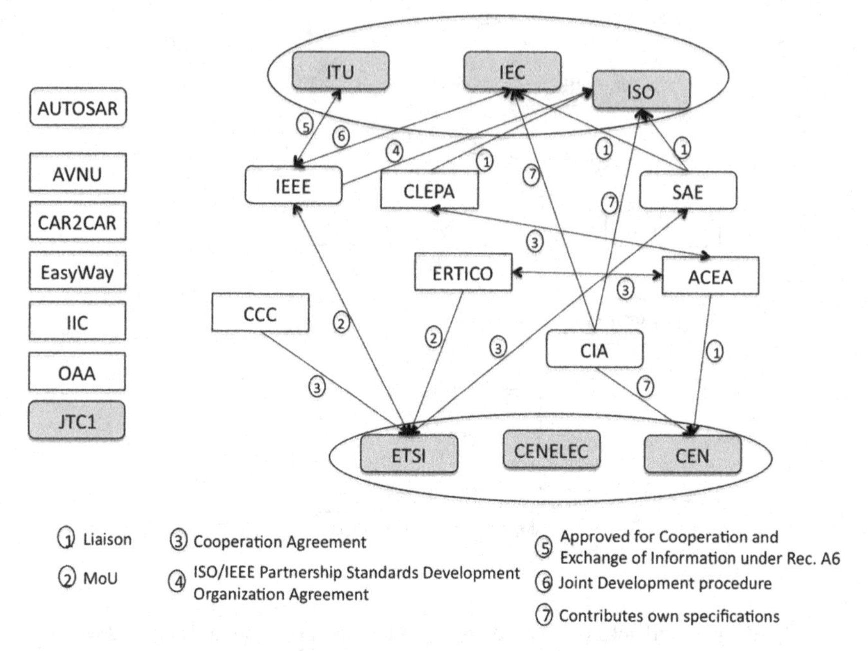

Figure 7.5 Entities and links between them in ITS (automotive) standardization today (well-defined links exist within and between the).

Working Groups in the telecommunication sector focusing on communication support for ITS services and applications. This development was not least triggered by the diffusion and increasingly advanced functionalities offered by mobile communication systems.

All in all, it seems safe to say that the ITS sector is quite mature by now. While the coordination between SSOs could certainly be improved, quite a few reasonably well-established links and coordination mechanisms are in place.

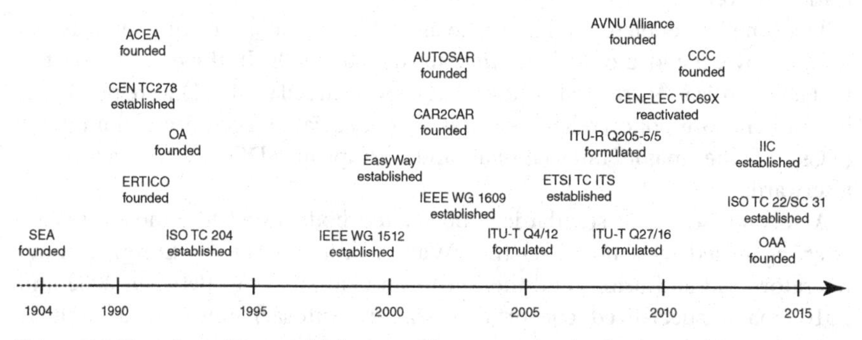

Figure 7.6 Time line of the establishment of important standardization entities in the ITS (automotive) sector.

7.4.2 Smart Manufacturing

The idea of "smart manufacturing" emerged in the late 1980s, became more popular as a research topic in the late 1990s, and got off the ground with the advent of the German "Industrie 4.0" initiative in 2013 (GTAI, 2014). Smart manufacturing uses ICT to optimize the use of labor, material, and energy to produce customized, high quality products for on-time delivery and to be able to quickly respond to changes in market demands and supply chains (Lu et al., 2015).

For 1996, the standards setting environment for smart manufacturing is not too dissimilar from the one in the ITS sector—a comparably small number of entities with very limited links between them. Here as well, the situation today is very different from the one to be found back then. The number of important players has almost tripled and a number of different types of links have been established (see Figure 7.7). Almost all private consortia and alliances (eCl@ss being the exception) focus on communication aspects. The explanation might be that smart manufacturing simply is not going to happen without an adequate underlying communication infrastructure that needs to meet special

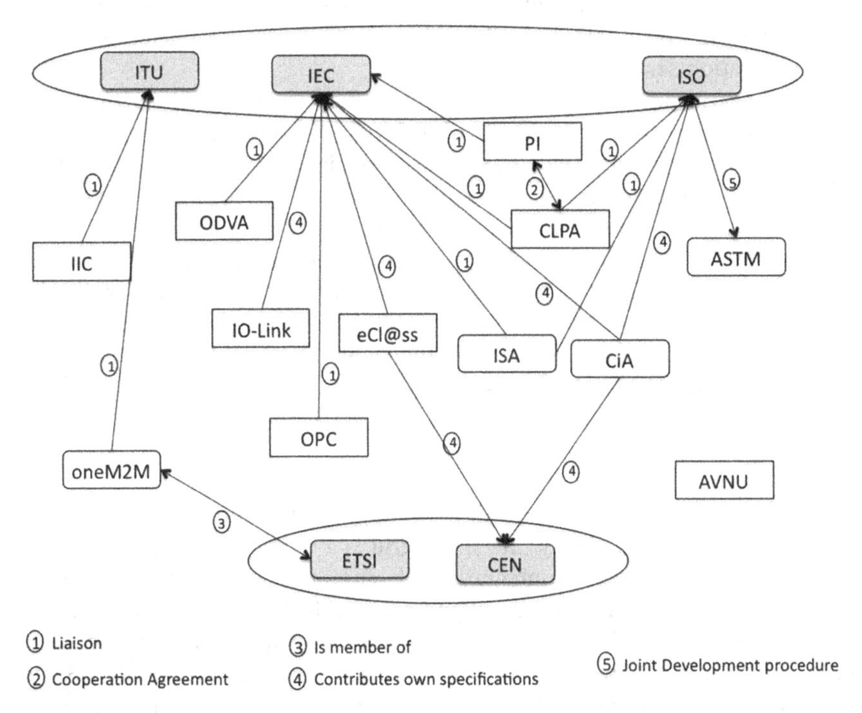

Figure 7.7 Links between SSOs in the smart manufacturing area today (well-defined links exist within and between the ⬭).

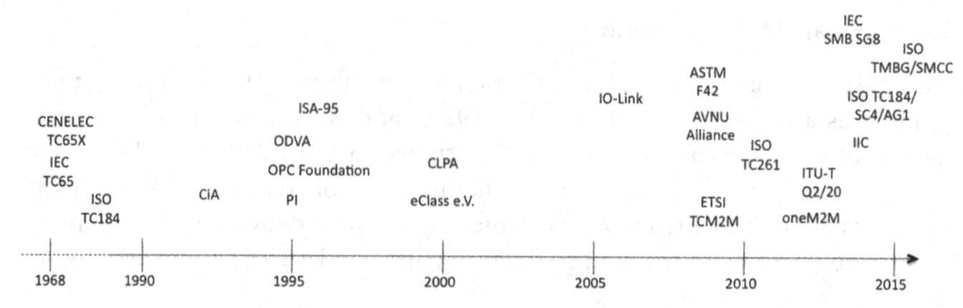

Figure 7.8 Time line of the establishment of important standardization entities in the smart manufacturing sector.

requirements concerning, for example, latency, resilience, reliability, and predictability (see, for example, Fettweis et al. (2014)). Accordingly, standards for this infrastructure will be much more widely used than those for the applications sitting on top of it. At least during the early days of development, they thus represent a much more lucrative field than application standards.

However, SSOs from the telecommunication sector (ITU, ETSI, oneM2M) appear to be isolated from the other entities. This is a bit of a surprise, since today supply chains are becoming global and the same may be expected for future manufacturing (IEC, 2015). Against this background the need for global communication services appears obvious.

Looking at the time line again, some older entities (ISO/TC 184 and IEC TC 65 and its European mirror entity, CENELEC TC65X, working on process automation) predate all others by quite a while (see Figure 7.8). In the mid-1990s a number of entities emerged, most of which had specified and developed products for plant floor communication. Many of their specifications found their way into the IEC process and were eventually formally standardized. In the nearer past, a number of SDOs established TCs or other groups that specifically focus on smart manufacturing. The agglomeration of newly found entities in 2013–2015 is quite remarkable and highlights the increasing importance recently assigned to smart manufacturing.

7.4.3 Smart Grid

First initiatives toward a more intelligent power supply system started in the late 1980s (Werbos, 2011). Today, the smart grid is a modern electric power grid infrastructure with smooth integration of renewable and alternative energy sources, through automated control and modern ICT (Gungor et al., 2010).

In 1996, the standardization arena relating to the smart grid was even thinner populated than in the two cases above. It is little wonder that mostly those SSOs active in the field of electrical engineering were around back then, working primarily in the field of power distribution. Links between these SSOs were nonexisting.

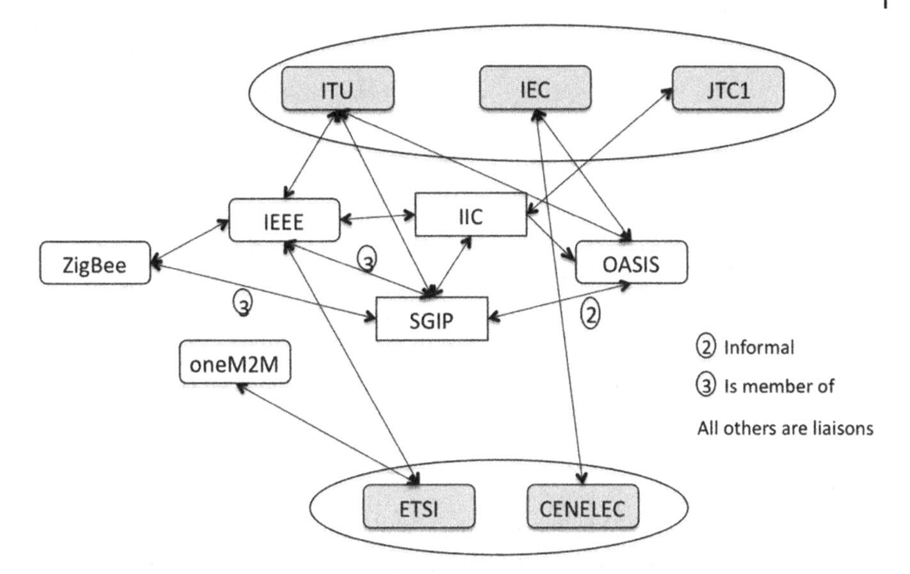

Figure 7.9 Links between SSOs in the smart grid sector today (well-defined links exist within and between the ⬭).

Again, similar to the development in the other two sectors already discussed, the web of SSOs working on smart grids (see Figure 7.9) has become much more complex today, involving many more players with a much more elaborate network of cooperations between them.

Compared to the other two sectors, the number of private consortia contributing to smart grid standardization is limited to the Smart Grid Interoperability Panel (SGIP) and the Industrial Internet Consortium (IIC), both of which also work on grid-specific aspects, not just on communication problems (as in smart manufacturing). Here, standards setting is mostly done by the SDOs (led by IEC), which have been active in the "traditional" fields of energy supply, electrical accessories, power systems management, or communication systems for decades (see also Figure 7.10).

In a way, the time line of the creation of entities working on standards for the smart grid further extends the development that could be observed for smart manufacturing (Figure 7.8) and, to a lesser degree, for ITS (Figure 7.6). In the latter case, the establishment of new entities was more equally distributed over time. For the former, a certain accumulation may be observed for the past 7 years or so. For the smart grid, this accumulation is much more pronounced. Here, we see the formation of numerous "specialized" entities starting in 2008. Prior to that point, only more "generic" aspects were covered (power supply, communication). That is, the smart grid may be considered as a novel application area, certainly in terms of associated standardization.

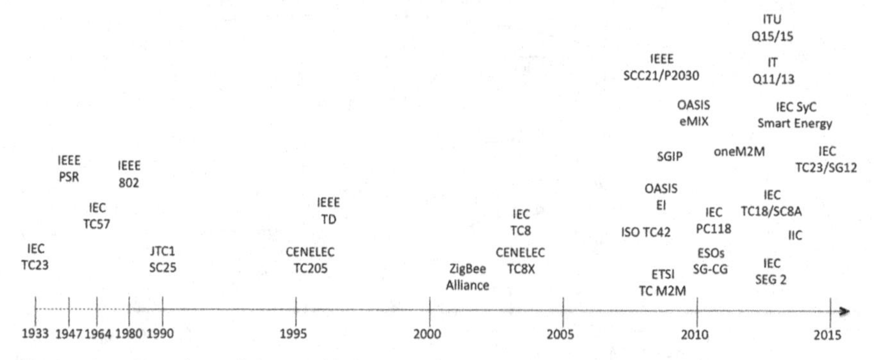

Figure 7.10 Time line of the establishment of important standardization entities for the smart grid.

7.4.4 Smart Cities

"Smart cities" are arguably the broadest application area. A smart city will, among others, comprise smart buildings, utilize the smart grid, provide smart transport facilities and e-health services, and also incorporate smart production sites. It will, therefore, be a particularly complex construct. Nevertheless, a survey by TU Vienna identifies 90 "smart" cities with 300,000–1,000,000 inhabitants and 77 with 100,000–300,000 inhabitants[9]. According to Navigant (2013), there are many pilot projects and small-scale developments but no examples of a smart city on a large scale, in line with the model depicted in Figure 7.11 (which is quite similar to the model used by TU Vienna). Truly "smart" cities are still closer to science fiction than to real life.

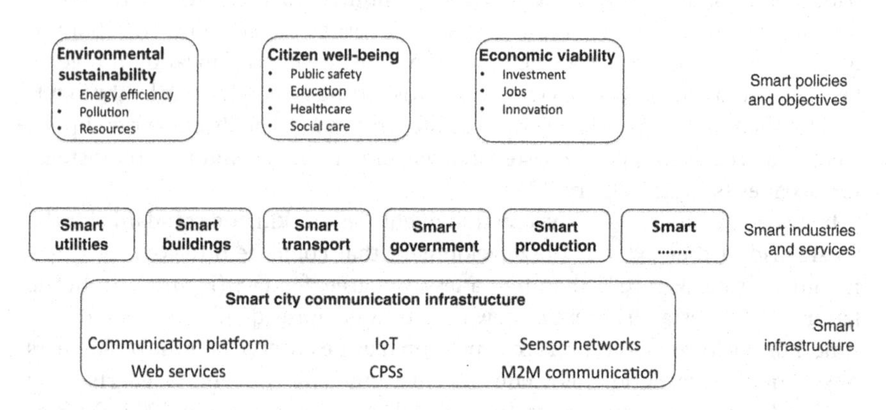

Figure 7.11 Smart city model according to and adapted from Navigant (2011).

9 http://www.smart-cities.eu/.

As can be seen, (virtually) all other applications contribute to the objectives of smart cities. In a way, a smart city represents a superset of "smart" applications. In addition, standards for "smart policies and objectives" (see Figure 7.2) provide guidance to city leadership for the development of an overall smart city strategy, the identification of priorities, the development of a practical implementation roadmap, and for an effective approach to monitoring and evaluating progress (BSI, 2015). Figure 7.12 shows the late 2016 status of SSO entities whose activities specifically focus at smart cities (again, those which develop more generic technologies or services that may also be deployed in a smart city context are not included) and the links that exist between them.

So far, the ITU has assumed a leading role in smart city standardization (not unlike IEC has done for the smart grid). Figure 7.12 also suggests that "smart cities" is a fairly new topic for standardization. Only a very limited number of players are active in this field and the links between them are neither particularly close nor numerous. Moreover, only a minority of entities within the individual SSOs focus on actual technical standardization work. Most are charged with high-level tasks like requirements identification as well as survey, road mapping, and/or coordination activities. However, the latter only relates to internal coordination within one SSO. In conjunction with the fairly young age of the activities, this explains the limited links between the entities.

Figure 7.12 also shows the complete absence of private standards consortia. This may be explained by the fact that smart city standardization focuses rather more on the strategic level (i.e., the top level in Figure 7.11), as opposed to the technical one. It seems fair to assume that little money is to be made form such activities, so consortia would stay clear off them.

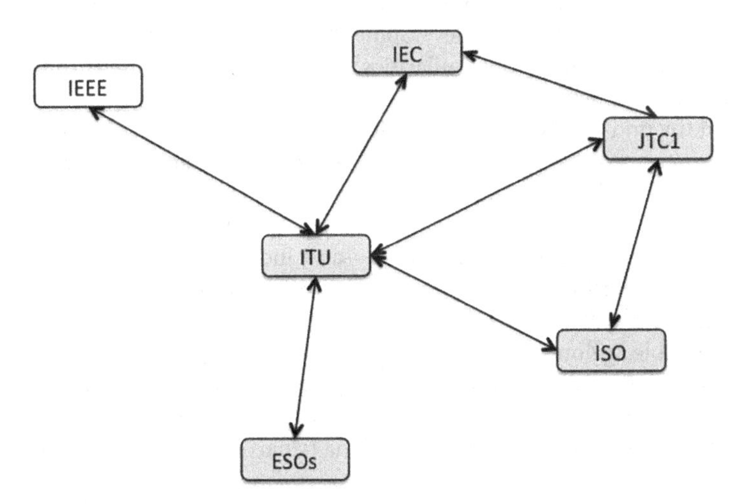

Figure 7.12 Entities and links between them in "smart city" standardization today.

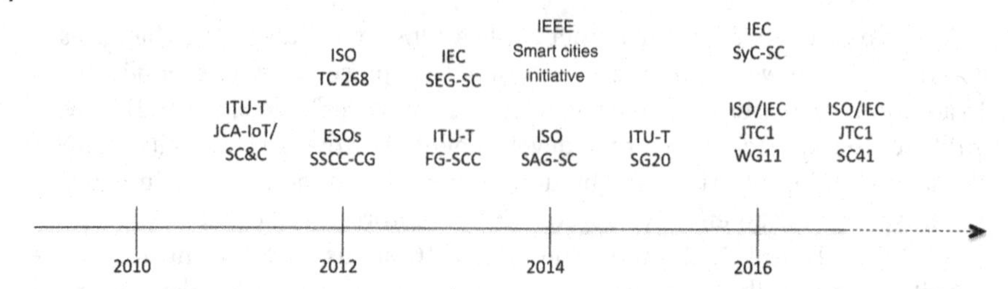

Figure 7.13 Time line of the establishment of important standardization entities in the "smart cities" sector.

As can be seen from the time line depicted in Figure 7.13, "smart cities" is indeed a very new field[10]. The first entity, ITU-T's Joint Coordination Activity on Internet of Things and Smart Cities and Communities (JCA-IoT and SC&C; now disbanded), was established in 2011.

The situation is similar to the one observed for smart manufacturing and the smart grid in that here as well we see a wave of newly founded standardization entities over the past 5 years. However, a major difference to the other sectors is that smart cities do not have any predecessor technologies—all standardization activities started from scratch in the 2010s.

7.5 Discussion and Some Speculation

So far, the chapter has looked at the standardization of the IoT and four (e) merging application areas. Their major common characteristic is the convergence of different technologies. The developments over time of the respective standardization environments show similarities, but also differences. Regarding the former, all environments emerged from comparably humble beginnings starting around the mid-1990s to fairly complex webs of SSOs today. The main reason for this increased complexity is the emergence of new and sometimes highly specialized SSOs. Likewise, in all cases, formal SDOs initially led the way; consortia and other "nontraditional" entities joined at a later stage (if at all, so far). It could be argued that their emergence—and increasing importance—eventually triggered the foundation of numerous new entities by the SDOs (Technical Committees, Working Groups, etc.).

Moreover, a notable agglomeration of SSOs may be observed over the past 5–7 (SM, SG, and SC) or 10 years (ITS). For SM and SG, this *might* be interpreted as the attempt of the industrialized, high-wage countries to improve their competitiveness versus emerging economies like China (smart manufacturing) and

10 Please note the scale of the x-axis, which is much larger than in the previous timelines.

to reduce their dependence on fossil energy sources (smart grid and, to a lower degree, ITS). Here as well, smart cities represent a superset of these and other "smart" applications.

The comparably more "homogeneous" development of the ITS area *may* also be attributed to the fact that lobbying entities and an overarching EU program preceded (and perhaps helped trigger) the first "wave" of SSOs in the early 1990s. This area also seems to have the most advanced integration of the telecommunication sector, which *may* also be considered a sign of greater maturity. In contrast, this integration is largely nonexistent in SM. The rather dominant role of SDOs for smart grid standardization is also worth noting. The fact that this is a highly regulated area *may* at least be part of the explanation.

Smart cities are a very new development without any preceding technologies. This is reflected in both the time line (which starts in 2011) and their sparsely populated and thinly linked web of SSOs. However, this is little wonder. After all, smart cities rely on the services provided by the other "smart" applications (located in the middle layer in Figure 7.13); the term "system of systems" is frequently used to highlight this characteristic (see, for example, IEC (2014)).

Comparing these time lines with the one for wireless communication systems (the most important part of their underlying infrastructure) we find a similar picture, but shifted to the left on the time line (see Figure 7.14).

The largest wave of SSO foundations occurred in the mid/late 1990s, when several major entities were formed. It predated those in SM and SG by more than 10 years. This is not such a big surprise since "smart" applications depend on an established ICT infrastructure. Today, all four application areas discussed require a wireless communication infrastructure as it offers much greater flexibility than a wired one (De Pellegrini et al., 2006; ITU, 2011; AIOTI,

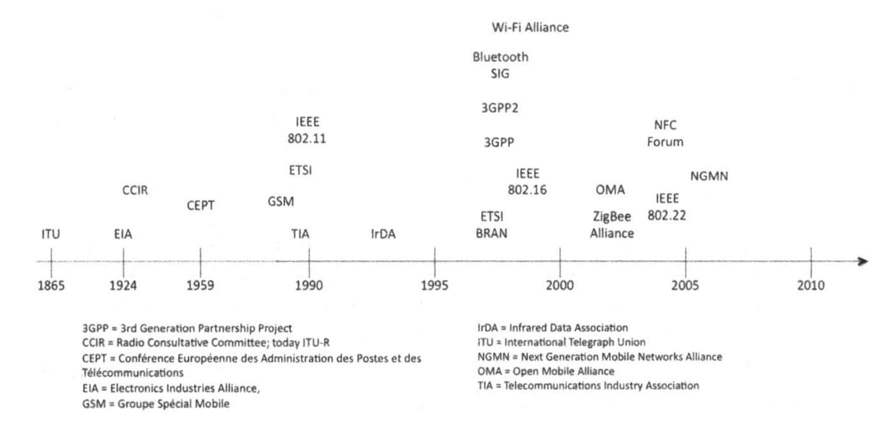

Figure 7.14 Time line of the establishment of important standardization entities for wireless communication.

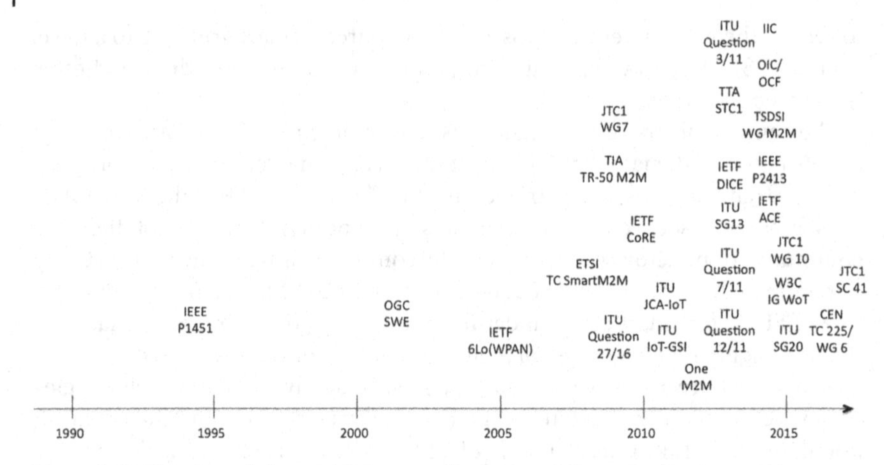

Figure 7.15 Time line of the establishment of important IoT standardization entities.

2015b); of course, the Internet still represents the communication backbone. Nevertheless, initially standardization in at least one "smart" application area, ITS, could apparently proceed without underlying standard-based wireless communication protocols and services. The beginning of standardization activities for wireless communication preceded that for ITS by only a few years.

The time line of the establishment of IoT-specific standardization entities depicted in Figure 7.15 shows a replication of this development. The establishment of the vast majority of dedicated IoT standards setting entities (new SSOs or specialized subunits of existing ones) occurred over the past 9 years. It got off the ground in 2009, with a peak in the period from 2013–2014. In 2015, JTC1 established its WG 10 "Internet of Things." This WG, together with WG 7 "Sensor Networks" has been moved to the newly established JTC1/SC 41 "Internet of Things and related technologies." This consolidation and, more so, the "elevation" to subcommittee status further highlights the importance ISO assigns to this field of work. That is, standardization activities for a dedicated IoT infrastructure started at around the same time as smart city standardization and the second "wave" of standardization of the smart grid and smart manufacturing. So, here again, application-specific standardization started without a dedicated infrastructure. Such a large mutually independent standardization of the IoT and its applications is a somewhat dangerous development. For one, a dialogue between the ICT world (developing the IoT-based infrastructure) and the world of applications (which will utilize this infrastructure) is crucial. After all, the latter needs to provide requirements for the former, and experts from the infrastructure side need to make clear any (technical) restrictions that may apply and potentially impact the applications. Moreover, and also due to an inadequate level of coordination between individual standardization activities (see Section 7.3), there is a real risk of creating new silos. Coordination largely occurs *within* individual SSOs, which may lead to the emergence of SSO-specific silos in which only standards from this SSO can seamlessly interoperate. This would then come in addition to proprietary silos. Furthermore, and considering the sheer number of standards setting entities that are active in the IoT domain and in the different application areas, one could be tempted to wonder if global interoperability can be achieved at all.

After all, the already quite considerable number of IoT-related standards may well contribute to interoperability issues in the field (silos) (Meddeb, 2016). The creation of a cross-SSOs coordination entity should be a next step. In addition, entities that actually try to address concrete application-related problems from a multidisciplinary perspective (see Table 7.3) will eventually be required. It remains to be seen if improved coordination helped get the mushrooming of standards and specifications under control.

7.6 Conclusion

This chapter has discussed some aspects of the standardization of the IoT and four of its major application areas. It turned out that the standardization environment has changed considerably over the past 20 years. Specifically, the number of SSOs working on standards relevant to the IoT and its applications has mushroomed during this period, as have the links that exist between them. A closer look at the time line of these developments reveals that the establishment of standardization entities relevant for the IoT, ITS, smart manufacturing, the smart grid, and smart cities show a very similar pattern. A first "wave" of new standardization entities occurred in the early–mid 1990s, followed by a second one around 2010 in the areas of SM, SG, and SC, when new entities were established at virtually explosive rate. For ITS, this happened a bit earlier and slightly less pronounced. The same picture emerges for mobile communication systems, where the peaks were in the late 1980s and mid-1990s, respectively. For the IoT, the time line looks slightly different, with one fairly massive wave between 2008 and 2014. In all cases, we can observe a considerable drop in the creation of new entities after 2014.

Considering the above, a look into a crystal ball suggests that the number of specialized standardization entities (SSOs, TCs, etc.) working on individual aspects of (e)merging applications and the IoT will continue to increase for a while, albeit a much slower pace (a fairly steep decline in the number of newly founded standardization entities may be observed for the past 2 years). It will probably reach saturation—at a fairly high level—in the not too distant future.

Today, the problem is that each SSO coordinates internally and may get updates from the other entities through more or less loose liaisons. Specifically, there is no cross-SSO coordination[11]. This once again reinforces the urgent need for some sort of effective and efficient cross-SSO coordination beyond what currently exists. This also holds for the coordination between the IoT-based infrastructure and the applications deploying it. However, it remains to be seen whether or not such overarching coordination will actually be established. This is highly unlikely to happen anytime soon, though.

Acknowledgments

Work on this chapter was funded by the Excellence Initiative of the German federal and state governments.

References

AIOTI (2015a) IoT LSP Standard Framework Concepts – Release 2.0. Available at https://ec.europa.eu/digital-agenda/sites/digital-agenda/files/discussions/aioti_wg3_sdos_alliances_landscape_-_iot_lsp_standard_framework_concepts_-_release_2_0.pdf (accessed November 18, 2016).

AIOTI (2015b) Report on Smart Mobility. Available at http://ec.europa.eu/newsroom/dae/document.cfm?action=display&doc_id=11822 (accessed November 18, 2016).

AIOTI (2015c) High Level Architecture (HLA) – Release 2.0. Available at http://ec.europa.eu/newsroom/dae/document.cfm?action=display&doc_id=11812 (accessed November 18, 2016).

Atzori, L., Iera, A., and Morabito, G. (2010) The Internet of Things: a survey. *Computer Networks*, **54**(15), 2787–2805.

Baheti, R. and Gill, H. (2011) Cyber-physical systems, in Samad, T. and Annaswamy, A.M. (eds.), *The Impact of Control Technology*, IEEE.

Biddle, B., White, A., and Woods, S. (2010) How Many Standards in a Laptop? (And Other Empirical Questions). In Beyond the Internet? Innovations for future networks and services. ITU-T Kaleidoscope Academic Conference. Available at http://www.itu.int/dms_pub/itu-t/opb/proc/TPROC-KALEI-2010-PDF-E.pdf (accessed November 18, 2016).

Blind, K. (2009) *Standardisation: A Catalyst for Innovation.* Inaugural Address Series, Rotterdam School of Management, Erasmus University, Rotterdam. Available at http://repub.eur.nl/res/pub/17558/EIA-2009-039-LIS.pdf (accessed November 18, 2016).

---

11 Private communication with Hanri Barthel, a long-standing and high-ranking expert in the field.

Blind, K., Jungmittag, A., and Mangelsdorf, A. (2011) The economic benefits of standardization. DIN German Institute for Standardization, Berlin. Available at http://www.din.de/blob/89552/68849fab0eeeaafb56c5a3ffee9959c5/economic-benefits-of-standardization-en-data.pdf (accessed November 18, 2016).

BSI (2015) Smart cities overview – Guide. PD 8100:2015. Available at http://shop.bsigroup.com/upload/Shop/Download/PAS/30313208-PD8100-2015.pdf (accessed November 14, 2016).

Cargill, C. F. (1995) A five-segment model for standardization, in Kahin, B. and Abbate, J. (eds.), *Standards Policy for Information Infrastructure*, MIT Press, Boston, pp. 79–99.

De Pellegrini, F., Miorandi, D., Vitturi, S., and Zanella, A. (2006) On the use of wireless networks at low level of factory automation systems. *IEEE Transactions on Industrial Informatics*, **2**(2), 129–143.

DIN (2000) *Economic Benefits of Standardization: Summary of Results.* Beuth Verlag GmbH and Deutsches Institut für Normung. Available at http://www.iec.ch/about/globalreach/academia/pdf/academia_governments/economic_benefits_standardization.pdf (accessed November 18, 2016).

EC (1996) Communication from the Commission to the Council and the Parliament on 'Standardization and the Global Information Society: The European Approach', COM (96) 359. Available at http://eur-lex.europa.eu/LexUriServ/LexUriServ.do?uri=COM:1996:0359:FIN:EN:PDF (accessed November 18, 2015).

Egyedi, T. (2014) the impact of competing standards: on innovation and interoperability for e-government. *PIK-Praxis der Informationsverarbeitung und Kommunikation*, **37**(3), 211–215.

Fettweis, G. et al. (2014) The Tactile Internet. TU-T Technology watch report. Available at http://www.itu.int/dms_pub/itu-t/oth/23/01/T23010000230001PDFE.pdf (accessed November 18, 2015).

GTAI (2014) Industrie 4.0 – Smart Manufacturing for the Future. Available at http://www.plattform-i40.de/I40/Redaktion/EN/Downloads/Publikation-gesamt/zukunftsbild-industrie-4-0.pdf?__blob=publicationFile&v=6 (accessed 10 November 2016).

Gunes, V., Peter, S., Givargis, T., and Vahid, F. (2014) A survey on concepts, applications, and challenges in cyber-physical systems. *KSII Transactions on Internet and Information Systems (TIIS)*, **8**(12), 4242–4268.

Gungor, V. C., Lu, B., and Hancke, G. P. (2010) Opportunities and challenges of wireless sensor networks in smart grid. *IEEE Transactions on Industrial Electronics*, **57**(10), 3557–3564.

Ho, J. Y. and O'Sullivan, E. (2015) Dimensions of standards for technological innovation–literature review to develop a framework for anticipating standardisation needs. EURAS Proceedings 2015: The Role of Standards in Transatlantic Trade and Regulation, 157–174.

IEC (2014) Orchestrating infrastructure for sustainable smart cities. Available at http://www.iec.ch/whitepaper/pdf/iecWP-smartcities-LR-en.pdf (accessed November 14, 2016).

IEC (2015) White Paper – Factory of the future. Available at http://www.iec.ch/whitepaper/pdf/iecWP-futurefactory-LR-en.pdf (accessed November 18, 2016).

IoT-A (2013) Internet of Things – Architecture – Deliverable D1.5 – Final architectural reference model for the IoT v3.0. Availabe at http://www.iot-a.eu/public/public-documents/d1.5/at_download/file (accessed November 18, 2016).

ISO (2014) ISO/IEC JTC 1: Smart cities. Preliminary report 2014. Available at http://www.iso.org/iso/smart_cities_report-jtc1.pdf (accessed November 14, 2016).

ITU (2011) Activities in smart grid standardization. Available at http://www.itu.int/en/ITU-T/focusgroups/smart/Documents/smartgrid_repository-V2.pdf (accessed November 18, 2016).

Jakobs, K. (2017) IC Standardisation, in Khosrow-Pour, M. (ed.), *Encyclopedia of Information Science and Technology*, 4th edn, IGI Global.

Jakobs, K., Wagner, T., and Reimers, K. (2010) Standardising the IoT – Collaboration or Competition? Proceedings of the e-Democracy, Equity and Social Justice—EDEM2010, IADIS.

Keoh, S. L., Kumar, S. S., and Tschofenig, H. (2014) Securing the Internet of Things: a standardization perspective. *IEEE Internet of Things Journal*, **1**(3), 265–275.

Lehr, W. (1992) Standardization: understanding the process. *Journal of the American Society for Information Science*, **43**(8), 550.

Li, S., Da Xu, L., and Zhao, S. (2015) The Internet of Things: a survey. *Information Systems Frontiers*, **17**(2), 243–259.

Li, S., Tryfonas, T., and Li, H. (2016) The Internet of Things: a security point of view. *Internet Research*, **26**(2), 337–359.

Lu, Y., Morris, K. C., and Frechette, S. (2015) Standards landscape and directions for smart manufacturing systems. 2015 IEEE International Conference on Automation Science and Engineering (CASE), pp. 998–1005.

Meddeb, A. (2016) Internet of Things standards: who stands out from the crowd? *IEEE Communications Magazine*, **54**(7), 40–47.

Navigant (2011) Smart City report. Navigant Research.

NIST (2013) Strategic R&D Opportunities for 21st Century Cyber-Physical Systems. http://www.nist.gov/el/upload/12-Cyber-Physical-Systems020113_final.pdf (accessed November 18, 2016).

NIST (2015) Draft Framework for Cyber-Physical Systems, Release 0.8. Available at https://s3.amazonaws.com/nist-sgcps/cpspwg/pwgglobal/CPS_PWG_Draft_Framework_for_Cyber-Physical_Systems_Release_0_8_September_2015.pdf (accessed November 18, 2016).

Sherif, M. H. (2001) A framework for standardization in telecommunications and information technology. *IEEE Communications Magazine*, **39**(4), 94–100.

Sicari, S., Rizzardi, A., Grieco, L. A., and Coen-Porisini, A. (2015) Security, privacy and trust in Internet of Things: The road ahead. *Computer Networks*, **76**, 146–164.

Swann, G.M.P. (2010) The Economics of Standardization: An Update. Report for the UK Department of Business, Innovation and Skills (BIS). Available at https://www.gov.uk/government/uploads/system/uploads/attachment_data/file/461419/The_Economics_of_Standardization_-_an_update_.pdf (accessed November 18, 2016).

Werbos, P. J. (2011) Computational intelligence for the smart grid-history, challenges, and opportunities. *Computational Intelligence Magazine, IEEE*, **6**(3), 14–21.

Part III

Security Issues and Solutions

8

Security Mechanisms and Technologies for Constrained IoT Devices

Marco Tiloca and Shahid Raza

RISE SICS Security Lab, Kista, Stockholm, Sweden

8.1 Introduction

The *Internet of Things* (IoT) emerging paradigm is commonly identified as the growing information technology trend toward a pervasive network society where all devices that can benefit from a connection will be connected with one another. This means that, in addition to traditional commodity devices and PCs, all kinds of electronic equipments that can benefit from a connection are expected to be available and accessible online.

In such a massively interconnected scenario, *security* has been immediately acknowledged as a dramatically important requirement to fulfill. On one hand, the well-known security threats, vulnerabilities, and attacks from traditional *Information Technology* (IT) systems are naturally inherited. This is why communication and management protocols proposed and adopted in the IoT already provide a number of *basic* fundamental security services. On the other hand, many additional security attack vectors are enabled against easier to target IoT devices, essentially because of two main reasons.

First of all, the long-term IoT vision assumes that the majority of devices are going to be directly connected to the Internet, in order to be directly accessible. This also makes them directly exposed to several kinds of security attacks, especially denial-of-service (DoS), that would be otherwise more difficult and less convenient for an adversary to carry out. Second, a typical IoT environment is expected to include a considerably large number of resource-constrained, often battery-powered, devices. This not only makes a large set of IoT devices considerably more vulnerable and less capable to cope with security attacks, but also results in the additional challenge of designing, implementing, and adopting security solutions that are also affordable for constrained IoT devices and do not significantly impact on performance and scalability of the overall network environment and IoT application scenario.

Internet of Things A to Z: Technologies and Applications, First Edition. Edited by Qusay F. Hassan.
© 2018 by The Institute of Electrical and Electronics Engineers, Inc. Published 2018 by John Wiley & Sons, Inc.

This chapter overviews the most important and crucial practical aspects concerning security in the IoT. In particular, this chapter first goes through the most important IoT security protocols and mechanisms, which are adopted in a typical IoT communication stack and are currently considered in ongoing standardization activities or have already been released as standards. They include, for instance, CBOR Object Signing and Encryption, OAUTH 2.0, Object Security for CoAP, DTLS, and compressed IPsec. This chapter also presents a number of specific security-related issues in the IoT, and discusses possible solutions to address them based on recent results from the scientific community and activities in standardization bodies. Particular attention is devoted to counteracting DoS attacks against IoT protocols and networks, preserving and improving high performance and scalability in the presence of security services, efficient and scalable protocols for management of cryptographic keys, and detection and counteraction of security attacks.

8.2 Security in IoT Protocols and Technologies

The IoT has the potential to transform the way we work and live on a daily basis. If there is one reason that can limit this transition, it will be the sage in security and privacy. While offering strong security is already hard in the Internet, it is in fact much more challenging in the IoT, since the "things" are going to support the Internet Protocol version 6 (IPv6), be globally reachable and extremely heterogeneous (comprising sensor devices, smartphones, standard computers, and even cloud environments), and usually deployed in physically unprotected, unattended environments. Moreover, most of them do not provide any conventional user interface, such as a display or keyboard. At the same time, IoT-constrained environments inherit the limitations typical of conventional wireless sensor networks (WSNs), for instance, limited processing and energy resources, multihop communication topologies, and lossy wireless links.

In order to meet these challenges, different protocols are being developed and standardized for IoT devices and networks. This section highlights the typical IoT security protocols and mechanisms currently adopted in IoT application scenarios and use cases. In particular, it goes through a typical network stack according to a top-down approach, and presents security services that IoT protocols rely on or directly provide. A quick summary of the presented protocols is provided in Table 8.1 (Section 8.2.7).

Figure 8.1 shows a resource-constrained IPv6-connected IoT network, also referred to as IPv6 over Low-power Wireless Personal Area Network (6LoW-PAN). It also sketches a typical IoT stack with different security protocols and technologies at different layers within an IoT device. At the application layer, OAuth 2.0 is being customized for IoT devices, with the aim to bring fine-grained access control mechanisms in the IoT; object security is being

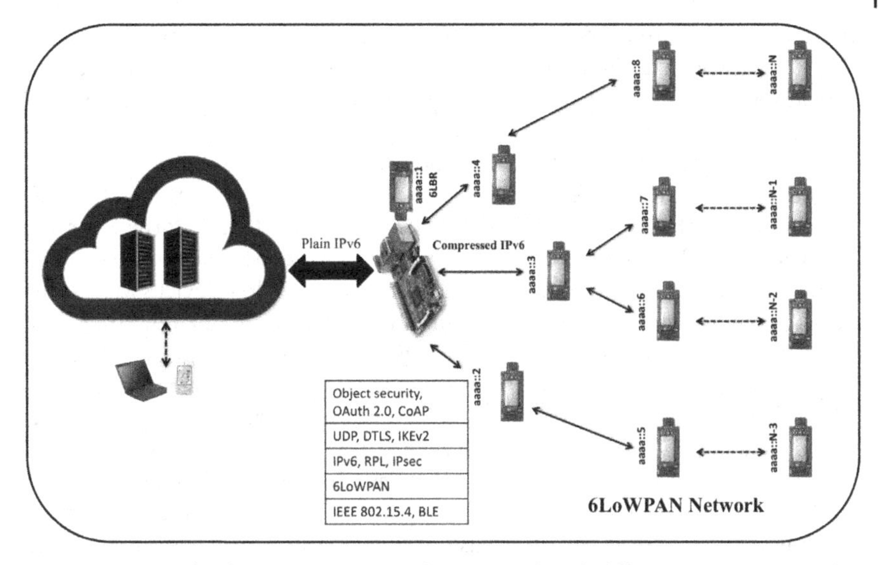

Figure 8.1 Example of resource-constrained IoT network with different security protocols and technologies at different layers.

standardized to protect individual data items; and the Constrained Application Protocol (CoAP) has asserted itself as the new web standard for IoT that can carry Object security and OAuth messages. At the transport layer, connectionless UDP is favored for IoT devices, and the *Datagram Transport Layer Security* (DTLS) protocol is bound with CoAP to form the secure CoAP (CoAPs) protocol. Alternatively, it is possible to use the Internet Key Exchange version 2 (IKEv2) protocol to dynamically establish and manage security associations and related key material when the adoption of IPsec (or network security) is preferred. Furthermore, IPv6 displays a potentially unlimited address space, and is the de facto networking and addressing protocol for the IoT, while the IPv6 Routing Protocol for Low-Power and Lossy Networks (RPL) is the standardized solution for packet routing within constrained, low-power, and lossy networks (Winter et al., 2012). Also, 6LoWPAN has been standardized in order to provide compression/decompression and fragmentation/reassembly mechanisms, and thus make larger packets fit in the IEEE 802.15.4 frames (or similar ones). Finally, there are a number of possibilities at the link and physical layer. Since there is a great potential in IEEE 802.15.4 as well as Bluetooth Low Energy (BLE), security in their presence is also discussed.

8.2.1 Lightweight Formats for Cryptosecurity Protocols

Since many IoT devices are expected to be resource constrained, there is the need to provide security services in a particularly efficient way, and hence reduce the amount of information to be stored, transmitted, and processed. This means that not only plain application-level content, but also encrypted content, digital signatures, and even cryptographic keys are required to be represented and encoded in a compact and efficient way.

To this end, CBOR Object Signing and Encryption (COSE) aims at creating cryptographic formats based on Concise Binary Object Representation (CBOR) (Bormann and Hoffman, 2013). CBOR is a succinct binary format for the serialization of data structures, suitable for resource-constrained devices (Bormann et al., 2014). A major share of envisioned billions of IoT devices will consist of battery-powered or energy-harvested devices. For these IoT devices, COSE has a huge potential to produce lightweight CBOR-based representations of cryptographic keys, encryption, message hashes, and digital signatures.

COSE is inspired by JavaScript Object Signing and Encryption (JOSE) that has already standardized cryptographic formats using JavaScript. Although COSE tries to reuse functionalities from JOSE, they are not directly compatible for straightforward adoption. One of the major differences that CBOR brings compared with JOSE is the direct use of binary data format with no need to first convert it into, for example, a string encoded in base64. The CBOR Encoded Message Syntax details the basic COSE structure and common COSE headers (Schaad, 2017). Furthermore, it elaborates the COSE objects by means of dedicated cryptographic key materials, as well as through cryptoalgorithms for data encryption/authentication and for computation of Message Authentication Codes (MACs). There already exist proposals for encoding parameters, keys and outcomes of RSA (Jones, 2017), and other algorithms (Schaad, 2016) as COSE messages. Also, lightweight COSE representation of cryptographic data structures and their implementation and evaluation, when compared with JOSE, could be an interesting research topic.

8.2.2 CoAP and DTLS

CoAP is a novel web transfer protocol especially designed for resource-constrained devices and networks. It relies on request/response interactions between endpoints and can easily integrate with the widely adopted Hyper Text Transfer Protocol (HTTP) for integration with the classical Web (Fielding and Reschke, 2014a, 2014b). Although the CoAP protocol has been designed for constrained devices and networks (Bormann et al., 2014), on the other hand it does not provide itself any particular primitives to secure communication (Shelby et al., 2014). Instead, CoAP messages can be protected either by means of object security, as discussed in Section 8.2.3 or through actual secure communication protocols. In such a case, the CoAP specification recommends to adopt the DTLS protocol (Rescorla and Modadugu, 2012). In particular, Shelby et al. (2014) specify the CoAP binding to DTLS as a set of deltas to plain unsecure CoAP. From a practical perspective, if a CoAP message is secured using DTLS, the *coaps://*URL scheme is considered, rather than the *coap://* scheme for unsecure communication. In particular, the three security modes *PreSharedKey*, *RawPublicKey*, and *Certificate* are available. That is, the *PreSharedKey* mode relies on preshared cryptographic keys, each of which include a list of devices allowed to use that key for secure communication. If the *RawPublicKey* is considered, a device owns a pair of asymmetric private–public keys with no associated certificate, hence entrusting validation to out-of-band mechanisms. In addition, the device maintains an identity computed from the public key, as well as a list of devices it is allowed to communicate with. Finally, in case it considers the *Certificate* mode, a device stores an asymmetric private–public key pair, together with an *X.509* certificate. The certificate binds the public key to its subject, and includes a digital signature from a well-known trust root (Cooper et al., 2008). In particular, the device additionally maintains a list of root trust anchors, which can be referred to in case certificate validation has to be performed.

Practically, DTLS allows two devices to establish a security association, authenticate with one another, and exchange protected CoAP messages. The Internet Engineering Task Force (IETF) has recently released a profile for Transport Layer Security (TLS) and DTLS 1.2. The profile provides communication security for resource-constrained devices employed for collecting data by means of sensors or for controlling actuators in applications for industrial control systems, home automation, smart cities, and other IoT deployments (Tschofenig and Fossati, 2016).

DTLS has been voluntarily conceived to be very similar to the TLS protocol (Dierks and Rescorla, 2008), and in fact fulfills equivalent security requirements. That is, it makes it possible for applications running on clients and server units to exchange secure messages, namely, DTLS *records*, especially preventing tampering, forgery, and eavesdropping of such messages. At the same time, DTLS displays a number of differences compared to TLS, so making it possible to provide secure message exchange on top of UDP and other unreliable datagram transport protocols. For instance, RC4 and other stream ciphers are not allowed to be used. Besides, a sequence number value is explicitly included in each DTLS message. So doing, DTLS messages result to be not related with one another, thus making it possible for recipients to correctly process them, even when out-of-sequence delivery occurs. Furthermore, DTLS addresses packet loss through message retransmission and local timeouts. Upon the reception of an invalid message, the related DTLS session may still be preserved, and that message can simply be silently discarded.

In order to exchange protected messages over DTLS, two *client* and *server* devices perform a *handshake* process in order to establish a secure session, as depicted in Figure 8.2. Handshake messages transmitted at a same step are grouped together as part of one DTLS *Flight*. Optional or situation-dependent messages are indicated in square brackets and not always sent.

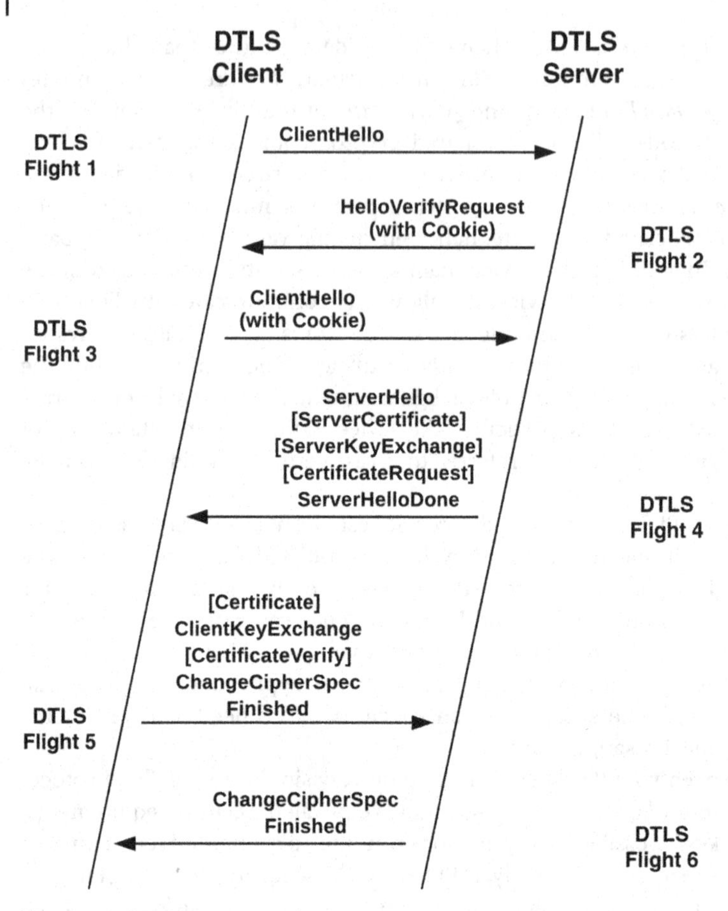

Figure 8.2 DTLS message exchange.

The client typically takes the initiative and sends a *ClientHello* message to the server in order to start establishing the DTLS session. DTLS admits a stateless cookie value that can be optionally exchanged between the server and the client by means of *Fligth2* and *Flight3*, with the intent to address DoS against the server. Then, client and server exchange *Flight4* and *Flight5* in order to authenticate each other, negotiate a cipher suite and security parameters, and establish cryptographic key material. Finally, *Flight6* confirms that a DTLS session has been fully established on both sides, and thus client and server can consider it for secure communication. The reader can refer to Dierks and Rescorla (2008) and (Rescorla and Modadugu), (2012) for additional information concerning the handshake and the specific format of DTLS Flights.

Before starting a handshake, two DTLS peers have to be provided with preliminary security material. This includes, for instance, preinstalled cryptographic keys that are employed to establish a *premaster secret* during the handshake. Then, client and server use this premaster secret together with self-generated random values in order to derive a *master secret*, out of which the actual security material for the DTLS session is finally derived. In order to distribute preinstalled keys, there are two main approaches that can be considered. The first one leverages asymmetric keys. Additionally to the typical usage of *X.509* certificates (Cooper et al., 2008), there are also available *raw* public key profiles, where private–public keys are provided without a certificate. In particular, such keys can be generated by device manufacturers and then installed before the device deployment (Wouters et al., 2014). In such a case, a device validates raw public keys received from other peers by means of out-of-band techniques. In particular, it usually stores a list of peers it is allowed to communicate with. Instead, an alternative approach leverages symmetric *pre-shared* keys (PSKs) (Eronen and Tschofenig, 2005). That is, the client stores a number of symmetric keys, each of which shared with a server that the client is allowed to communicate with. Then, during the handshake, the client specifies the particular key to use, indicating the related *PSK identity*. Finally, the client and server derive the DTLS premaster secret out of the symmetric key that they have been sharing. Adopting the approach based on preshared key is especially convenient for environments that are closed and within which one can easily provide cryptographic keys to the deployed devices. Furthermore, it does not require sending and receiving public certificates, or to engage in asymmetric cryptography operations, which are costly and particularly important to limit on resource-constrained DTLS servers. Besides, it makes key management operations considerably easier, particularly in IoT deployments within which manual and early configuration of connections is often common practice, and hence the provisioning of certificates often results to be a nonpreferable or even non-feasible option.

8.2.3 Object Security for Constrained RESTful Environments

The CoAP standard specifies the possible use of *proxies* as intermediary entities between CoAP clients and servers, to improve efficiency and scalability in a number of network and application scenarios (Shelby et al., 2014). At the same time, as previously discussed in this chapter, CoAP refers to the DTLS protocol to provide secure communication (Rescorla and Modadugu, 2012). This represents a problem in the presence of intermediary proxies, since the latter must be able to access, and possibly manipulate, specific portions of a CoAP message, in order to perform the intended proxy functionality. As a consequence, proxy operations on CoAP messages require that DTLS communication from a CoAP client terminates at the proxy. On the other hand, it is still very important to prevent the proxy from accessing and/or manipulating portions of the CoAP messages that are not strictly necessary for the correct performance of the intended proxy functionality. In particular, this would also contribute to reduce and relax the security and trust assumptions on the proxy units deployed in the network. To this end, it would be convenient to rely on a secure communication protocol that is able to protect CoAP messages in an *end-to-end* fashion, even when intermediary nodes are present.

At the time of writing, the document (Selander et al., 2018) describes a relevant proposal named *Object Security for Constrained RESTful Environments* (OSCORE). In principle, OSCORE is a data object-based security protocol enabling the exchange of CoAP messages that are end-to-end protected across intermediary nodes. OSCORE uses COSE objects (Schaad, 2017) in order to provide integrity, end-to-end encryption, and replay protection of CoAP messages. The usage of OSCORE is signaled in the first place by including a newly defined *Object Security* CoAP option in the secured CoAP messages. As a possible alternative, message protection can be limited to the payload only of individual messages, according to an akin approach named *object security of content* (OSCON). The end-to-end security services provided by OSCORE are also particularly desirable in constrained network settings where solutions like DTLS can not be supported. On the other hand, OSCORE can actually be combined and used together with DTLS. That is, OSCORE can enable end-to-end security of CoAP messages between a CoAP server and a CoAP client, together with protection of the entire CoAP messages (i.e., header included) in a hop-by-hop fashion provided by DTLS during the message transport between an endpoint and an intermediary node like a proxy.

More practically, OSCORE assumes that a security context has been pre-established and agreed upon between a CoAP client and a CoAP server considered as endpoints of the secure communication. In particular, a security context is the set of information and parameters that the two OSCORE endpoints use to perform the specific cryptographic operations. Furthermore, OSCORE is very similar to DTLS, and borrows a number of its mechanisms such as key derivation and construction of nonce values. However, OSCORE relies on COSE (Schaad, 2017) rather than the DTLS record layer, and thus is able to use security context identifiers and sequence numbers of variable length.

8.2.4 Compressed IPsec

The 6LoWPAN standard (Hui and Thubert, 2011) defines how heavyweight IPv6 datagrams can be transmitted over low-power and lossy networks based on IEEE 802.15.4 (IEEE Computer Society, 2011). To achieve this, 6LoWPAN introduces a number of schemes for header compression, which are able to considerably reduce the size of the UDP headers, IP datagrams, and IP extensions. In IP networks based on IEEE 802.15.4, also referred to as 6LoWPAN networks, security is particularly important due to the high exposure to the Internet and the resulting network vulnerability. Currently, IP security (IPsec) (Kent and Seo, 2005; Jankiewicz et al., 2011) is the standard security solution for

IPv6. In particular, IPv6 hosts on the Internet should implement it and be able to handle and process IPsec-protected packets. In particular, the *transport mode* of IPsec provides secure end-to-end communication between two nodes on the Internet. Hence, it is expedient to adapt 6LoWPAN in order to enable IPsec communications among IPv6-enabled *things* (e.g., sensor nodes) in 6LoWPAN networks and generic IPv6 nodes on the Internet.

A number of different proposals have been presented for compressing headers of IPsec packets. The majority of compression schemes is adoptable to generic Internet hosts, and does not specifically consider 6LoWPAN networks composed of resource-constrained devices. Migault *et al.* have proposed Diet-ESP (Migault and Guggemos, 2015; Migault et al., 2016) that specify how to perform compression of IPsec packets, but needs a number of changes and adoptions in conventional hosts on the Internet. Furthermore, RObust Header Compression (ROHC) relies on a flexible and efficient concept for header compression, but it is intended to generic hosts on the Internet and hence is not specifically intended for 6LoWPAN networks (Sandlund et al., 2010; Ertekin et al., 2010). Therefore, Raza et al. (2011) propose the 6LoWPAN-compressed IPsec, which is primarily targeted for resource-constrained IoT devices and network.

The schemes Diet-ESP and ROHC together with Generic Header Compression (Bormann, 2014) are accessory to the compressed IPsec described in Raza et al. (2011), where mechanisms for header compression specifically target 6LoWPAN networks, work with generic Internet hosts using IPsec, and do not result in any change of the standards Encapsulated Security Protocol (ESP) and IPsec Authentication Header (AH).

Currently, the IPsec's ESP protocol that provides both confidentiality and integrity of the ESP payload is mostly used. For the IoT, a number of use cases are envisioned where it is useful that ESP, as well as AH, protects the integrity of IPv6 headers in addition to the IPsec payloads. Because of different reasons, the usage of AH in the Internet is currently limited. That is, AH is not compatible with Network Address Translation (NAT), which modifies the IP source address while in transit, thus making the packet integrity check fail and ultimately causing the packet rejection at the recipient IPSec host. Moreover, ESP is able to provide both encryption and integrity protection of IPsec packets, but it makes it possible only to secure the application data and the ESP header, while leaving the IP header unprotected. Practically, this is fine in case the *tunnel mode* of IPsec is used, since the inner application data, ESP header, and IP header are both confidentiality and integrity protected.

Since IoT scenarios relying on IPv6-connectivity do not comprise NATs, it is possible and favorable to use the transport mode for providing end-to-end security. Thus, using AH along with ESP makes particularly more sense in the IoT. In particular, since the AH integrity check takes into account the IP address, IPsec can be effectively used to provide protection against IP spoofing, which is among the most common and likely security attacks mounted against resource-constrained devices communicating over IPv6. In addition, despite the stateless autoconfiguration of IPv6 addresses has been proposed, this is not considered to be an actual requirement to fulfill. That is, resource-constrained nodes in 6LoWPAN networks are assigned with IPv6 addresses at deployment time, and likely maintain the very same address during their lifetime, except in case of manual intervention by means of firmware/software updates. On the other hand, it is practically infeasible to address autoconfiguration in 6LoWPAN networks in a way that ensures end-to-end connectivity, unless a suitable and efficient mechanism specifically targeting 6loWPAN networks is designed and developed; this is in fact an interesting research challenge to take. Note that even if only a single application runs on a 6LoWPAN node, IPv6 potentially provides an unlimited address space. Therefore, one has the luxury to reserve many different IPv6 addresses for one single 6LoWPAN node. As a consequence, this enables the setup of unique IPsec security associations for each different application. In addition, in case IKEv2 (Bormann, 2015) is specifically considered as key management protocol, it is then possible to dynamically establish unique security associations on a per-application basis.

Besides, a number of use cases such as alarm actuating and simple environmental monitoring require only integrity protection, hence the usage of AH only is a convenient choice. At the time of writing, there is a pending IETF standard draft that defines the actual encoding of IPsec AH and ESP (Raza et al., 2016). The detailed implementation and evaluation of this draft are also available in Raza et al. (2011, 2014).

8.2.5 IEEE 802.15.4 and Bluetooth Low Energy

The IEEE 802.15.4 standard provides specifications for the physical and the Media Access Control (MAC) layer intended for wireless low-rate personal area networks. It entrusts other protocols like ZigBee or WirelessHART for additional services at the higher layers, or can alternatively rely on 6LoWPAN as an adaptation layer to operate together with standard Internet protocols.

In addition, the MAC layer of IEEE 802.15.4 directly provides a number of security services, that is, data authenticity, data confidentiality, and replay protection on a per-packet basis (IEEE Computer Society, 2011). In particular, the standard refers to a cryptosuite relying on the *Advanced Encryption Standard* (AES) 128 bits symmetric-key cryptography (National Institute of Standards and Technology, 2001). In case secure communication is enabled, an *Auxiliary Security Header* (ASH) is included next to the standard MAC header of each packet, which is secured upon its transmission based on what specified in the ASH. The recipient node receives and correctly unsecures the packet according to the same information.

In particular, IEEE 802.15.4 provides three different *security modes*, that is, CTR (encryption only), CBC MAC (authentication only), and CCM (encryption and authentication). The CCM and CBC MAC modes use a *Message Integrity Code* (MIC), which can be 4, 8, or 16 bytes in size. Security operations in IEEE 802.15.4 make use of a fresh nonce value. In particular, a nonce is generated at random on the sender side, and provides protection from replay attacks. The different available security modes and the different available configurations make it possible to trade specific application constraints with the security and performance requirements.

Finally, the IEEE 802.15.4 standard does not describe how to establish and distribute the key material in the network, or how to address device authentication. Instead, it assumes that both such security services are provided and enforced by the higher layers. As a consequence, the standard assumes that a given pair of sender and recipient nodes has successfully agreed on the same security settings and shared common key material, before they can start to securely communicate.

The same security services are provided in the enhanced standard IEEE 802.15.4e, which extends the previous standard IEEE 802.15.4 in order to specifically target embedded applications with critical requirements (e.g., healthcare or industrial applications) (IEEE Computer Society, 2012). In particular, it defines the MAC behavior mode *Time Slotted Channel Hopping* (TSCH), which combines multichannel and channel hopping functionalities with time slotted access. This not only makes it possible to provide better performance in terms of high reliability, large network capacity, energy efficiency, and predictable latency, but also contributes to make networks more robust against security attacks mounted at the physical layer, such as jamming and radio interference, thanks to the channel hopping mechanisms.

While on one hand IEEE 802.15.4 is currently the de facto standard covering the physical and link layers for 6LoWPAN networks, other new technologies are evolving as well. For instance, BLE, practically marketed as Bluetooth Smart, is among the energy efficient communication technologies currently available, and does represent an appealing alternative. In particular, BLE has asserted itself as a lightweight alternative for resource-constrained devices, compared to Classic Bluetooth. The Bluetooth 4.0 standard now also includes the BLE specifications, which comprise the broadcast communication mode in addition to the power-efficient connections among Bluetooth devices. Bluetooth 4.2 was published in December 2014, and provides a number of novel features which do make BLE a promising and valuable technology for the IoT (Bluetooth Special Interest Group, 2014).

The most important addition that enriches Bluetooth with IoT capabilities is the support for the Internet Protocol Support Profile (IPSP) (Bluetooth Special Interest Group, 2016). IPSP adds IPv6 support in the Bluetooth peripheral device and in the central master device acting as network coordinator, and specifies how devices are supposed to discover one another and to establish a connection at the link layer with each other. Besides, the BLE Generic Attribute

Profile (GATT) can be employed in order to determine whether a device has IPSP support enabled, which in turn enables the exchange of IP packets by means of the Bluetooth L2CAP Credit Based Flow Control Mode.

Furthermore, Nieminen et al. connect BLE with 6LoWPAN by specifying the standard way for transmission of compressed IPv6 packets on top of BLE (Nieminen et al., 2015). In particular, they propose using the header compression mechanisms of 6LoWPAN on top of BLE. On the other hand, they do suggest to not use the fragmentation mechanisms of 6LoWPAN, but rather to rely on the ones already outlined in Bluetooth L2CAP.

Also, one of the most relevant improvements brought by Bluetooth 4.2 is the increased size of packets, that is, the BLE payload at the link layer is increased from 27 to 251 bytes. Also, Bluetooth 4.2 improves the throughput up to 2:5 times. These enhanced capabilities enable efficient IPv6 communication over BLE, frequent and speedy updates of device firmware, as well as fast uploading of data from sensor devices to backend units. Also, it makes it possible to run even in the IoT a number of complex security protocols, for example, DTLS (Rescorla and Modadugu, 2012) and IPsec (Kent and Seo, 2005; Jankiewicz et al., 2011).

Prior to Bluetooth 4.2, the Bluetooth standard provides stronger security in comparison with BLE, mostly for achieving energy efficiency. However, it also displays a small packet size of 27 bytes, and hence does not allow devices to use asymmetric-key cryptographic protocols. Instead, Bluetooth 4.2 LE brings BLE to the same security level of the standard Bluetooth, by recommending to rely on Elliptic Curve Cryptography (ECC). Specifically, the National Institute of Standards and Technology (NIST) has recommended to use elliptic curves, whereas the Federal Information Processing Standards (FIPS) has recommended to use AES-CCM symmetric cryptography for encryption and integrity protection.

In addition, Bluetooth 4.2 also brings enhanced privacy by adding private address resolution both in the Bluetooth Host as well as in the controller segments of a Bluetooth node. Also, Bluetooth 4.2 specifies the controller to maintain a white list of private addresses at the link layer. So doing, the controller is able to generate and resolve private addresses at the link layer with no need to interact with the host, and to reject or accept incoming connection according to the maintained white list. This significantly reduces the rate according to which the host has to "wake up," therefore, considerably limiting the energy consumption of the BLE chip. The enhanced security and privacy with new low-power features and the fact that BLE is supported out of the box in most commodity smartphones makes BLE a valuable and promising technology for IoT deployment that involves mobility and allows a 6LoWPAN network to be directly connected to a smartphone, such as in deployments involving wearables. The standard IEEE 802.15.4 is still a preferable and more suitable choice for static deployments, such as smart homes.

8.2.6 OAUTH-Based Authorization in the IoT

According to the *Internet Security Glossary*, authorization is defined as the process for granting approval to a *client* to access a resource (Shirey, 2007). A typically adopted approach consists in managing the authorization of users, services, and their devices by means of dedicated *Authorization Server* (AS) entities. The *OAuth 2.0* authorization framework relies on this fundamental approach, and has asserted itself as one of the most widely adopted standard for managing authentication processes (Hardt, 2012). In fact, it makes it possible to address all typical issues of alternative approaches based on credential sharing, thanks to the introduction of an authorization layer, and the separation of the role of the client from the role of the actual *resource owner*. In principle, the OAuth 2.0 authorization framework allows a client entity (i.e., a host, a process, a user) to request and obtain a specific, regulated, and limited access to a resource available at a *resource server* (RS) by enforcing the permission of the related resource owner.

Specifically, the resource owner grants authorization to the client by means of an intermediary AS, which provides the client with the actual authorization-related information specified in an *access token*. More specifically, an access token is practically represented as a string, which encodes an authorization decision issued to the specific client, and is typically opaque to that client. Among other things, an access token specifies the duration and scope of authorized resource accesses, which are ultimately determined by the AS and enforced by the RS. Furthermore, the AS ensures that possible unauthorized parties are not able to generate, modify, or guess any issued access tokens for producing valid access tokens. Then, the client can contact the RS in order to access the intended resource, according to the credentials specified in the access token. That is, the RS verifies that the access token is valid and, in such a case, proceeds with serving the request received from the requesting client. Before starting to perform the authorization protocol, the client must have registered at the AS, and all access token credentials must be always transferred through a secure communication channel, that is, both from the AS to the client and from the client to the RS. Also, the AS must verify the identity of the resource owner upon every authentication request on behalf of a client, and is required to have preestablished a trust relation with the involved RS hosting the resources to access. Note that the same AS can be associated with, and issue access tokens to be used with, multiple resource servers.

Having said that, it is more than reasonable to assume that end consumers and enterprises will want to adopt the same kind of approach for managing authorization as well as resource access control in IoT application scenarios. Also, this will become more and more likely with the rapidly increasing number of services provided by applications hosted on IoT devices organized in large-scale deployments. This has led to an IETF proposal (Seitz et al., 2018) aimed at reusing the OAuth 2.0 framework to extend authorization to IoT devices with different kinds of constraints, by using the basic OAuth 2.0 mechanisms where possible, while providing implementers with additional guidance, profiles, and extensions for using it in a secure and a friendly way with privacy. In particular, Seitz et al. (2018) describe a framework for authorization in the IoT that combines together a number of building blocks—(i) the original OAuth 2.0 authorization framework, (ii) the transfer protocol CoAP (Shelby et al., 2014), and (iii) the application layer security services, based on object security to CBOR-encoded data (Selander et al., 2018)—required when transport layer security provided by protocols like DTLS is not sufficient, adequate, or convenient. The DTLS protocol as well as object security to CBOR-encoded data have been discussed in more detail in the previous sections of this chapter.

8.2.7 Overview of Security Protocols and Mechanisms in the IoT

Table 8.1 provides a compact overview of the security mechanisms and protocols discussed throughout Section 8.2, highlighting the layer that benefits from their contributions.

8.3 Security Issues and Solutions

Protocols and mechanisms presented in Section 8.2 are typically adopted to fulfill primary fundamental security requirements in the IoT, such as secure communication at different layers, authorization for resource access, and compact encoding for cryptographic formats. Nevertheless, even when such basic security warranties are ensured, IoT systems are prone to a number of security vulnerabilities and exposed to specific security attacks, exploiting weaknesses and inefficiencies of security protocols, as well as poor scalability and efficiency of typically adopted management schemes. Moreover, such attack vectors can be particularly easy to exploit in the IoT for two main reasons. First, the majority of IoT devices are supposed to be directly connected to the Internet in order to be directly reachable, and hence fully exposed to adversaries motivated to carry out, for instance, DoS attacks. Second, many IoT devices are resource constrained, as equipped with a limited amount of memory, computing, and energy resources. This means that attacks that are otherwise fairly feasible to deal with can result in a much more severe impact. Besides, even related security counteractions have to be designed in order to be feasible and adoptable for such classes of constrained devices.

This section presents a number of specific security issues related to IoT systems, protocols, and technologies, and discusses possible solutions to address them based on recent results from the scientific community and activities in

Table 8.1 Overview of protocols and mechanisms.

Protocol/ mechanism	Stack layer	Purpose	Reasons to adopt
CBOR Object Signing and Encryption	Application	Serialization of data structures for cryptographic formats	Efficient encoding of messages secured at the application layer
OSCORE	Application	Secure end-to-end communication between a CoAP client and a CoAP server	Required in the presence of intermediary entities (e.g., proxies) that can legitimately inspect and selectively alter CoAP messages
DTLS	Transport	Secure hop-by-hop communication between a DTLS client and a DTLS server	Required to entirely protect CoAP messages
Compressed IPsec	Network	Secure hop-by-hop communication between two IPsec nodes	Required to entirely protect transport packets. Compression by 6LoWPAN to deal with limitation of lower layers
IEEE 802.15.4 security services	MAC	Secure hop-by-hop communication between two 802.15.4 nodes	Required to entirely protect network packets and especially routing control traffic
Bluetooth Low Energy	MAC	Secure hop-by-hop communication between two Bluetooth nodes	Security level equal to that in standard Bluetooth. Possible white lists of private link-level addresses
OAuth 2.0	Application	Issue and enforcement of authorizations for resource access on remote server	Based on an Authorization Server, it totally separates clients and resource owners

standardization bodies. Particular focus is put on different forms of DoS attacks, including (selective) jamming performed at the physical layer, while considering also the need for efficient and scalable protocols for management of cryptographic keys, secure communication in multicast groups of IoT devices, and the adoption of Intrusion Detection Systems (IDSs) and firewalls.

8.3.1 Denial-of-Service Against CoAP

Energy-constrained, even battery-powered, IoT devices are substantially exposed to a number of security attacks, especially if they are directly connected to the Internet with no particular protection implemented through dedicated gateway or firewall boxes. A particular well-known class of security attacks is DoS, which in turn may be mounted with the specific intent to keep IoT devices constantly busy in receiving and processing invalid messages, so preventing them from switching to any energy-saving *sleep mode* and quickly draining their battery energy supply (see Figure 8.3).

This specific kind of DoS attacks is often referred as *denial-of-sleep*, and is considered a particularly severe security threat against constrained, battery-powered, IoT devices (Stajano and Anderson, 1999; Martin et al., 2004; Raymond et al., 2009; Gehrmann et al., 2015). In addition, their occurrence cannot be avoided altogether. That is, an active adversary that continuously transmits bogus messages can always induce a victim recipient IoT device to receive them and perform processing operations at different layers. This means that a practical countermeasure should instead focus on *reducing* the attack impact as much as possible, with particular reference to the energy consumption due to the useless message processing. However, counteracting this class of attacks in an effective and efficient way is definitely a challenging task, particularly in the presence of IoT devices that are directly connected to the Internet, that is, globally accessible to enable remote management and configuration, operation requests, and information retrieval.

Nowadays, *CoAP* has asserted itself as the de facto application layer protocol for the IoT, with particular reference to resource-constrained devices (Shelby et al., 2014). However, when it comes to providing secure communication, CoAP does not provide any particular security primitives or countermeasures against DoS attacks. Instead, CoAP suggests adopting the DTLS protocol (Rescorla and

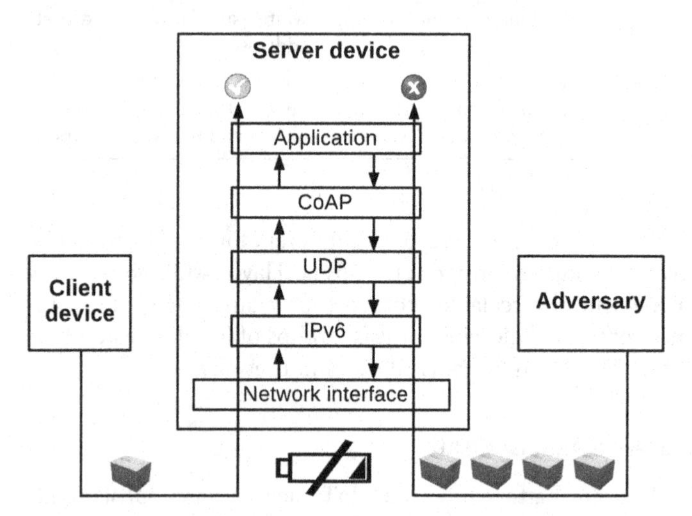

Figure 8.3 Denial-of-service attack against a server IoT device.

Modadugu, 2012), which relies on a costly and complex handshake process in order to setup a secure session, and results in a nonnegligible overhead to provide a whole set of security assurances, some of which may even not be necessary in some application scenarios.

A number of works have proposed detection techniques against denial-of-sleep attacks, typically considering traffic analysis or other anomaly detection mechanisms. For instance, Bhattasali and Chaki (2012) proposes a probabilistic model to detect denial-of-sleep attacks, based on Absorbing Markov Chains. Instead, Buennemeyer et al. (2007) describe B-SIPS, a system based on an innovative Dynamic Threshold Calculation algorithm that provides alerts upon detecting power changes. As pointed out in Raymond et al. (2009), these solutions based on intrusion detection techniques typically require to capture and analyze a considerable large amount of network traffic, and thus can be difficult or even unfeasible to deploy and run them on resource-constrained IoT devices.

Different proposed approaches consider traditional security services, especially at the link layer. For instance, Martin et al. (2004) propose a power-secure architecture, which processes only requests that deserve a high level of trust, but does not describe any concrete scheme to improve battery drain robustness. Instead, Raymond and Midkiff (2007) present an approach for rate limiting that relies on host-based intrusion detection mechanisms enforced at the link layer, considering messages to be legitimate only if both authenticated and not replayed, and taking actions to mitigate the attack effects. Also, Raymond et al. (2009) propose a framework based on classic link layer authentication, replay protection, and jamming identification. These detection approaches relying on traditional link layer security are however likely to result in a considerable overhead, especially as to energy consumption (Daidone et al., 2011). Also, they imply that validity of messages must be checked by every network node, along the path from the source to the destination device. This evidently produces a considerable additional impact on network performance as a whole. Furthermore, these approaches require to establish secure trust relations among all devices in the communication path (e.g., by distributing pairwise cryptographic keys), thus likely failing in efficiently providing protection against denial-of-sleep attacks from an end-to-end perspective.

More recently, Gehrmann et al. (2015) have proposed a security service relying on a short *Message Authentication Code* directly embedded in the CoAP message header, in order to early and efficiently detect invalid messages upon their reception. That is, it makes it possible to quickly determine whether received messages are valid or not, namely, whether they have been possibly transmitted by illegitimate sources. In such a case, a victim IoT device can early discard invalid messages and avoid performing useless additional parsing and processing operations. This considerably reduces the impact of denial-of-sleep attacks as to energy consumption, and thus preserves battery life in constrained IoT devices.

8.3.2 Denial-of-Service and Scalability Issues in DTLS

The DTLS handshake process displays two relevant security and performance issues that mostly affect devices acting as DTLS server.

First of all, as also stressed in the DTLS specification, a device acting as DTLS server is particularly vulnerable to DoS attacks mounted during the handshake process (Rescorla and Modadugu, 2012). Specifically, an active adversary has the possibility to continuously transmit *ClientHello* messages to the DTLS server, and induce it to start a great amount of DTLS handshakes. This in turn forces the server to start the setup of new (invalid) DTLS sessions. From a DTLS server point of view, this means allocating network and memory resources, and performing a number of resource-demanding operations. Moreover, this may exhaust available network and memory resources, so making the server less responsive or, in the worst case, even unavailable to process legitimate requests sent by DTLS clients. As a further drawback, attacks mounted through spoofed valid IP addresses force the DTLS server to send reply messages to "innocent" devices, so displaying a well-known *amplification* effect. The specification of DTLS provides a tentative solution against the DoS attack mentioned above, which leverages the stateless and optional exchange of a piece of information, namely, *cookie*, occurring during the first phases of the handshake between a DTLS client and server (Rescorla and Modadugu, 2012). Specifically, the DTLS server may reply to the first *ClientHello* message with an additional *HelloVerifyRequest* message, which includes a cookie as a locally generated value. After that, the DTLS client replies with another *ClientHello* message, which must include exactly the same cookie. However, the cookie exchange process only complicates the considered DoS attack, while it does not fundamentally provide any real protection against it. More in detail, as shown in Figure 8.4, an adversary able to intercept messages sent by the DTLS server during the handshake can still be able to perform the DoS attack, and induce the server to set up a considerable number of invalid, half-open DTLS sessions. It follows that the countermeasure based on the cookie exchange performed between client and

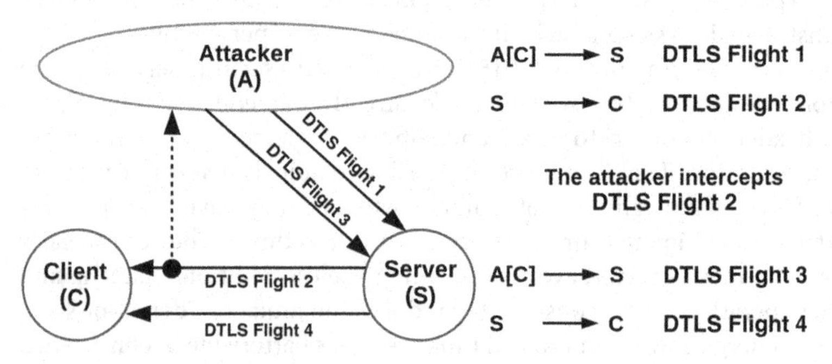

Figure 8.4 Denial-of-service attack targeting a DTLS server.

server is actually not a good solution against DoS attacks mounted by a resourceful and well determined on-path adversary.

Furthermore, if the handshake is based on the PSK approach (Eronen and Tschofenig, 2005), the DTLS server may require to store and manage a non-negligible number of symmetric keys associated with DTLS clients. As shown in Figure 8.5, this means that a DTLS server stores in the worst case one preshared key for every possible DTLS client. Of course, this approach scales poorly with the number of DTLS clients in the network. Moreover, it considerably complicates key provisioning and management operations, especially in dynamic scenarios. Nevertheless, the PSK scheme is very useful, convenient, and widely adopted in several dynamic IoT environments that involve resource-constrained devices as well as end users with no capability to securely manage a more complex public key infrastructure.

In Tiloca et al. (2017c), a security architecture has been proposed in order to address both issues discussed already. In particular, the proposed approach makes it possible for the DTLS server to detect an invalid *ClientHello* message upon its reception, and then immediately terminate the handshake at its very beginning, so practically neutralizing the attack and substantially limiting its impact. Also, it provides an alternative PSK scheme that greatly reduces the number of preshared keys stored at the server to one, thus preventing scalability and management issues. Furthermore, the proposed solution displays a number of advantages. First, it relies on a standardized method to extend *ClientHello* messages, and thus it is not required to modify the DTLS standard. Second, the DTLS client and server are not required to perform any further message exchange, and even the cookie exchange is not necessary anymore. Besides, the computing overhead on the client and server is not significantly affected, as the same computational complexity is preserved for the handshake process. Finally, it can be easily readopted in TLS, again with no required changes.

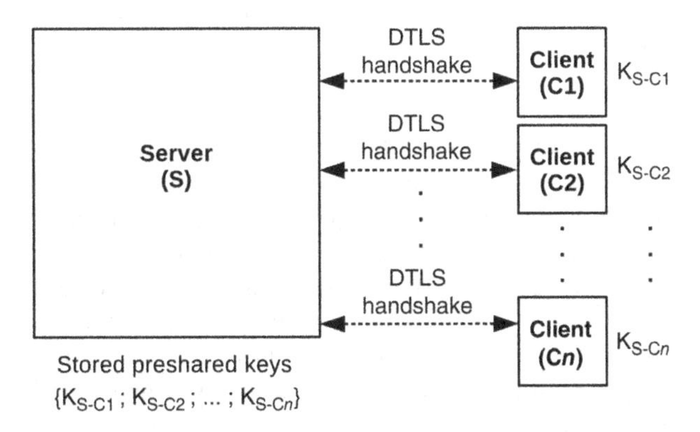

Figure 8.5 Poorly scalable storage of preshared keys on a DTLS server.

8.3.3 DTLS-Based Group Communication

As discussed in Section 8.2.2, the CoAP protocol (Shelby et al., 2014) mandates the adoption of the DTLS protocol (Rescorla and Modadugu, 2012) to secure unicast communications between a given pair of IoT devices. At the same time, in several IoT deployments involving *groups* with IoT devices as members, using multicast communication over IPv6 has several advantages, particularly as to scalability and performance. This is the reason why the standard (Rahman and Dijk, 2014) has explicitly defined group communication based on the CoAP protocol, with reference to a number of different use cases where the group communication model is particularly beneficial to IoT devices. For instance, group communication is highly convenient, or even inevitable, in IoT application scenarios such as firmware and software updates, lighting control, configuration and parameter updates, integrated building control, emergency broadcast, and commissioning of 6LoWPAN networks. In many IoT deployments, these use cases display a number of specific security requirements to be fulfilled, including the secure exchange of multicast messages among IoT devices within the same group. However, at the time of writing this chapter, CoAP does not provide support, or recommend an interoperable way, for securing multicast group communications.

At the same time, DTLS is the de facto security protocol to secure IoT unicast communications. Therefore, it is especially convenient to enhance DTLS so that it can also support secure group communication in the IoT. This choice would also be particularly convenient if compared with possible different alternatives. For instance, opting for IPsec multicast (Weis et al., 2008) would also require to use the heavyweight IPsec (Kent and Seo, 2005; Briscoe, 2010), and likely even additional supporting protocols, such as IKEv2 (Kaufman et al., 2014). Furthermore, DTLS is currently supposed to be used for protecting unicast communications in the IoT, hence opting for IPsec solutions would be even less practical and convenient in real-world application scenarios.

So far, there have been only a few initiatives to provide DTLS-based secure group communication for CoAP, in the remit of the standardization activities carried out within the IETF. In particular, the first approaches toward this direction were presented in the two documents Keoh et al. (2014) and Kumar and Struik (2015), which propose to protect communications in a multicast group of IoT devices by conveniently adapting the DTLS record layer, but were questioned because of a number of reasons. Among them it was pointed out that source authentication is a fundamental security requirement to be fulfilled, as group authentication alone may not be enough for several IoT application scenarios. Also, the protection of group response messages to multicast request messages was provided by means of traditional, pairwise, DTLS sessions. However, this can result in a number of practical issues and can display considerable degradation of performance, particularly in highly dynamic and large-scale groups of IoT devices.

On the other hand, the method described in Tiloca et al. (2015, 2017b) and Tiloca (2014) elaborates on the previous approaches, and proposes how to provide *two-way* secure group communication based on DTLS, while addressing the limitations displayed by the previously proposed approaches. Specifically, all the members of the multicast group consider the same group key material or efficiently derive individual key material. Group members rely on the overall key material in their possession to protect both requests sent to the group as multicast messages as well as possible responses sent back as unicast messages. The common key material is provided to a new group member upon its joining, and it is not necessary to perform DTLS handshakes among group members to enable secure communication in the group. Finally, the proposed approach ensures either source authentication or group authentication of messages sent within the group, according to the requirements of the specific application scenario.

8.3.4 Efficient and Scalable Group Key Management

As already mentioned, group communication is a convenient approach to adopt in many IoT application scenarios, in order to improve performance, responsiveness, and energy efficiency. However, it also requires a complex management of the group key material, as to generation, revocation, and (re) distribution. In general, the group key material comprises the secrets shared among the group members that enable secure communication only among devices in that group, at a given communication layer. For simplicity, the following refers to the group key material as a single symmetric *group key*, which all the group members commonly share and use to secure communications within the group at the application layer.

According to the group communication model, an IoT device explicitly joins a multicast group in order to become an active member. From then on, the device is able to receive (send) broadcast messages from (to) other devices in the same group. At some point in time, the device may want to leave the group, or can be forced to leave, in case it is compromised or suspected so. In order to allow only the actual group members to participate to the communications in the group, it is necessary to revoke and renew the current group key, that is, *rekeying*, according to the following policies. When a new device becomes member of the group, it must be prevented from decrypting and accessing messages exchanged before its joining, even if it has recorded them (backward security). Also, when a device decides to leave the group, or is forced to as compromised or suspected so, it must be prevented from decrypting and accessing any further future communication in the group (forward security). Practically, upon a change in the group membership, it is necessary to revoke the current group key and to distribute a new one to the current/remaining devices in the group. Therefore, it is particularly important that rekeying is performed in an efficient and highly scalable way, especially in highly dynamic and large-scale groups where joining and leaving can be very frequent events.

Furthermore, it is of particular importance to effectively deal with collusion attacks, which occur when multiple devices in the group are compromised, share their security material, and take part to the rekeying procedure anyway. Thus, these compromised nodes cooperate to regain knowledge of the group key, and would not be effectively removed from the multicast group. There are no group rekeying protocols that are not exposed to collusion attacks altogether, albeit they are characterized by different levels of resilience. However, only a few provide countermeasures to address collusion attacks. Moreover, several rekeying schemes recover the group from collusion attacks by means of a total member reinitialization, that is, each noncompromised device in the group has to be individually reinitialized. As a consequence, the collusion recovery displays a communication overhead that linearly grows with the group size, and can very negatively impact on the network scalability and performance.

Protocols for group key management have been typically classified as centralized, distributed, and decentralized (Rafaeli and Hutchison, 2003). Centralized approaches are particularly preferable in case of highly dynamic and large-scale groups, and especially good at minimizing communication, storage, and computing overhead. Centralized schemes use a set of administrative key material, which is logically organized for providing scalable rekeying upon node leaving (Dini and Savino, 2011; Dini and Tiloca, 2013; Gu et al., 2009; Tiloca and Dini, 2016; Wallner et al., 1999; Wong et al., 2000).

Furthermore, *Logical Key Hierarchy* (LKH) is a centralized protocol that relies on a hierarchical logical tree to organize administrative keys (Wallner et al., 1999). Specifically, the group key is the root of the tree, the group members' individual keys are the leaves of the tree, and the additional administrative keys are the internal tree nodes. In a group composed of n members, and assuming the key material to be organized in a balanced tree, both the storage overhead at the device side as well as the leave communication overhead grow as $O(log\ n)$.

Figure 8.6 shows the key material managed by LKH in a group composed of eight devices, namely, $u_1, u_2, \ldots, u_8$. In particular, each device owns the keys along the path from its related leaf node in the logical tree, up to the root node associated to the group key K_G. For instance, device u_3 owns the individual key K_3, the additional administrative keys K_{1-4} and K_{3-4}, and the group key K_G.

Other protocols deriving from LKH do not achieve better performance (Dini and Savino, 2011; Gu et al., 2009; Wong et al., 2000). Furthermore, the centralized protocols *HIghly Scalable Scheme* (HISS) (Dini and Tiloca, 2013) and *Group REkeying Protocol* GREP (Tiloca and Dini, 2016) leverage logical subgrouping in order to provide highly scalable and efficient group rekeying, display the same storage and computing overhead, that is, $O(\sqrt{n})$, and distribute a new group key by using a number of rekeying messages that is constant, small, and independent of the group size, that is, $O(1)$. In addition, the GREP protocol

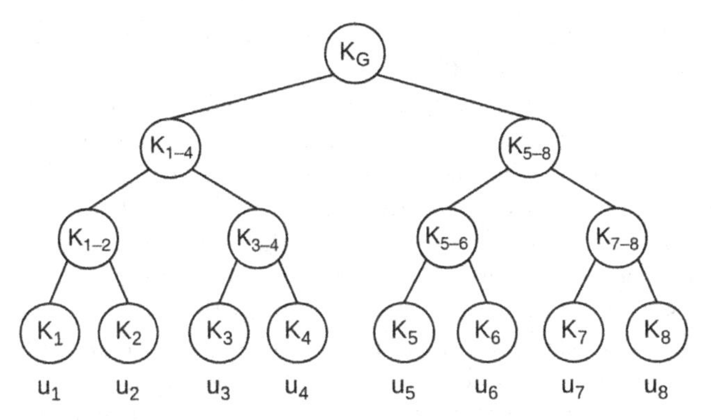

Figure 8.6 Example of LKH key material in the presence of eight group members.

relies on the novel idea of *member join history*, and in fact leverages it in order to perform collusion recovery in a considerably more scalable and efficient fashion, with no recourse to a total member reinitialization. Specifically, HISS displays a recovery overhead that always grows as $O(n)$. In contrast, the recovery overhead due to GREP gradually increases with the severity of the specific collusion attack and smoothly grows as $O(\sqrt{n})$, albeit only in the very unlikely worst case conditions.

On the other hand, distributed protocols rely on key management actions performed in a distributed way, involving the group members themselves or relying on additional key (sub)managers. However, they are harder to implement, are often less scalable, and raise security concerns. Moreover, some do not provide recovery from collusion attacks (Cheung et al., 2007; Fan et al., 2002), while others always require a total member reinitialization (Wang et al., 2008; Younis et al., 2006) or trade performance with collusion resistance (Chorzempa et al., 2005).

8.3.5 Selective Jamming in Wireless Networks

Jamming attacks are well known to be among the most common and severe kinds of DoS attacks against networks relying on wireless communication technologies (Lazos et al., 2009; Lu et al., 2014; Mustafa et al., 2012; Stojanovski and Kulakov, 2015; Xu et al., 2004, 2005, 2006). Specifically, jamming attacks consist in deliberately interfering with the operational frequencies in the network, and have been typically classified as *constant, random, deceptive,* and *reactive* (Xu et al., 2005).

On one side of the spectrum, constant jamming is extremely easy to perform. Its main objective consists in corrupting *all* transmitted network packets, by performing continuous overlapping transmissions of random signals. Note that this jamming strategy can be well considered as "always-on," since it relies on the constant presence of a high level of interference, which thus makes the attack much easier to detect (Xu et al., 2004, 2005). On the other side of the spectrum, reactive jamming denotes a particularly smart and power-efficient approach, according to which an adversary actively jams communications in the network only during transmissions from other devices. Moreover, it is very likely that reactive jamming is actually confused with normal collisions of network packets, which makes it much more difficult to detect than other forms of jamming attack. Even more specifically, *selective jamming* is a specific type of reactive jamming, and it aims at interfering with communications between network devices, in accordance to well defined and specific objectives and criteria. With respect to other kinds of jamming attacks and strategies, selective jamming is more power efficient to perform, and even more difficult to counteract and detect, due to the reduced exposure of the adversary. A common criterion considered for selective jamming consists in disrupting all communication of one particular device in the network. In particular, this attack is particularly easy to perform in a wireless network using Time Division Multiple Access (TDMA). Specifically, TDMA divides time into a periodic sequence of *superframes*, each of which is composed of a fixed amount of transmission *slots*. However, such slots are typically assigned to network devices in a way that every device can remain active only during its own slot(s), and it can instead go to sleep mode during the other slots, so limiting its energy consumption. On the other hand, a network device typically retains exactly the same slot(s) for several consecutive superframes. As a consequence, an adversary has to simply monitor network communications, determine the slots accessed by its victim devices, and jam them to completely thwart the devices' communications. Note that such an attack is also very efficient from a power consumption point of view, since the adversary requires to keep her radio active only during the transmission slots assigned to her victim devices.

The countermeasures proposed to counteract jamming attack have been typically divided into *physical-layer* and *cyber* solutions. In principle, physical-layer solutions try to prevent an adversary from successfully interfering with the frequencies used in the network. The most important techniques belonging to this category have been overviewed in Mpitziopoulos et al. (2009), and are often based on *Frequency-Hopping Spread Spectrum* (FHSS) (Pickholtz et al., 1982), that is, a method that selects a different carrier from the available frequency channels, in accordance to an algorithm shared by the receiver and transmitter device. However, physical-layer approaches are not actually capable to fundamentally neutralize jamming attacks. Conversely, cyber solutions assume the adversary as always able to interfere at the physical layer with regular transmissions in the network. Hence, cyber solutions contrast jamming attacks by means of appropriate security schemes. Most of cyber countermeasures are specialized to address constant jamming (Cagalj et al., 2007; Raymond and Midkiff, 2008; Xu et al., 2004; Xu et al., 2007). Instead, only a few solutions have been specifically designed to target selective jamming attacks. For instance, Wood et al. (2007) describe *Defeating Energy-Efficient JAMming in IEEE 802.15.4-based wireless networks* (DEEJAM), that is, a MAC protocol that provides protection against selective jamming attacks mounted by means of hardware supporting IEEE 802.15.4. In principle, DEEJAM tries to hide packets from an adversary jammer node, in order to evade its search and thus limiting the impact due to packets that are corrupted anyway. However, DEEJAM is specifically tailored to IEEE 802.15.4 networks, and results in a considerable computational overhead and energy consumption. Furthermore, Ashraf et al. (2012) proposed the low-overhead framework Jam-Buster, which leverages equal size of packets, multi-block payloads, and randomization wake up time of devices. By doing so, it tends to reduce predictability of transmission times on network devices and to eliminate differentiation of packet types. As a consequence, the adversary has to transmit additional jamming signals, and hence is required to consume additional energy to remain effective, and is faster and easier to detect as source of the jamming attack. However, Jam-Buster does not fundamentally counteract the attack by means of an actual solution against jamming, but instead tries to make selective jamming less efficient and convenient to perform. Proano and Lazos (2012) consider a particular kind of selective jamming, where only specifically important types of packets are jammed upon their transmission, and proposes some techniques to mitigate the attack effects, based on cryptographic primitives. While this can be a good solution to counteract packet classification, it also requires that the packets in their entirety are encrypted, so preventing receiver devices from early aborting the reception of packets not addressed to them. In Daidone et al. (2014), the authors propose a solution that counteracts selective jamming in IEEE 802.15.4 networks and use the mechanism Guaranteed Time Slot (GTS). In the GTS mode, a central coordinator can allocate at most seven reserved time slots to sensor nodes in the network, at each superframe. The proposed solution addresses the GTS severe vulnerability to selective jamming previously described in Sokullu et al. (2009), and entirely relies on the central coordinator node, which computes and randomly changes the utilization pattern of GTS slots at each superframe. This makes it possible to reduce the effectiveness of the attack to 1/7, but at the same time makes the central coordinator a single point of failure. Moreover, the proposed countermeasure specifically considers the IEEE 802.15.4 standard and thus is not general. Finally, Tiloca et al. (2013, 2015) present JAMMY, a dynamic and distributed solution that counteracts selective jamming in general TDMA wireless networks. In principle, JAMMY permutes the slot utilization pattern of network devices in a random way, at each superframe. It follows that the slots associated with every network device unpredictably change on a superframe basis. As a consequence, the adversary has the only option left to jam slots picked at random, hoping to guess the ones that the intended victim device uses for transmission at that superframe. In case only one slot per device is used at every superframe, then a successful selective jamming attack has a probability of *1/N*, given *N* the number of slots composing the superframe. More important, JAMMY does not rely on any central entity and is totally distributed, that is, every network device determines the slots to use for transmission in the next superframe in an autonomous way (i.e., with no data exchange with other devices) and in a consistent fashion (i.e., with no resulting collisions with other devices). Furthermore, JAMMY is dynamic, since multiple devices can join and leave the network at any time. Finally, it is in principle usable in any TDMA wireless network, regardless the specific technology.

8.3.6 Intrusion Detection Systems and Firewalls

Although communication security can protect against attacks that target confidentiality and integrity of sensitive data, there are a number of attacks aimed at disrupting sensor networks as a whole (Karlof and Wagner, 2003), such as DoS attacks. Especially for mission-critical networks, it is therefore necessary to adopt countermeasures against such kinds of attacks.

There are many approaches that can protect sensor networks against DoS attacks, by detecting and possibly counteracting intrusions and other anomalies. These intrusion detection systems (IDS) for sensor networks can be categorized as *watchdog-based* where a designated component is used for the detection of selfish nodes and malicious attackers (Marti et al., 2000; Rong et al., 2011); *signature or rule-based* that detect known types of intrusions by recognizing bad patterns (da Silva et al., 2005; Bhuse and Gupta, 2006); *anomaly-based* that detects deviations from a model of normal or good behavior (Onat and Miri, 2005; Loo et al., 2006); or *cluster-based* detection where a group of nodes elect a leader that participates in the detection process (Lazos et al., 2005; Rong et al., 2011). These conventional approaches target wireless sensor networks and are not directly applicable in the IoT. In fact, they assume that no central controller or management entity is available, messages within the sensor network are not secured, and nodes are not globally identifiable. Unlike a traditional sensor network, a 6LoWPAN network includes a 6LoWPAN Border Router (6BR) that is always reachable, sensor nodes are globally identified by means of an IPv6 address, and communication security is a fundamental requirement to fulfill. In addition, building an IDS that is suitable to the IoT remains a very challenging task, because 6LoWPAN devices are supposed to be globally reachable, are connected through lossy wireless links, are mostly resource-constrained units, and use not well tested new IoT protocols, for example, CoAP, 6LoWPAN, and RPL.

By exploiting these opportunities and threats, an IDS suitable to the IoT named SVELTE has been developed (Raza et al., 2013). SVELTE primarily defends against routing attacks against 6LoWPAN networks running the RPL protocol. A crucial decision in an IDS design concerns the placement of IDS modules in the network. SVELTE is a hybrid system where heavyweight detection modules are hosted in a 6BR, and lightweight notification functions are added in sensor nodes. SVELTE has two centralized modules hosted in 6BR: (i) the 6LoWPAN Mapper (6Mapper) that collects RPL-specific information from individual sensor nodes and rebuilds the RPL network in the 6BR and (ii) detection modules that analyze the mapped data and then perform the actual intrusion detection. The module 6Mapper has a corresponding function in each individual node that provides the mapping information to the 6BR.

Although SVELTE can protect 6LoWPAN networks against in-network routing attacks, it is fundamental that the sensor devices are safeguarded against global, much powerful, intruders. For example, a host on the Internet can easily perform DoS attacks aimed to deplete a sensor node's limited resources. A firewall is typically used to screen messages from Internet hosts destined to local area networks. For IoT deployments where end-to-end message integrity and confidentiality is enforced between a sensor device and a host on the Internet, a firewall cannot scrutinize the encrypted contents. SVELTE features a distributed mini-firewall for 6LoWPAN networks. Both the 6BR and each constrained node have a firewall module. In addition to offering classic blocking functionalities against external attackers specified by a system administrator, the firewall can filter and block malicious external hosts as notified by the legitimate nodes in the 6LoWPAN network in real time.

8.4 Conclusion

This chapter has overviewed the most notable security protocols and technologies currently available for adoption in the IoT, at different layers of the typical communication stack. Furthermore, this chapter has discussed specific security issues in the IoT and solutions to address them leveraging recent research and standardization efforts. More work is expected to be carried out and further challenges must be taken to achieve a de facto secure IoT, and ultimately pave the way to its pervasive and wide adoption. The following trends and topics are especially going to play a crucial role in the immediate research and standardization agenda about security in the IoT.

First, more and more application scenarios require explicit support for secure group communication, built-in at different stack layers. This encompasses both actual message protection and management of security material, and is expected to focus especially at the application layer (through approaches based on object-security) and at the network layer (through the IPsec and IKEv2 protocols). Second, a comprehensive set of profiles for the IoT has to be specified, in order to define how to take benefit of authorization frameworks based on OAuth 2.0, both for handling the actual authorized access to remote resources and for coassisting the contextual establishment of secure channels with IoT resource servers. Finally, novel approaches are required to improve robustness of IoT units against denial-of-services attacks aimed at thwarting their availability and even exhausting their power resources. These approach should better leverage context-aware information and be adaptive to mutable attack conditions.

Last but not the least, these open issues require efficient solutions that display a sustainable impact on performance and user experience, and that can be realistically deployed and easily maintained even in large-scale, highly dynamic deployments of resource-constrained devices.

References

Ashraf, F., Hu, Y.-C., and Kravets, R. H. (2012) Bankrupting the jammer in WSN. The IEEE 9th International Conference on Mobile Adhoc and Sensor Systems, 317–325.

Bhattasali, T. and Chaki, R. (2012) AMC Model for Denial of Sleep Attack Detection. *Journal of Recent Research Trends (JRRT)*. doi: arXiv:1203.1777v1.

Bhuse, V. and Gupta, A. (2006) Anomaly intrusion detection in wireless sensor networks. *Journal of High Speed Networks, Special issue on trusted internet workshop (TIW)*, **15**(1), 33–51.

Bluetooth Special Interest Group. (2014) Bluetooth Specification - Version 4.2

Bluetooth Special Interest Group. (2016) Internet Protocol Support Profile: Bluetooth Specification.

Bormann, C. (2014) *6LoWPAN-GHC: Generic Header Compression for IPv6 over Low-Power Wireless Personal Area Networks (6LoWPANs)*, RFC 7400, Internet Engineering Task Force.

Bormann, C. (2015) Signature Authentication in the Internet Key Exchange Version 2 (IKEv2), RFC 7427, Internet Engineering Task Force.

Bormann, C., Ersue, M., and Keranen, A. (2014) Terminology for Constrained-Node Networks. RFC 7228, Internet Engineering Task Force.

Bormann, C. and Hoffman, P. (2013) Concise Binary Object Representation (CBOR), RFC 7049, Internet Engineering Task Force.

Briscoe, B. (2010) Tunnelling of Explicit Congestion Notification. RFC 6040, Internet Engineering Task Force.

Buennemeyer, T. K., Gora, M., Marchany, R. C., and Tront, J. G. (2007) Battery exhaustion attack detection with small handheld mobile computers. The 2007 IEEE International Conference on Portable Information Devices (PORTABLE 2007), 1–5.

Cagalj, M., Capkun, S., and Hubaux, J.-P. (2007) Wormhole-based antijamming techniques in sensor networks. *IEEE Transactions on Mobile Computing*, **6**, 100–114.

Cheung, L., Cooley, J. A., Khazan, R., and Newport, C. (2007) *Collusion-Resistant Group Key Management Using Attribute-Based Encryption*, International Association for Cryptologic Research.

Chorzempa, M., Park, J.-M., and Eltoweissy, M. (2005) SECK: survivable and efficient clustered keying for wireless sensor networks. IEEE International Performance, Computing, and Communications Conference 2005, 453–458.

Cooper, D., Santesson, S., Farrell, S., Boeyen, S., Housley, R., and Polk, W. (2008) Internet X.509 Public Key Infrastructure Certificate and Certificate Revocation List (CRL) Profile. RFC 5280, Internet Engineering Task Force.

da Silva, A. P. R., Martins, M. H. T., Rocha, B. P. S., Loureiro, A. A. F., Ruiz, L., Linnyer, B., and Wong, H. C. (2005) Decentralized intrusion detection in wireless sensor networks. The 1st ACM International Workshop on Quality of Service and Security in Wireless and Mobile Networks, 16–23.

Daidone, R., Dini, G., and Tiloca, M. (2011) On experimentally evaluating the impact of security on IEEE 802.15.4 networks. The 3rd International Workshop on Performance Control in Wireless Sensor Networks (PWSN 2011), 20–25.

Daidone, R., Dini, G., and Tiloca, M. (2014) A solution to the GTS-based selective jamming attack on IEEE 802.15.4 networks. *Wireless Networks*, **20**(5), 1223–1235.

Dierks, T. and Rescorla, E. (2008) The Transport Layer Security (TLS) Protocol Version 1.2. RFC 5246, Internet Engineering Task Force.

Dini, G. and Savino, I. M. (2011) LARK: a lightweight authenticated rekeying scheme for clustered wireless sensor networks. *ACM Transactions on Embedded Computing Systems*, **10**(4), 1–35.

Dini, G. and Tiloca, M. (2013) HISS: a highly scalable scheme for group rekeying. *The Computer Journal*, **56**(4), 508–525.

Eronen, P. and Tschofenig, H. (2005) Pre-Shared Key Ciphersuites for Transport Layer Security (TLS). RFC 4279, Internet Engineering Task Force.

Ertekin, E., Jasani, R., Christou, C., and Bormann, C. (2010) Integration of Robust Header Compression over IPsec Security Associations. RFC 5856, Internet Engineering Task Force.

Fan, J., Judge, P., and Ammar, M. H. (2002) HySOR: group key management with collusion-scalability tradeoffs using a hybrid structuring of receivers. IEEE 2002 International Conference on Computer Communications and Networks, 196–201.

Fielding, R. T. and Reschke, J. (2014a) Hypertext Transfer Protocol (HTTP/1.1): Message Syntax and Routing. RFC 7230, Internet Engineering Task Force.

Fielding, R. T. and Reschke, J. (2014b) Hypertext Transfer Protocol (HTTP/1.1): Semantics and Content. RFC 7231, Internet Engineering Task Force.

Gehrmann, C., Tiloca, M., and Höglund, R. (2015) SMACK: short message authentication ChecK against battery exhaustion in the Internet of Things. IEEE International Conference on Sensing Communication and Networking 2015 (SECON 2015), 274–282.

Gu, Q., Liu, P., Lee, W.-C., and Chu, C.-H. (2009) KTR: an efficient key management scheme for secure data access control in wireless broadcast services. *IEEE Transactions on Dependable and Secure Computing,* **6**(3), 188–201.

Hardt, D. (2012) The OAuth 2.0 Authorization Framework. RFC 6749, Internet Engineering Task Force.

Hui, J. W. and Thubert, P. (2011) Compression Format for IPv6 Datagrams over IEEE 802.15.4-Based Networks. RFC 6282, Internet Engineering Task Force.

IEEE Computer Society. (2011) IEEE Standard for Local and Metropolitan Area Networks, Part 15.4: Low-Rate Wireless Personal Area Networks (LR-WPANs).

IEEE Computer Society. (2012) 802.15.4e-2012 - IEEE Standard for Local and Metropolitan Area Networks–Part 15.4: Low-Rate Wireless Personal Area Networks (LR-WPANs) Amendment 1: MAC sublayer.

Jankiewicz, E., Loughney, J., and Narten, T. (2011) IPv6 Node Requirements. RFC 6434, Internet Engineering Task Force.

Jones, M. B. (2017) Using RSA Algorithms with CBOR Object Signing and Encryption (COSE) Messages, RFC8230, Internet Engineering Task Force.

Karlof, C. and Wagner, D. (2003) Secure routing in wireless sensor networks: attacks and countermeasures. *Ad Hoc Networks,* **1**(2), 293–315.

Kaufman, C., Hoffman, P., Nir, Y., Eronen, P., and Kivinen, T. (2014) Internet Key Exchange Protocol Version 2 (IKEv2). RFC 7296, Internet Engineering Task Force.

Kent, S. and Seo, K. (2005) Security Architecture for the Internet Protocol. RFC 4301, Internet Engineering Task Force.

Keoh, S., Kumar, S., Garcia-Morchon, O., Dijk, E., and Rahman, A. (2014) DTLS-based Multicast Security in Constrained Environments, Internet Engineering Task Force.

Kumar, S. and Struik, R. (2015) Transport-layer Multicast Security for Low-Power and Lossy Networks (LLNs), C Internet Engineering Task Force.

Lazos, L., Liu, S., and Krunz, M. (2009) Mitigating control-channel jamming attacks in multi-channel ad hoc networks. The second ACM conference on Wireless network security, 169–180.

Lazos, L., Poovendran, R., Meadows, C. A., Syverson, P. F., and Chang, L. W. (2005) Preventing wormhole attacks on wireless ad hoc networks: a graph theoretic approach. IEEE 2005 Wireless Communications and Networking Conference, vol. 2, 1193–1199.

Loo, C. E., Ng, M. Y., Leckie, C., and Palaniswami, M. (2006) Intrusion detection for routing attacks in sensor networks. *International Journal of Distributed Sensor Networks,* **2**(4), 313–332.

Lu, Z., Wang, W., and Wang, C. (2014) Modeling, evaluation and detection of jamming attacks in time-critical wireless applications. *IEEE Transactions on Mobile Computing,* **13**(8), 1746–1759.

Marti, S., Giuli, T. J., Lai, K., and Baker, M. (2000) Mitigating routing misbehavior in mobile ad hoc networks. In the 6th annual international conference on mobile computing and networking (MobiCom '00), vol. 6, 255–265.

Martin, T., Hsiao, M., Ha, D., and Krishnaswami, J. (2004) Denial-of-service attacks on battery-powered mobile computers. The Second IEEE International Conference on Pervasive Computing and Communications (PERCOM 2004), 309–318.

Migault, D. and Guggemos, T. (2015) Implicit IV for AES-CBC, AES-CTR, AES-CCM and AES-GCM, Internet Engineering Task Force.

Migault, D. and Guggemos, T., and Bormann, C. (2017) ESP Header Compression and Diet-ESP, draft-mglt-ipsecme-diet-esp-05 (work in progress), Internet Engineering Task Force.

Mpitziopoulos, A., Gavalas, D., Konstantopoulos, C., and Pantziou, G. (2009) A survey on jamming attacks and countermeasures in WSNs. *IEEE Communications Surveys Tutorials*, **11**(4), 42–56.

Mustafa, H., Zhang, X., Liu, Z., Xu, W., and Perrig, A. (2012) Jamming-resilient multipath routing. *IEEE Transactions on Dependable and Secure Computing*, **9**(6), 852–864.

National Institute of Standards and Technology. (2001) Federal Information Processing Standards, Specification for the Advanced Encryption Standard (AES).

Nieminen, J., Savolainen, T., Isomaki, M., Patil, B., Shelby, Z., and Gomez, C. (2015) IPv6 over BLUETOOTH(R) Low Energy, RFC 7668, Internet Engineering Task Force.

Onat, I. and Miri, A. (2005) An intrusion detection system for wireless sensor networks. In the IEEE 2005 International Conference on Wireless and Mobile Computing, Networking and Communications (WiMob 2005), 253–259.

Pickholtz, R. L., Schilling, D. L., and Milstein, L. B. (1982) Theory of spread-spectrum communications: a tutorial. *IEEE Transactions on Communications*, **30**(5), 855–884.

Proano, A. and Lazos, L. (2012) Packet-hiding methods for preventing selective jamming attacks. *IEEE Transactions on Dependable and Secure Computing*, **9**(1), 101–114.

Rafaeli, S. and Hutchison, D. (2003) A survey of key management for secure group communication. *ACM Computing Surveys*, **35**(3), 309–329.

Rahman, A. and Dijk, E. (2014) Group Communication for the Constrained Application Protocol (CoAP), RFC 7390, Internet Engineering Task Force.

Raymond, D. R. and Midkiff, S. F. (2007) Clustered adaptive rate limiting: defeating denial-of-sleep attacks in wireless sensor networks. IEEE 2007 Military Communications Conference (MILCOM 2007), 1–7.

Raymond, D. R. and Midkiff, S. F. (2008) Denial-of-service in wireless sensor networks: attacks and defenses. *IEEE Pervasive Computing*, 7(1), 74–81.

Raymond, D. R., Marchany, R. C., Brownfield, M. I., and Midkiff, S. F. (2009) Effects of denial-of-sleep attacks on wireless sensor network MAC protocols. *IEEE Transactions on Vehicular Technology*, **58**(1), 367–380.

Raza, S., Duquennoy, S., Höglund, J., Roedig, U., and Voigt, T. (2014) Secure communication for the Internet of Things: a comparison of link-layer security and IPsec for 6LoWPAN. *Journal of Security and Communication Networks*, 7(12), 2654–2668.

Raza, S., Duquennoy, S., and Selander, G. (2016) Compression of IPsec AH and ESP Headers for 6LoWPAN Networks.

Raza, S., Duquennoy, S., Voigt, T., and Roedig, U. (2011) Demo abstract: securing communication in 6LoWPAN with compressed IPsec. The 7th IEEE International Conference on Distributed Computing in Sensor Systems (DCOSS '11).

Raza, S., Wallgren, L., and Voigt, T. (2013) SVELTE: real-time intrusion detection in the Internet of Things. *Ad Hoc Networks*, **11**(8), 2661–2674.

Rescorla, E. and Modadugu, N. (2012) Datagram Transport Layer Security Version 1.2. RFC 6347, Internet Engineering Task Force.

Rong, C., Eggen, S., and Cheng, H. (2011) An efficient intrusion detection scheme for wireless sensor networks. *Secure and Trust Computing, Data Management, and Applications*, Springer, 116–129.

Sandlund, K., Pelletier, G., and Jonsson, L.-E. (2010) The RObust Header Compression (ROHC) Framework. RFC 5795, Internet Engineering Task Force.

Schaad, J. (2016) *CBOR Encoded Message Syntax: Additional Algorithms*, Internet Engineering Task Force.

Schaad, J. (2017) CBOR Object Signing and Encryption (COSE), RFC8152, Internet Engineering Task Force.

Seitz, L., Selander, G., Wahlstroem, E., Erdtman, S., and Tschofenig, H. (2018) Authentication and Authorization for Constrained Environments (ACE), draft-ietf-ace-oauth-authz-10 (work in progress), Internet Engineering Task Force.

Selander, G., Mattson, J., Palombini, F., and Seitz, L. (2018) Object Security for Constrained RESTful Environments (OSCORE), draft-ietf-core-object-security-09 (work in progress), Internet Engineering Task Force.

Shelby, Z., Hartke, K., and Bormann, C. (2014) Constrained Application Protocol (CoAP). RFC 7252, Internet Engineering Task Force.

Shirey, R. W. (2007) Internet Security Glossary, Version 2. FYI 36, RFC 4949, Internet Engineering Task Force.

Sokullu, R., Korkmaz, I., and Dagdeviren, O. (2009) GTS Attack: an IEEE 802.15.4 MAC layer attack in wireless sensor networks. *International Journal On Advances in Internet Technologies*, **2**(1), 104–114.

Stajano, F. and Anderson, R. J. (1999) The resurrecting duckling: security issues for ad-hoc wireless networks. The 7th International Workshop on Security Protocols, SpringerVerlag, 172–194.

Stojanovski, S. and Kulakov, A. (2015) Efficient attacks in industrial wireless sensor networks. *ICT Innovations 2014*, vol. 311, Advances in Intelligent Systems and Computing, Springer International Publishing, 289–298.

Tiloca, M. (2014) Efficient Protection of Response Messages in DTLS-Based Secure Multicast Communication. SIN 2014, Glasgow, UK, 466–472.

Tiloca, M. and Dini, G. (2016) *GREP: a Group REkeying protocol based on member join history,* IEEE ISCC 2016, Messina (Italy), 326–333.

Tiloca, M., De Guglielmo, D., Dini, G., and Anastasi, G. (2013) SAD-SJ: a self-adaptive decentralized solution against selective jamming attack in wireless sensor networks. IEEE ETFA 2013, Cagliari (Italy), 1–8.

Tiloca, M., De Guglielmo, D., Dini, G., Anastasi, G., and Das, S. K. (2017) JAMMY: a distributed and dynamic solution to selective jamming attack in TDMA WSNs. *IEEE Transactions on Dependable and Secure Computing,* **14**(4), 392–405.

Tiloca, M., Gehrmann, C., and Seitz, L. (2017) On improving resistance to denial of service and key provisioning scalability of the DTLS handshake. *International Journal of Information Security,* **16**(2), 173–193.

Tiloca, M., Nikitin, K., and Raza, S. (2017) Axiom: DTLS-based secure IoT group communication. *ACM Transactions on Embedded Computing Systems,* **16**(3), 1–29.

Tiloca, M., Raza, S., Nikitin, K., and Kumar, S. (2015) Secure Two-Way DTLS-Based Group Communication in the IoT, Internet Engineering Task Force.

Tschofenig, H. and Fossati, T. (2016) Transport Layer Security (TLS)/Datagram Transport Layer Security (DTLS) Profiles for the Internet of Things. RFC 7925, Internet Engineering Task Force.

Wallner, D., Harder, E., and Agee, R. (1999) Key Management for Multicast: Issues and Architectures. RFC 2627, Internet Engineering Task Force.

Wang, W., Ma, J., and Moon, S.-J. (2008) CRMS: A collusion-resistant matrix system for group key management in wireless networks. IEEE 2008 International Conference on Communications, 1551–1555.

Weis, B., Gross, G., and Ignjatic, D. (2008) Multicast Extensions to the Security Architecture for the Internet Protocol. RFC 5374, Internet Engineering Task Force.

Winter, T., Thubert, P., Brandt, A., Hui, J. W., Kelsey, R., Levis, P., Pister, K., Struik, R., Vasseur, J.P., and Alexander, R. K. (2012) RPL: IPv6 Routing Protocol for Low-Power and Lossy Networks. RFC 6550, Internet Engineering Task Force.

Wong, C. K., Gouda, M., and Lam, S. S. (2000) Secure group communications using key graphs. *IEEE/ACM Transactions on Networking,* **8**(1), 16–30.

Wood, A. D., Stankovic, J. A., and Zhou, G. (2007) DEEJAM: defeating energy-efficient jamming in IEEE 802.15.4-based wireless networks. Proceedings of the 4th Annual IEEE Communications Society Conference on Sensor, Mesh and Ad Hoc Communications and Networks, 60–69.

Wouters, P., Tschofenig, H., Gilmore, J., Weiler, S., and Kivinen, T. (2014) Using Raw Public Keys in Transport Layer Security (TLS) and Datagram Transport Layer Security (DTLS). RFC 7250, Internet Engineering Task Force,

Xu, W., Ma, K., Trappe, W., and Zhang, Y. (2006) Jamming sensor networks: attack and defense strategies. *IEEE Network*, **20**(3), 41–47.

Xu, W., Trappe, W., and Zhang, Y. (2007) Channel surfing: defending wireless sensor networks from interference. Proceedings of the 6th International Conference on Information Processing in Sensor Networks, 499–508.

Xu, W., Trappe, W., Zhang, Y., and Wood, T. (2005) The feasibility of launching and detecting jamming attacks in wireless networks. The 6th ACM International Symposium on Mobile Ad Hoc Networking and Computing, 46–57.

Xu, W., Wood, T., Trappe, W., and Zhang, Y. (2004) Channel surfing and spatial retreats: defenses against wireless denial of service. The 3rd ACM Workshop on Wireless Security (WiSe 2004), 80–89.

Younis, M. F., Ghumman, K., and Eltoweissy, M. (2006) Location-aware combinatorial key management scheme for clustered sensor networks. *IEEE Transactions on Parallel and Distributed Systems*, **17**(8), 865–882.

9

Blockchain-Based Security Solutions for IoT Systems

Göran Pulkkis, Jonny Karlsson, and Magnus Westerlund

Department of Business Management and Analytics, Arcada University of Applied Sciences, Helsinki, Finland

9.1 Introduction

IoT security issues have many commonalities with IT security in general. However, much more sensitivity and confidentiality are required for IoT systems when these systems enter and digitalize the private lives of individuals. Some of the identified key privacy concerns for IoT related to the collection of individuals' data are unauthorized surveillance, uncontrolled data generation and use, inadequate authentication, and information security risks (Caron et al., 2016). The sensitivity of IoT technologies stems from security requirements such as confidentiality, integrity, authenticity, privacy, availability, and regulation. IoT security has physical and logical issues. The physical issues entail the limited capabilities of IoT devices in terms of computational power and memory and usually also in terms of energy, since most IoT devices are battery powered. The logical issues include authentication, privacy, protection against malware, the standardization of policies, and monitoring (Madakam and Date, 2016).

IoT deployment raises many security issues related to the characteristics of IoT devices, such as the need for lightweight cryptographic algorithms in terms of processing and memory capabilities, and the use of standard protocols, such as the necessity to minimize the size of data exchanged between nodes (Cirani et al., 2013). IoT devices are more vulnerable to security threats than Internet-connected computers, because of the limited processing capabilities and memory resources aggravating the implementation of protection. The current Internet networking protocol transition from IPv4 to IPv6 means that a growing number of IoT devices have global IP addresses, which facilitates the identification of these devices as targets of security attacks. Security attacks are also facilitated by the autonomous operation and communication of IoT devices. Thus, there is an urgent need for new and more robust security solutions for IoT systems.

Internet of Things A to Z: Technologies and Applications, First Edition. Edited by Qusay F. Hassan.
© 2018 by The Institute of Electrical and Electronics Engineers, Inc. Published 2018 by John Wiley & Sons, Inc.

The Internet and its technology stack have been in development for roughly four decades. During that time, the centralized client–server architecture has been fundamental in building current platforms and services. From an IoT point of view, this may also be troublesome, for example, when a myriad of wireless sensors need to submit their data back to a centralized service or when a monolithic service should be able to distribute security updates to a decentralized or distributed sensor network. These sensor networks would often benefit from a decentralized communication architecture being self-governing to a large extent. One issue that traditionally has been a hindrance in creating a decentralized architecture is the trust of other actors. The introduction of the cryptocurrency Bitcoin assumed that no trust is needed between two parties. This was achieved by integrating a distributed consensus mechanism as proof of the validation of new transactions while honoring the earlier transaction history. This was consequently extended to the design of generalized transactions outside the realm of cryptocurrencies. Today, this generalized mechanism often goes under the name of blockchain or decentralized ledger technology.

This chapter surveys blockchain-based security solutions for IoT and provides an insight into current research trends. Blockchain technology is briefly discussed, and its use in IoT environments is presented. Additionally, some current efforts to make blockchain technology suitable for IoT are described.

9.2 Regulatory Requirements

Recent attention from the regulator, particularly in the EU, has prompted an increased focus on security and privacy in the IoT sector. The adoption of blockchain technology as a viable solution for future IoT systems in fulfilling regulatory demands offers great potential. Concerning regulatory requirements regarding the design of IoT devices, new directives and regulations have recently been adopted by the EU Parliament. These requirements can be considered among the most stringent in the world and apply to the manufacturers of devices as well as service and platform providers anywhere if they deliver to the European Union and/or handle the personal data of EU residents. In addition, EU member states also provide some sector-specific regulation for areas handling sensitive information, such as health care and financial services. The United States lacks a general data protection or privacy law and instead relies mainly on a rather minimal sector-specific privacy-related legislation. The US approach makes drawing common conclusions more difficult for upholding a certain privacy level when designing information systems. For example, although the same IoT system may be used in different areas, the lack of a common privacy requirement or definition suggests that the manufacturer must at least, to a certain degree, anticipate the intended use of the design and maybe limit the areas of acceptable use when entering the US market. On the other hand, one can treat the EU regulatory requirements as the baseline for obligations that need to be fulfilled when dealing with either personal data or dealing with certain essential infrastructure operators. The two EU legal acts guiding the development and maintenance of information systems[1] are the General Data Protection Regulation (GDPR)[2] and the Directive on Security of Network and Information Systems (the NIS directive)[3]. The GDPR may have certain minor differences between the member states but lays the foundation for a unified digital single market within the European Union. As a directive, the NIS will likely be adopted differently by member states, although it defines what can be considered a minimum level of security responsibilities for information systems.

9.2.1 General Data Protection Regulation

The intention of the GDPR is to strengthen and clarify rights in regard to the personal data of natural persons—that is, individuals residing in the European Union. The GDPR affects organizations globally if their intended users are originating from the European Union and personal data are processed. A company found in violation of the regulation may face considerable fines and compensation fees to data subjects (GDPR Articles 82 and 83). The legislation separates the role of the controller and processor as different legal entities. The controller collects the initial consent or contract and cannot defer responsibilities in regard to the data subject to a third party. Even if the processor resides in a country outside the European Union, the processor is still bound by the GDPR while processing any personal data on an individual within EU borders. This is a measure ensuring that transnational data transfers do not violate the fundamental rights of the individual. Provided a processor is determined by the controller to follow the GDPR and that sovereign laws do not conflict with the EU laws, personal data can be transported and processed outside the EU.

The GDPR defines two types of data: personal and special category (sensitive) data. An identifiable natural person is one who can be identified, directly or indirectly, in particular by reference to an identifier such as a name, an identification number, location data, an online identifier, or one or more special category factors. Special categories in Article 4(1) of the GDPR are specific to the

1 A third act, Privacy and Electronic Communications Directive 2002/58/EC, is in its existing form of lesser importance to the IoT sector. However, a draft of a proposed change in this directive has been introduced that may have considerable consequences for the IoT sector in the future. See official EU communication available at https://ec.europa.eu/digital-single-market/en/proposal-eprivacy-regulation (accessed June 26, 2017).

2 General Data Protection Regulation (GDPR) (EUR-Lex Regulation [EU] 2016/679) . Available at http://eur-lex.europa.eu/eli/reg/2016/679/oj (accessed May 31, 2017).

3 Directive on Security of Network and Information Systems (the NIS Directive) (EUR-Lex Directive [EU] 2016/1148). Available at http://eur-lex.europa.eu/eli/dir/2016/1148/oj (accessed May 31, 2017).

physical, physiological, genetic, mental, economic, cultural, or social identity of that natural person. This separation between data classification may at times result in difficult design choices. For example, an image is considered under special categories when used for identification and/or authorization, while in documenting a care story the same image may be considered as personal data. Assuming the data subject has manifestly made the image public outside the immediate family, the system may in some cases be allowed to process the image at its own discretion.

Two important concepts in the legislation are, according to Article 5(1) A of the GDPR, that personal data shall be processed fairly and in a transparent manner in relation to the data subject. For the adoption of IoT, this may add additional system challenges compared to today's environment. The GDPR stipulates the following three main considerations on how to achieve compliance:

- *Design Requirements.* Requiring freely given consent or; by contract or for the performance of a contract; data minimization; data protection by design and default.
- *Handling of Personal Data.* Access; rectification; portability; transparency of usage; erasure.
- *Limitations on Processing.* Notification; restriction; security.

Let us consider the method for integrating IoT hardware and services/platforms as a traditional data analytics process. Data quality from data generators, such as sensors, must, in light of the GDPR, be accurate and reliable. Meta information describing, for example, data source, access rights, and justification for lawful processing should be recorded along with the data when it becomes associated with a natural person. During feature engineering, efforts must be made such that descriptive statistics do not infringe on the data subject's integrity or introduce new features that can be considered sensitive. An example of new feature is clustering based on location and additional information such as place of worship that can be considered to infer either religion and/or race. If such statistics are needed for determining factors and/or causes in ascertaining a hypothesis, then these efforts need to be disclosed to the data subject beforehand.

The need for transparent processing and transparent data transfers/manipulations/removals may be best met by an immutable data storage solution, such as a ledger-based technology like blockchain. Treating each corresponding operation as a transaction and defining a smart contract for that operation offers an auditable record for forensic purposes and an approach for showing compliance to potential users and the regulator.

9.2.2 Directive on Security of Network and Information Systems

The NIS directive is somewhat more restrictive in its application than the GDPR. The intention of NIS is that it should not place additional burden on small and medium-sized enterprises but rather be a recommendation for dealing with security and security incidents. It should be noted that the handling of security incidents involving personal data or sector-specific data may be further specified in the GDPR or other relevant sector-specific acts. Incidents are defined as any event having an actual adverse effect on the security of network and information systems. Enterprises that fall within the definition of the directive and thus must fulfill its obligations are operators of essential services, such as energy, water, transport, banking, financial market infrastructures, health care, and digital infrastructure. The requirements of these enterprises in terms of security measures are as follows[4]:

- *Preventing Risks.* Technical and organizational measures that are appropriate and proportionate to the risk.
- *Ensuring IT Security.* Ensure a level of security of the network and information systems appropriate to the risks.
- *Handling Incidents.* Prevent and minimize the impact of incidents on the IT systems used to provide the services.

The other category defined by NIS, in addition to the essential operators, comprises digital service providers such as online marketplaces, cloud computing services, and search engines. The obligations for companies belonging to this category are somewhat enlarged, and in addition to the requirements presented above for essential operators, these additional measures apply business continuity management; monitoring, auditing, and testing; and compliance with international standards.

9.3 Blockchain Technology

Blockchain technology was introduced in Nakamoto (2008) as a platform for the Bitcoin cryptocurrency. A blockchain is a distributed database for storing a continuously growing list of records called blocks. A blockchain is replicated in a peer-to-peer network of nodes, where each node stores a copy of the entire database. The topology of a blockchain is a chain of blocks since each block except the first block, the so-called Genesis Block, contains a link to the preceding block implemented as a hash of the preceding block. Each block in a blockchain is also time stamped. The basic structure of a blockchain is depicted in Figure 9.1.

A blockchain user is the owner of a node in the blockchain network for operations on the blockchain and must also own a unique key pair in public key

4 European Commission. (2016) Directive on Security of Network and Information Systems. Available at http://europa.eu/rapid/press-release_MEMO-16-2422_en.htm (accessed July 13, 2017).

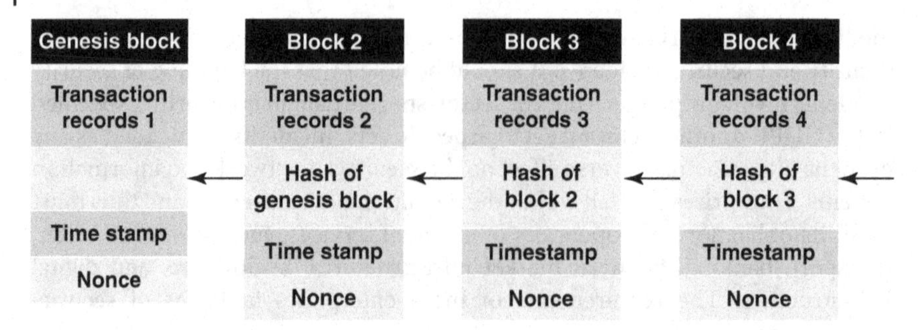

Figure 9.1 Basic blockchain structure.

cryptography. The public key identifies a blockchain user. A blockchain user operates on a blockchain from a node in a blockchain network. An operation on a blockchain is a transaction initiated by a blockchain user. A transaction creates a record, for example, about the transfer of Bitcoins, data, physical property, or some other asset from a blockchain user to another user of the same blockchain. The transaction record is signed by the blockchain user who initiated the transaction, and is sent to all blockchain network nodes. Each blockchain network node tries to validate the received transaction record with the public key of the initiator of the related transaction. Transaction records that cannot be validated by all the blockchain network nodes are considered invalid and get discarded. Special blockchain network nodes called mining nodes collect validated transaction records and store them as lists in time-stamped candidate blocks. A mining node performs a distributed computational review process, called mining, on its candidate block before it can be linked to the blockchain.

There are several implementations of mining for blockchains. In the Bitcoin blockchain, mining is based on proof-of-work (PoW). This means that each mining node repeatedly creates a hash of the concatenation of the last blockchain block and a new nonce (a randomly chosen value) until a hash of required difficulty is created. Created hashes are different because the nonce is different in each one of them. The mining node, which first creates a hash of required difficulty, can link its candidate block to the blockchain. The required difficulty is defined by the number of leading zero bits in a created hash. For example, if the difficulty is 10 zero bits, then on the average 2^{10} attempts are needed until a hash of the required difficulty is created. Thus, the mining node with the highest processing power will link its candidate block to the blockchain with the highest probability. In the Bitcoin blockchain, the PoW difficulty is increased after 2 weeks to control the blockchain growth rate.

A blockchain can be public, permissioned, or private. All software of a public blockchain is open source. Anyone can use a public blockchain, for example, the Bitcoin blockchain, by joining the blockchain network. A user joins a blockchain network by copying the entire blockchain and installing the blockchain software on the own node. Any blockchain user can also own a mining node, for example, by installing the mining software on the own node in a blockchain network.

More recent blockchain implementations that extend the functionality offered by the Bitcoin blockchain are denoted as Blockchain 2.0. One of the interesting features in Blockchain 2.0 is the support for smart contract, which was introduced as a concept by Szabo (1997). A smart contract is a computer program that encompasses contractual terms and conditions that enable the verification, negotiation, or enforcement of a contract. An example of a blockchain platform focused on smart contracts is Ethereum[5], whose cryptocurrency is called Ether. A smart contract in Ethereum is called DApp (Decentralized App), which can be executed on a server or directly on an Ethereum node. See, for example, Tapscott and Tapscott (2016) for a detailed description of blockchain technology and blockchain applications.

The security of a blockchain is based on the hash-interdependence between successive blocks in combination with the distribution of copies of the entire blockchain to all nodes in a blockchain network. A blockchain is in practice tamperproof—that is, resistant to modification attempts—since a block cannot be altered without alteration of all the subsequent blocks and participation of the entire network to verify and log the change. Moreover, a blockchain is not controlled by any single centralized authority, which could be an attack target, since complete blockchain copies are stored in all nodes of a peer-to-peer network. However, if an attacker can achieve control of a sufficient number of nodes in a peer-to-peer blockchain network, including some mining nodes, then there can be data losses and/or the insertion of corrupted data in the attacked blockchain. (Conoscenti et al., 2016)

Examples of security attacks against blockchains are the selfish mining attack (Eyal and Sirer, 2013), the history-revision attack (Barber et al., 2012), the eclipse attack (Heilman et al., 2015), and the stubborn mining attack (Nayak et al., 2016). Malicious mining nodes do not transmit all mined new blocks for validation in the blockchain network in a selfish mining attack. A malicious miner having more than twice as much computational power than all honest mining nodes have together can insert corrupted blocks in a blockchain in a history-revision attack. All incoming and outgoing connections of a target node in a blockchain network are controlled in an eclipse attack. A stubborn mining attack combines selfish mining with an eclipse attack.

9.4 Blockchains and IoT Systems

IoT devices generate enormous amounts of data, which must be stored and processed. Each CRUD (Create, Read, Update, or Delete) operation on IoT data

5 https://www.ethereum.org/ (accessed July 13, 2017).

can be registered as a transaction record in a blockchain block. Unauthorized operations on stored IoT data can, therefore, be detected. An identity credential for an IoT device can be registered as a transaction record in a blockchain block when the device is manufactured and retrieved later when the device is being used. Access rules for IoT devices can be specified and enforced by smart contracts. Unauthorized operations on stored IoT data can, therefore, also be detected. No centralized authority, such as a cloud storage provider, is needed for the protection of IoT data, which can be safely stored in different network nodes, when a blockchain guarantees data authenticity and prevents unauthorized access. The deployment history of an IoT device can be stored as transaction records in a blockchain. Transaction records can be stored in blockchain blocks when the IoT device is produced, delivered to an owner, installed, updated, and delivered to another owner, and so on. A blockchain can also enable secure messaging between IoT devices. Message exchanges between IoT devices can be treated similar to financial transactions in a Bitcoin network. Message exchanges between IoT devices can be enabled with smart contracts, which implement agreements between two parties. A fundamental capability of a blockchain is that a duly decentralized, trusted ledger of all transactions in a network is maintained. This capability can enable the many compliances and regulatory requirements for IoT systems, for example, in the GDPR. (Banafa, 2016; Conoscenti et al., 2016)

9.5 Examples of Blockchain-Based Security Solutions for IoT Systems

The decentralized and autonomous features of a blockchain make it a nearly ideal component in IoT security solutions. Blockchain use can enable an IoT security level, which otherwise would be difficult or even impossible to achieve. This section presents some of the recently proposed IoT security solutions that are based on blockchain technology.

9.5.1 Secure Management of IoT Devices

The management of an IoT device includes the control of configuration settings and operation modes as well as ensuring uninterrupted operation. Blockchain-based control of configuration settings and operation modes can prevent unauthorized access attempts and also provide protection against denial-of-service attacks.

In Huh et al. (2017), the authors proposed control and configuration of IoT devices using Ethereum as a blockchain platform. An identification credential for an IoT device can be implemented by a unique key pair (i.e., a private key and a public key) in public key cryptography. The private key is stored in the

IoT device while the public key is registered as a transaction record in an Ethereum block. An IoT device can then be addressed on Ethernet via its public key. Ethereum was chosen as the blockchain platform because its smart contracts enable the execution of programs on the blockchain. IoT device behavior can, therefore, be programmed in smart contracts. For a proof of the proposed concept, a simulation took place on a system composed of three IoT devices: an electricity meter, a LED light bulb, and an air conditioner. A policy requiring the air conditioner and the light bulb to switch to an energy-saving mode if the meter measurement exceeds 150 kW was set up by a smartphone. For the meter, a smart contract was programmed for sending measurement values and identity credentials (i.e., its public key and signature) to Ethereum. Smart contracts were also programmed for the air conditioner and the light bulb. These contracts retrieved measurement values with associated identity credentials from Ethereum. The identity credentials verified that the retrieved measurement value was a meter value and a transition to the energy-saving mode occurred when the threshold of 150 kW of the retrieved value was exceeded.

9.5.2 Secure Firmware Updates in IoT Devices

IoT device vendors remotely update the firmware of delivered devices in order to install new functionality and to patch discovered vulnerabilities. These updates are usually downloaded based on client requests from a repository server containing precompiled firmware binaries secured by public key infrastructure (PKI)-signed message digests. The signed message digest and the public signing key are attached to a downloaded firmware file. The firmware update on an IoT device starts only if a security check with the downloaded public key succeeds. However, this client–server firmware update protocol creates too much network traffic if millions of IoT devices simultaneously request updates.

A solution utilizing blockchains has been proposed for secure firmware updates in IoT devices where the global network traffic to a server is replaced by mostly local peer-to-peer communication between blockchain network nodes (Lee and Lee, 2016). In this solution, an IoT device manufacturer stores the hashes of released firmware versions on a blockchain that is accessible to all delivered IoT devices. Christidis and Devetsikiotis (2016) suggested that using a preinstalled smart contract with a condition to repeatedly check after a preset time interval has elapsed if a new firmware release is available, then an IoT device can autonomously find out about new firmware releases. The IoT device can retrieve the hash of the released firmware from the blockchain[6] and use it to

---

6 Case Study 1: Chain of Security. (2016) Chain of Things. Available at http://www.chainofthings.com/cs1chainofsecurity/ (accessed December 26, 2016).

securely download the new firmware version from a distributed peer-to-peer file system consisting of the manufacturer's node and IoT devices with firmware versions installed. The firmware hashes stored on the blockchain can also be used to verify that firmware installed on IoT devices is untampered.

In the aforementioned solution, all IoT devices delivered by the same vendor are normal blockchain nodes. Other blockchain nodes are verification nodes, which are operated by the firmware vendor through a secure network connection to maintain updated firmware and firmware metadata. An IoT device broadcasts firmware update requests to the blockchain nodes. If a response is first received from a verification node, and the firmware on the IoT device is up-to-date, then the firmware is verified. Otherwise, the latest firmware is downloaded from the responding verification node. If a response is first received from a normal blockchain node that has the same firmware version as the IoT device requesting a firmware update, then the firmware version is verified by a lightweight PoW mining procedure in which six verification log responses suffice. Otherwise, the up-to-date firmware is downloaded from a verification node to the IoT device—a normal blockchain node—whose firmware version is older. Each blockchain block consists of a header and a verification field, which consists of a verification counter, hashes of all stored transactions, a verification log, the name of the blockchain network node, the current firmware version, and a hash of the firmware file. The verification log contains time-stamps representing verification times and IDs of network nodes that request firmware updates, and IDs of network nodes that respond to such requests.

9.5.3 Trust Evaluation of a Trusted Computing Base in IoT Devices

A Trusted Computing Base (TCB) is the set of hardware, firmware, and/or software components that ensure the security of a computer system. This means that to break the security, an attacker must subvert one or more of these components. A TCB can, therefore, be a part of an IoT device, which is a tiny computer system. A TCB is trustworthy if all TCB components are unchanged by faults and untampered by adversaries. A TCB measurement, which creates hashes of all TCB components, is carried out to evaluate the trustworthiness of a TCB. If these hashes are securely stored then they can be later used to verify the TCB. TCB measurements are carried out when an IoT device is connected to the Internet and each time its TCB is updated. Trustworthiness has been compromised if a TCB measurement cannot be verified. A verifier performs remote attestation by issuing a cryptographic nonce and signing the concatenation of a verified TCB measurement and the nonce to assure that the TCB of an IoT device is trustworthy.

In Park and Kim (2017), a protocol called TM-Coin (TCB Measurement-Coin) has been designed for the trustworthy management of TCB measurements in

IoT devices. TM-Coin creates transaction records of verified TCB measurements and stores these records in blockchain blocks. TM-Coin uses the Trusted Execution Environment (TEE) provided by ARM TrustZone[7] as the TCB in IoT devices for the secure generation of transaction records for the blockchain. The TM-Coin protocol consists of two blockchain transaction types: registration and update. A registration transaction stores a record of the verification of TCB measurement in a blockchain block when an IoT device is connected. After a code update of at least one TCB component in an IoT device, an update transaction also stores a record of the verification of TCB measurement in a blockchain block. The miner nodes in the blockchain perform remote attestation of the TCB in an IoT device during a transaction. Data sensed by an IoT device are assured to be trustworthy after remote attestation by an external verifier. The attestation function is

$$\text{sign}(\text{hash}\,(\text{TCB\_M},\ \text{D},\ \text{N}))$$

where TCB_M is the latest TCB measurement of the IoT device obtained from the blockchain, D is the data sensed by the device, and N is a nonce issued by the verifier.

9.5.4 IoT Device Identity Validation

A secure identity of an IoT device can be implemented as a private key in an embedded public key cryptography chip. The corresponding public key is stored in a blockchain block by the IoT device manufacturer (Lombardo, 2016). A network node starts accessing an IoT device with a random challenge message, which is returned by the IoT device with a signature. The accessing network node can thereafter validate the IoT device identity with the public key that can be retrieved from the blockchain. An IoT device identity that is validated using a blockchain enables an almost perfectly secure authentication of an IoT, makes identity spoofing nearly impossible, and ensures the integrity of data captured from IoT devices because of the tamper-resilience of a blockchain.

An IoT device identity that is validated using a blockchain has been proposed to be used to create a blockchain based identity log capturing the device ID, its manufacturer, lists of available firmware updates, and known security issues (Manning, 2017). The history of a device with a securely validated identity can also be tracked by a blockchain. The history starts when the manufacturer stores the identity—the public key—of a manufactured IoT device in a blockchain block. Identities that are validated using a blockchain are being developed for IoT devices such as surveillance cameras.

---

7 ARM Security Technology Building a Secure System using TrustZone Technology. (2009) White paper. http://infocenter.arm.com/help/topic/com.arm.doc.prd29-genc-009492c/PRD29-GENC-009492C_trustzone_security_whitepaper.pdf (accessed May 30, 2017).

9.5.5 Secure Data Store System for Access Control Information

Current standard solutions for access control to network-connected devices are based on access control lists (ACLs). However, it would be impossible to maintain an ACL for each and every IoT device and rely on centralized access control servers when IoT scales to billions of devices and millions of device owners. To put these IoT device owners in control of the data generated by their devices, blockchain deployment is a possible solution that excludes dependence on centralized third parties.

A blockchain-based secure data store system for access control information has been introduced as a component in a proposed solution for the protection of IoT device owners' access control to data generated by their IoT devices (Hashemi et al., 2016). The other components are a data management protocol and a messaging service. The proposed solution implements role and capability-based access control. When a party with a defined role sends an access control message to another party also with a defined role, then the message is delivered to the messaging service. The messaging service sends the message to the data store system, where it is stored as a transaction record in a blockchain block. After that, the receiving party fetches the message from the blockchain block in the data store system through the messaging service. An overview of the proposed solution is illustrated in Figure 9.2.

Four roles are defined, as is shown in Figure 9.2: data owner, data source, data requester, and endorser. A data owner owns and grants access to the data generated by their IoT devices (i.e., data sources). A data requester (e.g., an IoT device) requests access to IoT data and an endorser validates such requests. The data management protocol is a message exchange protocol for capability-based access control. A capability permits a data requester to access a data owner's data object on a data source. The data management protocol consists of five access control message types: data source ticket generation message, data request message, ticket exchange message, data access message, and access announcement message. The data store system is a blockchain similar to the Bitcoin blockchain. The messaging service stores messages received from a sender as transaction records in blockchain blocks in the data store system and implements a publisher–subscriber protocol—as shown in Figure 9.2—for the delivery of messages stored as transaction records in blockchain blocks in the data store system.

9.5.6 Blockchain-Based Security Architecture for IoT Devices in Smart Homes

A blockchain-based architecture has been proposed for local networks in smart homes (or some other local environments) with a number of connected IoT devices like smart thermostats, smart bulbs, and IP cameras (Dorri et al., 2016,

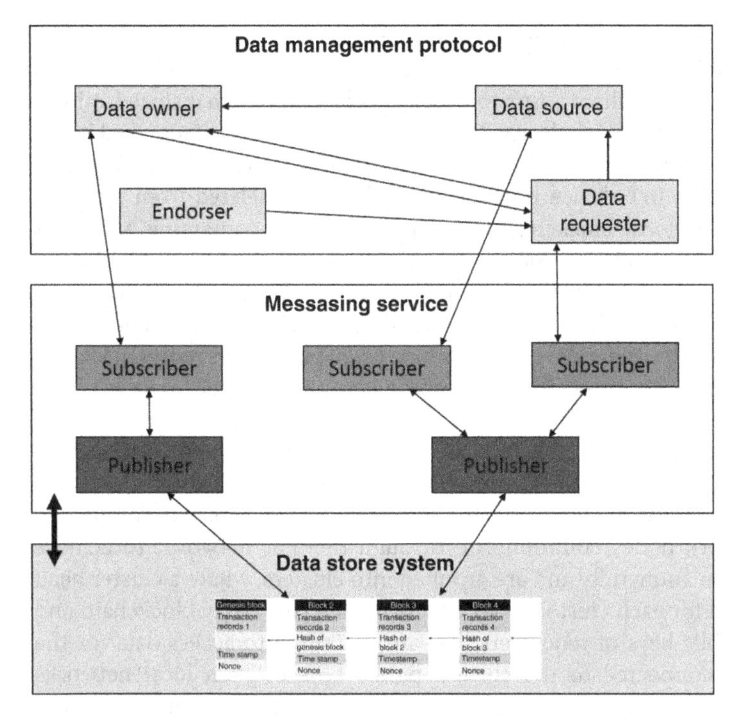

Figure 9.2 An overview of the proposed solution for secure access control messaging.

2017a, 2017b). The architecture has three tiers, namely, the local networks in smart homes, an overlay network, and cloud storage. In each tier, entities use blockchain transactions for communication with each other. The transaction types include genesis transactions, store transactions, access transactions, and monitor transactions. The architecture, which is illustrated in Figure 9.3, has

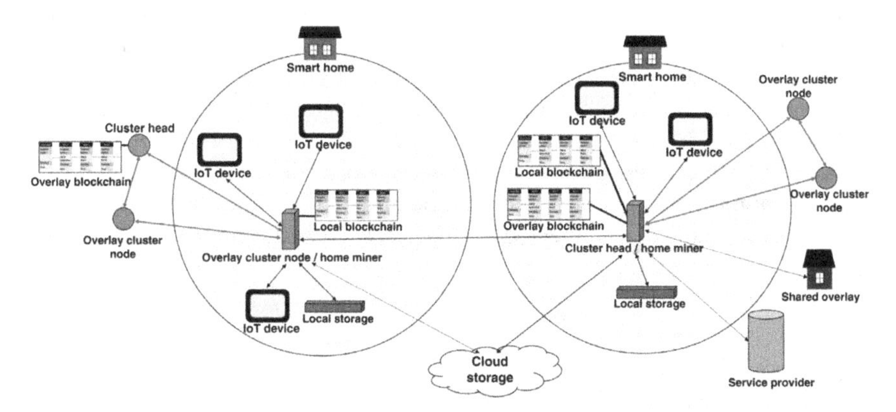

Figure 9.3 An overview of the proposed blockchain-based architecture for smart homes.

strong protection against denial-of-service attacks, modification attacks, dropping attacks, mining attacks, appending attacks, and linking attacks.

In the local network, there is a private local blockchain, which is stored, mined, and managed by at least one device. When a new IoT device is connected to the local network, a genesis transaction record is stored in a local blockchain block. When an existing IoT device is removed, its ledger is deleted from the local blockchain. This local blockchain has a policy header containing an access control list, which enables a local network owner's control of all blockchain transactions in the local network. Communication between IoT devices is encrypted by preshared Diffie-Hellman[8] keys. The local network can have local storage for data. The miner device maintains a list of public keys representing the digital identities of entities, which can be given permissions to access local network data from outside.

The overlay network resembles the peer-to-peer Bitcoin network (Nakamoto, 2008) and is composed of local network miner devices, other local network devices, and/or local network owners' smartphones or personal computers. Overlay network nodes communicate through the Tor network[9] to achieve communication anonymity and are grouped into clusters, where a cluster head (CH) is elected for each cluster. Each CH maintains an overlay blockchain and three lists: public keys of requesters that are permitted to access data for the smart homes connected to the cluster, public keys of such local networks connected to the cluster that can be accessed from outside, and transactions sent to other CHs (Dorri et al., 2016). Transaction records with more than one signature and access transaction records are stored in overlay blockchain blocks. Several local networks having the same owner can be managed together as a shared overlay consisting of a shared blockchain with a common miner and shared storage.

IoT devices in a local network can store their data locally or in cloud storage. Cloud storages are blockchains, where IoT data is stored in identical blocks with unique block numbers. A local network device is authenticated in cloud storage by the given block number and a hash of the stored data. A store transaction, illustrated in Figure 9.4, is initiated by an IoT device in a local network to store generated data, for example, by a thermostat to store temperature measurements in cloud storage. An access transaction, illustrated in Figure 9.5, can be initiated by a local network owner or a service provider to retrieve stored IoT data. A monitor transaction, which can be initiated by a local network owner, retrieves the status of a connected IoT device, such as the current temperature value of a thermostat.

8 Diffie-Hellman Protocol. 2017. Wolfram MathWorld. http://mathworld.wolfram.com/Diffie-HellmanProtocol.html (accessed January 3, 2017).

9 Tor. 2016. https://www.torproject.org/ (accessed December 28, 2016).

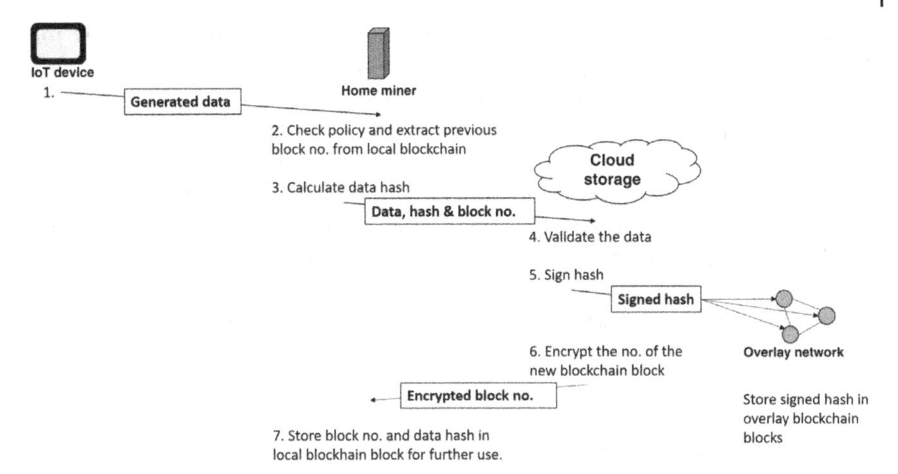

Figure 9.4 Handling a store transaction.

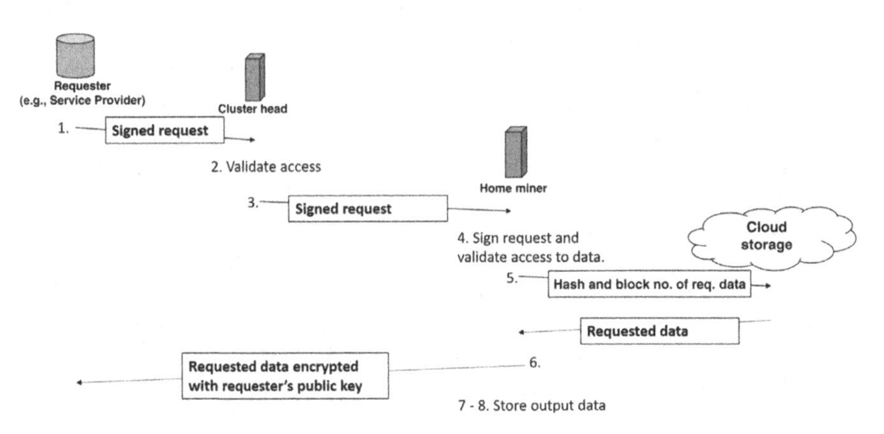

Figure 9.5 Handling an access transaction.

9.5.7 Improved Reliability of Medical IoT Devices

Medical IoT devices are subject to the same security concerns of other IoT devices. User safety in medical IoT-based systems is a top priority. A user must be protected from any system malfunction caused by a device fault or security incident. A medical IoT device must function reliably and resist security attacks. Also, the integrity and user privacy of data generated by medical IoT systems must be granted. Blockchain use in the management of medical IoT devices can provide protection against malicious tampering of device settings and modes of operation. An immutable blockchain ledger of records from management events can mitigate the risk of device malfunction.

Nichol and Brandt (2016) proposed the use of blockchain technology for device management to improve the reliability of medical IoT devices. When a medical IoT device is manufactured, a hash of a unique device identifier together with other relevant information, such as company name, is stored in a blockchain block. Later, this data can be updated with patient data, hospital, doctor, emergency contacts, and patient care directives. The patient and the caregivers can be automatically notified about device service needs, approaching battery expiration, and detected patient health irregularities through a set of smart contracts. The risk of a catastrophic device failure is, therefore, reduced by smart contracts sending preventive maintenance information to the patient and caregiver.

9.6 Challenges and Future Research

The implementation of proposed blockchain-based security solutions in IoT is a highly relevant topic for future research. First, IoT applications potentially benefiting from blockchain security features should be rigorously surveyed. An example of a relevant domain is health care IoT applications, where the tampering of health-related measurement data could be disastrous. When the applications and their security requirements have been established, the next step is to evaluate how blockchain technology could be implemented. Critical simulations and experimental evaluations of the security and performance of blockchain-based security solutions are needed before these solutions can be implemented in real applications. At the same time, new tamper-resistant blockchain-based security solutions providing detailed forensics about security attacks are required. Much work is also needed in the development of standards for the design of IoT hardware, IoT firmware, and other IoT software supporting implemented and verified blockchain-based security solutions.

A significant issue with using blockchain solutions in IoT systems is the limited processing capabilities of most of the used devices. As blockchain technology uses cryptography extensively for hashing, digital signing, and encryption, more research on lightweight cryptographic algorithms is needed for the practical implementation of blockchain-based security solutions.

9.7 Conclusions

Although IoT systems have existed in various forms for a long time, security challenges are emerging and will continue to emerge in the foreseeable future. General IT security methods and tools do not fulfill all specific requirements for secure IoT deployment. Therefore, the identification of emergent methods suitable for IoT security solutions is important. Blockchain technology may improve the ability of automatic response to a security incident. This is especially important for IoT systems, since decentralized IoT networks are expected to function reliably and securely without human supervision.

A relevant security risk for all IT systems, and therefore also for IoT systems, is the possibility of attackers tampering with the software and/or data of a security solution. Blockchain-based IoT security solutions mitigate this risk because they are “virtually” tamper resistant and because of their ability to perform real-time audits on transactions. However, there are also disadvantages and shortcomings of blockchain use in security solutions for IoT systems. The major disadvantages for resource-constrained IoT devices are the increasing processing power requirements of PoW mining and the increasing storage requirements in blockchain nodes when the size of a blockchain ledger grows. To mitigate these disadvantages, some other mining techniques should be deployed and only a device-dependent transaction ledger should be maintained when a device is too resource constrained (Banafa, 2016; Kolias et al., 2016).

It should be noted that none of the presented examples of blockchain-based IoT security solution proposals gives full protection against all possible security threats and attacks. Moreover, the practical deployment of these security solutions is still a future issue. A practical security solution for an IoT system can, therefore, combine a choice of blockchain-based security solutions with a set of other security solutions. The current rapid increase of IoT implementations and the IoT security incidents that have hitherto occurred emphasize the necessity of continued research on improving decentralized security measures and reliability of IoT systems.

References

Banafa, A. (2016) A Secure Model of IoT with Blockchain. OpenMind. https://www.bbvaopenmind.com/en/a-secure-model-of-iot-with-blockchain/ (accessed May 16, 2017)

Barber, S., Boyen, X., Shi, E., and Uzun, E. (2012) Bitter to better – how to make Bitcoin a better currency, in *Lecture Notes in Computer Science*, vol. 7397, Springer, Switzerland, 399–414.

Caron, X., Bosua, R., Maynard, S. B., and Ahmad, A. (2016) The Internet of Things (IoT) and its impact on individual privacy: an Australian perspective. *Computer Law & Security Review*, **32**(1), 4–15.

Christidis, K. and Devetsikiotis, M. (2016) Blockchains and smart contracts for the Internet of Things. *IEEE Access*, **4**, 2292–2303. doi: 10.1109/ACCESS.2016.2566339

Cirani, S., Ferrari, G., and Veltri, L. (2013) Enforcing security mechanisms in the IP-based Internet of Things: an algorithmic overview. *Algorithms*, **6**, 197–226. doi: 10.3390/a6020197

Conoscenti, M., Vetro, A., and De Martin, J. C. (2016) Blockchain for the Internet of Things: a systematic literature review. Proceedings of the 13th International Conference of Computer Systems and Applications (AICCSA), IEEE, 1–6.

Dorri, A., Kanhere, S. S., and Jurdak, R. (2016) Blockchain in Internet of Things: Challenges and Solutions. Available at https://arxiv.org/ftp/arxiv/papers/1608/1608.05187.pdf (accessed November 13, 2016).

Dorri, A., Kanhere, S. S., Jurdak, R., and Gauravaram, P. (2017) Blockchain for IoT security and privacy: the case study of a smart home. Proceedings of the International Conference on Pervasive Computing and Communications Workshops (PerCom Workshops), IEEE, 618–623.

Dorri, A., Kanhere, S. S., and Jurdak, R. (2017) Towards an optimized BlockChain for IoT. Proceedings of the 2017IEE/ACM Second International Conference on Internet-of-Things Design and Implementation IoTDI'17, IEEE, 173–178.

Eyal, I. and Sirer, E. G. (2013) Majority is not Enough: Bitcoin Mining is Vulnerable. Available at https://arxiv.org/pdf/1311.0243v5.pdf (accessed January 14, 2017).

Hashemi, S. H., Faghri, F., Rausch, P., and Campbell, R. H. (2016) World of empowered IoT users. Proceedings of the First International Conference on Internet-of-Things Design and Implementation (IoTDI), IEEE, 13–24.

Heilman, E., Kendler, A., Zohar, A., and Goldberg, S. (2015) Eclipse attacks on Bitcoin's peer-to-peer network. Proceedings of the 24th USENIX Security Symposium, Washington: Usenix, 129–144.

Huh, S., Cho, S., and Kim, S. (2017) Managing IoT devices using Blockchain platform. Proceedings of the 19th International Conference on Advanced Communication Technology (ICACT), IEEE, 464–467.

Kolias, K., Stavrou, A., Bojanova, I., Voas, J., and Grance, T. (2016) Leveraging Blockchain-based protocols in IoT systems. NIST National Institute of Standards and Technology. Available at http://csrc.nist.gov/groups/SMA/ispab/documents/minutes/2016-06/1_iot_stavrous.pdf (accessed December 27, 2016)

Lee, B. and Lee, J.-H. (2016) Blockchain-based secure firmware update for embedded devices in an Internet of Things environment. *The Journal of Supercomputing*, 1–6. doi: 10.1007/s11227-016-1870-0

Lombardo, H. (2016) Blockchain serves as tool for human, product and IoT device identity validation. Available at https://inform.tmforum.org/nfv-it-transformation/2016/11/blockchain-serves-tool-human-product-iot-device-identity-validation/ (accessed December 26, 2016).

Madakam, S. and Date, H. (2016) Security mechanisms for connectivity of smart devices in the Internet of Things, in Mahmood, Z. (ed.), *Connectivity Frameworks for Smart Devices*, Series: Computer Communications and

Networks, Springer International Publishing, Switzerland, 23–41. doi: 10.1007/978-3-319-33124-9_2

Manning, J. (2017) Factom Receives Second DHS Grant For Blockchain IoT Project. ETHNews. Available at https://www.ethnews.com/factom-receives-second-dhs-grant-for-blockchain-iot-project (accessed June 27, 2017).

Nakamoto, S. (2008) Bitcoin: A Peer-to-Peer Electronic Cash System. Available at https://bitcoin.org/bitcoin.pdf (accessed December 27, 2016).

Nayak, K., Kumar, S., Miller, A., and Shi, E. (2016) Stubborn mining: generalizing selfish mining and combining with an eclipse attack. 2016 IEEE European Symposium on Security and Privacy (EuroS&P), 305–320.

Nichol, P. B. and Brandt, J. (2016) Co-Creation of Trust for Healthcare: The Cryptocitizen Framework for Interoperability with Blockchain. doi: 10.13140/RG.2.1.1545.4963

Park, J. and Kim, K. (2017) TM-Coin: trustworthy management of TCB measurements in IoT. Proceedings of the International Conference on Pervasive Computing and Communications Workshops (PerCom Workshops), IEEE, 654–659.

Szabo, N. (1997) The Idea of Smart Contracts. Available at http://www.fon.hum.uva.nl/rob/Courses/InformationInSpeech/CDROM/Literature/LOTwinterschool2006/szabo.best.vwh.net/idea.html (accessed June 24, 2017).

Tapscott, D. and Tapscott, A. (2016) *Blockchain Revolution: How the Technology Behind Bitcoin Is Changing Money, Business, and the World*, Penguin Publishing Group, USA.

10

The Internet of Things and IT Auditing

John Shu,[1] Jason M. Rosenberg,[2] Shambhu Upadhyaya,[2] and Hejamadi Raghav Rao[1]

[1]*Department of Information Systems and Cyber Security, University of Texas San Antonio, San Antonio, TX, USA*
[2]*University at Buffalo School of Management, The State University of New York at Buffalo, Buffalo, NY, USA*

10.1 Introduction

The growth and proliferation of Internet and Communication Technologies (ICTs) within the last decade has resulted in a giant web of interconnected devices. Gartner research currently estimates that there are about 3 billion connected devices in consumer home environments. It is, however, predicted that by the year 2020 we would have nearly 21 billion devices connected and talking to each other (Boorstin, 2016). This growth is not limited to the traditional consumer environments but also comprises the millions of devices that are continually being installed and deployed industrially in facilities such as plants, warehouses. Currently, there are huge cost-saving advantages being realized by the prowess of IoT in industries such as health care, transportation, and retail and logistics in almost every facet of their operation. However, as with the adoption of any burgeoning technology, IoT comes with a gamut of issues such as acceptance, usability, security, among which security usually turns out to be one of the most critical. This is coupled with the fact that on average an individual can be associated with up to three different IoT-related devices, which increase the chances and opportunities for malicious individuals to gain illegal access. A report published by the Gartner Research Group demonstrated an increase of over 300 million dollar in security-related spending on IoT worldwide between 2014 and 2018 as seen in Figure 10.1 (Van der Meulen, 2017)

This tremendous increase in the spending forecast for the security of IoT is indicative of the potential vulnerability associated with these technologies. Hence, this calls for detailed policies on the use of these technologies in

Internet of Things A to Z: Technologies and Applications, First Edition. Edited by Qusay F. Hassan.
© 2018 by The Institute of Electrical and Electronics Engineers, Inc. Published 2018 by John Wiley & Sons, Inc.

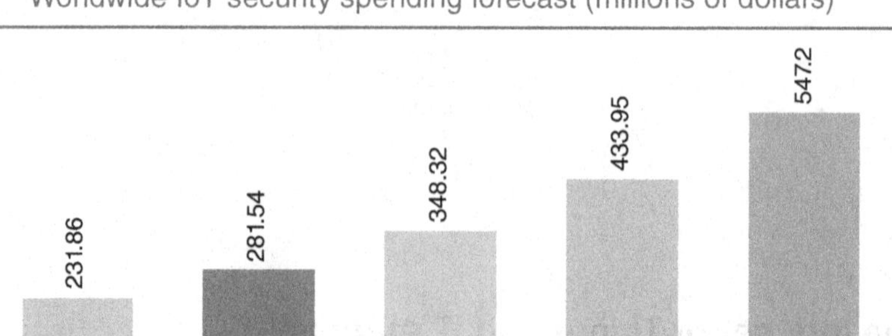

Figure 10.1 Worldwide IoT security spending forecast (in millions of dollars).

organizations, but more importantly, the implementation of comprehensive routine audit procedures. The implementation of such routine audits will ensure a consistently robust and resilient infrastructure regardless of the changes/upgrades in hardware, software, or patches. This chapter, therefore, discusses and elaborates on the somewhat novel concept of audits in relation to IoT.

10.2 Risks Associated with IoT

In their definition, the International Telecommunication Union (ITU) described IoT as (i) a global infrastructure for Information Systems in the Society that (ii) enables and facilitates advanced services by way of physical and virtual objects interconnection (iii) empowered by existing and evolving interoperable information and communication technologies. (Wortmann et al., 2015). The second point in this definition presents a very interesting and unique scenario hitherto not observed—the concept of interconnecting physical and virtual objects in an interoperable fashion. This essentially adds a layer of interconnectivity that was practically nonexistent as objects, which have traditionally not been considered online capable devices are now obtaining Internet capabilities. Objects such as fridges, thermostats, health devices, and a number of previously Internet technology incapable objects are coming online. The issue with the rapid pace with which these devices are coming alive on the Internet is the lack of security built into the typical IoT devices. Security is sometimes an afterthought once the original communicability objective of the device is attained. HP published an Internet of Things Risk report based on a security analysis of IoT devices in 2015 with some intriguing statistics as seen in Figure 10.2. The study revealed that on the lists of commonly used IoT devices, more than 70% of these devices are highly vulnerable to attacks not being properly secured. A

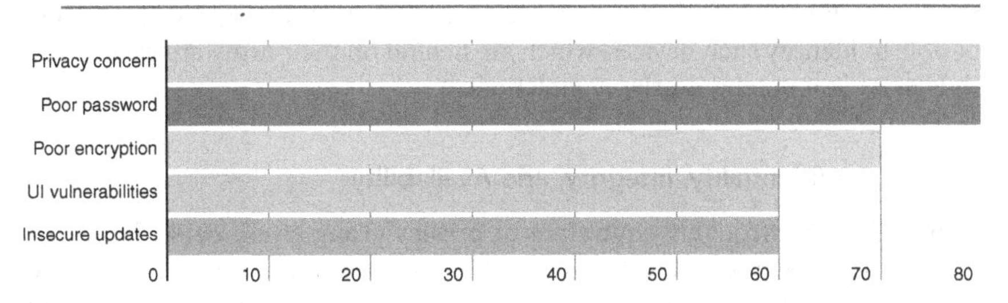

Figure 10.2 Internet of Things risks.

significant number of these devices use unencrypted network services whose connections to the Cloud and associated apps are exposed. This makes them vulnerable to simple exploits such as man in the middle and the likes of it. HP also found that 80% of passwords used in these devices were either commonly used passwords or the same password repeated over multiple devices and accounts creating more opportunities for attackers among other things (see Figure 10.2) (HP News, 2016). These vulnerabilities are a serious concern in any business model where these devices are used and may have serious repercussions if not addressed in a timely manner.

This all goes to show that while the IoT carries a lot of potential and possibilities, there are crucial issues related to risks, threats, security, and auditing that have to be taken into consideration. Here, discussed further are some of the risks that arise and, therefore, need to be considered when considering IoT and auditing.

10.2.1 Privacy

The number one issue in dealing with the IoT is privacy. This is the case because these sensors and devices have come into very close proximity of our lives and our personal spaces like homes, cars, and even our bodies, and hence, maintaining privacy with all the data collected becomes a paramount issue. A vivid example will be that of the hacking of a baby monitor that took place in April of 2014. An intruder was able to gain access to a baby monitor remotely while the family was asleep. He was actually able to swivel the camera to any direction he wanted and was also heard screaming at the baby and yelling very expletive words (Wagstaff, 2014).

This is a glaring example of a traumatizing violation of the privacy of this family. In an organizational setting, the intruder could have used this avenue to monitor the organization. Information such as who is the last to leave or the first to come into the office and the hours spent at the facility could be easily obtained. Not only that but they might be able to obtain passcodes to open doors either visually or by the beep tones produced as the door is unlocked. A statement by the manufacturing company of the baby monitor indicated that the model in question was 3 years old and had not gotten a firmware update, hence creating the avenue for this attack. A routine audit of devices in an organization would be able to identify such devices, which are behind on their firmware update and thus avert any potential privacy violations.

10.2.2 Confidentiality, Integrity, and Availability

Confidentiality is roughly equivalent to privacy (Yang et al., 2010). However, here we are very much concerned with the measures that need to be taken to prevent sensitive information from reaching wrong people. Integrity deals with maintaining consistency, accuracy, and trustworthiness of the data throughout its life cycle (Yan, 2016). Devices connected to a network can be hijacked if the network is breached and the integrity of the data in that network can be compromised or falsified. Availability here deals with maintaining functionality when needed by the user.

In 2014, a study by the SANS institute indicated that 375 health organizations in the United States were compromised all within a month. The intruder infiltrated a set of new and improved radiological imaging units in tandem with the network and was able to gain access to patients files and confidential information (Filkins, 2014). The patients' information was at the mercy of the intruder who could as well have changed or manipulated the data thus violating its integrity. They could also have introduced a bug to the system to hamper the proper functioning of the imaging units denying medical professionals its availability at crucial moments. An audit performed on this category of devices in such medical establishments will reveal the potential areas intruders might exploit and offer recommendations on the best course of action. Admittedly, some manufacturers might not immediately see the need for integrated security in devices such as components of an imaging unit. However, once the industry/organizations start putting standard audit procedures into place, manufactures will be forced to think about how best to integrate security into their products in order to be audit compliant.

10.2.3 Identity Management

Identity management issues might arise if the user uses their social media account to access their IoT service online. In a scenario where their social media account gets compromised, the IoT account/device could also be compromised. For example, an intruder can take control of your home security once they breach your social media account that is connected to a smart watch or Fitbit. The intruder could also perpetrate identity theft once they have access to details from the social media account as well as other personal information available on the devices, for example, financial information used by NFC apps present in phones and also smart watches/bands. Standard audit procedures within any organization would either prevent vulnerable devices from joining the network or notify the network administrators of devices in need of patch or firmware update.

10.2.4 Physical Attacks

Attackers can stage targeted physical attack on the smart network in many ways. Cutting off power or tampering with circuit breakers, installing signal jamming devices on communication lines, and so on, are all possible physical attacks that can debilitate the network. Another facet of physical attacks in scenarios where proximity permits could also include resetting the passwords, changing settings, and redirecting traffic to a server controlled by the hacker. From their servers, attacks can be launched in a number of different ways, for example, studying the firmware of the device and exploiting unmitigated vulnerabilities. Local attacks can also occur over Wi-Fi/Ethernet.

10.2.5 Cloud Infrastructure Attacks

Allowing the users to use weak passwords, not locking out users after unsuccessful attempts, missing two-factor authentication (2FA), unsecured password recovery, and, in general, not enforcing standard security procedures offers an easy target for lurking attackers. Scenarios such as these invariably attract attacks such as brute force attacks, blind SQL injection attacks, and other targeted account-harvesting attacks. Eventually, any successful attack will allow the attacker gain access to the device(s) and personal data. With some of the attacks, such as blind SQL injection attacks, the hacker can end up with read access to the database of the console and obtain the login credentials of the other users connected to the cloud IoT infrastructure.

10.2.6 Malware Attacks

Malware bearing software accidentally downloaded to any device could easily tell the attacker about the devices in the network and perform the previously mentioned attacks. It would just be a matter of time before the attacker can use the device as well as other connected devices to perform attacks like connected toasters that mine crypto currencies or smart TVs that are held ransom by Malware. IoT auditing of these devices will also be to identify anomalous' in the logs and generate alerts to that effect.

10.3 IT Auditing

The purpose of Information Technology (IT) auditing is to examine and scrutinize the management controls in the IT infrastructure. Typically, after the scrutiny, a decision is then made as to whether or not the IT systems are properly espousing the three information assurance tenants (confidentiality, integrity, and availability) while still being properly aligned with the organization's objectives. IoT auditing evidently follows much of already standardized procedures in auditing but requires additional steps to properly ensure a truly multilayered/tiered secured system infrastructure.

10.3.1 IoT Auditing

A proper understanding of the challenges facing IoT devices will elucidate that these IoT devices need a certain degree of security controls and standards. This is evident as the evolution and progress of most of these devices have been disparate. Thus, bringing a huge number of functionally disparate devices into a single network could potentially make for a network with gapping loopholes that is susceptible to cyberattacks. Given their geographically dispersed originating factories, these devices are not always manufactured with the necessary security protocols and standards. A lot of manufacturers already involved in the manufacture of other Internet capable or peripheral devices might use the same procedures and standards to build IoT devices. As with the common Internet capable peripheral devices, such as routers, switches, gaming consoles, manufacturers pressed for time or to meet up with demand surges could ramp up production without necessarily enforcing security. This was the case with Sony games and the recently hacked Jeep Cherokee car replete with IoT-like technology (Newcomb, 2016; Martin, 2016). Hence the result of devices manufactured and built with varying standards could very well imply that they will be vulnerable to attacks common to any device connected to the Internet and possibly other newly developed attacks. An added risk that ensues it that since these devices eventually become part of the network, there is a high probability that these "weak links" in the system can potentially provide a gateway to attack the rest of the network as well as the other devices connected to the network. The vulnerabilities discussed earlier shown in Figure 10.2 would be a serious concern in any business model where these devices are used and may have serious repercussions if not addressed in a timely manner (Rawlinson, 2017).

10.3.2 Need for Auditing

With the proliferation of IoT, billions of devices are to be continually connected to a vastly expanding network all in a bid to improve the quality of peoples' lives, change business processes and models, and reinvent entire industries. On the other hand, IoT also has the potential to provide entrance points for cybercriminals into personal and corporate networks and data storage units. This unequivocally poses a problem that warrants auditing procedures. The loss related to these sorts of attacks have historically been significant when you consider examples like Target, Sony, Home Depot, and Ashley Madison (Keith, 2017).

Clearly the main challenge in today's and future implementations of the IoT is ensuring we have not made any compromises on the security aspect. Lack of proper security measures could provide chances for intruders to access and use personal information that is collected and transmitted to or from a device. Personal information can be misused by the unauthorized person and may result in identity theft or fraud. This may also create risks to physical and public safety in some cases (Alexandra Carmichael, 2011; Tollefson, 2015). To achieve desirable levels of security, IoT systems must adopt and evolve a multilayered security checks and balances, which will be evaluated during auditing. The device, the software, the communication channels must all be tamper proof and ensure data confidentiality. Security should not be an afterthought whereby a layer of protection is wrapped around a finished product. Industry standardization and best practices should push for a "security by design" approach where security is built into the different layers of the device presenting several walls obstructing access to any intruder. This can take several forms, for example, 2F authentication (already commonly used today) or proximity authentication, which will block out most remotely staged attacks. Regardless of the security and assurance technique used, provisions have to be made by the manufacturer to allow for audits within the organization.

Speaking of the organization, incidentally, most of the IoT devices are not actually included in the security audits, as it is currently. An internal audit function can educate the managing body on the competitive edge that a properly functioning IoT implementation can bring to the enterprise. It will also elaborate on the importance, benefits, and potential cost saving advantages in that respect. On the other hand, potential security loopholes and malpractices can be identified and the associated risks dealt with. Moving forward, preventive, corrective, and detective measures and controls could also be implemented to reenforce the IoT infrastructure. This practice of auditing becomes a very important routinely needed exercise especially with the constant progress of the field because the associated risks and vulnerabilities also change with the technology's rapid evolution.

Performing internal audits can be very beneficial as it has the potential to offer strategic advice to the organization's management on the importance, the benefits, and the competitive edge that the IoT could offer the organization. A competent audit can demonstrate to the organization's management how IoT can be effectively implemented in daily operational procedures such as the automated tracking of inventory. These can range from inbound logistics, sales, and marketing all the way to product disbursement. The internal auditing process also permits constructive recommendations and advice to management on how to implement preemptive preventive, detective, and effectively corrective measures. Given the incredible pace at which the IoT is advancing, the inherent risks are a major looming concern, as evidently not enough time will be devoted to the security evaluation of these systems (Salman, 2015).

Furthermore, due to the lack of security audit in IoT, there is no way for an organization to ascertain the source and the type of attack. An organization will be ill-prepared for such an incident and this would affect the business continuity of the organization. To mitigate the risks involved with the use of IoT devices, an organization has to perform a risk-based assessment of all the assets included under the IoT umbrella and perform an end-to-end security audit at appropriate intervals along with the documentation, testing, and reporting of business continuity procedures. Organization can also perform controlled self-assessment (CSA) that would aid in seamless audits. Controlled self-assessment is an internal control assessment technique that has been used in industry for identifying and managing aspects of risk and exposure within an organization. Strong arguments have been made in its favor as it also identifies and highlights areas in the organization with potential opportunities (Ahmed et al., 2003).

10.3.3 Risk Identification and Assessment

Every IT security audit begins with a thorough risk identification and risk assessment along with a holistic validation of the impact of the systems to the goals of the organization. This process essentially starts with risk identification where potential risks to the system are recognized and described. Risk identification is followed by risk assessment, where the likelihood and the consequences of each risk is determined and documented. Control risks, detection risks, inherent risks, and overall audit risk are considered.

After a thorough risk assessment, the auditor must define the scope of the audit by holistically validating the business function to be audited. Typically, prior approval from the senior management is obtained and authority is delegated from the board of directors before the audit process is initiated. An auditor will usually have to consider the points described in Figure 10.3 before auditing the IoT system.

Figure 10.3 outlines some very important points to consider before any auditing procedure can begin within the organization. Primarily, the value that the IoT system generates to the business or organization is key. A system that is centralized and directly integrated to the production or manufacturing arm of the organization would be very critical as it essentially forms an integral part of the organizations driving engine. This calls for a more critical assessment to ensure that the manufacturing or production engine is robust enough to withstand attacks that can bring the whole system down. IoT systems that are more peripheral in nature or decentralized might not necessarily need such scrutiny. Another important aspect directly related to the value of the IoT system in the organization is the threat environment. Not every IoT system is vulnerable to a particular attack. NFC and Bluetooth-based systems might not

Prior to audit, consider:	The value that the IoT system generates to the business and the business processes that are supported.
	Thorough understanding of the threat environment and plans of mitigation.
	Access control list of users using the IoT system.
	Evaluation of risk scenarios and anticipated business impact.
	Privacy and legal issues with the use and users of the IoT systems.
	Type of information that is collected from these IoT systems.
	Validating the risks related to the confidentiality, integrity and availability of data collected by these systems.

Figure 10.3 Points to consider before the audit procedure.

be necessarily vulnerable to remotely staged attacks, such as SQL injection, but could be vulnerable to attacks requiring close proximity. An understanding of this threat environment and plans for mitigation is therefore required. It is also worth mentioning that in recent times the damages caused by insider threats call for a closer scrutiny of people on the systems' access control list as they also could be a part of the threat environment. From instances such as that of Snowden and the NSA, we have come to understand that the list of users with access control privileges effectively constitute the threat environment. These insiders are capable of using any kind of IoT device and exfiltration method to siphon data out of the organization that effectively constitutes the threat environment.

Some other important points to consider as specified by Figure 10.3 include the evaluation of risk scenarios and anticipated business impact, privacy and legal issues that arise with the use of the IoT systems, type of information that is collected from these IoT systems, and the damages that can result if this data is obtained by intruders. All of these will permit the auditor draw up a more focused assessment plan for an audit that will better serve the organization. Organizations whose IoT systems are more customer centric will be worried more about privacy and legal issues; whereas, with more manufacturing or production centric organizations, they might be concerned more with risk scenarios and their related impact on business. After considering these points, an effective audit strategy can be developed based on what the expectations are with regard to the result of the audit.

10.3.4 Audit Strategy

An auditor must keep the organizations interest in mind while auditing the IT systems. An auditor's independence is of crucial importance so that he/she is not to be influenced by any factors that could jeopardize the audit. The audit can essentially begin by focusing on the following aspects of the IoT system:

- *Security.* As the name implies, the IoT devices generally have some built-in Internet connectivity capabilities, and hence, become just as susceptible to attacks from cyber criminals and hacktivists as laptops, notebooks, and other Internet capable devices. A thorough vulnerability assessment of the IoT systems must be conducted and potential risk factors and internal controls have to be identified. These vulnerabilities, threats, and controls have to be documented and periodically tested. The documentation is essential to strengthen the controls for IoT systems. Security of systems provided by third parties must also be considered and audited at frequent intervals. A thorough analysis of the encryption used in IoT systems must also be considered in the audit. Moreover, auditors must also ensure that these devices follow the basic security standards and protocols that have been defined by an appropriate security framework (Kessinger and Duffer, 2017; Hare-Brown, 2017).
- *Health and Safety.* Of all the risks posed by the IoT devices, risks associated with human life and safety are indispensable. Health and safety are of utmost importance in industries like health care, chemical industries, manufacturing unit, laboratories, where smart devices are employed. Examples of these health devices include pacemaker, defibrillators, or other vital signs tracking devices. These systems must be thoroughly tested before they are deployed into these business units. In addition to that, control measures are needed to ensure that the requisite testing procedures are completed before major overhauls such as upgrades, patches, and other changes are made to IoT systems. This is very critical where health and safety-related faults pose a significant risk (Crossman and Liu, 2015; Kessinger and Duffer, 2017).
- *Resilience.* Since IoT devices are used in crucial systems that are prone to attacks, an auditor must assess the existence of controls that could recover systems in the event of a failure. An auditor must elucidate the importance of business continuity, disaster recovery, and incident response to the senior management and actively participate in the design and testing of these procedures. These procedures are crucial to identify the organizations preparedness in the event of a mishap. All crucial systems must be considered while testing these scenarios and appropriate documentations must be in place to guide a smooth transition in the event of a change. Performing testing to ensure the continuity of these procedures are of prime importance to identify their concomitance to the RPO (recovery point objective) and RTO (recovery time objective) (Kessinger and Duffer, 2017).
- *Monitoring.* Akin to any other access-based system, there is a dire need for controls measures that can monitor the functioning of the IoT systems.

Frequent testing has to be performed to ensure that the controls are operating as expected. Any exception or error that occurs in the system must be successfully recorded. These recordings can assume the form of any kind of logging available to the system. Logging obviously has been in the past and continues to be a tremendous asset during audits. It has been compared to an administrative partner that is always at work, never complains, never gets tired, and is always on top of things. If properly instructed, such a partner can provide extensive details on the time and place of every event that has taken place in the network or system (Tuli and Sahu, 2013). The SANS Institute identifies different logging levels such as Debug and Informational, Notice, Warning, Error, Critical, Alert, and Emergency in that order of severity (GadAllah, 2003). Considering a more proactive stance, preventive controls need to be consistently maintained and can be tested with penetration tests to ensure their operability. Likewise, detective controls need to log any illegal access to the system and corrective controls must successfully restore data if lost (Hare-Brown, 2017).

- *ASSET Management.* An auditor must give sufficient importance to the procurement and classification of IoT assets that are used in an organization. A holistic risk-based assessment must be performed while classifying these assets and the data that they transmit. These devices must also contain sufficient amount of encryption to the point where the loss of encrypted data does not pose serious risk to the organization (Hassan, 2016). This is of immense importance as recently U.S. HealthWorks suffered from a data breach via an unencrypted laptop that was lost (Lewis, 2017). Tightening up security measures in asset management evidently should be a major priority.
- *CHANGE Management.* While upgrading/changing a system from a legacy to an improved system, care must be taken to ensure a smooth transition. The newly employed system must mitigate the risks that possibly plagued the legacy systems while also not compromising on critical controls. As second generation IoT devices begin rolling off assembly lines and factories, it will be imperative to ensure that their integration into the organization does more in the way of mitigating existing loopholes and vulnerabilities. Due to pressing schedules, some legacy SCADA (Supervisory Control and Data Acquisition) systems undergo limited amounts of testing and fail to achieve a compromise between concrete security measures and smooth daily functioning. The security features turn out to be either too stringent and slow down smooth functioning or not stringent enough to promote robust functionality, hence allowing security loopholes and vulnerabilities. An example of such a failed attempt was observed with Windows 7's attempt on enforcing system-wide privacy and security. A thorough assessment of any new IoT devices and systems in general will, therefore, be needed before they can be deployed company wide.

10.4 Use Cases of IoT in IT Auditing

The rapid innovations made possible by IoT are consistently pushing the boundaries of how we interact with technology. A direct result of this is the fact that organizations are seeing new nonstandardized forms of technology entering into their networks. This brings about new scenarios having security implications for which the organization is not adequately prepared to handle, as there have been no prior organizational standards in these areas. Listed below are three use cases in which the use of IT auditing would be useful in preventing or detecting possible security lapses in the IoT.

10.4.1 Bring Your Own Devices

The first use case will deal with a Bring Your Own Device scenario, specifically wearable devices due to their growing popularity and expanding capabilities. Depending on the brand, smart watches can browse the web, sync up to your email, write notes, take voice recordings, and even take pictures. The enhanced functions of these wearable devices would allow for easier corporate espionage due to the ability to carry out small bits of data hidden on the watch. Another potentially dangerous scenario would be if the smart watch was hacked, either through the web function or through Bluetooth-based attacks. A virus could unwittingly be brought into the business office where the virus was then allowed to spread after the watch connected to your work computer or other devices in the office.

10.4.2 Electronic Utility Meter Readers

The next use case will deal with the idea of electronic utility meter readers. An electronic meter makes it easier to keep track of the utility costs in a company, but comes with its own hazards. If an outsider is able to gain access to those meters, they will be able to monitor traffic throughout the building or company. A malicious agent can figure out when a certain area will be least populated and then utilize social engineering to get through that area. It is a lot easier to fool one or two individuals with social engineering techniques than a whole group of people. In addition to that an agent who is out to cause lethal damage could redirect the flow of gas to concentrated areas within the building, which could end up in a fire hazard. Finding ways to cut off the flow of air to these regions of the buildings could have similar lethal effects on human lives.

10.4.3 Smart Parking Meter Interfaces

Another use case can involve smart parking meters and their connections to buildings' or organizations Wi-Fi. Smart parking meters in a particular building will be connected to a main interface that can provide incoming drivers with information on exactly where parking is available in that building. In such a scenario, an incoming driver can quickly query the building's parking interface to obtain information on the parking available in that building. For instance, a vehicle might have just pulled out of Level 1 parking spot 27, making that spot available on the building's parking interface. An incoming vehicle would not have to go to level 5 to find a parking spot. An additional benefit of this interface would be the automated tagging of employees' cars. So, employees would no longer need a physical tag to park but could use RFID tags or possibly have the car computer system connected to the smart parking grid network, which could as well be hosted through the organizations intranet. The first issue that could arise with this is some sort of denial-of-service (DoS) attack where an attacker or malicious agent can breach the system and tag empty lots as occupied thereby denying legitimate users parking service. The situation becomes more critical if the attacker can connect to the organization's intranet and extract valuable information on the organization or trivial information such as what vehicle the CEO drives. There are a number of different ramifications that could come with this particularly the potential loss of valuable information. Organizations who own or share smart parking grid interfaces would have to work together to establish common standards to enhance security and consequently audit procedures.

10.5 Protecting the Business Network

The government created the Federal Trade Commission (FTC) for the purpose of protecting the consumer in their purchase of products and services[1]. They are an independent agency and as such, do not have direct authority or the upper hand in the enforcement of their ideas in a particular industry. Instead, they come up with their version of a best practice solution, such as in the case of securing the IoT and then would recommend that the industry adopt these practices to handle security and privacy issues all in a bid to protect the consumer (Ohlhausen, 2014). The FTC understands that IoT, for instance, has great potential for communication innovation and would like to see the network grow, but they also understand that users have to believe in the network for them to use it. The apparent dilemma here is that even though they believe in the importance of this security, they lack direct enforcement ability upon companies in the industry. As such they resolve to be more persuasive in their approach by releasing reports that lay out best practices and host on workshops to spread their ideas.

1 https://www.ftc.gov/

While this seems great, some of the ideas that the FTC has come up with are actually pretty basic (Federal Trade Commission and others, 2015). The fact that these solutions were not already in place is a symptom of the new attention being placed on the IoT network. For example, one of the solutions is to have security be part of the first step of product design and not just put in as an afterthought. It would be easy for us to just blame the producer for not doing enough to protect the privacy of their customers. However, the consumer will not find many products on the market that were made to be tamper proof.

10.5.1 Traditional Security Measures

Another best practice for a company is to minimize the data collected by this network or to notify consumers to make them fully aware of this collection of data. People are wary of any data collection due to "Big Data" being a major buzzword in the media. Unless all players in a certain grouping (e.g., wearables) get together, and all promise to disclose the extent of data collection, this solution will not be taken seriously. If one brand announces the collection of data, consumers will jump ship from that brand to another brand that has not announced collection, even though most likely their some data be produced for a hacker to steal.

The solutions that the FTC is pitching need to be disseminated further before they can be accepted as serious answers. So, while the FTC may only be offering basic ideas, this may eventually create enough public attention to get these ideas implemented in future products to increase security moving forward. The FTC cannot just have workshops attended by industry insiders and expect instant change to current practices. They do have their reports and information on their site, but realistically, if a consumer cared about this topic, they would most likely already have collected all of the relevant and meaningful information from another source. The average consumer is just aware of the benefits they stand to gain from being able to connect to more devices around their house.

There are numerous other big, high-tech firms providing their thoughts on this topic, and there does not seem to be one magic answer as to how to infallibly secure this network is. In the meantime, we just keep growing the network because we do not believe that hackers will crack into our fridges or other such devices. As with credit card theft, we never think that it will happen to us, until it does. And even then, we are already so ingrained into this system that we do not know how to act different so we hope that an issue does not happen again. With this in mind, it calls into question the Federal Trade Commission's warning of doom for the growth of the IoT in relation to consumer buy-in and privacy concerns. By getting the word out to the average consumer about potential dangers and the impact they can have on a company's bottom line if security is not made a top priority, the FTC can help ensure change.

10.5.2 New Policies to Address New Threats

Industries that are keen on implementing these IoT devices must be prepared to efficiently manage IoT devices to gain maximum rewards from it. They must have prepared to mitigate any risks that IoT poses by following specific guidelines and standards. A few recommendations for organizations planning to implement IoT are as follows:

- Designing security into IoT systems from the bottom-up. Security must not be added to these systems after their implementation, but rather, they must be incorporated from the initial stages of development. In other words, security controls must not be a value added to the IoT systems, but an essential integrated feature.
- Understanding vital assets and values and investing in their protection. Health companies focus on the well-being of the patient while commercial organizations focus on great products and sales maximization. These assets and values have to be the central focus when planning on IoT implementations.
- Collecting sufficient amount of data that is required and encrypting the sensitive data.
- Partnering with appropriate vendors on elements of security like identity management, access control management intelligence analytics, and patch management.
- Conducting a comprehensive security audit of the IoT systems including privacy, risks, and fraud assessment.
- Sufficient testing before implementing or changing the IoT systems.
- Training the organization staff on the risks related to IoT systems and reiterating it.
- Creating a security awareness program and educating all the members of the organization on the importance of security practices related to IoT systems (Hare-Brown, 2017).

10.6 Conclusion

The next generation of technology belongs to network-interfaced devices that perform intelligent and complex tasks in order to enhance human lifestyle experiences. The evolution of these devices now allows them to exchange copious amounts of data, process this data, and obtain results, which allow them to make decisions very often without any human intervention.

Unfortunately, this luxury does not come without its drawbacks as these networks teeming with data pose as a very attractive bed for intruders and other ill-intentioned minds. This chapter has highlighted and discussed some of the possible ensuing vulnerabilities and demonstrated the need for routine auditing. The onus, however, does not only lie with the organizational auditing bodies, but manufacturers must find a way for end-to-end security to be incorporated in IoT devices and IoT systems. This should be done in concert with factory level audits to ensure compliance with designated standards. In a nutshell, security at both the device and system levels should be an integral part of their build process followed by recurring audits to ensure standards are met.

Establishing audit procedures for IoT devices might seem far-fetched as these devices cut across a wide array of categories, as seen in the chapter. There however already exist basic tenets for auditing devices that exchange data in today's world. In like manner, rigorous auditing routines akin to the commonly known and widely accepted such as with penetration testing routines or with BYOD infrastructure could easily be referenced. Adapting and modifying these already existing technologies will doubtlessly ensure compliance at every level of society ranging from homes to job sites. As it is, the available technology already possesses the tools and capabilities for built-in security or at the very least periodic audits. To this end, priority must not be placed on investing new technologies and gadgets. The immediate objective must be directed toward conveying the present best in class IT security controls, streamlined for this new and complex ecosystem of technology that is driving the IoT.

Acknowledgments

The research of the second, third, and fourth authors is funded in part by the National Science Foundation through the Scholarship for Service program under grant #1241709. The first author is funded by the University of Texas at San Antonio.

References

Ahmed, A. M., Yang, J. B., and Dale, B. G. (2003) Self-assessment methodology: the route to business excellence. *The Quality Management Journal*, **10**(1), 43. Available at http://search.proquest.com/openview/0697fb8426fc3c99f60970e174cb1d9b/1?pq-origsite=gscholar.

Alexandra Carmichael. (2011) HIT – Health Internet of Things. *Quantified Self*. February 14. Available at http://quantifiedself.com/2011/02/hit-%e2%80%93-health-internet-of-things/.

Boorstin, J. (2016) Humans hooked on 21 billion of these by 2020. *CNBC*. February 1. Available at http://www.cnbc.com/2016/02/01/an-internet-of-things-that-will-number-ten-billions.html.

Crossman, M. A. and Liu, H. (2015) Study of authentication with IoT testbed. 2015 IEEE International Symposium on Technologies for Homeland Security

(HST), pp. 1–7. Available at http://ieeexplore.ieee.org/xpls/abs_all.jsp?arnumber=7225303.

Federal Trade Commission, and others. (2015) Internet of Things: Privacy & Security in a Connected World. Washington, DC: Federal Trade Commission.

Filkins, B. (2014) Health Care Cyberthreat Report. Widespread Compromises Detected, Compliance Nightmare on Horizon. *SANS Institute*. http://www.redwoodmednet.org/projects/events/20150731/docs/Norse-SANS-Healthcare-Cyberthreat-Report2014.

GadAllah, S. M. (2003) *The Importance of Logging and Traffic Monitoring for Information Security*. Available at https://pdfs.semanticscholar.org/b1bd/427cb53b8ccc7a9b7630dcec77abcbb27c2b.pdf.

Hare-Brown, N. (2017) How to mitigate security risks associated with IoT. *ComputerWeekly*. Available at http://www.computerweekly.com/opinion/How-to-mitigate-security-risks-associated-with-IoT (accessed July 10, 2017).

Hassan, M. K. A. L. (2016) Governance, risk and compliance "GRC" for Internet of Things" IOT. *International Journal of New Technology and Research*, **2**(3), 148–152. Available at https://www.ijntr.org/download_data/IJNTR02030038.pdf.

HP News (2016) HP News—HP Study Reveals 70 Percent of Internet of Things Devices Vulnerable to Attack. Available at http://www8.hp.com/us/en/hp-news/press-release.html?id=1744676#.V2ByEuYrKgQ (accessed June 14).

Keith, C. (2017) A Quick Guide to the Worst Corporate Hack. *Bloomberg.com*. Available at http://www.bloomberg.com/graphics/2014-data-breaches/ (accessed July 10, 2017).

Kessinger, K. and Duffer, J. (2017) Internet of Things: Risk and Value Considerations. Available at http://www.isaca.org/knowledge-center/research/researchdeliverables/pages/internet-of-things-risk-and-value-considerations.aspx (accessed July 10).

Lewis, D. (2017) US healthworks suffers data breach via unencrypted laptop. *Forbes*. Available at http://www.forbes.com/sites/davelewis/2015/06/01/us-healthworks-suffers-data-breach-via-unencrypted-laptop/ (accessed July 10).

Martin, L. (2016) PlayStation hacked: what to do when your PSN account gets hacked. *Express.co.uk*. Available at http://www.express.co.uk/pictures/galleries/7133/PlayStation-Store-Summer-Sale-PS4-discounts.

Newcomb, A. (2016) Jeep hackers' are back with a scary new trick. *NBC News*. Available at http://www.nbcnews.com/tech/tech-news/jeep-hackers-are-back-scary-new-trick-n623756.

Ohlhausen, M. K. (2014) Privacy challenges and opportunities: the role of the federal trade commission. *Journal of Public Policy & Marketing*, **33**(1), 4–9. Available at http://journals.ama.org/doi/abs/10.1509/jppm.33.1.4.

Rawlinson, K. (2017) HP study reveals 70 percent of internet of things devices vulnerable to attack. Available at http://www8.hp.com/us/en/hp-news/press-release.html?id=1744676#.V2ByEuYrKgQ (accessed July 10).

Salman, S. (2015) Auditing the Internet of Things. Internal Auditor, October 29. Available at https://iaonline.theiia.org/2015/auditing-the-internet-of-things.

Tollefson, R. (2015) Healthcare data at risk: Internet of Things facilitates healthcare data breaches. Third Certainty, January 9. Available at http://thirdcertainty.com/news-analysis/internet-things-facilitates-healthcare-data-breaches/.

Tuli, P. and Sahu, P. (2013) System monitoring and security using keylogger. *International Journal of Computer Science and Mobile Computing,* **2**(3), 106–111. Available at http://d.researchbib.com/f/8nq3q3YzydL3AgLl5wo20iMT9wpl9jLKOypaZiGJSlL2tlZQRmY1LlFGZlZQRmZwVhpTEz.pdf.

Van der Meulen, R. (2017) Gartner says worldwide IoT security spending to reach $348 million in 2016. Available at http://www.gartner.com/newsroom/id/3291817 (accessed July 10)

Wagstaff, K. (2014) Man hacks monitor, screams at baby girl. NBC News, April 28. Available at http://www.nbcnews.com/tech/security/man-hacks-monitor-screams-baby-girl-n91546.

Wortmann, F., Flüchter, K., and others. (2015) Internet of Things. *Business & Information Systems Engineering,* **57**(3), 221–224. Available at http://search.proquest.com/openview/ac6643ce2897d57bbad03f45ba9436e1/1.pdf?pq-origsite=gscholar&cbl=816386.

Yan, W. Q. (2016) *Introduction to Intelligent Surveillance,* Springer.

Yang, G., Xu, J., Chen, W., Qi, Z.-H., and Wang, H.-Y. (2010) Security characteristic and technology in the Internet of Things. *Journal of Nanjing University of Posts and Telecommunications (Natural Science),* **30**(4).

Part IV

Application Domains

11

The Industrial Internet of Things

Alexander Willner[1,2]

[1]*Fraunhofer FOKUS, Software-based Networks (NGNI), Berlin, Germany*
[2]*Technische Universität Berlin, Next Generation Networks (AV), Berlin, Germany*

11.1 Introduction

Within industrial use cases, computers were introduced over the last decades, mainly to fulfill specific requirements, such as meeting hard real-time response times or operating reliably in very rough environments. Their task was, and still is, to automate physical control loops, to process input signals, and trigger actuation signals based on this collected information. These systems are part of the Operational Technology (OT). Respective fields of application include energy, health care, manufacturing, smart cities, and transportation. This development significantly enhanced the efficiency of local processes within these and other application domains and their benefits cannot be argued away.

Nowadays, however, we live in a connected world. Networks of devices, processes, and services constantly exchange data with each other and enable the cooperation for a common task. Under the umbrella of the Internet of Things (IoT) (Ashton, 2009), the number of interconnected devices is expected to grow exponentially toward 30 billion devices until 2020 (Markit, 2016). As described in the former chapters, this development will be a large driver for economic growth within the foreseeable future. For example, Woodsite Capital Partners estimated that IoT-related value-added services will grow from 50 billion USD in 2012 to 120 billion USD in 2018, attaining around 16% compound annual growth rate (CAGR) in the forecast period (Woodside Capital Partners, 2015).

Arguably, the Industrial Internet of Things (IIoT) (Jeschke et al., 2017) will be the biggest driver of productivity in the future. This concept, that is, the usage of IoT technologies within industrial domains is also called the Industrial Internet and the related market value is estimated to reach 124 billion USD by 2021 at a high CAGR (IndustryARC, 2016). Therefore, in Germany, for example, 80% of

Internet of Things A to Z: Technologies and Applications, First Edition. Edited by Qusay F. Hassan.
© 2018 by The Institute of Electrical and Electronics Engineers, Inc. Published 2018 by John Wiley & Sons, Inc.

all industry corporations will already have their value chain digitized by 2020 (PricewaterhouseCoopers, 2014) to participate in this paradigm shift. A countermeasure to mitigate a development that might inspire the reader to examine the topic of digitization in more detail: A 40% share of worldwide manufacturing is already held by developing countries and they have doubled their share in the last two decades (Roland Berger Strategy Consultants, 2014); Western Europe, on the other hand, has lost over 10% of its manufacturing share.

Following the definition of Gartner,[1,2] OT causes a change through direct monitoring and control of physical devices. OT is traditionally associated with industrial environments using nonnetworked embedded proprietary technology that usually does not generate data for the enterprise. Information and Communication Technology (ICT), on the other hand, inherently covers the entire spectrum of technologies for information processing and open communications. Therefore, OT and ICT systems have historically chosen different technological approaches, which makes the application of IoT mechanisms a challenging task. Nevertheless, in order to enable a digital transformation across the industrial value chains, both worlds have to converge. A key aspect in this regard is the interoperability between systems. Starting with technical aspects, such as connectivity mechanisms and communication protocols, this further includes syntactical and semantic conformity as well as organizational interoperability (van der Veer and Wiles, 2008). In order to coordinate efforts, to discuss various economic and technical aspects, and to reach agreement on common concepts, a number of alliances, initiatives, and Standards Developing Organizations (SDOs) work together on different layers.

This chapter gives a general overview on the subject and provides the reader with an overall motivation behind the development of the IIoT and a classification of related technologies. Not only the most relevant use cases with their predicted market values are described, but also technological challenges and candidates to realize the IIoT vision are identified. Finally, the work of the two most important alliances is illustrated. They aim at digitizing the whole industrial value chain across domain boundaries to enhance efficiency and enable new and disruptive business models.

11.2 Market Overview

The aforementioned expected growth of the Industrial IoT market will facilitate the invention of creative business models; it will be accompanied by the development of new and the adoption of existing IoT technologies in more and more fields of application, and will finally enable the digital

1 http://www.gartner.com/it-glossary/it-information-technology
2 http://www.gartner.com/it-glossary/operational-technology-ot

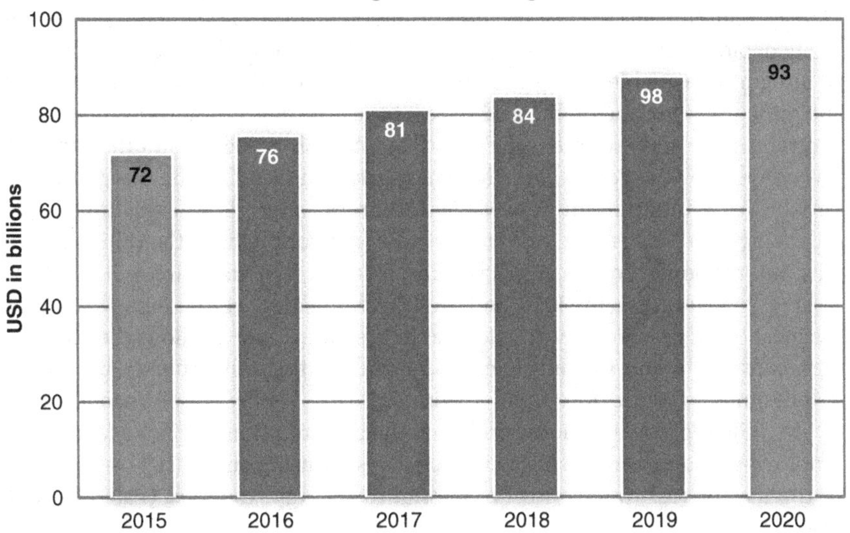

Figure 11.1 Value of smart energy market, global 2015–2020 (Frost & Sullivan, 2016c).

networking of the whole value chain across multiple domains. In this section, a deeper insight into five related use cases within the most important verticals is provided. As with all attempts to look into the future, the following market forecasts should be taken with a grain of salt.

11.2.1 Energy

The global revenue for the smart energy segment amounted to 72 billion USD in 2015 (Frost & Sullivan, 2016c) and as depicted in Figure 11.1, the revenue is expected to show a CAGR, between 2015–2020, of 5.3% resulting in a market volume of approximately 93 billion USD in 2020. Leading technologies will be Advanced Metering Infrastructures (AMIs), Demand Response (DR), Distribution Grid Management (DGM), and Advanced Transmission Technology (ATT), while DGM will be the dominant segment with a 64% of the market share by 2020.

As highlighted in Chapter 14, the energy market is evolving to a more efficient, cleaner, and flexible ecosystem. For example, the aim of the Paris Agreement[3], entered into force on November 4, 2016 with 116 partner ratifications, is to strengthen the global response to the threat of climate change. Energy generation accounts for 68% of the shares of global anthropogenic greenhouse gas

3 http://unfccc.int/paris_agreement/items/9485.php

(GHG) (International Energy Agency, 2016), therefore, it is necessary to shift to a cleaner and efficient energy production market. Renewable energy plants are being deployed all over the world, but nevertheless, one of the biggest challenges of integrating variable energy sources, like Photovoltaic (PV) or wind energy, is the difficulty in balancing the grid in real time. Moreover, renewable plants are erected where the resource (solar, wind, biomass) is available, and they are not always close to the consumer. The smart grid will facilitate the integration of variable and intermittent renewable resources, allow load adjustment and balancing, and distribute power over the network efficiently (ITU, 2012). The International Energy Agency (IEA) foresees that its share will reach at least a 26% increase in 2020 and IIoT technologies will change the utility business models. AMIs will allow a bidirectional power flow; hence, the customer will be able not only to consume but also to produce power, becoming a "prosumer" (World Resources Institute, 2016). Demand side management (DSM) will improve the energy grid from the consumption side, for example, by employing smart energy tariffs with incentives for using energy at a certain time of the day, or real-time control of distributed energy resources (Palensky and Dietrich, 2011).

11.2.2 Health care

As depicted in Figure 11.2, the global revenue in the health care market will grow from 86 billion USD in 2015 to 233 billion USD in 2020 and the projected CAGR is around 21% (Little, 2016).[4] With a market share of 44% by 2020, the wireless health segment will be the most relevant one mainly driven by wireless sensors, handheld devices, and eHealth applications. The Organization for Economic Co-operation and Development (OECD) reported that in 2014, 9.945% of the world gross domestic product (GDP) was spent on health, up to 0.144% since 2005.[5] Circulatory, digestive, cancer, and mental health conditions represent almost 60% of the current health spending and, likewise, chronic diseases account for 60% of the causes of death.[6] The World Health Organization (WHO) and its member states endorsed health care as a cost-effective and secure approach to strengthen the health care systems (WHO, 2005), and governments are focusing on making them more efficient and sustainable health care (Frost & Sullivan, 2016b). For instance, European Union health care policies pursue making health care tools useful and widely accepted by involving health care professionals and patients in the strategy, design. and implementation.[7]

4 https://solarcity.com
5 http://data.worldbank.org
6 http://stats.oecd.org
7 https://ec.europa.eu/health/ehealth/policy

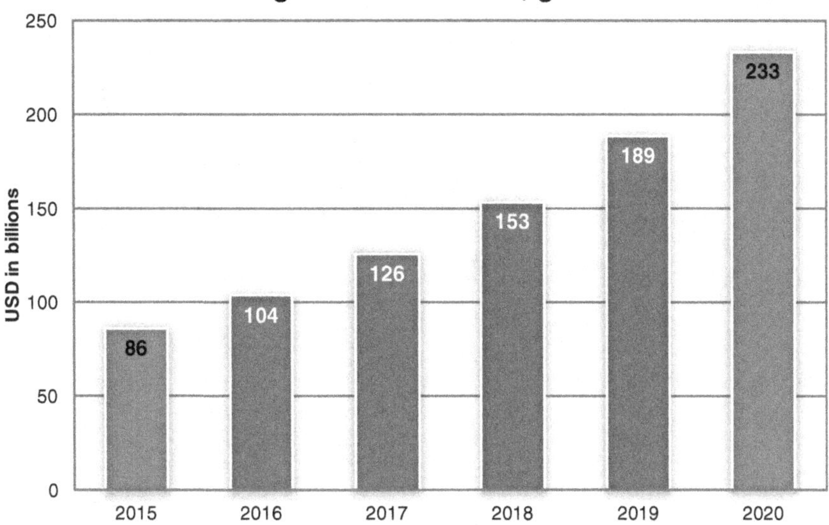

Figure 11.2 Value of health care market, global 2015–2020 (Little, 2016).

Devices such as heart rate monitors, pulse oximeters, blood pressure monitors, pedometers, smartwatches, smartphones apps, and so on, are being used to measure health conditions and activities. When this information is exchanged between the device and a health care platform, patients benefit not only from self-monitoring but the information could also be used for different purposes such as detection, prevention, treatment of diseases, supporting a rehab process, and so on. Seamless communication aids patients that need remote assistance, thus, reducing costs for them and the insurance system. This specific application of IoT technologies in the health care domain is further described in Chapter 16. The IIoT will help to improve access to comprehensive health care services, quality of medical services, decrease medical errors, and improve patients' quality of life. Moreover, real-time monitoring, control, and automation empower assisted living to provide personal safety and health care management at home. Additionally, one of the main benefits of health care is a patient's empowerment by providing more autonomy and increasing their treatment.

11.2.3 Manufacturing

As shown in Figure 11.3, the global revenue in the manufacturing market will grow from 39 billion USD in 2015 to 62 billion USD in 2020 and the projected average CAGR is 9.7% for the global market (Mordor Intelligence, 2017). The smart manufacturing domains include automotive, chemical and petrochemical, oil and gas, pharmaceuticals, aerospace, defense, mining, among

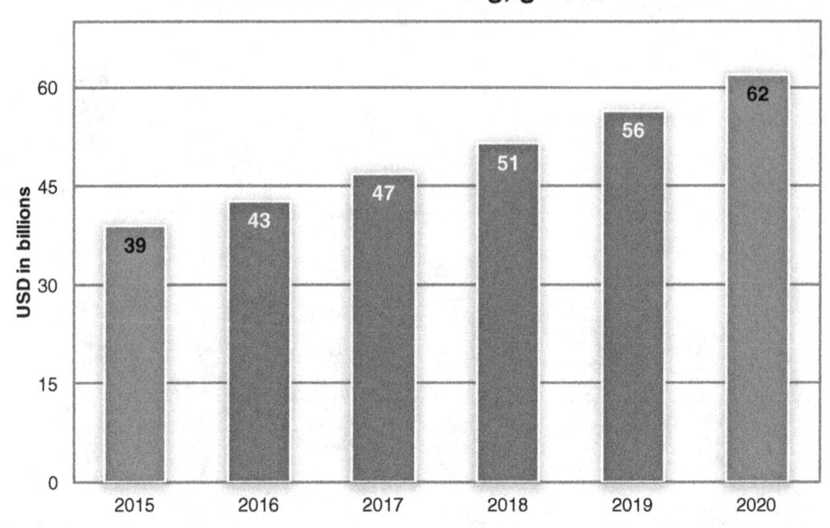

Figure 11.3 Value of smart manufacturing market, global 2015–2020 (Mordor Intelligence, 2017).

others. The chemical and petrochemical industries hold the major share (23%) while the oil and gas market is expected to grow at a higher CAGR.

Within this major IIoT application domain, digital technologies will be used to move toward resource-saving and more efficient manufacturing. For example, Cyber-Physical Systems (CPSs) and prescriptive analytics will enable automated decision-making at the topological edge of a network to allow for timely maintenance measures and an extended lifecycle of machines while optimizing the overall production at the same time. Every digital device will be able to provide their real-time status (the so-called digital shadow), thus allowing other devices to react on this information. It is foreseen that the application of IIoT technologies in the manufacturing sector will lead to process optimization (possibly enabling efficient lot-size one productions) and supports the prioritization of workloads, and as a result will significantly reduce needed quality inspections, surveillance, and operational expenditures in the industrial manufacturing sector (Frost & Sullivan, 2014, 2016a). The development of a virtual factory will provide a holistic, scalable, and virtual representation of a manufacturing facility and allow synchronization, dynamic configuration and thus, enable cost savings of manufacturing facilities (Ghielmini, 2013).

However, these potential cost savings are also offset by risks. Downtimes in the production are very expensive, which is why reliability has been a top priority in automation technology over the last 40 years. Installations are also often operated over many years to decades without the need to install updates, as

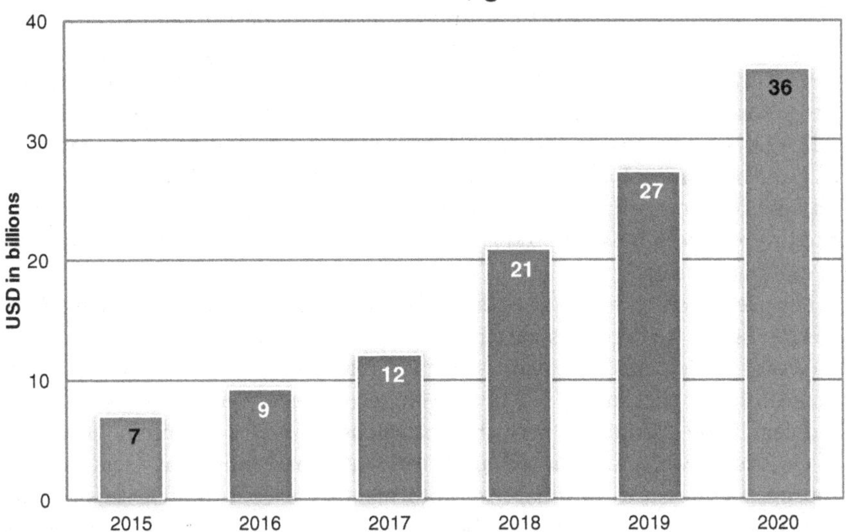

Figure 11.4 Value of smart cities market, global 2015–2020 (Markets and Markets, 2016).

it is common (and required for at least security reasons) in IT infrastructures. Therefore, the continuous merging of OT and IT in this context faces both high potential and great challenges.

11.2.4 Smart Cities

The global smart cities value in 2015 was approximately 312 billion USD and is expected to reach 758 billion USD by 2020 with a CAGR of 19.4% (Markets and Markets, 2016) (see Figure 11.4). The building segment is projected to grow at the highest CAGR, on top of transportation, energy, and smart citizen services such as education, health care, and security. According to the United Nations (UN), urban areas represent approximately 70% of energy-related global emissions, and by 2050 more than half of the world's population will live in cities, mainly in African and Asian regions (United Nations, 2014). Nevertheless, energy efficiency and GHG emissions are not the only matters of concern. With urban population increasing, challenges such as security, balance public expenditure, transportation, health care, and education have to be considered. A city is a complex network of people and infrastructure that interacts, expands, and transforms continuously. Traditionally, the infrastructure and services of the cities are operated as verticals or domains, with little or no interaction: transportation, energy, health care, buildings, industry, and so forth.[8]

8 http://english.gov.cn/2016special/internetplus/

Each vertical is evolving to a smarter and more efficient version of itself, and cities must take advantage of those improvements. A smart city should be able to integrate the current infrastructure with ICT to operate more efficiently while improving the quality of life of its citizens. According to the International Telecommunication Union – Telecommunication Standardization Sector (ITU-T), a smart sustainable city is defined as "an innovative city that uses ICT and other means to improve quality of life, efficiency of urban operation and services, and competitiveness while ensuring that it meets the needs of present and future generations with respects to economic, social, environmental as well as cultural aspects."[9]

The use of IIoT technologies will enable the efficient use of resources in urban areas; however, to become smarter, a city needs its municipality, industries, and society to participate. Some use case scenarios include Blackout Prevention, that is, the utility applies smart "self-healing" to reconfigure itself whenever there is a problem in the distribution network and, whenever there is an imminent cut-off of electric power inform, in advance, residential and industrial users to take appropriate measures; air quality monitoring, that is, collaborative sensing will help to determine contaminants before they reach a dangerous level and to identify the source and their impact on transportation, industry or energy generation industries; or smart parking, that is, buildings, streets, and parking lots are connected to determine available parking spaces to save time, make efficient use of resources (gas, diesel, and public spaces), and minimize stress as well as emitted pollutants. Further details can be found in Chapter 12.

11.2.5 Transportation

As shown in Figure 11.5, the global revenue for the smart transportation segment amounted to 10 billion USD in 2015 and its market is expected to show a CAGR between 2015–2025 of 18.7%, resulting in a market volume of 24.5 billion USD in 2020 (Zpryme Agency, 2015). The smart transportation ICT segment includes hardware, software, communications and networking, and sensors and Intelligent Electronic Devices (IEDs). Smart transportation or intelligent transport systems (ITSs) are those systems that enable connection, integration, and automation of the transportation network to improve experience for travelers and system operators (users) by enhancing vehicles and infrastructure (U.S. Department of Transportation, 2015). Therefore, the scope of smart transportation is not only limited to connected cars, but also to car/bike sharing systems, pay as you drive (users); smart roads, road pricing, parking systems, traffic management, backhaul communications, fleet management (infrastructure); connected car, automated vehicles, public transportation (vehicles), just to mention a few. Although in 2015 more than 69 million passenger

9 http://www.itu.int/en/ITU-T/ssc/Pages/info-ssc.aspx

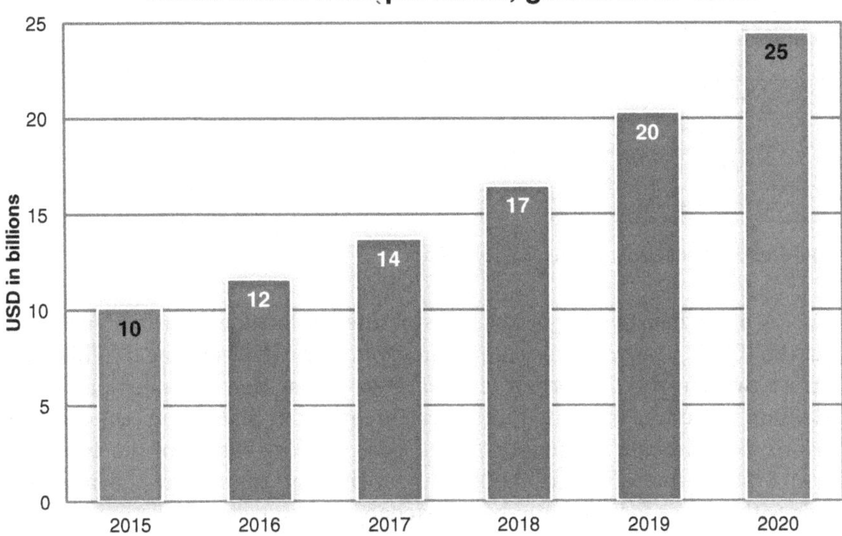

Figure 11.5 Value of smart transportation market, global 2015–2020 (Zpryme Agency, 2015).

cars were produced, the automobile industry is experiencing changes. City policies are discouraging private vehicles (McKinsey & Company, 2016) and today there are more than 80K car-sharing vehicles in operation with more than 6 million users (BCG Perspectives, 2016).

The use of IIoT technologies in the transportation industry will allow proactive maintenance and prevent failures through predictive analytics. Safer vehicles and roads will improve crash avoidance by developing vehicle-to-infrastructure cooperative systems. Advanced sensing technologies and high-bandwidth connectivity will enable real-time applications that interwork with different domains. For example, Advanced Traffic Management Systems (ATMSs) will improve the flow of vehicles, thereby decrease traffic commuting time and CO_2 emissions across urban areas. Moreover, fleet management (from rental cars to freight transport) will be supported by ubiquitous and affordable mobile communications as well as location systems to maximize customer service and productivity (GSMA, 2015). Finally, a transformation from current cars to driverless cars is expected and will be based on the IIoT (PWC, 2016).

11.3 Interoperability and Technologies

There are clear indications that a digital transformation will take place across all industrial domains and at the same time a number of technical challenges will

Application
Presentation
Session
Transport
Network
Data link
Physical

Figure 11.6 Connectivity on the OSI layer stack.

arise. A nonexhaustive list of key areas of interest include security, Quality of Service (QoS), connectivity, communication, and data exchange. While the former two are very important cross-layer concerns that will need particular attention in the area of the IIoT, the latter three directly build upon each other to ensure cross-domain interoperability. Based on an extension of the common Open Systems Interconnection (OSI) model, this starts at the physical layer and ends with a layer for semantic-based exchange of knowledge.

11.3.1 Connectivity

As shown in Figure 11.6, overall connectivity between objects involve the first four layers of the OSI model. They build the basis for IIoT devices to connect with each other, and within single, noninterconnected use case domains, a crucial prerequisite is the use of interoperable wired or wireless links. Depending on the requirements of the application area, proprietary technologies, such as PROFIBUS[10] or Modbus[11] fieldbus systems, might be used. However, two main trends can be observed: First, wired networks are in the transition to become mainly Ethernet based, with the IEEE Time-Sensitive Networking (TSN) standards as one notable development. Second, where possible low-power wireless technologies are applied for connectivity. For short ranges, mainly Bluetooth Low Energy (BLE), RFID technologies, such as Near Field Communication (NFC), and IEEE 802.15.4-based approaches like ZigBee are being deployed. For Low-Power Wide-Area Networks (LPWANs), LoRa, Sigfox, nWave, and Neul are popular technologies in unlicensed bands. At the same time, the 3rd Generation Partnership Project (3GPP) is standardizing LTE-M, NB-IoT, and EC-GSM-IoT for licensed bands.

As data exchange between multiple verticals is one particular characteristic of the IIoT, a joint routable network layer is a second prerequisite. Despite its specification of IPv6 already in the end of the twentieth century, many networks are still using IPv4. However, given the number of interconnected devices and

---

10 http://www.profibus.com
11 http://www.modbus.org

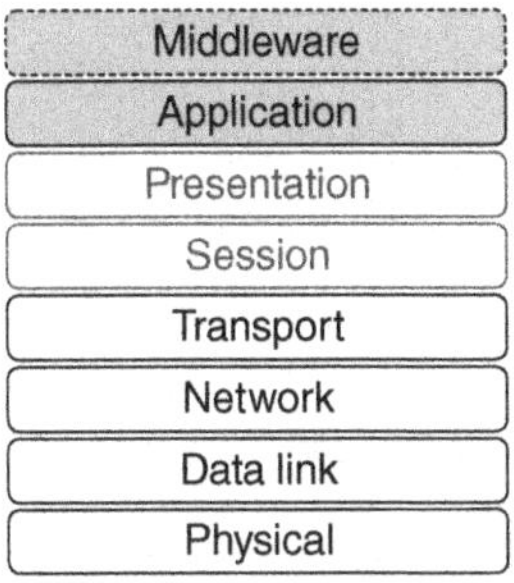

Middleware
Application
Presentation
Session
Transport
Network
Data link
Physical

Figure 11.7 Communication on the OSI layer stack.

their often-limited capabilities, the Internet Engineering Task Force (IETF) working group IPv6 over Low-Power Wireless Personal Area Networks (6LoW-PAN) defined respective concepts to use IPv6 as the common networking layer, which will also be predominant in the IIoT context. Depending on the requirements, typical Transmission Control Protocol (TCP) and User Datagram Protocol (UDP)-based transports will be used on top of IPv6.

11.3.2 Communication

As indicated in Figure 11.7, in modern networks, the next and final OSI layer is the application. However, for interoperable data exchange, an additional layer is needed that we call "middleware" for convenience. First, the Hyper Text Transfer Protocol (HTTP), the Advanced Message Queuing Protocol (AMQP), the Message Queue Telemetry Transport (MQTT), the Constrained Application Protocol (CoAP), or WebSockets are typical examples of standard application-level protocols that are used within the IIoT context. Next, to enable standard Remote Procedure Calls (RPCs) and data exchange, mechanisms such as XML-RPC or the Simple Object Access Protocol (SOAP) are being used as well as Representational State Transfer (REST) concepts. Finally, depending on the use case and the geopolitical area, a variety of IIoT specific middleware systems are then deployed. The following three examples are currently under discussion within different IIoT verticals.

The oneM2M[12] (Swetina et al., 2014) alliance is a partnership of international standards bodies. While historically being focused on the telecommunications industry, the architecture aims to cover smart building, smart factory, and smart power grid use cases. In the current release 2.0 of the middleware specification, messages are allowed to be sent via HTTP, MQTT, CoAP, and WebSockets. Further, a number of Common Service Entities (CSEs) are defined that are often used in Machine-To-Machine Communication (M2M) (Wu et al., 2011) environments that can be invoked by Application Entities (AEs); and Network Service Entities (NSEs) provide respective services to the CSEs.

12 http://onem2m.org

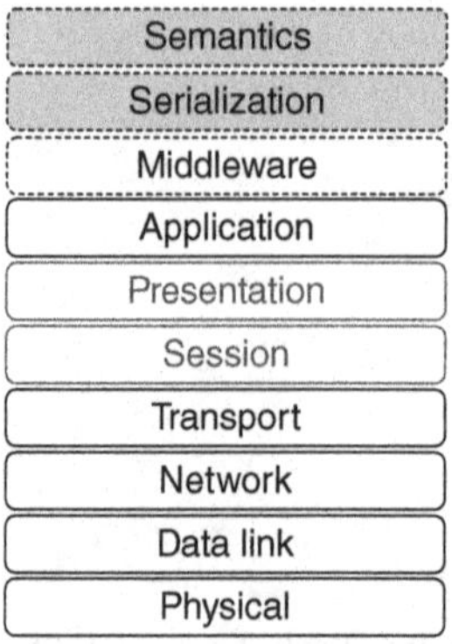

Semantics
Serialization
Middleware
Application
Presentation
Session
Transport
Network
Data link
Physical

Figure 11.8 Data exchange on the OSI layer stack.

Developed within the international OPC Foundation, the Open Platform Communications Unified Architecture (OPC UA) (IEC, 2016) middleware is, as a successor to the former Object Linking and Embedding for Process Control (OPC) architecture, mainly being used in the automation industry. The two different communication types are either directly exchanging binary data using raw TCP sockets or exchanging XML data via SOAP and HTTP over TCP. Since June 2016, further application-level publish/subscribe protocols, such as AMQP, are being evaluated.[13] The standards further define common base services such as (historical) data access, alarms and conditions, and programmability, and a common object-oriented meta model for describing exchanged information.

Finally, the Data Distribution Service (DDS) (Pardo-Castellote, 2003) was developed within the Global Information Grid (GIG) project and standardized by the Object Management Group (OMG). Based on the DDS Interoperability Real-time Publish-Subscribe Wire Protocol (DDSI-RTPS), the DDS Application Programming Interface (API) offers via TCP or UDP, access to a data-centric publish/subscribe system with potentially multiple hierarchical control domains. Since 2017, it has been further extended with RPC capabilities.[14]

11.3.3 Data Exchange

As shown in Figure 11.8, the aspect of data exchange is located on top of the afore discussed middleware layer. Following the European Telecommunications Standards Institute (ETSI) white paper on technical interoperability (van der Veer and Wiles, 2008), at least three different layers have to be considered. First, all aspects from the network layer up to the middleware have to be considered

---

13 https://opcfoundation.org/news/opc-foundation-news/opc-foundation-announces-support-of-publish-subscribe-for-opc-ua/
14 http://www.omg.org/spec/DDS-RPC/1.0/

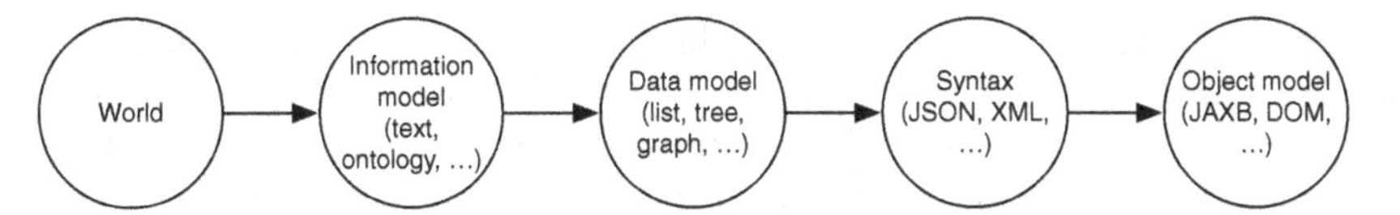

Figure 11.9 Information modeling based on Pras and Schoenwaelder (2003) and Pras et al. (2007).

for technical interoperability. Due to the heterogeneity of the involved systems in IIoT environments, the application of a homogeneous set of protocols is unlikely. Therefore, implementations need to be abstracted from specific APIs. To implement such a Separation of Concerns (SoC) (Martin, 2003; 2012), a term coined by Edsger W. Dijkstra in 1974 (Dijkstra, 1982), numerous architectural design patterns can be applied. Examples are the classic Model View Controller (MVC) (Krasner et al., 1988), Entity, Boundary, Interactor (EBI) (Jacobson et al., 1992) or Data, Context and Interaction (DCI) (Coplien and Bjørnvig, 2010), in addition to more modern Microservices-based architectures (Newman, 2015; Fowler and Lewis, 2014; Thones, 2015).

Second, assuming the usage of appropriate design patterns or gateways to allow IIoT systems to exchange data with each other, this data has to either be serialized using the same syntax or unambiguous mapping rules have to exist. As indicated in Figure 11.9, to exchange information between distributed systems in general, world concepts are abstracted into an information model, often in form of a human-readable text document such as a Request for Comments (RFC). To transmit the information over a network, the derived data model is then serialized using a syntax such as the Extensible Markup Language (XML) or the JavaScript Object Notation (JSON). Within an application, this string is then deserialized again by functional code into an object for either document- or stream-based processing, depending on the size of the data and the way the recipient application is designed. To decrease transmission and deserialization time, more efficient serializations such as Protocol Buffers (protobuf) (Varda, 2008) can be used as well.

Finally, assuming the same syntax is being used, the meaning of the exchanged data has to be understood by each involved component. This in particular holds true in IIoT environments in which heterogeneous devices across multiple application domains shall negotiate interactions autonomously in order to further automate processes. Typically, a tree-based data model along with structure- and identifier-implied semantics are being used, for example, based on schema definitions known within the specific use case domain. This approach, however, does not scale with cross-domain IIoT-wide scenarios, as, due to their heterogeneous nature, it would result in involving n different approaches to encode information in a tree, which leads to a combinatorial problem of n^2 required conversions using functional code. A formal information model of types, properties, and relationships of objects within specific domains, also known as ontologies (see Figure 11.9), is needed instead. This would allow for semantic reasoning over the information to infer logical consequences such as transitivity, symmetry, or equality of specific data; in other words, it would allow machines to understand each other. One common approach is the Semantic Web (Berners-Lee et al., 2001), along with its canonical graph-based data model Resource Description Framework (RDF) (Cyganiak et al., 2014), ontology languages such as the Resource Description Framework Schema (RDFS) (Dan and Guha, 2014) and the Web Ontology Language (OWL) (Herman et al., 2012), and other related concepts. As indicated in Figure 11.10, the respective stack of technologies for semantic interoperability

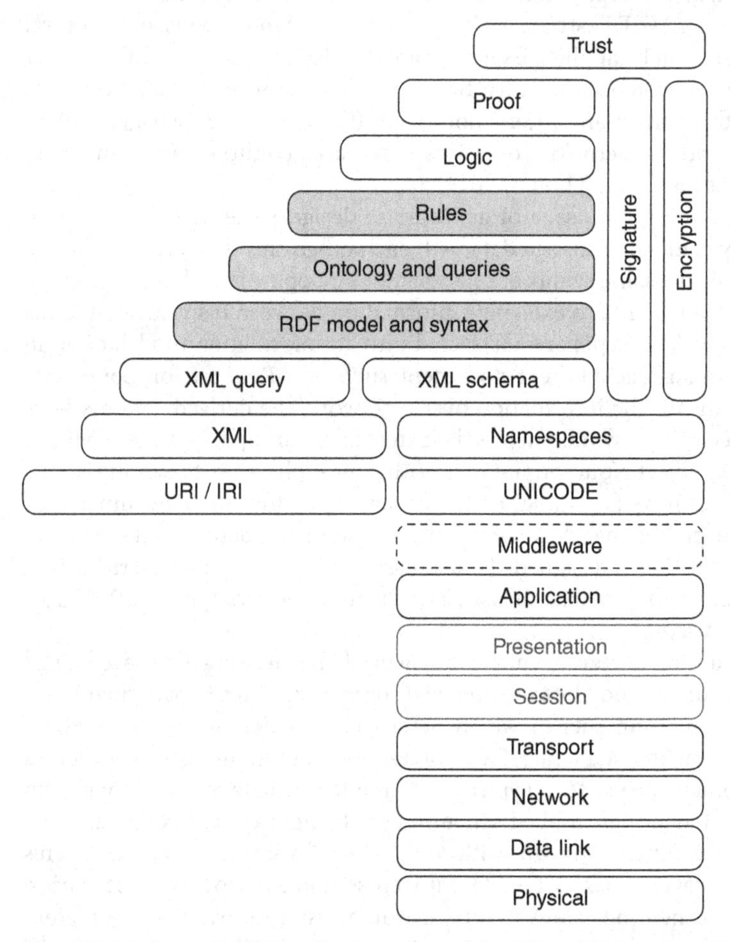

Figure 11.10 Semantic Web layer cake on top of the OSI model based on Berners-Lee (2003).

is located above the typical OSI model and middleware systems. Being mainly independent of the protocols and serializations used within different application areas, interoperability on this layer will probably be one of the most important aspects within the envisioned IIoT. For performance reasons, binary serializations, such as Header, Dictionary, Triples (HDT) (Fernández et al., 2011), can also be used.

11.4 Alliances

The importance and development of standardization within the overall IoT is highlighted in Chapter 7. Within the IIoT context alone, the Alliance for Internet of Things Innovation (AIOTI) Working Group for IoT Standardization (WG3)[15] is listing over 60 relevant SDOs and alliances. This overview covers building, manufacturing, transportation, health care, energy, cities, and farming use cases as well as horizontal telecommunication aspects.

11.4.1 Industrial Internet Consortium

The one initiative that stands out is the Industrial Internet Consortium (IIC)[16] as it covers almost every vertical domain. While not being an SDO, the IIC is an open membership organization bringing together government, academia, and the industry. It aims at gathering use case requirements and coordinating standardization efforts across the whole IIoT ecosystem. One notable outcome of this effort is the Industrial Internet Reference Architecture (IIRA) (Industrial Internet Consortium, 2015). Its main purpose is to provide a common basis for heterogeneous stakeholders to design IIoT systems by presenting an architectural overview. Following the concepts and terminology introduced in ISO/IEC/IEEE 42010:2011 (ISO/IEC/IEEE, 2011), the overall architecture is decomposed into four different viewpoints. For each of them, concerns and models from different stakeholder perspectives are described in more detail. As the names imply, within the business viewpoint, commercial and regulatory aspects are identified; the usage viewpoint describes matters related to components or humans interacting with the IIoT system; the functional viewpoint focuses on the overall internal and external interactions and activities; and finally, the implementation viewpoint covers specifics for carrying out the described concerns in the other viewpoints.

From a technical perspective, the specification of the functional viewpoint provides the most in-depth details. In particular, the ongoing convergence between local control systems of traditional OT and the globally interconnected

15 http://www.aioti.org/workinggroups/
16 https://www.iiconsortium.org

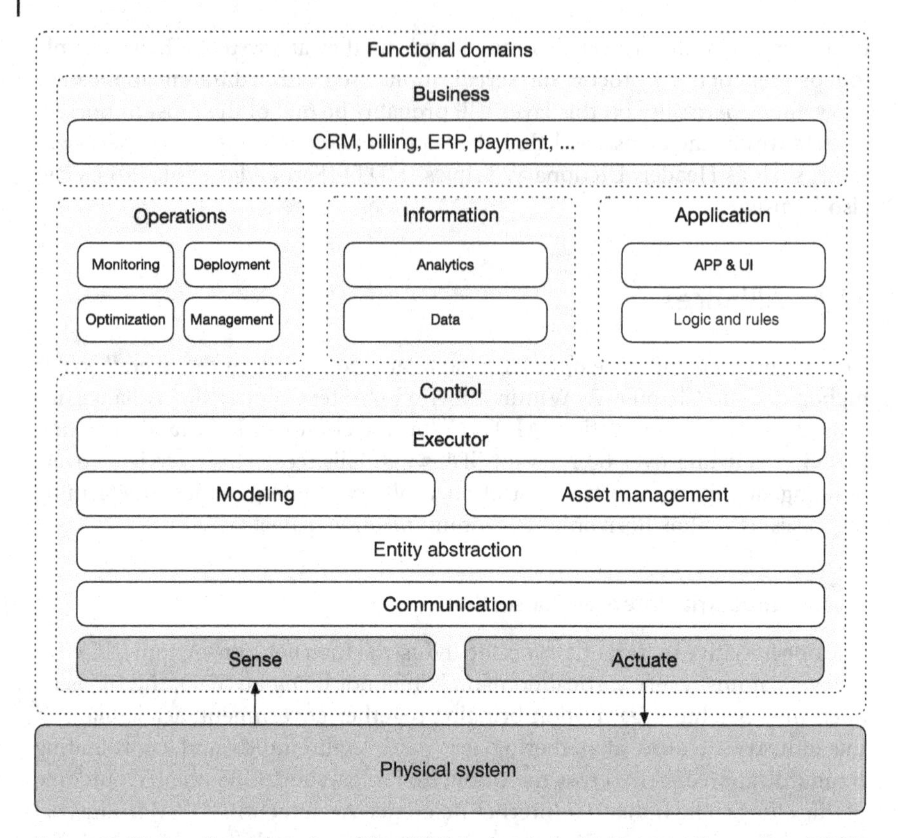

Figure 11.11 Functional domains of the IIRA based on Industrial Internet Consortium (2015).

Information Technology (IT) is a major concern. As a result, a number of functional domains have been identified (see Figure 11.11) to form a concrete functional architecture based on interconnected CPSs. The *control* domain directly interacts with the physical system by sensing information and applying closed-loop logic through actuation. This domain further includes communication with external entities, data abstraction and analytics, and asset management. Within the *operations* domain, a number of functionalities are contained that are required to manage the systems under a single *control* domain. These functionalities include provisioning, deployment, monitoring, diagnostics, prognostics, and optimization. The *information* domain contains data and analytics functionalities that are complementary to those within the control domain. Its purpose is to transform, process, persist, distribute, and analyze data for systemwide, long-term optimizations. Next, the *application* domain holds both global use case specific logic and rules as well as interfaces for humans or applications to interact with the logic. Finally, within the *business* domain traditional functionalities such as enterprise resource planning (ERP), customer relationship management (CRM), or manufacturing execution system (MES) reside.

Another source for technical details on how to design IIoT systems is the description of the implementation viewpoint. It describes the architectural design patterns that are applied across use case domains. The most general abstraction is the three-tier architecture that defines three different tiers: the *edge*, *platform*, and *enterprise* tier. On the highest level, the enterprise tier mainly contains domain applications with their rules and controls, and the platform tier in the center holds more generic data aggregation and flow analytics functionalities. While outsourcing these activities to third parties to meet organization needs in an efficient manner (Hassan, 2011), communicating data to a centralized cloud over the Internet introduces latency and jitter. Therefore, functionalities can also be placed close to the devices within the lowest edge tier (also known as edge computing (Lopez et al., 2015)) to be connected to the potentially deterministic and real-time capable, access network. Missing in this view is the fog computing (Vaquero and Rodero-Merino, 2014; Yi et al., 2015) paradigm, which is further described in Chapter 4. To further reduce latency and at the same time to improve scalability, security, and data sovereignty, in this concept, functionalities of the Control domain are, depending on the resource constraints of the devices, either running on the device or on nodes that are attached directly to the devices. This architecture further allows nodes to actuate autonomously without dependency on the network and above all presents the basis to implement the aforementioned concept of CPSs. Finally, the decision where to place specific virtualized functionalities in a topology is always a tradeoff between available resources, network performance, data privacy, and device autonomy.

11.4.2 Plattform Industrie 4.0

In the mid-eighteenth century, the mechanical production was powered by water by over 80% in Great Britain (Minchinton, 1989). This changed gradually with the invention of the steam engine, until over 98% of the required power was supplied by steam in the beginning of the twentieth century. This development changed the way of production dramatically for the first time and is, therefore, called the first industrial revolution. After this, in the beginning of the twentieth century, the introduction of assembly line work changed the overall process of production again. While it took over 720 min to manufacture a Ford Model T in 1911, the time dropped below 90 min only 3 years later (Minchinton, 1989), thanks to this second industrial revolution.

Until 1968, dedicated controllers, relays, and fixed circuits were used to automate the production process in factories. As a result, the process to update such facilities was very time consuming, expensive, and error-prone. The invention of the Programmable Logic Controller (PLC), by Dick Morley, started the third industrial revolution. With its input/output (I/O) modules for field devices, it built the foundation of the modern five-layer automation pyramid. Multiple PLCs, remote terminal units (RTUs), and human machine interfaces (HMIs) are interconnected over Supervisory Control and Data Acquisition (SCADA) fieldbus systems, and on top of this a MES that monitors the most important key performance indicators (KPIs) and finally, the information can be integrated into the ERP system.

The current convergence between this OT and ICT toward self-managed CPS-based automation is called the fourth industrial revolution. The concept makes use of virtual representations of physical objects for smart factories of the future; also called Industry 4.0, based on the German term "Industrie 4.0." As a union of the most relevant companies and associations in Germany, the "Plattform Industrie 4.0" aims at developing recommendations for the implementation of smart factories of the future. Hence, and in contrast to the IIC, the initiative is mainly focusing on modeling next generation manufacturing systems while focusing on the economic impact of interconnected cross-domain value chains.

The Plattform Industrie 4.0 specified the Reference Architecture Model Industrie 4.0 (RAMI) (Adolphs et al., 2015; Deutsches Institut für Normung, 2016). This three-dimensional layer model is the basis to systematically classify related technologies and builds upon standards defined by the International Electrotechnical Commission (IEC), namely, IEC 62890 (lifecycle management for systems and products used in industrial process measurement, control, and automation), IEC 62264 (enterprise control system integration), and IEC 61512 (batch control). Analog to the IIC IIRA, the RAMI defines six different layers and beginning in 2016, both initiatives agreed on a cooperation and started to map functionalities of both architectures.[17] As shown in Figure 11.12, from the lowest to the highest level, the RAMI layers correspond to the IIRA domains as follows: The *asset* layer with the physical system, the *integration* layer with the control domain, the *communication* layer with the communication part of the control domain, the *information* layer with the information domain, the *functional* layer with the operations and application domains, and finally the *business* layer with the business domain.

The most important concept in this model is the so-called Asset Administration Shell (AAS) (Plattform Industrie, 2016), as depicted in Figure 11.13. It can be seen analog to the aforementioned control domain to implement a CPS. It wraps a software component around physical objects holding a digital representation, a so-called Digital Twin or Avatar, of the asset to which other systems communicate with. As such, it encapsulates an existing Asset/physical system to

17 http://blog.iiconsortium.org/2016/03/the-industrial-internet-is-important-new-technologies-and-new-business-opportunities-will-disrupt-industries-on-many-level.html

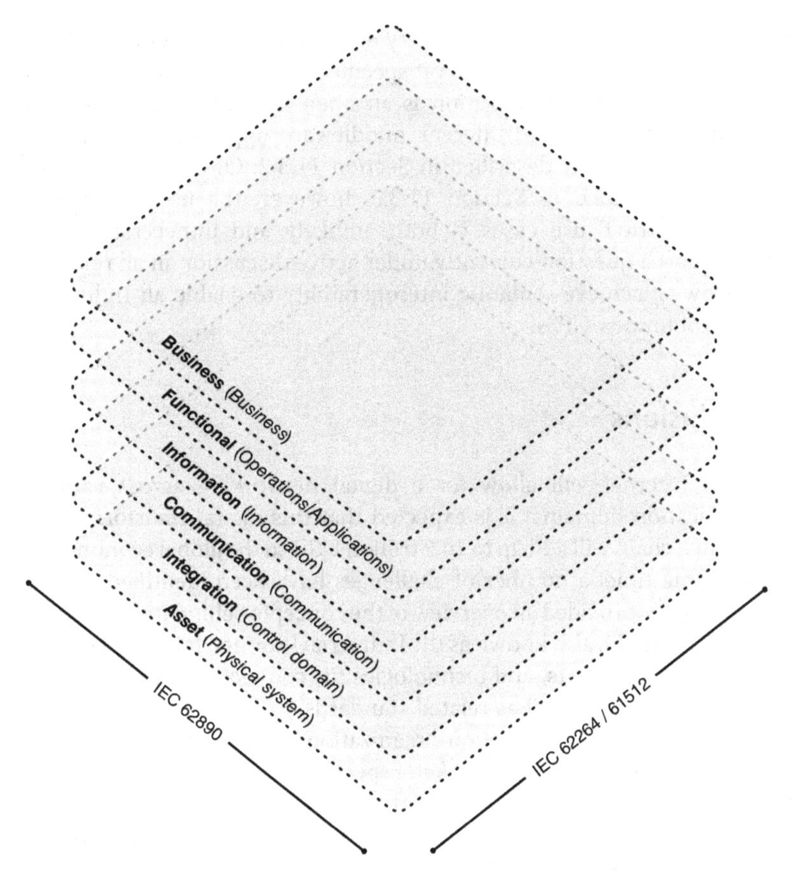

Figure 11.12 Functional domains of the IIRA based on Adolphs et al. (2015).

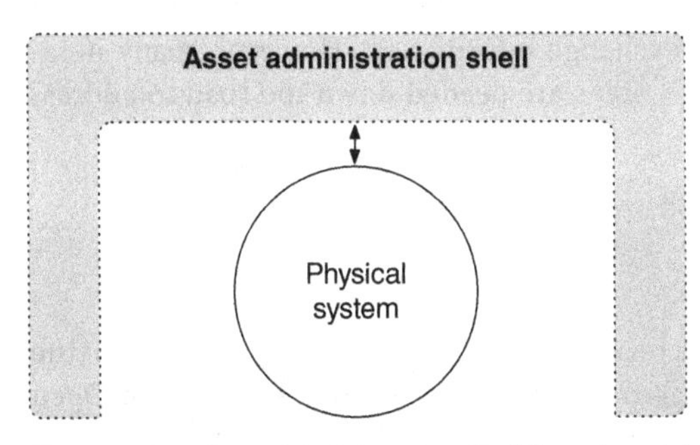

Figure 11.13 Asset Administration Shell based on Adolphs et al. (2015).

integrate it into Industrial Internet/Industry 4.0 environments. As neither the Plattform Industrie 4.0 nor the IIC has yet specified how these concepts should be implemented, the actual realization is an open and interesting research topic. Currently, mainly three different middleware approaches are under evaluation that were briefly described in Section 11.3.2: OPC UA, oneM2M, and DDS. As we learned in Section 11.3.3, however, the use of a single middleware in all IIoT use cases is both, unlikely and unnecessary. The challenging research question currently under active discussion in all relevant alliances is, how to achieve semantic interoperability to enable an Industrial Internet of autonomous CPSs.

11.5 Conclusions

The Industrial Internet will allow for a digital networking across various industrial application domains. It is expected that this digital transformation of the entire value chain will add up to 14.2 trillion USD to the global economy by 2030. At the same time, a number of challenges have been identified in this context. This chapter provided an overview of the concepts behind the Industrial Internet of Things (IIoT), also known as the Industrial Internet. It focused on use cases, interoperability aspects, and technologies for connectivity, communication, and data exchange as well as related standards.

In summary, various standardization organizations and alliances are aiming for harmonization in this emerging market. Especially, the IIC is proposing with its IIRA a concept that covers all relevant verticals at the same time. As a specific use case example, the Plattform Industrie 4.0 initiative defines a Reference Architecture Model Industrie 4.0 (RAMI) to push forward the concepts behind smart factories that will be operated by autonomous cyber-physical systems (CPSs). The overall vision draws an exciting future of the next industrial revolution characterized by highly interconnected, autonomous CPSs that continuously communicate and exchange information. However, many iterative, consecutive, and evolutionary steps are needed down the road to address multiple hurdles on the way.

Acknowledgments

This chapter would not have been possible without the dedicated support of the whole Industrial Internet of Things groups at the Fraunhofer Institute for Open Communication Systems (FOKUS) and the Technical University Berlin (TUB). I would like to give special thanks to Alejandra Escobar Rubalcava for her valuable input and Birgit Francis and Richard Figura for restless proofreading.

References

Adolphs, P. et al. (2015) Reference Architecture Model Industrie 4.0 (RAMI 4.0). Technical report ZVEI/VDI/Plattform Industrie 4.0.

Ashton, K. (2009) That 'Internet of Things' Thing. *RFiD Journal*, p. 4986.

BCG Perspectives. (2016) What's ahead for car sharing? the new mobility and its impact on vehicle sales. Technical report, February.

Berners-Lee, T. (2003) WWW Past and Future. Presentation at the Royal Society, London. World Wide Web Consortium (W3C), September.

Berners-Lee, T., Hendler, J., and Lassila, O. (2001) The semantic web. *Scientific American* **284**(5), 34–43.

Coplien, J. O. and Bjørnvig, G. (2010) *Lean Architecture: for Agile Software Development*, John Wiley & Sons, Ltd, Chichester, p. 376.

Cyganiak, R., Wood, D., and Lanthaler, M. (2014) Resource Description Framework (RDF) 1.1 Concepts and Abstract Syntax. Recommendation. World Wide Web Consortium (W3C), February.

Dan, B. and Guha, R. V. (2014) Resource Description Framework (RDF) Schema Specification 1.1. Recommendation. World Wide Web Consortium (W3C), February.

Deutsches Institut für Normung. (2016) *Referenzarchitekturmodell Industrie 4.0 (RAMI4.0).* Technical report SPEC 91345. DIN, April.

Dijkstra, E. W. (1982) On the role of scientific thought. *Selected Writings on Computing: A Personal Perspective*, Springer, New York, NY, pp. 60–66.

Fernández, J. D. et al. (2011) Lightweighting the web of data through compact RDF/HDT. *Advances in Artificial Intelligence* (14th Conference of the Spanish Association for Artificial Intelligence (CAEPIA)). Lecture Notes in Computer Science, vol. 7023, Springer, pp. 483–493.

Fowler, M. and Lewis, J. (2014) *Microservices.*

Frost & Sullivan. (2014) Technology Convergence Enabling Smart Factories (Technical insights): Factories with Advanced Intelligent Capabilities. Technical report, December.

Frost & Sullivan. (2016) Advanced Manufacturing Techvision Opportunity Engine. Technical report, August.

Frost & Sullivan. (2016) Future of the Smart Grid Industry. Technical report, May.

Frost & Sullivan. (2016) Technologies Empowering Smart Healthcare. Technical report, October.

Ghielmini, G. (2013) Virtual factory manager for semantic data handling. *CIRP Journal of Manufacturing Science and Technology*, **6**(4), 281–291.

Global System for Mobile Communications Association (GSMA). (2015) Mobilizing Intelligent Transportation Systems (ITS). September.

Hassan, Q. (2011) Demystifying cloud computing. *The Journal of Defense Software Engineering*, **1**, 16–21.

Herman, I., Horrocks, I., and Patel-Schneider, P. F. (2012) OWL 2 Web Ontology Language Document Overview (Second Edition). Recommendation. World Wide Web Consortium (W3C), December.

International Electrotechnical Commission. (2016) OPC Unified Architecture - Part 1: Overview and Concepts. Technical report, TR 62541-1:2016. IEC.

Industrial Internet Consortium. (2015) Industrial Internet Reference Architecture. Technical Report Version 1.7. IIC, June pp. 1–101.

IndustryARC. (2016) Industrial Internet of Things (IIoT) Market Analysis - By Components (Sensors, Memory, Processors, RFID); By End Use Industry (Manufacturing, Transportation, Energy, Retail, Healthcare, Agriculture) - Forecast (2016–2021). Technical report, March.

International Energy Agency. (2016) CO_2 emissions from fuel combustion. Technical report, September.

ISO/IEC/IEEE. (2011) Systems and Software Engineering–Architecture Description. Technical report 42010:2011, May.

International Telecommunication Union (ITU). (2012) Boosting energy efficiency through Smart Grids. Technical report, April.

Jacobson, I., Christerson, M., and Jonsson, P. (1992) *Object-Oriented Software Engineering: A Use Case Driven Approach*, 4.A. ACM Press, Reno, USA.

Jeschke, S. et al. (ed.), (2017) *Industrial Internet of Things*, Springer Series in Wireless Technology, Springer International Publishing, Cham, Switzerland.

Krasner, G. E., et al. (1988) A description of the model-view-controller user interface paradigm in the smalltalk-80 system. *Journal of object oriented programming*, **1**(3), 26–49.

Little, A. D. (2016) Succeeding with Digital Health Winning Offerings and Digital Transformation. Technical report.

Lopez, P. G. et al. (2015) Edge-centric computing. *ACM SIGCOMM Computer Communication Review*, **45**(5), pp. 37–42.

Markets and Markets. (2016) Smart Cities Market by Solution and Services for Focus Areas - Global Forecast to 2020. May.

Markit, I.H.S. (2016) Complimentary Whitepaper: IoT Platforms: Enabling the Internet of Things. Technical report, March.

Martin, R. C. (2003) *Agile Software Development: Principles, Patterns, and Practices*, Prentice Hall PTR, Upper Saddle River, NJ.

Martin, R. C. (2012) The Clean Architecture.

McKinsey & Company. (2016) Automotive Revolution-Perspective Towards 2030. Technical report, January.

Minchinton, W. (1989) The Energy Basis of the British Industrial Revolution. Bayerl, G. (ed.), *Wind- und Wasserkraft. Die Nutzung regenerierbarer Energiequellen in der Geschichte*, VDI Verlag pp. 342–362. ASIN: B005K7SYNI

Mordor Intelligence. (2017) Global Smart Factory Market - Growth, Trends, and Forecasts (2015 - 2020). Technical report, February.

Newman, S. (2015) *Building Microservices*, O'Reilly Media, pp. 1–250.
Palensky, P. and Dietrich, D. (2011) Demand side management: demand response energy systems, and smart load. *IEEE Transactions on Industrial Informatics*, **7**(3), 381–388.
Pardo-Castellote, G. (2003) OMG data-distribution service: architectural overview. 2003 Proceedings of the IEEE 23rd International Conference on Distributed Computing Systems Workshops, pp. 200–206.
Plattform Industrie 4.0/ZVEI. (2016) Structure of the Administration Shell. Technical report PI4.0.
Pras, A. and Schoenwaelder, J. (2003) RFC 3444: On the Difference between Information Models and Data Models. *RFC 3444* (Informational). Internet Engineering Task Force (IETF).
Pras, A. et al. (2007) Key research challenges in network management. *IEEE Communications Magazine*, **45**(10), 104–110.
PricewaterhouseCoopers (2014) Wirtschaftsprüfungsgesellschaft Aktiengesellschaft. Industrie 4.0 - Chancen und Herausforderungen der vierten industriellen Revolution. Technical report, October.
PricewaterhouseCoopers (PWC). (2016) Auto Industry Trends. Technical report, October.
Roland Berger Strategy Consultants. (2014) INDUSTRY 4.0 - The new industrial revolution How Europe will succeed. Technical report, March.
Swetina, J. et al. (2014) Toward a standardized common M2M service layer platform: Introduction to oneM2M. *IEEE Wireless Communications*, **21**(3), 20–26.
Thones, J. (2015) Microservices. *IEEE Software*, **32**(1), 116–116.
U.S. Department of Transportation. (2015) ITS 2015-2019 Strategic Plan, Joint Program Office (JPO).
United Nations (UN). (2014) World's population increasingly urban with more than half living in urban areas. Technical report, New York, June.
van der Veer, H. and Wiles, A. (2008) *Achieving Technical Interoperability*, White Paper. European Telecommunications Standards Institute (ETSI).
Vaquero, L. M. and Rodero-Merino, L. (2014) Finding your way in the fog: towards a comprehensive definition of fog computing. vol. 44(5). ACM SIGCOMM Computer Communication Review.
Varda, K. (2008) Protocol buffers: Google's data interchange format. Google Open Source Blog.
Woodside Capital Partners. (2015) The Internet of Things: "Smart" Products Demand a Smart Strategy Using M&A for a Competitive Edge. Technical report, March.
World Health Organization (WHO). (2005) Fifty-Eighth World Health Assembly Resolutions And Decisions Annex. Technical report, Geneva, May.
World Resources Institute. (2016) The Rise of the Urban Energy "Prosumer". Technical report, March.

Wu, G. et al. (2011) M2M: from mobile to embedded internet. *IEEE Communications Magazine*, **49**(4), pp. 36–43.

Yi, S., Li, C., and Li, Q. (2015) A survey of fog computing. *Proceedings of the Workshop on Mobile Big Data (Mobidata)*, ACM Press, New York, pp. 37–42.

Zpryme Agency. (2015) Smart Transportation: Definitions and Market Outlook. Technical report, June.

12

Internet of Things Applications for Smart Cities

Daniel Minoli[1] *and Benedict Occhiogrosso*[2]

[1] *IoT Division, DVI Communications, New York, NY, USA*
[2] *Intellectual Property Division, DVI Communications, New York, NY, USA*

12.1 Introduction

In 2008, the world's population reached a 50–50 split in the distribution of populations between urban and nonurban environments. At this juncture, we are witnessing an expansion of cities, as populations accelerate the transition from rural and suburban areas into urban areas driven by economic opportunities, demographic shifts, and generational preferences. Seventy percent of the human population is expected to live in cities by the year 2050. The largest growth in urban landscapes is occurring in developing countries. There are now more than 400 cities with over one million inhabitants and there are 20 cities with over 10 million people (Staff, 2016). In most instances, especially in the Western world, cities have aging infrastructure, such as roads, bridges, tunnels, rail yards, and power distribution plants. Some locations have experienced tremendous real estate development in recent years, yet the roads, water mains, sewers, power grids, and sometimes even communication links have seen no, or extremely limited, upgrades. The physical infrastructure that is in place in many cities is aging and, going forward, the services provided by such infrastructure may, in fact, be subject to temporary rationing as necessary, even emergency, upgrades are made. Sometimes just closing a lane for a few days creates chaotic and dangerous traffic conditions and national headlines.

New technological solutions are needed to manage the increasingly scarce infrastructure resources, especially in view of the challenges imposed by population growth, typically limited financial resources, and perennial political inertia. IoT technologies and principles hold the promise of being able to improve the resource management of many assets related to city life, including the flow of goods, the movement of private and public vehicles, and the greening of the environment. When cities deploy on a broad scale state-of-the-art

Internet of Things A to Z: Technologies and Applications, First Edition. Edited by Qusay F. Hassan.
© 2018 by The Institute of Electrical and Electronics Engineers, Inc. Published 2018 by John Wiley & Sons, Inc.

Information and Communication Technologies (ICT), including, in particular, IoT technologies, they are referred to as being "smart cities." IoT technologies can be leveraged, for example, to optimize energy consumption and maximize life-activity efficiency, thus improving and optimizing urban quality of life (QoL). There is a consensus that IoT deployments in support of smart city environments are a clear benefit to society in terms of livability and to the environment in terms of efficient use of resources, such as energy; the concept of sustainability also comes into play.

At the technology level, there is interest in establishing how the evolving technology itself can be effectively used, the requisite architectures, the standards, and the critical requirements for cybersecurity. The applicable technology spans sensors, networking (especially wireless technologies for personal area networks, fogs, and cores, such as 5th Generation [5G] cellular), and analytics. Architecture deals with how things are assembled, including hierarchy. Standards relate to the ability to deploy the technology in a commodity fashion, assuring simple and reliable end-to-end interoperability—these also include Machine to Machine (M2M) standards and approaches. Cybersecurity, in canonical form, spans confidentiality, integrity, and availability.

Smart cities application areas include but are not limited to intelligent transportation systems (including smart mobility, vehicular automation, and traffic control), smart grids, smart buildings, goods and products, logistics (including smart manufacturing), sensing (including crowdsensing and smart environments), surveillance/intelligence, and smart services. Cities have been incorporating new technologies over the years, but recently the rate of technology adoption has increased, especially for, but not limited to, surveillance, traffic control, energy efficiency, and street lighting (Caragliu et al., 2011; Minoli, 2013). There is a very extensive body of literature on this topic; some recent references of interest include (Calvillo et al., 2016; Grodi et al., 2016; Lizzi et al., 2016; Martínez et al., 2016; Ramaswami1 et al., 2016; Romero et al., 2016; Soriano et al., 2016; Srivastava and Pal, 2016; Wewalaarachchi et al., 2016; Zanella et al., 2014).

This chapter aims at discussing a number of IoT issues and applications specific to smart cities. In particular, there is a plethora of issues, technical solutions, and technical challenges associated with a broad-based deployment of IoT in urban settings to enjoy the benefits of a smart city. We note only in passing (as depicted in Figure 12.1) that the IoT ecosystem, also in the context of smart cities, is comprised of many objects, many sensors, some relatively well-defined aggregation networks, and many (cloud-based) analytics and storage engines. Objects and sensors are both static and mobile, with mobility likely being more prevalent. Smart cities do not depend on any specific or unique IoT technology, but include all forms of IoT technologies, including appropriate sensors, appropriate networks, and appropriate analytics, all of which may depend on the particular vertical application under consideration. Additionally, there will also be actuator devices that are used to actually control the environment in

Figure 12.1 A logical view of an IoT ecosystem, also as applicable to smart cities.

response to a sensed set of data or some analytical computation—for example, changing the signs and the barriers on a road to reverse lanes at various times during the day or changing the operating value on a pump to control water or sewer flows.

Traditionally, even in smart cities applications, the IoT has been envisioned as supporting a large population of relatively low-bandwidth parameter-sensing devices, particularly in M2M environments, and generally in stationary locations (e.g., data-collecting meteorological weather stations, electric meters, industrial control, and the like). Nonetheless, video-oriented applications that have inbound streams ranging up to ultrahigh definition resolutions both in pixel density and in frame rate and video-oriented applications where the stream is outbound to the receiving devices (e.g., but not limited to smartphones) are now emerging. These emerging IoT applications are known as IoT-based multimedia (IoTMM) applications; these also have applicability to smart cities (Alvi et al., 2015; RWS, 2015).

In addition to an overview of the applicability of IoT to smart cities, this chapter provides two new research-oriented results for IoT deployment in smart cities environments: (1) the support of multimedia-oriented IoT applications, for example, for public safety and surveillance applications and (2) the application of the family of Mobile IPv6 (MIPv6) protocols for mobility-based applications, for example, for vehicular crowdsensing (Hu et al., 2014) (e.g., to improve traffic flows). These lines of investigation have not previously received the research exposure they perhaps deserve. Mobility and mobility management are underlying requirements of many smart city applications.

12.2 IoT Applications for Smart Cities

Cities are now constantly challenged "to do more with less"; thus, automation in general, and IoT in particular, are important tools to meet the requirements of the citizenry while managing costs. Indeed, many city challenges can be addressed or at least ameliorated by IoT principles. Livability, infrastructure and real estate management, traffic transportation and mobility, logistics, electric power and other city-supporting utilities, and security are perhaps the key aspects of a city from a substratum perspective. For some applications, the well-defined concepts of (ETSI-defined) M2M apply, especially for infrastructure monitoring applications (ETSI, 2011). Drones (a type of IoT device) will also play a role in many smart city applications. Table 12.1 identifies a number of key, specific urban challenges and related IoT-based solutions to these challenges; this table is clearly not exhaustive. Figure 12.2 depicts some of these applications graphically. Table 12.2 provides taxonomy of the requisite synthesis to achieve a pervasive broad-scale deployment of the IoT technology in smart cities environments. The United States, for example, has more than 19,000 cities and towns; thus, an IoT-based solution to some of the issues impacting cities offers a major market opportunity—that opportunity is clearly even more extensive worldwide. However, urban budgets are increasingly more constrained both in terms of the revenue collection as well as by the increasing system and infrastructure upgrade requirements, as cities age and/or populations swell. Some forecast the market opportunity for technology developers to be $1 trillion worldwide in 2020, while others size the market at several hundred billion dollars (Staff, 2016). The United States cities, such as Los Angeles, New York, Miami, San Francisco, Seattle, and Washington DC, are deploying service-enhancing technologies of the IoT/sensor type; an even more rapid trend seen around the world, in cities or countries such as Abu Dhabi, Amsterdam, Barcelona, Berlin, Dubai, Hong Kong, London, Melbourne, Milton Keynes, Paris, Qatar, Rio de Janeiro, Santiago, Saudi Arabia, Southampton, Stockholm, Singapore, Seoul, Sydney, and Tokyo. Table 12.2 provides a more extensive summary of smart city initiatives and a number of specific applications around the world as of press time—it should be noted that this list and the applications being deployed continuously evolves over time. Figure 12.3 provides a mapping to the continental regions.

In this context, the following four trends are seen as important by city planners: (i) demographic and workforce trends; (ii) infrastructure cost and financing; (iii) the growth of public and private mobility systems; and (iv) the availability of new modes of transportation (also including autonomous transportation) (Staff, 2015). In each of these cases, IoT has something to offer. However, at the current time, the smart cities IoT segment is fragmented into discrete vertical domains, multiple stakeholders, and disconnected Information and Communication Technology (ICT) systems. A set of streamlined technical standards and a usable multiservice architecture that support a "plug and play" mode of expansion would greatly enhance deployment and broad-based penetration. Smart city solutions to date have tended to be vendor–proprietary approaches (Machina Research White Paper, 2016).

Table 12.1 Key urban challenges and IoT-supported solutions.

Issue	Key challenges	IoT support/solutions
Livability	QoL, affordability, optimal access to services, efficient transportation, low delays, safety.	IoT can enhance urban life by supporting smart multimodal transportation, information-rich environments, environmental and safety sensing, smart parking and smart parking meters, smart electric meters, location-based services, and real-time connectivity to health-monitoring resources
Infrastructure and real estate management	Need to monitor spaces, buildings, transportation arteries, bridges, tunnels, railroad crossings, and street lighting.	Highly interconnected sensors optimized to the task at hand can provide real-time and historical trending data that enable agencies to provide enhanced capabilities and services economically. Drones (a type of IoT device) will also play a role
Traffic, transportation, and mobility	(Since recorded history) a transportation network is a platform for commerce and human interaction. Efficiency is paramount. There is typically a scarcity of thru-traffic choices that affects many (U.S.) cities today, also compound by unsynchronized street signals.	Driverless technology (in the U.S. California, Florida, Michigan, Nevada, and Washington D.C. have already passed legislation related to autonomous vehicles and more are anticipated soon.) Some expect to see driverless bus transit within the next 10 years. IoT sensors (and vehicle-based actuators) are needed to support this function. Dynamic transportation systems enable people to shift seamlessly between multimodal elements, depending on their needs. IoT can facilitate the process of enabling people to change modes in a predictable and efficient manner. Mobility is a basic aspect of population's growth in the urban environment, and technology is the mechanism that supports this growth. Initially app-based transportation models did not exist outside of major metropolitan areas, but many cities now have these capabilities, which are ultimately based on IoT. Drones (a type of IoT device) will also play a role
Logistics	The need to supply urban dwellers with fresh food as well as all the supplies and material necessities of life	IoT technologies can be utilized to streamline warehousing, transportation, and distribution of goods into the urban setting. Traffic management is a key element of such logistical support

(continued)

Table 12.1 (*Continued*)

Issue	Key challenges	IoT support/solutions
Power and other city-supporting utilities	A reliable flow of electric energy, gas, and water; efficient waste-management; sewer disposition, snow removal material; and gasoline storage and distribution are all critical	Smart grid solutions and sensor-rich utility infrastructure can address these critical urban challenges
Physical security	Open-space security in streets, parks, public transportation hubs, tunnels, bridges, trains, buses, ferries, and government building are increasingly more important.	IP-based surveillance video, plate reading, face recognition, gunshot detection, crowd control, biohazard, and radiological contamination monitoring based on IoT technologies will address these concerns in a more effective and standardized manner. Drones (a type of IoT device) will also play a role

Figure 12.2 Graphical examples of smart city IoT applications.

Another application entails the smooth interaction between the smart city infrastructure and evolving smart building IoT technologies (Minoli et al., 2017a).

Naturally, not all challenges faced by cities can be completely solved by IoT solutions—such IoT solutions basically provide data on resources (e.g., power usage), data on state (e.g., traffic), data on logistics (e.g., movement of goods), data on security (e.g., surveillance), and the ability to deliver advanced smart services (e.g., smart parking meters.) For example, in some advanced western countries, the so-called Millennials are avoiding using/purchasing cars (when they can) and showing preferences for other modes of transportation such as biking and walking. Increasingly, and as a consequence, offices are being located in proximity to these transportation options; such new real estate development urban blueprint is independent of IoT, but the subtending capabilities in support of the new urban layout can be facilitated by IoT principles, especially in the context of management of mobility.

Table 12.2 List of smart cities initiatives and cities with interest in smart cities technologies, January 2017.

City	Project name	Type	Description
Barcelona, Spain	Smart City Barcelona	Various	The city has deployed responsive technologies across urban systems that include public transit, parking, street lighting, pollution control, and waste management. Initiatives started in 2012 and continue to the present. The city has 670 Wi-Fi hotspots at a maximum distance of 300 ft. from point to point
Boston, MA	BOS:311	Smart services	The BOS:311 mobile application allows residents to instantly report city issues including potholes, blocked drains, and faulty street lights. As soon as a report has been submitted (along with pictures taken through a smartphone or tablet's camera), the app forwards the report automatically into the city's work order system
Charlotte, NC	Verizon: Envision Charlotte	Environmental monitoring	Charlotte is utilizing Verizon's Smart City Solutions to help businesses to cut back on greenhouse gases and increase energy efficiency. The system utilizes interactive kiosks and monitors located around the city that gather and analyze information relating to energy usage. The project partners include Charlotte Center City Partners, Duke Energy, and Verizon. The project runs through July 2018
Fujiwasa, Japan	Japan's Sustainable Smart Town	Energy, e-health	The Fujisawa Sustainable Smart Town (SST) deals with energy conservation, with additional focus on community, mobility, security, and health care. IoT sensors, controls, and networking support this initiative. The project partners include Panasonic, the Fujisawa City local government, Tokyo Gas, Accenture, and Yamato Transport
Kansas City, KS	The Smart City Corridor	Internet access/smart services	The Kansas Most Connected Smart City (KCMO) project is a group of IoT city initiatives funded through public–private partnership with Cisco, Sprint, and Think Big Partners. The city is expanding the "Smart City Corridor" that includes the KC Streetcar, a Wi-Fi-enabled tram, and 25 interactive kiosks along the streetcar line that provide access to city services, local business information, public digital art, and entertainment

London, UK	Digital Catapult: Things Connected	General	Digital Catapult provides technologies (networks) to act as a "sandbox" for businesses to test out emerging IoT technologies. Participants can test how IoT solutions will work over a low-power wide-area network (LPWAN). The project provides London-based businesses with 50 LoRaWAN base stations (more planned later) and support for additional LPWAN technologies. Partners include BT, Future Cities Catapult, Imperial College London, King's College London, UCL, and Queen Mary
Moscow, Russia	Telensa Smart Parking	Parking meter	Telensa offers a smart parking solution called PARKet that aims at finding parking spaces more efficiently and lower city congestion rates. The on-street parking monitoring system utilizes an array of magnetic sensors set into the road surface to detect when a car parks over one of the devices. When this pressure is detected, a flag is automatically sent to a central database through a low-power radio system, to update the status of the parking space. The system allows the administrators to keep an eye on parking availability, but consumers can also access this information through a web interface
New York City, NY	Sidewalk Labs: LinkNYC	Internet access/smart services	LinkNYC has recently installed interactive kiosks across New York City to replace payphones. The kiosks provide NYC residents free access to the Internet. The gigabit-speed Wi-Fi access is supported by revenue through advertisements shown in the kiosk terminals. The kiosks also allow visitors access services such as maps, advice, and emergency services. The kiosk has a touch screen tablet for browsing and offer mobile connectivity for smartphones, tablets, and laptops. LinkNYC will initially install 7500 Links across all five boroughs of NYC
New York City, NY	SST's ShotSpotter	Physical security	SST is working with the NYPD to roll out ShotSpotter, a gunfire detection system. The sensors working collaboratively can detect the geolocation of weapons fired employing a 360-degree coverage algorithm. ShotSpotter systems have been installed in seven Bronx precincts and 10 in Brooklyn. The data collected by the sensors can also be used to assess trends and crime hotspots in urban areas
Oslo, Norway; Dresden and Klingenthal Germany; Hong Kong; and, Paris	Green City Solutions' CityTree	Highly specific	Green City Solutions' CityTree combines plant life and IoT technology to improve air quality in tight urban spaces. The CityTree is a bioengineered vertical 12 ft. stand that is able to purify the air around it with a capacity equivalent to 275 trees—while taking up 99% less space. It utilizes a special moss culture that attracts pollutants and converts them to a biomass. Deployment started in 2014

(*continued*)

Table 12.2 (*Continued*)

City	Project name	Type	Description
San Diego, CA	Smart City Solutions	Various	San Diego has a number of IoT-based initiatives including (i) the introduction of 75,000 intelligent LED streetlights; (ii) the revitalization of the Port of San Diego; (iii) revitalization of the PETCO Park baseball stadium; and (iv) solar-to-electric vehicles (EVs) charging technology
Tel Avis, Israel		Smart services	With its smart city initiatives, Tel Aviv aims at eliminating or reducing the barriers between the municipality and the residents
Tirana, Albania		Smart services	The city plans to use digital technology as a way to connect more directly with the residents. An app was introduced that enables the public to report real-time problems and concerns to municipal staff
Worldwide list (265 cities listed)	*Africa:* Kenya—Nairobi; Malawi—Blantyre; Mauritius—Port Louis; Morocco—Casablanca; Nigeria—Abuja *Asia:* China—Beijing; Hebei; Pekin; Shanghai; Shenzhen; India—Ghaziabad; Gurgaon; Haryana; Ludhiana; Mumbai; Naiya Raipur; New Delhi; Iraq—Duhok; Israel—Haifa; Jerusalem; Ramat Gan; Rosh Haain; Tel Aviv; Japan—Fujiwasa; Fukuoka; Kobe; Kyoto; Yokohama; Jordan—Amman; Malaysia—Cyberjaya; Kuala Lumpur; Philippines—Manila; Qatar—Lusail; Singapore—Singapore; South Korea—Seongnam City; Seoul; Taiwan—Taipei; Turkey—Ankara; United Arab Emirates—Dubai *Europe:* Albania—Tirana; Andorra—Andorra la Vella; Austria—Graz; Vienna; Belgium—Antwerp; Brussels; Liège; Bosnia and Herzegovina—Mostar; Croatia—Dubrovnik; Denmark—Copenhagen; Kalundborg; Estonia—Tallinn; Tartu; Finland—Jyväskylä; Tampere; France—Andiran; Grenoble; Lannion; Levallois-Perret; Lille; Obernai; Paris; Paris La Défense; Germany—Berlin; Bonn; Cologne; Dresden; Hamburg; Hannover; Karlsruhe; Klingenthal; Leipzig; Ludwigsburg; München; Munich; Nürnberg; Wolfsburg; Greece—Thessaloniki; Ireland—Dublin; Italy—Bergamo; Firenze; Milan; Modena; Rome; Torino; Turin; Luxembourg—Luxemburg; Netherlands—Amstelveen; Amsterdam; De Meern; Eindhoven; Zeist; Norway—Drammen; Oslo; Portugal—Dafundo; Lisboa; Russia—Moscow; Spain—A Coruña; Alcobendas; Alicante; Almería; Ávila; Barcelona; Bellatera; Benalmádena; Bilbao; Boecillo; Calvià; Castellón; Cerdanyola del Vallès; Gavà; Gijón; Granada; Hospitalet del Llobregat; Las Palmas de Gran Canaria; L'Hospitalet de Llobregat; Logroño; Lorca; Madrid; Majadahonda; Murcia; Orihuela; Oviedo; Palafrugell; Palencia; Palma de		

	Malllorca; Pals; Paterna; Portillo; Pozuelo de Alarcón; Rivas Vaciamadrid; Salamanca; Sant Adrià del Besòs; Sant Boi De Llobregat; Sant Cugat del Vallès; Sant Feliu de Llobregat; Santa Coloma de Gramenet; Segovia; Sevilla; Valencia; Vic; Viladecans; Zaragoza; Sweden—Älmhult; Kista; Malmö; Stockholm; Switzerland—Carouge; Geneva; Gockhausen; Meyrin; Rolle; St. Gallen; United Kingdom—Banbury; Birmingham; Bristol; Cambridge; Cowes; Edinburg; Glasgow; Huntingdon; Isle of WIght; London; Royston; Southampton; Stirling; Swindon
	North America: Canada—Montréal; Ottawa; Toronto; Vancouver; Mexico—Culiacan; Guadalajara; Leon; Mérida; Mexico City; Mineral de la Reforma; Nogales; Puebla; San CIro de Acosta; Tepic; United States of America—Amherst; Atlanta; Berkeley; Boston; Brooklyn; Burlington; Cambridge; Chicago; Framingham; Kansas City; Lafayette; Lemont; Los Angeles; Milford; Milpitas; New York City; Orlando; Philadelphia; Purchase; Redmond; Salt Lake City; San Diego; San Francisco; San Jose; Silver Spring; Waltham; Washington DC; Wayland; Weehawken
	South America: Argentina—Buenos Aires; Córdoba; Mendoza; Parana; Santo Tomé; Brazil—Aparecida de Goiania; Brasilia; Brasília; Divinópolis; Goiania; Ilhabela; Manaus; Natal; Palmas-TO; Porto Alegre; Recife; Rio de Janeiro; Santos; Tres Rios; Vitoria; Chile—Santiago; Santiago de Chile; Colombia—Barranquilla; Bogotá; Bucaramanga; Buritica; Cajica; Calamar; Carmen de Carupa; Cartagena; Chiquinquirá; Cogua; Cota; Ibagué; Inirida; Manizales; Medellín; Montería; Oicata; Orito; Pasto; Pereira; Rio Quito; San Jacinto del Cauca; Soledad; Solita; Tabio; Tunja; Valledupar; Viterbo; Zipaquirá; Ecuador—Quito; Peru—Lima; Magdalena del Mar; Uruguay—Montevideo; Venezuela—Barcelona

Source: Some data based partially on Staff (2016) and on data from *Smart City Expo World Congress,* Av. Reina Mª Cristina s/n, 08004 Barcelona, Spain.

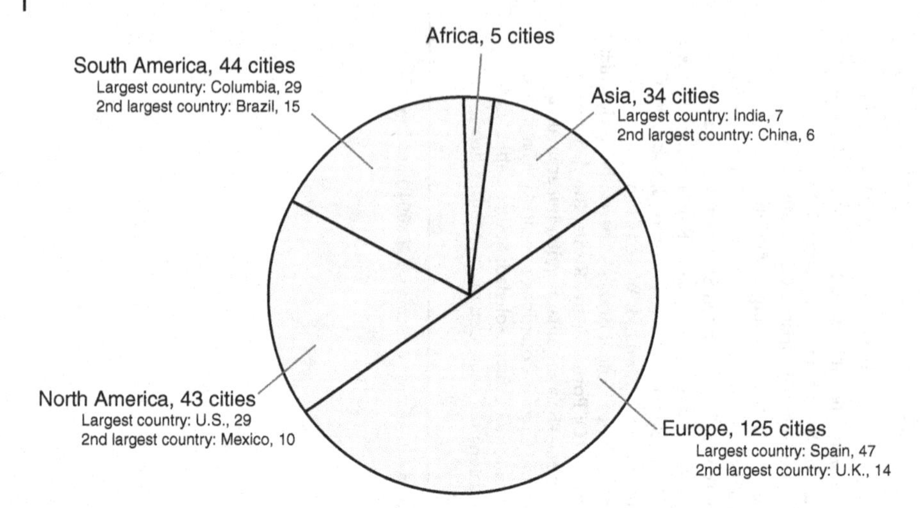

Figure 12.3 Global distribution of smart cities initiatives.

12.3 Specific Smart City Applications

The previous section identified a number of applications now being deployed in smart cities environments. This section describes five of these applications in some additional detail.

12.3.1 Driverless Vehicles

Advances in wireless communication techniques and location-aware sensor technologies are fueling the evolution of vehicle networks (VNs). In the next 5 years, driverless cars, buses, and trucks will begin to share the roadways with traditional vehicles. Additionally, one might see a significant expansion of autonomous transportation in fleet vehicles such as commercial trucks. Fleets are the initial area where driverless vehicle technology is expected to be widely deployed—driverless public transit (trains and buses) may also shift toward autonomous transportation. The U.S. National Highway Traffic Safety Administration has defined five levels of vehicle automation based on how active the driver is during operation.

- *Level 0—No Automation.* The operator is in control of all aspects of the vehicle's functions at all times, although the vehicle may have features that passively warn the driver of a potential collision or lane departure. IoT applications: basic internal in-vehicle monitoring of functionality.
- *Level 1—Function-Specific Automation.* The operator is in full control of the vehicle, but may use automated features that can affect control speed, braking,

or steering to assist with specific functions (cruise control, automatic braking, and lane keeping systems). IoT applications: more advanced internal in-vehicle monitoring of functionality.

- *Level 2—Combined Function Automation.* Automation allows the operator to be disengaged for some portions of the trip (e.g., the driver may be able to take his or her hands off the wheel and foot off the pedal), but the operator must still actively monitor the vehicle and be ready to take control at any time. IoT applications: internal in-vehicle monitoring of functionality in conjunction with onboard/off board signaling to detect environmental conditions.
- *Level 3—Limited Self-Driving Automation.* The operator no longer needs to be constantly monitoring the roadway, since the vehicle handles critical safety functions under certain conditions, while alerting the operator if there is an upcoming obstacle. IoT applications: advanced onboard/off board signaling to detect environmental conditions.
- *Level 4—Full Self-Driving Automation.* The operator no longer has any responsibility for safe operation of the vehicle, and is not expected to monitor road conditions or take control at any point during the trip. IoT applications: complete and exhaustive onboard/offboard signaling to detect environmental conditions.

A substantial amount of new ICT infrastructure will be required to achieve the goal of self-driving automation and broad deployment of Intelligent Transportation Systems (ITSs). This includes on-board sensors and actuators, road transponders to support the concept of a "digital road," and other actuators (for example to manage traffic lights and/or video surveillance for road conditions and situational awareness). Considerable vehicular-to-vehicular (device-to-device [D2D]) and vehicular-to-infrastructure communications mechanisms have to be deployed, configured, managed, and optimized. Significant technical work is underway to advance the science to the point where deployments can be made to support safe operations (Alam et al., 2016; Cheng et al., 2015; Ghazal et al., 2015; Hegyi and De Schutter, 2015; Nuss et al., 2014; Picone et al., 2015; Qureshi and Abdullah, 2013; The National Academies of Science, 2014; van der Mei et al., 2013; Vogel and Mueller, 2015). In particular, work on vehicle-to-vehicle (V2V) and vehicle-to-infrastructure (V2I) protocols, communication channels, transmission equipment, and analytics is in progress.

12.3.2 Crowdsensing

Crowdsensing allows a large population of mobile devices to measure phenomena of common interest over an extended geographic area, enabling "big data" collection, analysis, and sharing. It has major urban applications for the (active or passive) collection of traffic conditions, weather conditions, and even video images. There is a lot of current interest in this topic, especially in the context of urban environments (Bei et al., 2013; Cardone and Corradi, 2016; Ganti et al., 2011; Jamil and Basalamah, 2014; Károly and Lendák, 2015; Mikko et al., 2014; Vladimir and Gruteser, 2013; Wang et al., 2016a; Xiping and Liu, 2013; Xu et al., 2014). Crowdsensing leverages ubiquitous mobile devices and the increasingly more pervasive wireless network infrastructure to collect and analyze sensed data without the need to deploy a large set of static sensors: The mobile crowdsensing paradigm enables large-scale sensing opportunities at lower deployment costs than dedicated infrastructures by utilizing today's dedicated and/or stationary sensors. Thus, crowdsensing can be a major technical enabler to address the challenges associated with the urban-based paradigm shift discussed earlier. Crowdsensing entails massive collection of data and often the (geo)location of that data (i.e., location of the [mobile] sensor where the data is being collected) for aggregation and analysis. The term mobile crowdsensing (MCS) has been coined to describe a broad set of crowdsensing applications (Ganti et al., 2011; Mikko et al., 2014; Xiping and Liu, 2013). MCS applications tend to be community-type sensing where large-scale phenomena (e.g., vehicular traffic, air quality monitoring) are assessed when many individuals or other entities (automatically) supply localized information which is then aggregated for system-wide results.

Crowdsensing (with connected wearable and ubiquitous computing as backdrops) may ultimately become the most disruptive and transformative technology since the World Wide Web. In recent years, the widespread availability of sensor-provided smartphones has enabled the possibility of harvesting large quantities of data in urban areas exploiting user devices, thus enabling a suite of urban crowdsensing applications. Sensing devices include smartphones, music players, gaming systems with embedded sensors, smart watches, health/fitness devices, wearables, and in-vehicle sensors. Functional improvements in terms of capabilities of smart devices, including but not limited to smartphones, smartwatches, and other personal devices, and in terms of open-air hotspot communication technologies are now allowing crowdsensing solutions to emerge, for both data and multimedia streams, as a way to enhance environment sensing, particularly for smart cities applications. However, mobility management (e.g., retaining end-to-end connectivity while in motion) is critical to its practical and cost-effective use. Since on-board power is generally less of an issue for crowdsensing sensors, IPv6 protocols and MIPv6 mobility management techniques may be a useful technology to consider (see Section 12.5).

12.3.3 Smart Buildings

An evolution of the smart city concept is the application of these concepts to commercial building environments, also possibly including multibuilding campuses (Kyriazis et al., 2013). Commercial buildings have a wide gamut of monitoring, management, and resource optimization requirements and many of the applications for smart cities have applicability to building management; some of these applications include video surveillance, traffic/access control, surveillance, energy management (including lighting), indoor environmental and air quality/comfort control, and fire detection, among others (Corna et al., 2015; Ghayvat et al., 2015; Keles et al., 2015; Lizzi et al., 2016; Mauser et al., 2015; Minoli et al., 2017a; Oti et al., 2016; Srivastava and Pal, 2016; Sun et al., 2014; Tragos et al., 2015). The list that follows identifies typical elements and systems where energy is consumed, all of which benefit greatly from improved (IoT-based) sensing, automation, and management (list is not exhaustive):

- *Heating, Ventilation, and Air Conditioning (HVAC) Room.* modular boilers, air compressors, and chillers.
- *Server Room.* uninterruptible power supplies (UPSs); computer room air conditioners (CRACs); telecom closets; racks and virtualized/blade servers.
- *Office Space.* light-emitting diodes (LEDs) lighting; daylight sensors; thermostats (used in controlling HVAC systems and energy consumption); demand response mechanisms.
- *Cooling System Elements.* rooftop units (RTUs), cooling towers, and heat pumps.

In aggregate (across all commercial buildings), the key electrical energy consumption elements are as follows: cooling: 14.9%; ventilation: 15.8%; lighting 17.1%; refrigeration: 15.8%; office equipment: 4.1%; and computers 9.5% (U.S. Energy Information Administration, 2012).

Building management systems (BMSs) have traditionally been used to manage several building-related functions. A BMS is a computerized platform that allows for monitoring and controlling a building's mechanical and electrical equipment. A BMS is typically used to manage loads and enhance efficiency, thus reducing the energy needed to illuminate, heat, cool, and ventilate a building. BMSs interact with controls hardware in the various mechanical/electrical subsystems for real-time monitoring and controlling of the energy used; they are often used to implement demand response (DR) arrangements. Although current-generating BMSs typically focus primarily on electrical consumption, in the future BMSs are expected to cover all energy sources supporting a building, also including natural gas, renewable energy, water usage, and steam systems. Indoor environmental and air quality capabilities are also important. Lately, BMSs have migrated to IP-based networking. This allows remote monitoring by a centralized operations center, such as cloud-based analytics. Smart in-building lighting not only allows intelligent centralized (and/or remote) control and improves the inhabitants' experiences but also lowers energy consumption. IoT will take the current BMS capabilities to the next level. Inexpensive sensors are emerging, and user-friendly applications are becoming available, often as a Software as a Service (SaaS) cloud-provided service (U.S. Energy Information Administration, 2012). HVAC optimization and intelligent lighting controls are just two of the key areas that are facilitated by the IoT. These developments are now driving the deployment of the IoT in building applications, in the broader context of smart city applications.

12.3.4 Smart Campuses

There is a relatively small body of literature on the topic of a smart campus; a few key references include Alghamdi and Shetty (2016); Bates and Friday (2017); Gomes et al. (2016); Huang et al. (2016); Lazaroiu et al. (2016); Li et al. (2016); Lin and Chen (2016); Rha et al. (2016); Various Speakers (2016); Yu et al. (2011); and Zhang et al. (2016). In the context of infrastructure management, a subset of IoT applications apply to the physical continuum that spans a smart city, a number of institutional campuses, and independent smart buildings. A campus is typically comprised of several buildings under one administrative jurisdiction, such as a (private) university or college, or a hospital complex encompassing several structures in a small geographic area. Some also consider a stadium to be a campus. A campus can also be seen as a group of clusters in various regions, but all managed by an oversight entity, for example, a state university that may have a number of campuses (say two dozen or more) throughout the state, or a state prison system with a number of sites, each comprised of several buildings.

Typically, one building acts as an administration office where all the campus communications might terminate, for example, if there is a control center, and where possibly there is a wide area network (WAN) handoff. In multisite campuses, there may be one centralized site where the various data (environmental parameters, control, video, and so on) is centralized to one, say statewide or city-wide, control center for monitoring, processing, storage, or analysis. Even if in-campus connectivity is available (e.g., campus-wide fiber between buildings), it may not be usable for M2M/IoT applications for various administrative, security, or technological reasons. A newly built campus, for example, a business park may well have interbuilding fiber connectivity, but older campuses may not have such wired connectivity. Either way, a dedicated campus network for M2M applications may be needed; today it may typically be added as a wireless overlay to an existing campus as well as greenfield applications.

While the applications that are being considered for smart cities are fairly encompassing, as seen in Table 12.3, the applications that are typically considered for smart campus are somewhat more limited. These might include external campus surveillance; internal surveillance; building emergency generator, automatic transfer switch (ATS), and digital meter (DM—aka smart meter) monitoring and control; elevator monitoring and control; and HVAC monitoring and control. Other campus-related applications include remote door control, water leak detection, washing machine scheduler, smart parking, smart trash cans, light control, and emergency notification (Various Speakers, 2016). Energy efficiency and conservation are becoming more important, especially

Table 12.3 A taxonomy of the requisite synthesis to achieve broad-scale deployment of IoT in smart cities environments.

Area	Subdiscipline
IoT/M2M technologies for smart cities	Sensors
	Networking (especially wireless technologies for personal area networks, fogs, and cores, such as 5G cellular)
	Analytics
System architectures for smart cities	Proposed IoT architectures
	M2M architecture
	Architectures suited for smart cities
IoT/M2M standards applicable/ needed for smart cities	Layer 1, wireless
	Layer 2/3, IP, IPv6, MIPv6
	Upper layers
	Vertical specific
Cybersecurity for smart cities	Confidentiality
	Integrity
	Availability
IoT/M2M applications for smart cities	Intelligent transportation systems (including smart mobility, vehicular automation, and traffic control) For example, traffic monitoring, to assess traffic density and vehicle movement patterns, e.g., to adjust traffic lights to different hours of the day, special events and public safety (e.g., ambulances, police and fire trucks).
	Smart grids For example, Advanced Metering Infrastructure (AMI) and Demand Response (DR); also, natural gas distribution and other pipelines (e.g., steam in some cities)

(continued)

Table 12.3 (*Continued*)

Area	Subdiscipline
	Lighting management Control light intensity when area is empty or sparsely populated and/or when background light is adequate (e.g., depending on lunar phases, seasons, etc.)
	Smart building For example, building service management, specifically for city-owned real estate to remotely monitor and manage energy utilization
	Waste management For example, for disposition of public containers or city-owned properties
	Sensing (including crowdsensing, smart environments, and drones) Environmental monitoring, for example, sensors on city vehicles to monitor environmental parameters Also, crowdsensing, where the citizenry at large utilizes wearables, smartphones, and car-based sensors to collect and forward data for aggregation for a variety of visual, signal, and environmental information
	Water management For example, to manage water usage, for example, for sprinklers, considering rain events
	Surveillance/Intelligence
	Smart services As an example, the New York City Transit Department of Buses has recently designed a 700/800 MHz radio digital system to be deployed in up to 6500 city buses and 1500 nonrevenue generating vehicles. Applications include advanced computer-aided dispatch automatic vehicle location to track the position of buses in real time via global positioning system (GPS) using cellular overlay and provide advanced fleet management and next-bus time of arrival notification at bus stops throughout the region and on customer smartphones.
	Goods and products logistics (including smart manufacturing)

considering governmental mandates in many jurisdictions to reduce energy consumption by 20% by 2020 or 2025; IoT-based capabilities can facilitate the achievement of these goals (e.g., tracking the electricity used by various systems and appliances in the building by monitoring energy usage data from a smart meter). Clearly, the smart building IoT applications mentioned above (for example, occupancy, lighting, daylight harvesting, access control, fire safety, and so on) can also be considered to be part of smart campus applications (Minoli et al., 2017a). For campus wireless IoT connectivity, typically, one can make use of license-free industrial, scientific, and medical (ISM) bands. While a number of such bands exist, the ISM unlicensed radio band at 900 MHz (specifically at 902–928 MHz) is often optimally employed due to better weather-related performance and the reduced congestion from Wi-Fi and other devices (operating in the 2.4 or 5 GHz bands);900 MHz also supports "long distance links" that can span several miles. However, wireless services operating in the ISM band(s) must intrinsically accept potential interference from other users since there is no regulatory protection from other ISM devices' operation: the transmissions of nearby devices using ISM (e.g., including other similar radios, cordless phones, Bluetooth devices) can give rise to electromagnetic interference and disrupt radio communication utilizing the same frequency. Fortunately, there are power restrictions mandated by FCC to minimize interference and thus unlicensed low-power users are generally able to operate in these bands without being impacted by or causing problems to other ISM users. Traditional spread spectrum techniques, where a radio signal generated with a particular bandwidth is by design spread in the frequency domain into a signal with a much wider bandwidth, will also reduce the interference (however, spread spectrum system are slightly more expensive than normal transmitter–receivers.) Other bands available include the V-band (40 to 75 GHz) and the E-band (60 to 90 GHz); these bands, equating to millimeter wavelengths, are just now beginning to be broadly deployed; typically they support wireless Gigabit Ethernet in metropolitan areas and can substitute for physical fiber over short distances (of a few miles).

12.3.5 Smart Grid

Smart cities benefit from being served by smart grids (SGs). The SG is an evolution of the electricity network that integrates the activities of power consumers, power generators, distribution grid, and devices connected to the grid (e.g., substations, transformers, etc.). The goal of SG is to economically and efficiently deliver sustainable, reliable, and secure electricity supplies. The reliable and cost-effective delivery of power is obviously critical to cities; thus, SGs support the smart city paradigm. M2M/IoT technology is designed for automated data exchange between devices, and thus has applicability to SGs. M2M communications takes place between two or more mechanistic entities that routinely avoid direct human intervention. With M2M technology, organizations track and manage assets; inventories; transportation fleets; oil and gas pipelines; mines; widespread infrastructure; natural phenomena such as weather conditions, farm production, forestry condition, and water flows; and, as noted, SGs. Wireless communication is a staple of M2M. These wireless technologies range from unlicensed local (the so-called fog) connectivity to licensed 3G/4G/5G cellular to low earth orbit (LEO) satellites. All of these technologies are relevant to the SG.

Utilities have started to gradually support M2M and Supervisory Control And Data Acquisition (SCADA) systems over wireless and satellite links; these connectivity technologies are applicable to the SG for city and rural environments, respectively, particularly for the transmission and distribution space (T&D) sector. The intelligent integration of information from actions of users connected to the electricity grid—consumers, generators, and the distribution grid—are performed by the SG. Efficient, sustainable, economic, and secure delivery of electricity supplies is the main goal of SG. A SG encompasses the various stages of power generation, distribution, and consumption. The goal is to exploit the power of automation to better control distribution, green efficiency, and consumption. SG management technologies are needed to address these and related power management issues. The power industry is increasingly seeking to incorporate ICT in general and IoT in particular, into its operations, including at the edges; the thus-enhanced power grids are known as SGs. Three basic issues are of interest: monitoring of rural transmission systems, demand response (DR) support, and urban automated meter readers (AMRs). Well known examples of IoT usage in energy efficiency and interactions with the SG/AMI include the following: smart thermostats; smart appliances that can interwork with DR-based SG power management (with the use of appliance-level actuators); "plug-level" control of electrical outlets, where lower end devices can be turned on or off remotely; IoT-based LED lighting and daylight sensors for "smart lighting" that not only allow intelligent centralized (and/or remote) control but also lower energy consumption while improving the residents' experience; and, consumers' ability to generate green renewable power and sell it back to the SG. Refer to Chapters 14 and 15 for a more extensive discussion of SGs and energy applications for the IoT.

12.4 Optimal Enablement of Video and Multimedia Capabilities in IOT

Evolving video and multimedia-oriented IoT applications applicable to smart cities (for example video surveillance, crowdsensing, social networking, and even drone-based video/sensors) can make use of the newly developed High Efficiency Video Coding (HEVC) algorithm (ISO/IEC 23008-2 and ITU-T

Recommendation H.265 (Lazaroiu et al., 2016).) HEVC enjoys the same structure as MPEG-2 and H.264/AVC (advanced video coding), but it provides more efficient video compression. Specifically, it can provide the same image resolution as H.264/AVC but with a lower coding data rate, or it can provide better image resolution than H.264/AVC but at the same coding data rate. The lower data rate is useful in IoT not only for the practicality of transmission, especially when wireless networks are utilized, but also in terms of storage if the information is aggregated centrally to support analytics (Banitalebi-Dehkordi et al., 2014; Belhadj et al., 2015; Dey and Kundu, 2016; Diaz-Honrubia et al., 2014; Pham and Vafi, 2015; Schwarz et al., 2007; Shen et al., 2013; Su et al., 2014, 2016; Wiegand et al., 2003).The objective of HEVC is to achieve a compression gain of 50% over H264/AVC. A related tradeoff, however, is that the coder/decoder becomes more complex. The design objective was that the coder should have at most a maximum tenfold increase in complexity of the underlying DSP for the coding aspect and a maximum 66% increase in complexity for the decoder. The increased computational complexity requires more advanced chip sets for the equipment utilizing this algorithm and also an increase in power consumption (and decreased battery life (Minoli, 2008)).

While UltraHD entertainment applications may opt for the increase in quality for the same data rate as the older H.264/AVC, IoT multimedia applications applicable to smart cities will likely prefer to go the route of (i) a specified quality for reduced bandwidth; or (ii) depending on the application, reduced quality for reduced bandwidth; or (iii) even sticking with H.264/AVC in order to have less expensive and more energy efficient endpoints, especially in the case of mobile nodes, for example, in crowdsensing or wearable/ubiquitous computing applications (vehicle oriented applications such as V2V or V2I may not have these power restrictions (Guevara et al., 2012).)

Quality of Service (QoS) and Quality of Experience (QoE) are critical, especially as the number of endpoints increase and as these endpoints are in densely populated areas such as a city or even a stadium or conference center. The reliance on wireless technology also drives the requirement for efficient use of the channel bandwidth.

Efficient compression is a first step in addressing overall traffic management, under the thrust of increased deployment/usage, the evolution to multimedia, and the concentration experienced in large urban environments. It is well known that video frames have repetitive structures. Hence, a pixel's color can be derived from the color of its neighbors within the same frame (intraframe, also known as "intra") or from recent frames (interframe, also known as "inter")—particularly when the objects are moving in some normal, predictable fashion. The fundamental approach is to encode a block of pixels as a prediction plus a delta from that prediction; the delta information typically requires fewer bits than would be the case by coding the pixel values directly, hence achieving content compression. Compared to H.264/AVC, HEVC utilizes a recursive quadtree structure for frame partitioning, uses larger block transforms, uses more efficient motion vector prediction and motion compensation, makes use of additional sample adaptive offset filtering, and employs an enhancement to H.264's Context-adaptive Binary Arithmetic Coding (CABAC), known as Syntax-based Context-adaptive Binary Arithmetic Coder (SBAC) (Belyaev et al., 2006; Chien et al., 2016; Guo et al., 2016; Rojals, 2016).

Other QoS/QoE techniques for multimedia, which are particularly applicable to IoT, entail the proper priority management of the I-, P-, and B-frames, dropping some of the frames when needed (in case of congestion), as discussed in Ghazal et al. (2015). These techniques are particularly of interest due to power and bandwidth scarcity of IoT devices.

12.5 Key Underlying Technologies for Smart Cities IOT Applications

A number of fundamental IoT technologies are needed to support smart cities in an effective manner. Two of these technologies are discussed in this section: *MIPv6* and *5G wireless*. The expectation is that the IoT may broadly embrace IPv6 in the near future, because of the abundant IPv6 address spaces and because globally unique object (thing) identification and connectivity can be provided in a standardized manner without additional status or address (re)processing, this being its intrinsic advantage over IPv4 or other schemes (some of the proposed protocols, however, state that the object should have its own ID—which may not be unique within a certain application, but the entirety would have a unique address—these proposals wish to keep the ID separate from the address.) IPv6 offers a large 2^{128} address space; hence, the number of available unique node addressees is $2^{128} \sim 10^{39}$. In addition to its expanded addressing capabilities, other advantages of IPv6 that could prove useful for smart cities application include scalability, "plug and play" mechanisms that facilitate the connection of equipment to the network, and mobility management. Because of the large population of users (and sensors) in an urban environment, the use of IPv6 and the mobility extensions, for example, MIPv6 and Proxy MIPv6 (PMIPv6), makes good technical sense.

While a number of addressing schemes are available (and usable at different protocol layers), for example, the IEEE 48-bit Extended Unique Identifier (EUI-48) and the 64-bit Extended Unique Identifier (EUI-64) used for IEEE 802 MAC addresses, or the ITU-T E.164 address space (with up to 15 digits plus the international call prefix), a network layer address is needed for optimal, broad-scale deployment of IoT devices. In that context, the usable number of public addresses in IPv4 is less than about 3.8 billion (the theoretical maximum being in the range of 4.3 billion). Organizations have made use of network address translator (NAT) functionality or proxy server mechanisms to simultaneously build private intranets or closed user group extranets and also communicate with entities in the public Internet. Furthermore, recently IANA allocated the address block 100.64.0.0/10 (or about 4 million, addresses) for use in what has been called carrier-grade NAT; this is intended for use within carrier networks enabling carriers to uniquely identify their customer's access devices within a given carrier's area of operation in a large city, on the assumption that a customer only had one device (lately we have observed people who carry two smartphones, one for company operations and one for personal social media access and so on.) Such an addressing scheme could be used in the short term for fog-oriented (edge subnetworks) environments. However, the significantly larger address space afforded by IPv6 is ideal for smart cities applications, especially in the near future, when there could be a large set of endpoints (say, 5x the size of the population considering smart home applications, wearables, city-wide sensor systems, etc.) (Minoli, 2013; Wang et al., 2016b).

IPv6 was originally defined in 1995 in an IETF Request for Comments (RFC) 1883 and then further refined by RFC 2460, "Internet Protocol Version 6 (IPv6) Specification" (December 1998). A large body of additional RFCs have emerged in recent years (over 432 RFCs by press time) to add capabilities and refine the IPv6 concept (specifically, RFC 2460 is updated by RFC 5095, RFC 5722, RFC 5871, RFC 6437, RFC 6564, RFC 6935, RFC 6946, RFC 7045, RFC 7112). Mobility is supported with MIPv6 protocols, including the protocol defined in RFC 3775 "Mobility Support in IPv6" (June 2004), which provides basic network-layer mobility support in the IPv6 environment; this protocol offers an improved mechanism to support mobility management without having to rely on session reinitiation required by other mobility approaches. One example of an application for which the MIPv6 approach is ideal is crowdsensing in smart city/urban environments. There is an extensive area of active research on MIPv6 and related standards: over 70 IETF RFCs have been developed in the past 12 years (starting in 2004). Extensions to RFC 3775 include Network Mobility (NEMO) Basic Support Protocol (RFC 3963, January 2005); Hierarchical Mobile IPv6 Mobility Management (RFC 4140/5380, August 2005); Multihoming (RFC 4980, October 2007); Multicast Mobility Routing for Proxy Mobile IPv6 (RFC 7028, September 2013; Quality of Service for Proxy Mobile IPv6 (May 2014, RFC 7222), and Proxy Mobile IPv6 Extensions to Support Flow Mobility (RFC 7864, May 2016). Work has also recently been undertaken related to cybersecurity (e.g., Authentication, Authorization and Accounting – RFC 5637), functional optimization (various RFCs), and wireless use (RFC 7561). Additionally, transition mechanisms are methods used for the coexistence of IPv4 and/or IPv6 devices and networks. For example, an "IPv6-in-IPv4 tunnel" (6to4 tunnel) is a transition mechanism that allows IPv6 devices to communicate through an IPv4 network. Furthermore, some specific IPv6-oriented IoT protocols have also been developed. It is well known that IoT nodes have noteworthy design constraints, which impact both the operation at the physical layer, as well as functions at the upper layers. To that end, an IETF Working Group (WG) was chartered in 2005 to define IPv6 over IEEE 802.15.4, specifically the use of IPv6 over Low-Power Wide-Area Networks (LPWAN). The 6LoWPAN group defined encapsulation and header compression mechanisms that allow IPv6 packets to be transmitted over IEEE 802.15.4-based networks. The base specification developed by the 6LoWPAN IETF group is RFC 4944 (September 2007), further updated in 2011–2012 by RFC 6282 with header compression, and by RFC 6775 with neighbor discovery optimizations. It allows link-layer mesh routing under IP topology. 6LoWPAN can utilize 802.15.4 and other PHYs, and it facilitates seamless integration with other IP-based systems. While 6LoWPAN is generally low bandwidth (in the 250 kbps range), the use of H.265 compression methods allow its use in some IoT multimedia applications (e.g., video surveillance). Mobility intrinsically implies the use of wireless technologies, such as LPWAN type, cellular 3/4/5G network type, or satellite type.

Figure 12.4 depicts graphically the concept of mobility management, especially in a MIPv6 context and also in a smart cities context (e.g., for

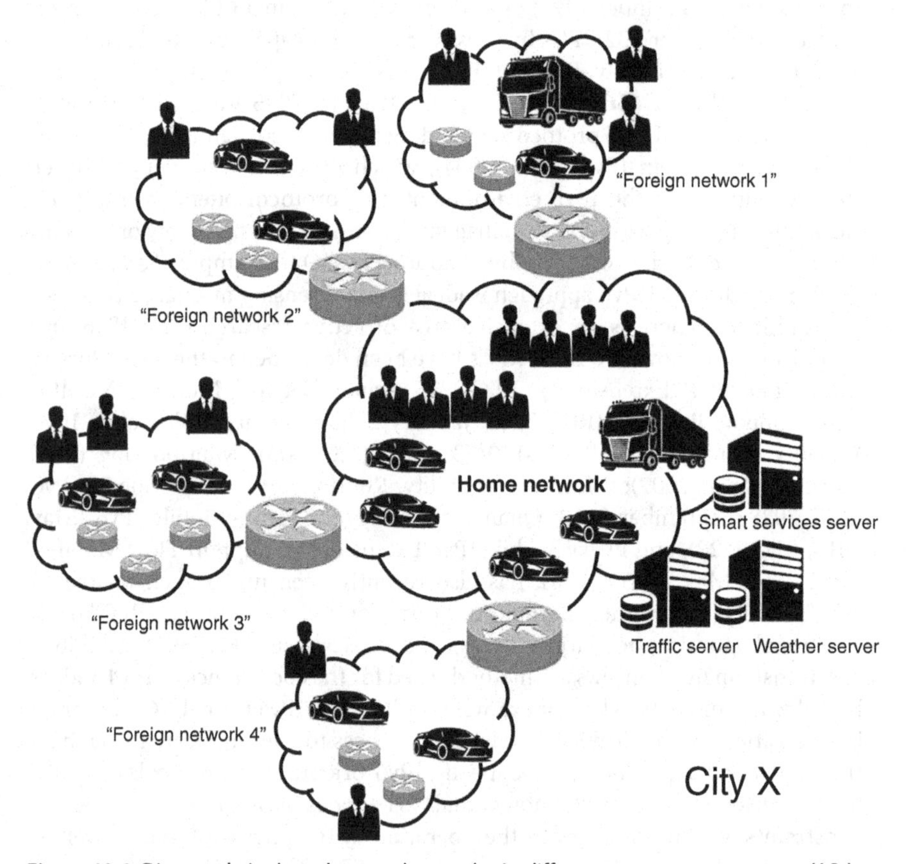

Figure 12.4 Dispersed city-based networks may be in different autonomous systems (ASs).

crowdsensing, video streaming, and traffic control applications.) Very likely IoT applications in smart cities environments will make extensive use of wireless technologies, including SDR (software-defined radios) and SON (self-organizing network) approaches. Therefore, there are two key drivers that animate traffic engineering optimization: (i) wireless channels are usually at a premium in terms of bandwidth (spectrum continues to be an issue); (ii) there is a desire to make major advancements in the multimedia arena; for example, the 5G requirements under consideration by 3GPP (for Release 14) include, but are not limited to, Multimedia Broadcast for Public Warning Systems, Location services, Mission Critical Video over LTE, LTE support for Video-to-Everything (V2X) services, drone video capture, emergency services over WLAN, and smart grid applications. Extending a concept broached in ZTE Corporation (2015), one can perceive what can be called an M^8 connectivity "world": *M*an–*M*an, *M*an–*M*achine, *M*achine–*M*achine, *M*obile, *M*IPv6.

The goal of enabling enhanced mobile broadband (eMBB) is one of three key 5G use cases ITU's IMT 2020 Vision. 5G goals are to ultimately provide 1 Gbps throughput in high-mobility environments (10 Gbps peak throughput for quasistationary environments).

Earlier, new results to optimize multimedia delivery in support of (smart cities) IoT applications (such as surveillance, social networking, crowdsensing) were discussed. That optimization can be done at the content level (applications residing above the OSI layer 7) as discussed earlier. However, a robust network-level support of QoS is needed. It is one thing to optimize content (by analogy think of "lightening up" a package to be sent via Fedex, to make it fit in a smaller box, which in theory could improve delivery), but that would be to no avail if Fedex did not plan out its airplane traffic (e.g., landing protocol), and the package got "stuck" along the way. The QoS mechanisms proposed and assessed in RFC 7222, Quality-of-Service Option for Proxy Mobile IPv6 (May 2014) and RFC 7561, Mapping Quality of Service Procedures of Proxy Mobile IPv6 (PMIPv6) and WLAN (June 2015), indeed provide important mechanisms going forward.

New Radio Access Technologies (RATs) are being analyzed for 5G that range from massive multiple input multiple output (MIMO) (with up to hundreds of antennas) to beamforming, from unlicensed spectrum and above C-band (6 GHz) to millimeter waves (targeting the 60–80 GHz region.) Complementing the complex and advanced methods to increase bandwidth availability with the use of advanced modulation and small cells, one way to support enhanced mobile broadband (eMBB) is to use the MIPv6 family of standards. MIPv6 will also enable 5G to support a stated migration goal from nomadic to even aeronautical applications.

3GPP made a strategic and very propitious leapfrog move a decade ago when they adopted a full IP approach to telephony (the defined goal of 3GPP in 1998 was to develop technical specifications for a 3G mobile system). 5G affords the opportunity to make another leapfrog move to an IPv6-baselined approach to multimedia communications with legacy support of other protocols. In fact, recently 3GPP has published specifications that utilize MIPv6. Mobility management under roaming is more likely to be needed and/or occur in a city environment than in the more sparse rural environment, thus having applicability to the smart city application.

3GPP TS 123 402 v13.5.0 (2016-04) notes that the S5 and S8 reference points in the Evolved Packet Core (EPC) architecture have been defined to support both a GTP (General Packet Radio System [GPRS] Tunneling Protocol) and PMIPv6 (Proxy Mobile IP version 6) protocol stack. The PMIPv6 stack is described in TS 23.402 (UMTS, 2016) and the GTP stack is described in TS 23.401 (GPRS, 2012).

3GPP TS 29.275, Proxy Mobile IPv6 (PMIPv6—also published as ETSI TS 129 275 V13.4.0, 2016-03) (PMIP, 2016)—specifies protocol mechanisms to be used over the PMIPv6-based S2a, S2b, S5, and S8 reference points as defined in 3GPP TS 23.402 (discussed later). Figure 12.5 depicts the PMIPv6-based protocol stack. The protocols are applicable to the Serving GW (Gateway), PDN (Packet Data Network) Gateway, ePDG (Evolved Packet Data Gateway), and Trusted Non-3GPP Access. Importantly it covers the following:

- Mobility management procedures
- Tunnel management procedures
- Path management procedures
- PMIPv6-based S5 and PMIPv6-based S8 description
- Trusted non-3GPP access over S2a description
- Untrusted non-3GPP access over S2b description
- S2a and S2b chaining with PMIPv6-based S8 description

The EPC environment defines very precisely the interfaces between the various functional elements that participate in the enhanced mobile broadband environment, and the allowed protocols over those interfaces, in order to support global interoperability. Figure 12.6 depicts one example of the EPC environment and the cited reference points. Table 12.4 defines the architectural

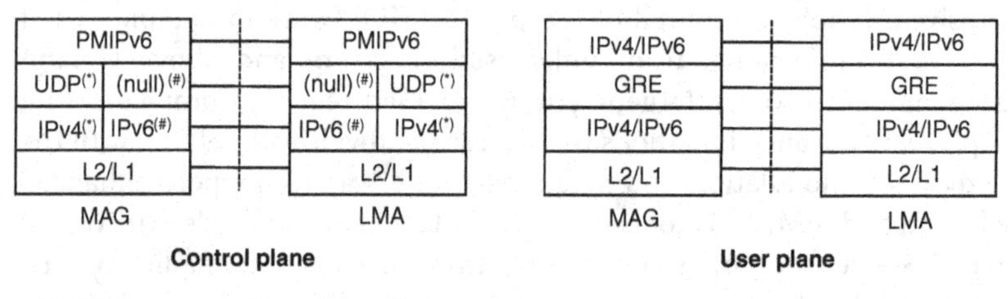

Figure 12.5 3GPP/ETSI protocol stack specification of PMIPv6 for enhanced mobile broadband.

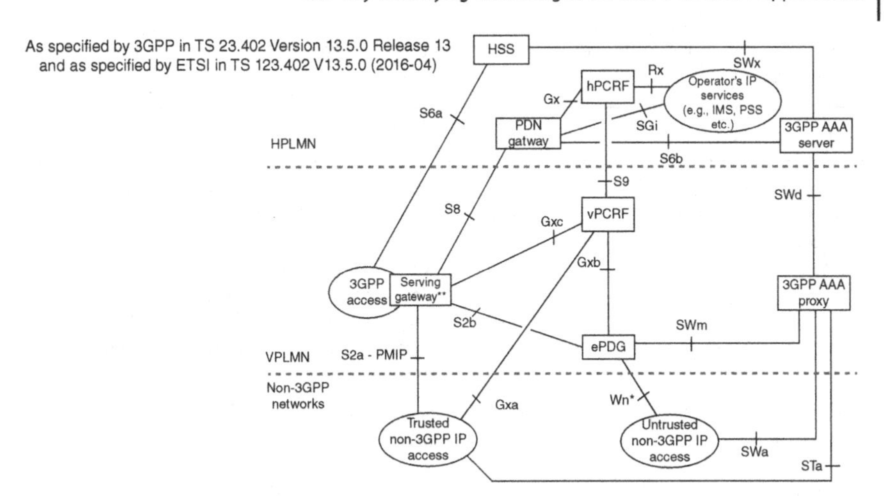

Figure 12.6 Example of PMIPv6 usage in 3GPP context.

interfaces described when MIPv6 (or more precisely PMIPv6) is used; while the technology is fairly complex, the takeaway for the reader is that MIPv6 is being applied to mobility management tasks in advanced 4G/5G cellular networks, which are likely to be critically important in smart city applications in the future.

QoS consideration for the MIPv6 environment have also been studied in the 3GPP environment (QoS, 2011). The application of the well-developed machinery of MIPv6 to mobility management, to enhanced mobile broadband, IoT, and smart cities in particular is expected to see deployment in the near future.

The high-density requirements of urban environments, particularly as driven by the plethora of new IoT-oriented applications will require the technical advancements afforded by 5G. While the 5G technology could take several service directions, it appears that the vision of a superfast mobile network where densely clustered small cells provide a contiguous urban coverage is the path that will be taken by the standards development teams and the implementers. METIS (Mobile and wireless communications Enablers for the Twenty-twenty Information Society—an integrated 2012-15 European Community effort considered to be the 5G European flagship initiative), has recommended that 5G will have to support the following:

- Far more stringent latency and reliability requirements for applications related to health care, security, logistics, automotive applications, or mission critical control.
- A wide range of data rates (up to multiple Gbps, and tens of Mbps) and to guarantee high availability and reliability.

Table 12.4 Reference interfaces in the evolved packet core (EPC) architecture that can use and PMIPv6

Reference points	Protocol
S2a provides the user plane with related control and mobility support between trusted non 3GPP IP access and the Gateway (GW).	S2a is based on Proxy Mobile IP version 6. For Trusted WLAN, S2a may also be based on GTP (General Packet Radio System [GPRS] Tunnelling Protocol). To enable access via Trusted Non 3GPP IP accesses that do not support GTP and PMIPvo\S2a also supports Ghent Mobile IPv4 FA mode
S2b provides the user plane with related control and mobility support between ePDG (evolved packet data gateway) and the GW.	S2b interface is based on GTP or PMTPv6 specification (RFC 5213)
S5 provides user plane tunneling and tunnel management between Serving GW and the PDN (packet data network) GW. It is used for Serving GW relocation due to UE (user equipment) mobility and in case the Serving GW needs to connect to a noncollocated PDN GW for the required PDN connectivity.	The S5 interface is based on the PMIPv6 specification.
PMIP-based S8 is the roaming interface in case of roaming with home-routed traffic. It provides the user plane with related control between Gateways in the VPLMN (visited public land mobile network) and HPLMN (home public land mobile network).	PMIPv6-based S8 interface is based on the PMIPv6 specification. The GTP variant interface is described in TS 23.401

Note: PMIPv6-based interfaces (S5, S8, S2a, and S2b) utilize Generic Routing Encapsulation (GRE), RFC 2784, along with the key field extension (RFC 2890).

- Network scalability and flexibility to support a large number of devices with very low complexity while having very long battery lifetimes.

5G is a technology for urban environments, thus a technology that can advance many of the smart cities IoT applications—more specifically, the support of IoT, M2M, and D2D services is intrinsically incorporated in the emerging 5G plans. The technical objectives for 5G are to develop technical solutions for a system that supports the following:

- 1000× higher mobile data volume per area
- 10–100× higher number of devices

- 10–100× higher user data rates
- 10× longer battery life for low-power IoT devices
- 5× reduced end-to-end latency

Observers make the case that societal development will lead to changes in the way wireless communication systems are used and these developments will lead to a significant increase in mobile and wireless traffic volume, predicted to increase a 1000-fold over the next decade. Applications driving wireless traffic include but are not limited to on-demand mobile information and high-resolution entertainment, augmented reality, e-health, and ubiquitous IoT rollouts.

Standardization work for 5G has been undertaken by international bodies, such as the ITU-R, with the goal of achieving one unified global standard. Many well-known research centers, universities, standards bodies, carriers, and technology providers are currently involved in advancing the development of the technology. Development work has been underway for several years, but gaining speed after the World Radio Conference of 2012 (WRC'12), where the exploration of new system concepts was given priority, followed by a period of optimization and standardization spanning the 2015–2018 timeframe, bookended by the WRC'15 and WRC'18. As of press time, countries and carriers worldwide were beginning to articulate potential rollout plans. Implementation is expected to start in 2018 in some advanced countries, although further developments on fundamentals will continue. In the United States, 5G commercial services are expected to start to appear in 2020; naturally, the current 4G/LTE and 5G are expected to coexist for many years (Lee, 2016; Nordrum, 2017). In the United States, Verizon Wireless and AT&T started 5G trials in 2017, with the first commercial deployments expected to materialize in 2020. Verizon Wireless became the first U.S. carrier to define (in 2016) some of the technical details on its 5G vision, a blueprint providing testing details for new multiantenna array processing techniques, carrier aggregation technologies, and wide bandwidth utilization; the company is undertaking trials, working with KT Corporation, Samsung, and Nokia. In Japan, NTT DoCoMo is researching the use of the millimeter wave spectrum and conducted its first 5G in 2016. In South Korea, KT Corporation is planning to unveil its 5G technology in time for the 2018 Olympics. In China, the Academy of Telecommunications Research has launched a 5G three-year research program. In the United Kingdom, the University of Cambridge published a comprehensive plan on 5G Mobile and Wireless Communications Technology and the government has reportedly made plans to free up 750 MHz of spectrum by 2022 and has set aside some of its budget for research and development. In Continental Europe, 20 telecommunications companies, including Nokia, Vodafone, BT, and Deutsche Telekom, have indicated that they are planning to make investments in 5G technologies by 2018, but only under certain regulatory relief (Triggs, 2016).

5G networks will be five times as fast as the highest current speed of today's 4G networks (with download speeds as high as 5 Gbps—4G offering only up to a maximum of 1 Gbps). These higher speed networks will use significantly higher frequency bands for operation than currently used by cellular systems, including millimeter-wave frequencies (these bands ranging from 28 to 73 GHz, specifically the 28, 38, 60, and 72–73 GHz bands). These targeted cellular frequencies thus overlap with K-band transmissions of communication satellites: Some of the proposed bands have been long in use for point-to-point microwave paths, Direct Broadcast satellite TV, and high throughput satellite systems. Some of the design details are a latency below 5 msec, support for device densities of up to 100 devices/m^2, reliable coverage area, integration of telecommunications services, including mobile, fixed, optical and Medium Earth Orbit/Geosynchronous Earth Orbit (MEO/GEO) satellites, and support for the IoT ecosystem. In 2016, the FCC approved spectrum for 5G, including the 28, 37, and 39 GHz bands. Initially, 5G will use the 28 GHz band, but higher bands will very likely be utilized later on. Until now cellular networks have used frequencies below 6 GHz. Lower frequencies are less subject to weather impairments, can travel a longer distance, and penetrate building walls more easily. Waves at higher frequencies do not naturally travel as far or go through objects as easily. However, a lot more channel bandwidth is available in millimeter-wave bands. Initial implementations, however, will support a maximum speed of 1 Gbps.

New mobile generations are typically assigned new frequency bands and wider spectral bandwidth per frequency channel (1G up to 30 kHz, 2G up to 200 kHz, 3G up to 5 MHz, and 4G up to 20 MHz.) The integration of new radio concepts such as massive MIMO antenna technologies, ultra dense networks, moving networks, device to device, ultra reliable, and massive machine communications will allow 5G to support the expected increase in the data volume in urban mobile environments opening the door for new applications. Developers foreseeing the need for "an innovative utilization of spectrum" and "small cell" approaches are required to address the scarcity of the spectrum, but at the same time cover the geography.

In addition to new bands, 5G technology is expected to use beamforming and beamtracking, where a cell's antenna can focus its signal to reach a specific mobile device and then track that device as it moves around. Massive MIMO is a transmission node (base station) equipped with a large number (hundreds) of antennas that simultaneously serve multiple users; with this technology, multiple messages for several terminals can be transmitted on the same time–frequency resource. Beamforming utilizes a large number (hundreds) of antennas at base stations to achieve highly directional antenna beams that can be "steered" in a desired direction to optimize transmission performance, including throughput.

12.6 Challenges and Future Research

As noted already, not all the issues that impact cities can be solved with IoT technologies. But even in those cases where the technologies could be applicable there are challenges. Some of the major challenges related to the broad scale introduction of IoT services in urban environments include the following:

- The lack of widely adopted IoT standards, especially at the upper layers; users are left with having to select a vendor's system and not only be "stuck" with that vendor, but unable to expand more broadly and interconnect with other systems.
- The various applications (e.g., traffic management, infrastructure management, power management, surveillance, and public transportation) are now and for the foreseeable future stand-alone solutions and technology-silos: administrators have to deploy separate and fragmented systems, rather than one all-encompassing system, a system that could share data and information among the discrete subsystems; interoperability is clearly an issue.
- Scalability remains a concern: as the urban user population gets larger, will the protocols, architectures, and analytics systems be able to keep up and/or grow smoothly?
- System reliability becomes a concern as people begin to rely more consistently on the services (e.g., smart services) offered by cities and municipalities.
- Security, confidentiality, and privacy issues will become more pressing as the IoT makes itself more pervasive in peoples' lives. For example, in crowdsensing applications, one may be able to track individuals and/or sensitive data they have collected on themselves, including fitness and exercise data. Any IoT device poses an attack surface that can result in a security risk, more so than the average computer on a network. Recent surveys have shown that only a fraction of IoT devices have robust security mechanisms built into them: most objects in the new connected world are developed with minimal security features, making them very vulnerable endpoints. It is thus incumbent upon the smart city network to provide a high level of security needed for each object (Avaya Inc., 2017; Minoli et al., 2017b).
- The use of wireless networks, especially in crowded urban settings, will tax the radio spectrum, and new frequency bands will have to be allocated; 5G technologies aim at addressing some of these issues.
- Many city applications have to be funded, procured, and administered through the bureaucratic efforts of city governments; these are not always known to be agile, well-funded, and "up to speed" with the latest technologies, especially in the context of other pressing problems often impacting cities (e.g., crime, unemployment, aging infrastructure).

In spite of these challenges, many are optimistic that the technology will progressively find its way into the fabric of a city. A number of the technology issues identified above are being aggressively researched in order to resolve residual concerns. For example, it has been recognized by the IoT researchers and developers that standards are critical for the success of smart cities deployment. The classical benefits of standardization include interoperability, replicability, reusability, and reduced costs. Studies have shown that a more rapid deployment (by around 30% acceleration) and a cost reduction in actual disbursement by city agencies (a reduction of 30% by 2025 compared with the vendor-proprietary status quo) can be achieved (Machina Research White Paper, 2016). Thus, it is important to continue to develop appropriate standards to cover the core, the fogs, and perhaps even the analytics (Moreno-Cano et al., 2015). New research into 5G is expected to lead to fundamental changes in the design of cellular systems, making bandwidth more cost effective and thus usable for IoT and smart city applications. The large bandwidth will also support the multimedia-oriented IoT smart city applications. Technologies that may lead to both architectural and component disruptive design changes include increased device intelligence, native support for machine-to-machine communications, device-centric architectures, massive MIMO, and even millimeter wave to microwave connectivity (Boccardi et al., 2014). Cybersecurity research is also critical for smart city applications, especially in the context of infrastructure management (e.g., power grids, water/sewer management, traffic flows, and surveillance) (Cerrudo, 2015; Chakrabarty and Engels, 2016; Minoli et al., 2017b). The treatment of Big Data generated by a city-wide system also needs to be further understood and addressed (Hu and Yang, 2015).

12.7 Conclusion

Numerous applications and opportunities are afforded by IoT deployment in support of smart cities. The promise exists to improve resource management; the flow of goods, people, and vehicles; improved urban QoL; the greening of the environment by optimizing energy consumption; and maximizing life-activity efficiency. City authorities generally have pressing concerns that prevents them from being "early adopters" as stand-alone entities: In practical terms, deployments (even development) of smart cities systems have been based on a public–private partnering approach. This rollout approach may well continue, but the need for additional standardization of the protocols and architectures exist. The gamut of applications of IoT in the smart city context is expected to continue to grow, especially as broadband city-wide connectivity becomes more broadly available under the 5G rollout planned for the next decade and such connectivity is, by design, optimized for M2M applications (while previous generations of the technology were optimized for people-to-people interactions.) Four billion people already live in cities at this time, therefore, the application of ICT and IoT to all aspects of city life is an indisputable imperative. As the

Department of Economic and Social Affairs of the United Nations recently noted, "As the world continues to urbanize, sustainable development challenges will be increasingly concentrated in cities, particularly in the lower-middle-income countries where the pace of urbanization is fastest. Integrated policies to improve the lives of both urban and rural dwellers are needed" (United Nations, 2014). IoT will prove to be an indispensable tool to address these many evolving urban challenges.

References

Alam, M., Ferreira, J., and Fonseca, J. (2016) Introduction to Intelligent Transportation Systems. *Intelligent Transportation Systems*, vol. 52, Studies in Systems, Decision and Control, Springer, pp 1–17.

Alghamdi, A. and Shetty, S. (2016) Survey toward a smart campus using the Internet of Things. IEEE 4th International Conference on Future Internet of Things and Cloud (FiCloud), August 22–24, 2016, Vienna, Austria.

Alvi, S.A., Afzal, B. et al. (2015) Internet Of multimedia Things: vision and challenges. *Ad Hoc Networks*, **33**, 87–111.

Avaya Inc., (2017) Avaya's 2017 tech trends to watch: the distinguished dozen, January. Available online https://www.avaya.com/blogs/archives/2017/01/2017-technology-trends.html.

Banitalebi-Dehkordi, A., Azimi, M. et al. (2014) Compression of high dynamic range video using the HEVC and H. 264/AVC standards. 10th International Conference on Heterogeneous Networking for Quality, Reliability, Security and Robustness (QShine), August 18–20, 2014. doi: 10.1109/QSHINE.2014.6928652.

Bates, O. and Friday, A. (2017) Beyond data in the smart city: learning from a case study of re-purposing existing campus IoT. IEEE Pervasive: Special Issue on Smart Buildings and Cities, January.

Bei, P., Zheng, Y., Wilkie, D., and Shahabi, C. (2013) Crowd sensing of traffic anomalies based on human mobility and social media. Proceedings of the 21st ACM SIGSPATIAL International Conference on Advances in Geographic Information Systems, pp. 344–353. ACM.

Belhadj, N., Bahri, N., and Marrakchi, Z. (2015) H. 264/AVC high definition intra coding implementation on multiprocessor system on chip technology architecture. *IET Computers & Digital Techniques*, **9**(5), 259–267. doi: 10.1049/iet-cdt.2014.0151.

Belyaev, E., Gilmutdinov, M., and Turlikov, A. (2006) Binary arithmetic coding system with adaptive probability estimation by "virtual sliding window." 2006 IEEE Tenth International Symposium on Consumer Electronics, p. 5.

Boccardi, F., Heath, R.W., Jr Lozano, A., Marzetta, T.L., and Popovski, P. (2014) Five disruptive technology directions for 5G. *Communications Magazine, IEEE*, **52**(2), 74–80.

Calvillo, C.F., Sánchez-Miralles, A., and Villar, J. (2016) Energy management and planning in smart cities. *Renewable and Sustainable Energy Reviews*, **55**, 273–287. doi: 10.1016/j.rser.2015.10.133.

Caragliu, A., Bo, C.D., and Nijkamp, P. (2011) Smart cities in Europe. *Journal of Urban Technology*, **18**(2) 65–82. doi: dx.doi.org/10.1080/10630732.2011.601117.

Cardone, G., Corradi, A. et al. (2016) ParticipAct: a large-scale crowdsensing platform. *IEEE Transactions on Emerging Topics in Computing*, (1), 21–32.

Cerrudo, C. (2015) Keeping Smart Cities Smart: Preempting Emerging Cyber Attacks in U.S. Cities, ICIT (Institute for Critical Infrastructure), Whitepaper, June

Chakrabarty, S. and Engels, D.W. (2016) A secure IoT architecture for smart cities. Consumer Communications & Networking Conference (CCNC), 13th IEEE Annual, January 9–12, 2016. doi: 10.1109/CCNC.2016.7444889.

Cheng, X., Yang, L., and Shen, X. (2015) D2D for intelligent transportation systems: a feasibility study. *IEEE Transactions on Intelligent Transportation Systems*, **16**(4), doi: 10.1109/TITS.2014.2377074.

Chien, W.-J., Rojals, J.S., and Karczewicz, M. (2016) Context Reduction For Context Adaptive Binary Arithmetic Coding, US 9277241 B2, March 1.

Corna, A., Fontana, L., Nacci, A.A., and Sciuto, D. (2015) Occupancy detection via iBeacon on Android devices for smart building management. Proceeding of DATE '15 Proceedings of the 2015 Design, Automation & Test in Europe Conference & Exhibition, Grenoble, France, March 09–13, pp. 629–632. ISBN: 978-3-9815370-4-8.

Dey, B. and Kundu, M.K. (2016) Enhanced macroblock features for dynamic background modeling in H. 264/AVC video encoded at low-bitrate. *IEEE Transactions on Circuits and Systems for Video Technology*, **PP**(99), 1–1.

Diaz-Honrubia, A.J., Martinez, J.L., and Cuenca, P. (2014) Multiple reference frame transcoding from H.264/AVC to HEVC, *MultiMedia Modeling: 20th Anniversary International Conference, MMM 2014, Dublin, Ireland, January 6–10, Proceedings, Part I*, Lecture Notes in Computer Science, vol. 8325 Springer, pp. 593–604. doi: 10.1007/978-3-319-04114-8_50.

ETSI, (2011) ETSI TS 102 690: machine-to-machine communications (M2M); functional architecture. October. Available at http://www.etsi.org/deliver/etsi_ts/102600_102699/102690/01.01.01_60/ts_102690v010101p.pdf.

Ganti, R.K., Ye, F., and Lei, H. (2011) Mobile crowdsensing: current state and future challenges. *IEEE Communications Magazine*, **49**(11), 32–39.

Ghayvat, H., Mukhopadhyay, S., Gui, X., and Suryadevara, N. (2015) WSN- and IOT-based smart homes and their extension to smart buildings. *Sensors*, **15**(5) 10350–10379; doi: 10.3390/s150510350.

Ghazal, A., Wang, C.-X., Ai, B., Yuan, D., and Haas, H. (2015) A nonstationary wideband MIMO channel model for high-mobility intelligent transportation systems. *IEEE Transactions on Intelligent Transportation Systems*, **16**(2), doi: 10.1109/TITS.2014.2345956.

Gomes, R., Pombeiro, H., Silva, C. et al. (2016) Towards a smart campus: building-user learning interaction for energy efficiency: the Lisbon case study. *Handbook of Theory and Practice of Sustainable Development in Higher Education*, World Sustainability Series, Springer, pp 381–398.

3GPP TS 23.401: General Packet Radio Service (GPRS) Enhancements for Evolved Universal Terrestrial Radio Access Network (E-UTRAN) access. Available at http://www.etsi.org/deliver/etsi_ts/123400_123499/123401/11.03.00_60/ts_123401v110300p.pdf (accessed February 22, 2018).

Grodi, R., Rawat, D.B., and Rios-Gutierrez, F. (2016) Smart Parking: Parking Occupancy Monitoring And Visualization System For Smart Cities, SoutheastCon, 30 March–3 April. DOI: 10.1109/SECON.2016.7506721.

Guevara, J., Barrero, F., Vargas, E., Becerra, J., and Toral, S.L. (2012) Environmental wireless sensor network for road traffic applications. *IET Intelligent Transport Systems*, **6**, 177–186.

Guo, L., Chien, W.-J., and Karczewicz, M. (2016) Context Optimization For Last Significant Coefficient Position Coding, US 9357185 B2, May 31.

Hegyi, A. and De Schutter, B. (2015) Introduction to the special issue on the 16th IEEE international conference on intelligent transportation systems (ITSC'13). *IEEE Transactions on Intelligent Transportation Systems*, **16**(1), doi: 10.1109/TITS.2015.2393172.

Hu, J., Yang, K. et al. (2015) Special issue on big data inspired data sensing, processing & networking technologies. *Ad Hoc Networks*, **35**, 1–2.

Hu, J., Min, G., and Xia, M. (2014) Special issue on Ubiquitous Computing and Communications Technologies. *Computers and Electrical Engineering*, **40**, 673–674.

Huang, Y., Ali, S., Bi, X. et al. (2016) Research on smart campus based on the Internet of Things and virtual reality. *International Journal of Smart Home*, **10**(12), 213–220.

Jamil, S.S., Basalamah, A. et al. (2014) Demonstrating Map++: a crowd-sensing system for automatic map semantics identification. 2014 Eleventh Annual IEEE International Conference on Sensing, Communication, and Networking (SECON), pp. 152–154. IEEE.

Károly, F. and Lendák, I. (2015) Simulation environment for investigating crowd-sensing based urban parking. 2015 International Conference on Models and Technologies for Intelligent Transportation Systems (MT-ITS), pp. 320–327. IEEE.

Keles, C., Karabiber, A., Akcin, M., Kaygusuz, A., Alagoz, B.B., and Gul, O. (2015) A smart building power management concept: smart socket applications with DC distribution. *International Journal Of Electrical Power & Energy Systems*, **64**, 679–688. doi: 10.1016/j.ijepes.2014.07.075.

Kyriazis, D., Varvarigou, T., White, D., Rossi, A., and Cooper, J. (2013) Sustainable smart city IoT applications: heat and electricity management & eco-conscious cruise control for public transportation. 2013 IEEE 14th International Symposium and Workshops on a World of Wireless, Mobile and Multimedia Networks (WoWMoM), June 4–7. doi: 10.1109/WoWMoM.2013.6583500.

Lazaroiu, G., Dumbrava, V., Costoiu, M. et al. (2016) Energy-Informatic-Centric Smart Campus. IEEE 16th International Conference on Environment and Electrical Engineering (EEEIC), June 7–10, 2016, Florence, Italy.

Lee, J. (2016) "(Plenary talk): 5G trial in 2018 PyeongChang winter olympics—technical challenges & preparations. International Symposium on Antennas and Propagation (ISAP), October 24–28, 2016, IEEE.

Li, H., Shou, G., Hu, Y., and Guo, Z. (2016) WiCloud: innovative uses of network data on smart campus. 11th International Conference on Computer Science & Education (ICCSE), August 23–25, 2016.

Lin, Y.-B. and Chen, J.-H. (2016) Keynote speech 2: implementing smart campus through IoTtalk. International Conference on Communication Problem-Solving (ICCP), September 7–9, 2016.

Lizzi, L., Ferrero, F., Danchesi, C., and Boudaud, S. (2016) Design Of antennas enabling miniature and energy efficient wireless IoT devices for smart cities. Smart Cities Conference (ISC2), 2016 IEEE International, September 12–15. doi: 10.1109/ISC2.2016.7580825.

Machina Research White Paper, (2016) Open Standards in IoT Deployments Would Accelerate Growth by 27% and Reduce Deployment Costs by 30%. May Published via Interdigital, 200 Bellevue Parkway, Suite 300, Wilmington, DE 19809.

Martínez, D., Gracia, T., Munoz, E., and García, A. (2016) Smart cities' challenge: how to improve coordination in the supply chain. *Sustainable Smart Cities,* Innovation, Technology, and Knowledge Management, Springer, pp. 129–142 doi: 10.1007/978-3-319-40895-8_10 ISBN: 978-3-319-40894-1.

Mauser, I., Feder, J., Müller, J., and Schmeck, H. (2015) Evolutionary Optimization of Smart Buildings with Interdependent Devices. *Applications of Evolutionary Computation,* vol. 9028, Lecture Notes in Computer Science Springer, pp. 239–251 doi: 10.1007/978-3-319-16549-3_20.

Mikko, R., Törmä, S., and Kratinov, D. (2014) Mobile crowdsensing of parking space using geofencing and activity recognition. 10th ITS European Congress, Helsinki, Finland, pp. 16–19.

Minoli, D. (2008) *IP Multicast with Applications to IPTV and Mobile DVB-H,* John Wiley & Sons, Inc. New York.

Minoli, D. (2013) *Building the Internet of Things with IPv6 and MIPv6,* John Wiley & Sons, Inc. New York.

Minoli, D., Sohraby, K., and Occhiogrosso, B. (2017a) IoT considerations, requirements, and architectures for smart buildings—energy optimization and next generation building management systems. *IEEE Internet of Things Journal*, 1–15. doi: 10.1109/JIOT.2017.2647881.

Minoli, D., Kouns, J., and Sohraby, K. (2017b) IoT security (IoTSec) considerations, requirements, and architectures. 14th Annual IEEE Consumer Communications & Networking Conference (CCNC 2017), Las Vegas, January 8–11.

Moreno-Cano, V., Terroso-Saenz, F., and Skarmeta-Gomez, A.F. (2015) Big data for IoT services in smart cities. Proceedings of the IEEE 2nd World Forum on Internet of Things (WF-IoT), Milan, Italy, December 14–16, 2015, pp. 418ff.

Nordrum, A. (2017) Here comes 5G—whatever that is [Top Tech 2017]. *IEEE Spectrum*, **54**(1), 44–45. doi: 10.1109/MSPEC.2017.7802747.

Nuss, D., Stuebler, M., and Dietmayer, K. (2014) Consistent environmental modeling by use of occupancy grid maps, digital road maps, and multi-object tracking. Intelligent Vehicles Symposium Proceedings, 2014 IEEE, June 8–11. doi: 10.1109/IVS.2014.6856516.

Oti, A.H., Kurul, E., Cheung, F., and Tah, J.H.M. (2016) A framework for the utilization of building management system data in building information models for building design and operation. *Automation in Construction*, **72**(2), 195–210 doi: 10.1016/j.autcon.2016.08.043.

Pham, H.D. and Vafi, S. (2015) Motion-energy-based unequal error protection for H. 264/AVC video bitstreams. *Signal, Image and Video Processing*, **9**(8), 1759–1766. doi: 10.1007/s11760-014-0641-8.

Picone, M., Busanelli, S., Amoretti, M., and Zanichelli, F. (2015) *Advanced Technologies for Intelligent Transportation Systems*, Springer, ISBN: 978-3-319-10667-0.

3GPP TS 29.275, Proxy Mobile IPv6 (PMIPv6) based Mobility and Tunnelling protocols; Stage 3, 3GPP TS 29.275 version 13.4.0 Release 13); also published as ETSI TS 129 275 V13.4.0 (2016-03). Available at http://www.tech-invite.com/3m29/tinv-3gpp-29-275.html.

3GPP TS 23.207, End-to-End Quality of Service (QoS) Concept and Architecture, Release 14,. Available at http://www.arib.or.jp/IMT-2000/V900Jul11/5_Appendix/Rel5/23/23207-5a0.pdf (accessed February 22, 2018).

Qureshi, K.N. and Abdullah, A.H. (2013) A survey on intelligent transportation systems. *Middle-East Journal of Scientific Research*, **15**(5), 629–642. doi: 10.5829/idosi.mejsr.2013.15.5.11215.

Ramaswami1, A., Russell, A.G., Culligan, P.J., Sharma1, K.R., and Kumar, E. (2016) Meta-principles for developing smart, sustainable, and healthy cities. *Science (AAAS)*, **352**(6288), 940–943. doi: 10.1126/science.aaf7160.

Rha, J.-Y., Lee, J.-M., Li, H.-Y. et al. (2016) From a literature review to a conceptual framework, issues and challenges for smart campus. *Journal of Digital Convergence*, **14**(4), 19–31. doi: 10.14400/JDC.2016.14.4.19.

Rojals, J.S. et al. (2016) Scan-Based Sliding Window IN Context Derivation For Transform Coefficient Coding, US 9363510 B2 June 7.

Romero, C.D.G., Barriga, J.K.D., and Molano, J.I.R. (2016) Big data meaning in the architecture of IoT for smart cities. *Data Mining and Big Data*, First International Conference, DMBD 2016, Bali, Indonesia, June 25-30. Proceedings, Lecture Notes in Computer Science, vol. 9714, pp. 457–465. doi: 10.1007/978-3-319-40973-3_46.

RWS,(2015) RWS-150002 Views on Next Generation Wireless Access 3GPP RAN Workshop on 5G, September, Lenovo, Motorola Mobility.

Schwarz, H., Marpe, D., and Wiegand, T. (2007) Overview of the scalable video coding extension of the H. 264/AVC standard. *IEEE Transactions on Circuits and Systems for Video Technology*, **17**(9), 1103–1120. doi: 10.1109/TCSVT.2007.905532.

Shen, T., Lu, Y., Wen, Z., Zou, L., and Chen, Y. (2013) Ultra fast H. 264/AVC to HEVC transcoder. Data Compression Conference (DCC), March 20–22, IEEE doi: 10.1109/DCC.2013.32.

Soriano, F.R., Samper, J.J., Martinez, J.J., Cirilo, R.V., and Carrillo, E. (2016) Smart cities technologies applied to sustainable transport. open data management. Telematics and Information Systems (EATIS), 2016 8th Euro American Conference on, April 28-29, doi: 10.1109/EATIS.2016.7520155.

Srivastava, S. and Pal, N. (2016) Smart cities: the support for internet of things (IoT). *International Journal of Computer Applications in Engineering Sciences*, **VI**(II), 5–7.

Staff, (2016) Population Statistics, Population Reference Bureau, Washington, DC 20009. Available at www.prb.org.

Staff, (2015) National League of Cities, Center for City Solutions and Applied Research, "City of the Future – Technology and Mobility", White Paper Washington, D.C. 20004. Available at www.nlc.org.

Staff (2016) The World's Top 10 Smart City Projects, Informa Telecoms & Media, Informa UK Limited, December London, SW1P 1WG.

Su, C.L., Chen, T.M., and Huang, C.Y. (2014) Cluster-based motion estimation algorithm with low memory and bandwidth requirements for H.264/AVC scalable extension. *IEEE Transactions on Circuits and Systems for Video Technology*, **24**(6), 1016–1024. doi: 10.1109/TCSVT.2013.2290385.

Su, P.C., Tsai, T.F., and Chien, Y.C. (2016) Partial frame content scrambling in H. 264/AVC by information hiding. *Multimedia Tools and Applications*, doi: 10.1007/s11042-016-3406-2.

Sun, Y., Wu, T.-Y., Zhao, G., and Guizani, M. (2014) Efficient rule engine for smart building systems. *IEEE Transactions on Computers*, **64**(6), doi: 10.1109/TC.2014.2345385.

The National Academies of Science, (2014) Engineering and Medicine, USDOT's Intelligent Transportation Systems (ITS): ITS Strategic Plan 2015–2019, J. Barbaresso, G. Cordahi, D. Garcia, C. Hill, A. Jendzejec (eds), Report/Paper Numbers: FHWA-JPO-14-145, Created Date: December 23.

Tragos, E.Z., Foti, M., Surligas, M., Lambropoulos, G. et al. (2015) An IoT based intelligent building management system for ambient assisted living. Communication Workshop (ICCW), 2015 IEEE International Conference on, June 8–12. DOI: 10.1109/ICCW.2015.7247186.

Triggs, R. (2016) 5G networks—state of the industry, August 18. Available at http://www.androidauthority.com/what-is-5g-705914/.

U.S. Energy Information Administration, (2012) Commercial Buildings Energy Consumption Survey (CBECS), Energy Usage Summary Edition. Available at (http://www.eia.gov/consumption/commercial/reports/2012/preliminary/).

3GPP TS 123 402 v13.5.0 (2016-04), Universal Mobile Telecommunications System (UMTS); LTE; Architecture Enhancements for non-3GPP Accesses (3GPP TS 23.402 version 13.5.0 Release 13). Available at http://www.etsi.org/deliver/etsi_ts/123400_123499/123402/13.05.00_60/ts_123402v130500p.pdf.

United Nations, (2014) World Urbanization Prospects, Department of Economic and Social Affairs, Document ST/ESA/SER.A/352, published by the United Nations, ISBN 978-92-1-151517-6, Available at https://esa.un.org/unpd/wup/.

van der Mei, R.D., Ottenhof, F., and Bots, M. (213), Digital road authority for coordination between in-car navigation systems and traffic control centres. ERCIM News, vol. 94, ISSN 0926-4981.

Various Speakers, (2016) Part I: Smart Campus And IoT, Collaborative Innovation Community Meeting—Internet2, 2016 Technology Exchange, September 25–28, Miami, FL.

Vladimir, C. and Gruteser, M. (2013) Crowdsensing maps of on-street parking spaces. 2013 IEEE International Conference on Distributed Computing in Sensor Systems (DCOSS) pp. 115–122. IEEE.

Vogel, A. and Mueller, G. (2015) Method For Creating A Directory Of Road Sections, Method For Ascertaining All Road Sections Within A Search Area, And Computer Program, US 9214097 B2, December 15.

Wang, Q., Wang, W., Shi, J., Zhu, H., and Zhang, N. (2016a) Smart media pricing (SMP): non-uniform packet pricing game for wireless multimedia communications. Proc. IEEE INFOCOM, 5th Workshop on Smart Data Pricing April.

Wang, W., Wang, Q., and Sohraby, K. (2016b) Multimedia Sensing As A Service (MSaaS): exploring resource saving potentials of at cloud-edge IoTs and fogs. *IEEE Internet of Things Journal.* **4**(2), 487–495.

Wewalaarachchi, B.J., Shivanan, H., and Gunasingham, H. (2016) Integration Platform To Enable Operational Intelligence And User Journeys For Smart Cities And The Internet Of Things, Patent US20160239767 A1, August 18.

Wiegand, T., Sullivan, G.J. et al. (2003) Overview of the H. 264/AVC video coding standard. *IEEE Transactions on Circuits and Systems for Video Technology,* **13**(7), 560–576. doi: 10.1109/TCSVT.2003.815165.

Xiping, H., Liu, Q. et al. (2013) A mobile crowdsensing system enhanced by cloud-based social networking services. Proceedings of the First International Workshop on Middleware for Cloud-enabled Sensing, p. 3. ACM.

Xu, C., Li, S., Zhang, Y., and Miluzzo, E. (2014) Crowdsensing the speaker count in the wild: implications and applications. *IEEE Communications Magazine*, **52**(10), 92–99.

Yu, Z., Liang, Y., Xu, B., Yang, Y., and Guo, B. (2011) Towards a smart campus with mobile social networking. Internet of Things (iThings/CPSCom), 2011 International Conference on and 4th International Conference on Cyber, Physical and Social Computing, October 19-22, Dalian, China.

Zanella, A., Bui, N., Castellani, A., Vangelista, L., and Zorzi, M. (2014) Internet of Things for smart cities. *IEEE Internet of Things Journal*, **1**(1), 22–32. doi: 10.1109/JIOT.2014.2306328.

Zhang, L., Oksuz, O., Nazaryan, L. et al. (2016) Encrypting wireless network traces to protect user privacy: a case study for smart campus. IEEE 12th International Conference on Wireless and Mobile Computing, Networking and Communications (WiMob), October 17–19, 2016.

ZTE Corporation (2015) Considerations on 5G Key Technologies and Standardization, ZTE Corporation, September 2015, 3GPP RAN 5G Workshop—The Start of Something, Phoenix, AZ, USA, September 19.

13

Smart Connected Homes

Joseph Bugeja, Andreas Jacobsson, and Paul Davidsson

Internet of Things and People Research Center and Department of Computer Science and Media Technology, Malmö University, Malmö, Sweden

13.1 Introduction

Nowadays, connectivity and smartphone ownership is pervasive in large parts of the world. This alongside technological innovations, availability, and affordability of smart devices has given momentum to the emerging paradigm of the Internet of Things (IoT). A smart connected home is an instance of this paradigm, inheriting all the aspects of connectivity of the involved devices.

A smart connected home is a residence equipped with sensors, systems, and devices that can be remotely accessed, controlled, and monitored, typically via the Internet (Balta-Ozkan et al., 2013). According to a study conducted by Statista,[1] the global smart home market in 2015 was valued at close to $9.8 billion and it is expected to reach about $43 billion in 2020. Similarly, a survey commissioned by August Home and Xfinity Home[2] reported that the smart home market is anticipated to double in the United States by 2017. Research from the company Icontrol Networks[3] indicate that personal and family security are the key drivers for householders purchasing a smart home system in the United States and Canada. The report also identifies altruistic aspects connected to smart home technologies. For instance, around 74% of the surveyed parents say they would sleep better at night if their parents or grandparents had a smart home so that they can keep an eye on them. Similarly, the report indicates that nearly half of the consumers interested in saving energy consumption costs were

1 https://www.statista.com/outlook/279/100/smart-home/worldwide#market-revenue (accessed September 4, 2017).

2 http://corporate.comcast.com/images/August_Xfinity-Safe-and-Secure-Study-Report.pdf (accessed September 4, 2017).

3 https://mysmahome.com/news/4496/icontrol-2015-report-reveals-security-remains-key-driver-for-smart-home-adoption-2/ (accessed September 4, 2017).

Internet of Things A to Z: Technologies and Applications, First Edition. Edited by Qusay F. Hassan.
© 2018 by The Institute of Electrical and Electronics Engineers, Inc. Published 2018 by John Wiley & Sons, Inc.

excited in using smart home technologies to conserve energy and thus help the environment.

Recently, commercial organizations have intensified their smart home activities to enter the market through hardware, software, and services. For instance, in 2015 Apple introduced HomeKit,[4] Amazon introduced Echo,[5] and Google introduced Brillo.[6] A number of survey studies have been published, for example, Alam et al. (2012), Chan et al. (2008), and De Silva et al. (2012). The mentioned studies vary widely in their approach and scope, but nonetheless, they were published preceding recent developments in the smart home industry. Latest progress includes, in particular, integration platforms, new stakeholders, increased usage of cloud services, voice-controlled devices, and widespread availability of sensor technologies in modern everyday appliances. Understanding these components is key to comprehending the modern smart connected home.

This chapter provides a contemporary overview of smart connected homes. Specifically, it looks at industrial and academic projects introduced over the past few years, identifying their underlying technologies, architectures, and offered services. The smart connected home brings various benefits to the householders and society, but it also introduces different technical and social challenges to smart home developers and researchers alike.

13.2 The Smart Connected Home Domain

Different from a traditional home that features appliances that are operated locally and manually (e.g., magnetic switches and push buttons), the smart connected home incorporates Internet-connected appliances and devices that can interact "intelligently" with the householders. Several efforts to promote smart connected home functionality have been produced over the last decades conveying different ideas, application areas, and utilities.

In the subsequent sections, a detailed account of the concept of the smart connected home is provided and the main stakeholders are identified.

13.2.1 The Concept of the Smart Connected Home

There is no commonly accepted definition of the smart connected home, but in general, a smart home may denote any kind of residence (e.g., apartment,

---

4 http://www.apple.com/ios/homekit/ (accessed September 4, 2017).

5 http://www.amazon.com/Amazon-SK705DI-Echo/dp/B00X4WHP5E (accessed September 4, 2017).

6 https://developers.google.com/brillo/ (accessed September 4, 2017).

cottage, and rented living space), which involves information and communication technologies allowing for remote control, monitoring, and access (Balta-Ozkan et al., 2013). Meanwhile, some researchers further add that a smart home needs to have ambient intelligence and automatic control allowing it to recognize and possibly make decisions on its own guided by the behavior of the residents (De Silva et al., 2012; Pedrasa et al., 2010; Zhang et al., 2013).

Consequently, it can be observed that there are three types of homes: smart home, connected home, and a smart connected home. The relation between these types is illustrated in Figure 13.1.

Smart home includes systems that allow the residents to operate home appliances, typically only locally from within the house. This type of smart home tends to rely on wireline-based standards such as KNX and it is not connected to the Internet. Commonly, it tends to focus on the automation of lighting, windows, and in-house entertainment, and is associated with building automation.

Connected home allows for remote control and management of appliances, typically over IP-based networks (e.g., the Internet). Additionally, this type of home usually provides services that promote and support, for example, security, health care, and energy management. Moreover, this type of house generally includes a central hub (gateway) from which the system can be controlled together with a user interface that can be operated typically through a smartphone.

Smart connected home includes functionality from both previous types, but also adds communication and service-exchange between related areas, such as the grid and electric vehicle and on-site microgeneration (e.g., rooftop solar panels) (Balta-Ozkan et al., 2013). Smart connected homes thus merge the functionality from both the connected home and the smart home. It may also include system capabilities, such as learning, prediction of, and response to the occupants' needs and lifestyle preferences in their home environment. Typically,

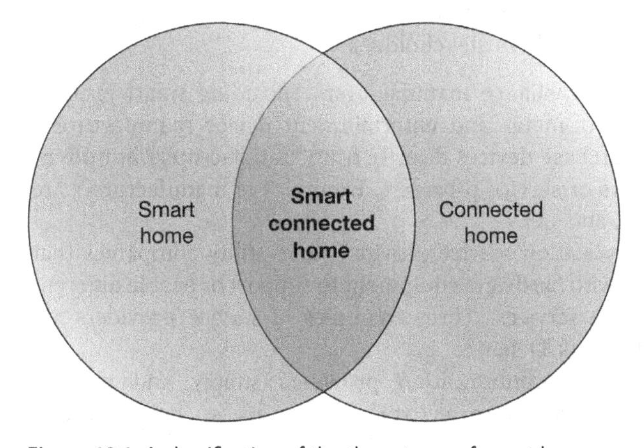

Figure 13.1 A classification of the three types of smart homes.

this type of home incorporates connected devices that exhibit some form of "intelligent" logic commonly based on machine learning algorithms for activity recognition and to perform some action automatically. In implementing this, cloud services are often used. In this case, cloud-based applications analyze the data collected and processed from the smart connected home and they are often able to take actions, sometimes autonomously. For example, a smart leak detection system may notify the home owner that there is a leak in the water heater system, and as well automatically turn off the water, gas, or electricity supply to prevent valuable resources and save money. Additionally, homes in this category may also feature advanced forms of interaction typically supporting voice detection, facial recognition, and gesture controls.

Conforming to the current home automation trends, the emphasis of this work is on the smart connected home. So, hereinafter, the term smart connected home is used to denote a residence incorporating a range of sensors, systems, and devices that can be remotely accessed, controlled, and monitored via a communication network such as the Internet. In the rest of this text, this is also associated with the concept of the do-it-yourself (DIY) home, where end users can build their own smart connected home through computing units (e.g., Arduino, Banana Pi, and Raspberry Pi) without relying on professionals.

13.2.2 Smart Connected Home Stakeholders

In IoT applications, such as smart homes or smart buildings, there are diverse stakeholders ranging from technology investors, technology developers, technology integrators, and more. Additionally, specific stakeholders may be involved depending on the actual smart home system. For instance, it is common to have a specific stakeholder such as a meter point operator in an energy-focused smart connected home, and a content provider in an entertainment-focused smart connected home.

In general, there are six main stakeholders that can be identified as described below. Figure 13.2 illustrates these stakeholders.

- *Device Manufacturers.* Appliance manufacturers, including smart product suppliers such as smart meter and entertainment device manufacturers. Householders may purchase devices directly from manufacturers but oftentimes through retailers or service providers. Examples of manufacturers are Samsung, Honeywell, and LG.
- *Service Providers.* Application service providers and utility companies that provide the end users with hardware equipment to support or enable different smart connected home services. Three examples of service providers are Verisure, AT&T, and Leak Defense.
- *Network Providers.* Telecommunication providers supply and manage network infrastructure like core network, radio access network, and

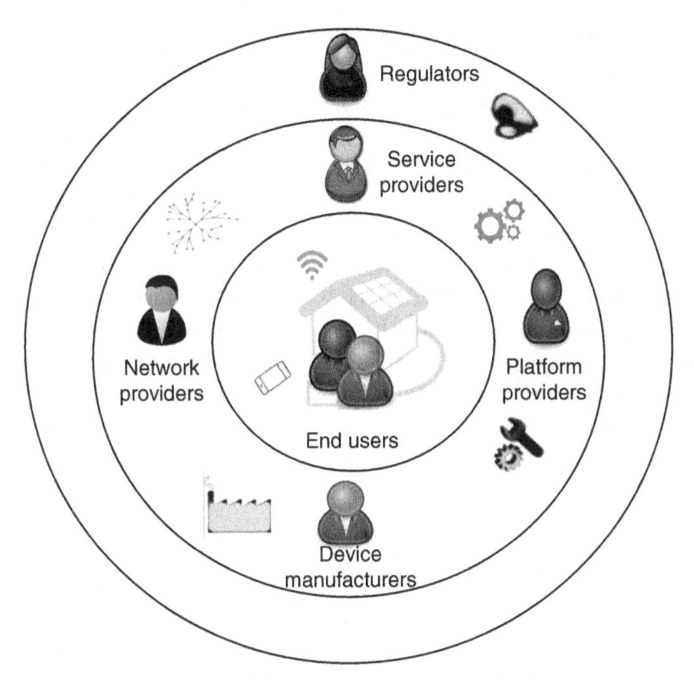

Figure 13.2 The main smart connected home environment stakeholders.

interconnectivity network to service providers that want to offer smart home services. Effectively they are the stakeholders that connect the householders to the Internet. An example of a network provider is, for instance, Verizon.

- *Regulators.* Regulators are external entities overseeing business services or specific industry sectors. This can include certification and accreditation bodies related to quality, security, and safety. An example of this could be a privacy regulator that develops laws to safeguard personally identifiable information from undue exploitation. Possibly, this entity may affect the entire spectrum of industry stakeholders.
- *Platform Providers.* Entities supplying mechanisms, tools, and frameworks assisting in overcoming integration challenges and facilitating easier assess, customization, and automation support. Commonly, different device manufacturers work with different platform providers for third-party integration. Some organizations serving this role are Apple, Google, and Amazon.
- *End Users.* The end user is the stakeholder that uses services. Typically, this represents the home residents that purchase and operate the different smart connected home devices and services.

13.3 Smart Connected Home Systems

The smart connected home can integrate different smart home services to provide a convenient, comfortable, and safe environment to the household members, as well as to help them perform their household tasks effectively. Broadly, smart connected home systems can be organized into four categories as systems that support energy, entertainment, health care, or security services (Badica et al., 2013). There are also systems that do not fit in any of these categories, and there are systems that overlap. However, these categories often help in clarifying the purpose and functionality of any given system.

In the succeeding sections, the smart connected home systems are grouped according to the aforementioned service categories. Recent systems from both academia and industry are described.

13.3.1 Energy

Energy systems are targeted to provide efficient energy consumption and management for the home. The energy domain commonly involves the use of smart meters, smart thermostats, and adaptive lighting systems. System architectures in this domain may utilize "intelligent" multiagent systems and control strategies to predict and automatically maximize energy efficiency and user comfort (Reinisch et al., 2011). Other projects put the onus of monitoring energy consumption to the householders. This can be assisted, for instance, by smart energy monitors that display in real time the energy consumption patterns, for example, the heating, hot water usage, and CO_2 emissions (Hargreaves et al., 2013; Qin et al., 2014). Two examples of commercial systems include Google Nest[7] and Ecobee[8] smart thermostats. Both of these systems allow for preventing a home from overheating or overcooling. They also include functionality to automatically adjust the house temperature to fit the occupants' preferences, and also allow for remote adjustment of the temperature.

The energy management domain appears to be an area where there is numerous research projects and funds especially related to reducing energy usage and optimizing its utilization in houses and commercial buildings. The driving forcing behind this tends to be associated with the rising energy costs, and reducing greenhouse gas emissions (arguably, a primary cause of global climate change). However, energy management is also important because it also contributes for peace of mind. For instance, by automatically turning on

7 https://nest.com/support/article/Nest-Learning-Thermostat-technical-specifications (accessed September 4, 2017).

8 https://www.ecobee.com/wp-content/uploads/2014/09/ecobee3_techspecs.pdf (accessed September 4, 2017).

lights to help discourage potential intruders while the residents are away from home.

13.3.2 Entertainment

Smart connected home systems tend to promote entertainment typically maximizing occupants' comfort and convenience by providing personalized amusement content and social communication services. The entertainment sector commonly involves game consoles, connected TVs, and smart speaker systems. Microsoft's Easy Living Project (Brumitt et al., 2000) is an architecture that uses cameras to tailor services according to the occupants' location within the house. This system is often mentioned as a case study in well-cited surveys (Balta-Ozkan et al., 2013; De Silva et al., 2012). More recent systems adopt smart home technologies allowing the residents to get an idea of outside weather conditions without having to leave the house. This can be done, for instance, through the use of smart cameras and temperature sensors (Wang et al., 2015). In addition to that, some systems also capture and try to alter the individuals' social emotions. One way of implementing this is by utilizing smart furniture such as tables augmented with cameras and motion sensors (Yu et al., 2012). These sensors analyze the viewer's facial expressions (e.g., eyes and mouth) and then play the user's favorite music and supporting sentences to alleviate his/her mood (Yu et al., 2012). Two recent industry examples are Samsung Smart TV[9] and Apple TV.[10] Both of these systems are connected to the Internet, allowing end users to stream content from services like Netflix, and access more services, for example, a web browser, through built-in apps.

It can be observed that the entertainment domain is emerging as one of the top reasons why people purchase a smart home system. Mostly, this is for the ability to remotely control or monitor TVs and sound systems.[11] Additionally, it can be noted that the technologies in this domain tend to rely on microphones and cameras and may support different forms of advanced interaction, most likely in the form of speech recognition.

13.3.3 Health Care

The health care service area is focused on providing mobile health care and fitness support, and aims to provide independent healthy living. In comparison to the other domains, the health care service also involves the use of wearable

9 http://www.samsung.com/us/system/consumer/product/un/24/h4/un24h4000afxza/H4000_SpecSheet_7_29_14_1.pdf (accessed September 4, 2017).

10 http://www.apple.com/tv/specs/ (accessed September 4, 2017).

11 https://mysmahome.com/news/4496/icontrol-2015-report-reveals-security-remains-key-driver-for-smart-home-adoption-2/ (accessed September 4, 2017).

sensors (e.g., wrist straps) in order to allow for possibly continuous monitoring of body signal parameters (e.g., cardiac diseases) even while not at home. Health care services can monitor the residents' personal health, generate tailored health reports, and may support remote diagnoses and chronic disease management. Generally, this includes connected devices such as wireless scales, fitness monitors, and physiological devices. Recent projects connected to the health care domain tend to utilize audio technologies, for example, speech recognition, to facilitate and ease daily living for elderly and frail persons (Amiribesheli et al., 2015; Ni et al., 2015; Portet et al., 2013). It is also common to have architectures that support continuous elderly monitoring, for example, by checking the health indicators of persons such as glucose levels. This can be implemented for instance by ontologies, web service technologies, and specialized sensors (Rivero-Espinosa et al., 2013). Notable examples from the industry include the in-home patient monitoring systems Vignet platform[12] and Philips Tele-Station.[13] These systems support connecting wireless measurement devices, such as physiological monitors, allowing individuals to be active participants in their health care.

The health care domain has been extensively researched by many authors and the amount of publications on connected health care is continuously increasing. Monitoring a person's cognitive and physical health is a precursor to ensuring a healthy society. On the flip side, it can be noted that this application involves the most privacy invasive sensors with technologies that can monitor the patient's intimate physiological conditions, such as blood glucose levels.

13.3.4 Security

Systems in the security category are often targeted to offer services, which are designed to monitor, detect, and control security and safety threats. Smart home security and safety systems typically range from remote entrance monitoring services to systems that automatically recognize physical threats, such as a fire or a burglary, and autonomously take the corresponding action. Such actions can include for instance starting the immediate fire sprinklers or automatically switching on or off appliances, such as digital locks or video cameras. Typically, this domain includes functionality that support smart door locks, cameras, and alarm systems. In the market, there are many instances of security systems, and two such examples are Verisure Securitas Direct System[14] and Alarm.com.[15] Both systems allow for remote monitoring and managing security systems in the

12 http://www.vignetcorp.com/node/26 (accessed September 4, 2017).

13 https://www.usa.philips.com/healthcare/product/HCNOCTN483/ecarecompanion-patient-app-your-patients-gateway-to-care (accessed September 4, 2017).

14 https://www.verisure.com/our-offer.html (accessed September 4, 2017).

15 https://www.alarm.com/get_started/select_features.aspx (accessed September 4, 2017).

home, and include 24/7 professional monitoring and emergency response solutions.

As a central observation, it can be noted that the security domain is one of the main motivations for consumers to purchase smart home systems,[16] despite the various privacy concerns related to in-house video surveillance (Brezovan and Badica, 2013). Smart connected homes security systems are especially popular among residents living in towns where there are reports of high criminal activities. However, their widespread adoption is also linked to the convenience factor being offered by such systems. Furthermore, it can be noted that this domain is tightly linked to surveillance systems, involving technologies such as cameras and motion sensors that are predominantly used in intelligent buildings.

13.4 The Smart Connected Home Technologies

A smart connected home system comprises a multitude of Internet-connected devices that serve different application areas. These devices are typically characterized by heterogeneous hardware and software, and support different communication technologies. Typically, gateways act as the core devices supplying connectivity to service providers and other external parties. Depending on the actual architecture, the use of cloud services may also be used as an enabling technology. Figure 13.3 provides a depiction of the generic smart connected home architecture.

The following sections review the technologies used in the home environment. Most of the analyzed technologies are extracted from the smart home project descriptions identified in the previous section.

13.4.1 Sensors and Actuators

Sensors are nodes that are used to sense, measure, and detect changes in the environment including the occupants. In smart connected homes, sensors that record the temperature of an object, for example, room, or body, are used across all the previously introduced systems. Similarly, some sensor types, for example, CO/CO_2, are used in multiple domains for different purposes. For instance, these may be used for tracking CO_2 emissions in the energy domain and to indicate possible CO leakages in the security domain. Some domains, in particular the health care domain, employ specialized sensor types, for example, physiological sensors, that are exclusively used in that service area. While physiological sensors tend to be mobile and are generally worn by householders, other sensor types are often fixed, for example, wall-mounted camera, or

16 https://mysmahome.com/news/4496/icontrol-2015-report-reveals-security-remains-key-driver-for-smart-home-adoption-2/ (accessed September 4, 2017).

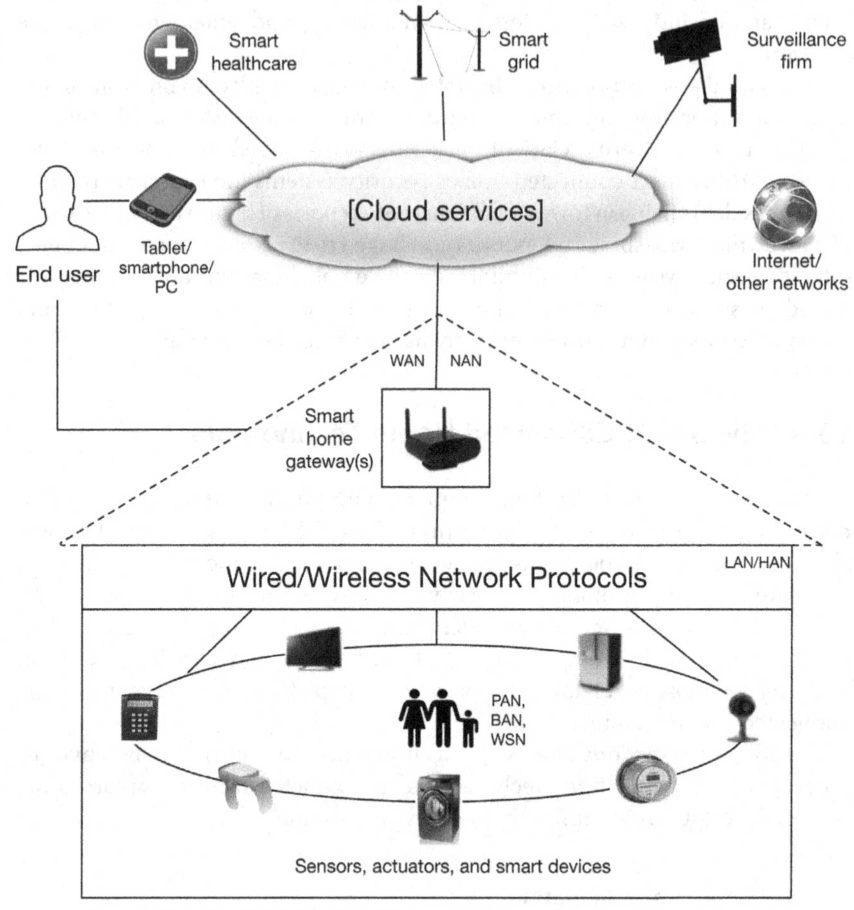

Figure 13.3 A generic architecture of the smart connected home.

attached to an object, for example, door contact sensor. In summary, Table 13.1 presents the different sensor types used by the different smart home systems introduced in Section 13.3. The sensor types are classified according to the type of data or parameters they measure (Amiribesheli et al., 2015).

Actuators provide the means to implement physical actions, such as turning on/off lights, raising alarms, and activating heating appliances. In smart connected homes, typical examples of these include motor controllers, switches, keys/locks, speakers, and displays. Actuators may influence sensors and can consequently cause the system or a user to activate the actuator. It can be argued that the more sophisticated and autonomous this type of interaction is, the more the connected home system is seen as smart.

Table 13.1 Categorization of the different sensor types and their corresponding data captured by the smart connected homes described in Section 13.3.

	Data captured/service area			
Sensor type	**Energy**	**Entertainment**	**Health care**	**Security**
Environmental	Ambient light, CO_2 emissions, Humidity, Luminosity, Rain, Temperature	Light intensity, Temperature	Smoke and gas leakage, Temperature	CO emission, Humidity, Water leakage, Smoke, Temperature
Physiological			Blood pressure and pulse, Blood oxygen, Blood glucose, Heart rate/rhythm, Peak flow, Temperature, Weight	
Multimedia		Audio, Video	Audio	Video
State	Motion, Near-field/far-field activity, Presence, Proximity	Accelerometer, Gyroscope, Motion	Presence	Motion, Presence

In practice, it is common to assemble different sensors and actuators in a single unit. For example, Canary,[17] a home security solution, has a built-in HD camera, microphone, temperature, humidity, and air quality sensors, as well as a siren.

13.4.2 Gateways

Smart connected home gateways are devices that compile, convert, and transmit information collected through sensors, sensor hubs, and commands from devices, of which some may be smart. The gateway acts as the component

17 https://canary.is/ (accessed September 4, 2017).

responsible for interfacing with the outside world allowing for routing internal network data in and out to the Internet as well as providing services to the residents. It can range from a dedicated device, smartphone, to a computing unit such as a Raspberry Pi.

This gateway tends to be connected to the home's broadband router. This is then typically connected to the Internet via a wired connection (e.g., Ethernet connection to a cable or DSL modem). Nonetheless, it is possible to have a gateway embedded with a wireless/cellular radio. In this case, the device may communicate directly to the Internet or a cellular carrier.

In setting the requirements for home gateways, the Home Gateway Initiative (HGI)[18] standardization organization is actively involved. This organization is made up of multiple telecommunication companies and device manufacturers, and it is actively involved in improving the interpretability of the gateway with smart home devices. HGI has published various guidelines, standards, and requirement documents to promote an open and modular deployment for home gateways to provide compatibility among various devices. Modern gateways would typically offer the following functions:

- *Notifications.* Most gateways implement some form of notification system supporting the routing of messages and alerts from connected devices to the end user. For instance, in the case of an emergency, when a washing machine is flooding, the house owner might get notified of this occurrence. Typically, such alerts are in the form of SMS text messages, emails, or prompts/pop-ups displayed through a mobile application installed on an end user device such as smartphone.
- *Automation.* Some gateways support an automation or scheduling system that allows for the creation of different rules for the connected devices. For instance, some systems allow for the creation of rules based on time, for example, sunrise and sunset rules, and others allow for more granular rules based on events and conditions. An example of this could be a rule that automatically turns on the lights when movement is detected in the home.
- *Local Control.* Gateways may also feature local control. This allows for manual control and execution of rules in case the Internet or cloud services are unavailable. This interface is implemented usually in the form of a local web browser, mobile, or desktop application. However, some devices also offer a built-in display panel.
- *Cloud Service.* Commonly, major gateways utilize a cloud service(s). This allows for remote device management, however, oftentimes the cloud is also used for storing home data, serving the user interface, and communicating with external systems such as weather, push messages, and SMS gateways.

---

18 http://www.homegatewayinitiative.org/ (accessed September 4, 2017).

- *Third-Party Integration.* With an increasing number of smart home devices and an ensuing variety of standards, there is a need to allow for connections between dissimilar products. Modern gateways facilitate this by offering integration with third-party services, such as, Amazon Echo, Apple HomeKit, and "If This Then That" (IFTTT[19]), that allow so. Amazon Echo is a cloud-based, voice-activated platform that makes it possible to control connected devices. Apple HomeKit is a platform designed to make different devices communicate with each other possibly with the help of compatible bridges or hubs. IFTTT is a web-based service allowing users to pull together multiple services and to program devices, through a construct known as "recipes," to run routines, react to triggers, or pass commands to other devices in the house. An example of an IFTTT recipe could be one that automatically powers on a smart TV after a certain amount of physical activity, for example, walking 10,000 steps, has been performed by the residents. An alternative application to IFTTT that is arguably more flexible in terms of automation support is Stringify.[20] This works similarly to IFTTT but instead of using "recipes" it uses a concept known as "flows." Through flows multiple applications and services can be chained together. As an example, a flow could be designed to improve the home comfort by automatically adjusting the house lighting and room temperature through voice commands. The actual implementation of this using Stringify may consist of three "things":[21] Amazon Alexa, Nest thermostat, and Philips Hue Bulb. Amazon Alexa voice service (supported for instance by Amazon Echo device) can be programmed to wait for the householders to trigger it, for example, by saying "Alexa, tell Stringify Night Mode." This trigger can then consequently adjust Hue to amber at 15% and Nest thermostat to set the room temperature to 20 °C. This setup is shown in Figure 13.4. Implementing this with IFTTT, especially if the technologies are supplied by different manufacturers, would commonly require multiple recipes.
- *API/SDK.* Typically, most hubs feature an Application Programming Interface (API) and/or Software Development Kit (SDK). These allow for interfacing and building customized functionality. For instance, an Alexa smart home ecosystem contains a Smart Home Skill API[22] that enables the creation of capabilities (e.g., ability to play music, answer questions, provide weather forecasts) to control cloud-connected devices. Another example could be a Smart TV App Store that has a SDK allowing for screen control, app control, and download and upload via network.

---

19 https://ifttt.com/ (accessed September 4, 2017).

20 https://www.stringify.com/ (accessed September 4, 2017).

21 The term "things" is used in Stringify to represent apps and services that can be linked together in a "flow."

22 https://developer.amazon.com/alexa-skills-kit (accessed September 4, 2017).

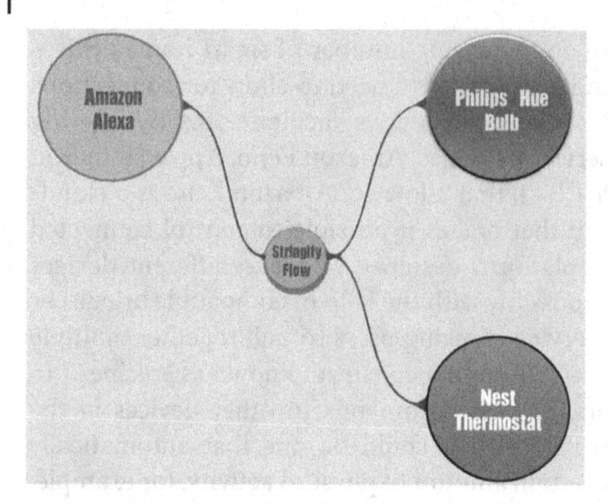

Figure 13.4 Integration between different smart home devices and a third-party service.

13.4.3 End User Client Devices

Typically, end users access, control, and monitor the smart connected home functionality through a user interface on an end user device. Taking advantage of mobile devices, the residents can check the information of the smart home system, schedule different tasks, and send immediate commands to be performed.

For instance, end users might issue commands to open their home door or to monitor their children remotely through an application available on their smartphones, tablets, desktop computers, or specialized devices such as smart remote control systems. Furthermore, these devices may recognize speech input, hand gestures, and perform user tracking.

Commonly, a specific application is required to manage and control different smart home devices issued by different manufacturers. For instance, to control smart locks, lights, and thermostat, one would commonly require separate applications for each gadget. However, the emergence of standards in the future may allow some integration between such systems/devices, and thus, use of less number of applications.

13.4.4 Cloud Services

Smart connected home systems may involve the use of cloud services that collect and analyze data readings from home devices. Such systems tend to be implemented over private, public, or hybrid backend infrastructures (clouds) and may include service application or delivery platforms operated and managed by service providers and device manufacturers (Hassan, 2011).

Commonly, the platforms or cloud service delivery models adopt the Platform as a Service (PaaS) or the Software as a Service (SaaS) cloud resource provisioning solution. The PaaS refers to the platforms that provide cloud computing services exposed in the form of APIs, programming languages, middleware, and frameworks, allowing for data storage functionality, device management, access control, and more. SaaS solutions, for example, IFTTT, tend to be centered on data mash-up using cloud computing facilities.

Chapter 4 provides more information about the different cloud service delivery models and the utilization of cloud computing in IoT.

13.4.5 Integration Platforms

Integration platforms or frameworks are technologies that allow for better interoperability and support across the diverse smart connected home environment. Examples of current platforms that are developed and supported by companies are Apple HomeKit, Google Brillo/Weave, Allseen AllJoyn,[23] Samsung SmartThings,[24] and Amazon Alexa. These smart home frameworks also provide a programming framework for developers to build apps that realize smart home benefits. Moreover, there are also open source home automation platforms built on open protocols and under open source license. Three such systems are openHAB,[25] OpenMotics,[26] and Domoticz.[27] Commonly, these connected home platforms are popular within the DIY communities. DIY users install and configure the smart home devices themselves instead of relying on professionals.

13.4.6 Communication Protocols and Models

The Internet protocol (IP) is the basis of the IoT, and hence, it is widely used in smart connected homes. Due to its easy interoperability, pervasive nature, widespread adoption, and recent research of lightweight interfaces, the IP is considered vital to the success of the smart connected home (Kailas et al., 2012). However, non-IP-based devices can also be part of the home.

Smart connected home systems often use many different types of communication protocols. These range from wired protocols (e.g., X10, HomePlug, and LonWorks) to wireless communication protocols (e.g., Wi-Fi, Z-Wave, and ZigBee), open standards (e.g., ZigBee) to proprietary (e.g., Z-Wave), and short-range protocols (e.g., NFC, RFID, and 6LowPAN) to long-range protocols (e.g., NWave, Sigfox, and LoRaWAN).

It is also important to distinguish between low-level and high-level protocols. Low-level protocols (e.g., ZigBee) are principally used for device-to-device

23 https://allseenalliance.org/opportunities/consumers (accessed September 4, 2017).
24 https://www.smartthings.com/ (accessed September 4, 2017).
25 http://www.openhab.org/ (accessed September 4, 2017).
26 https://www.openmotics.com/ (accessed September 4, 2017).
27 https://domoticz.com/ (accessed September 4, 2017).

networking within the house and in the case of long-range protocols (e.g., 4G) to communicate with service providers. High-level protocols are used to transmit sensor data or receive remote commands through the home gateway. Examples of these are MQTT, CoAP, and XMPP. Additionally, there are high-level domain-specific protocols such as the Smart Energy Profile 2.0[28] that connect the home energy devices to the smart grid.

Smart connected home devices can communicate with each other through heterogeneous communication protocols and by using different models. Typically, there are three communication models for IoT environments, such as, the smart connected home: device-to-device, device-to-cloud, and device-to-gateway (Rose et al., 2015). In the device-to-device model, devices exchange direct messages with each other through low-level protocols. An example of this could be a smart lock sending a signal to turn on the smart light when a door is opened. In the device-to-cloud model, devices are instead connected to a service provider to exchange and control messages. This model is used for instance by Nest smart thermostat, for example, to support remote control through a smartphone. Most consumer devices rely on the device-to-gateway model; in this model, the gateway acts as the mediator facilitating the interaction between the devices and cloud services. As an example, a smart scale may be connected to a smart watch indirectly through the help of a local gateway and cloud services. Such a setup can be used, for instance, to help householders stay motivated and on track regarding their weight goals even when not physically being in the proximity of the scale. Figure 13.5 illustrates the three identified communication models.

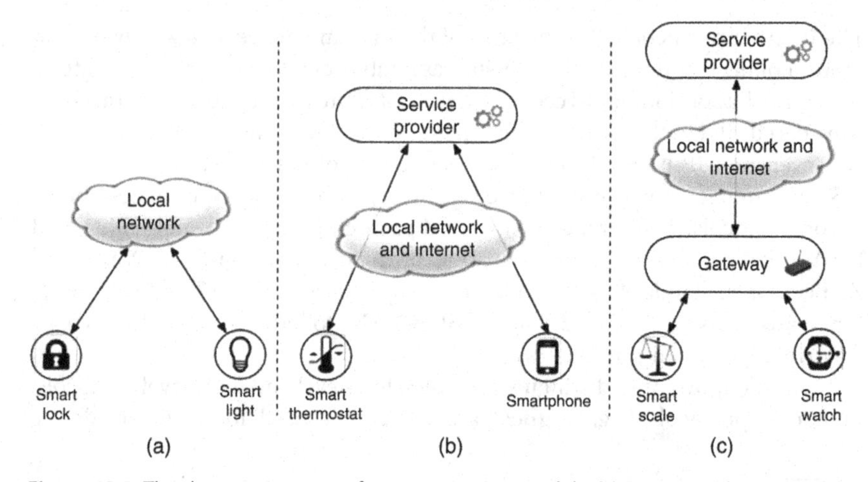

Figure 13.5 The three main types of communication models. (a) Device-to-device model. (b) Device-to cloud model. (c) Device-to-gateway model.

28 https://standards.ieee.org/develop/wg/SEP2.html (accessed September 4, 2017).

13.5 Smart Connected Home Architectures

There are two main architectural styles of smart connected homes: centralized and distributed. This distinction is based on where the intelligence is placed, that is, the provisioning of services and the cooperation between the devices to attain a common goal. Figure 13.6 illustrates the two approaches.

13.5.1 Centralized

In the centralized model, all the data are retrieved by a single central entity (Roman et al., 2013). This entity is oftentimes a dedicated local gateway that houses the application logic, stores data, and communicates with the IoT devices that are connected to it.

Another option could be to have the cloud acting as the central connection hub (Spender, 2015). In terms of information flow, the central node collects all the information from smart devices within the home network and acts on it. Consequently, end users need to connect typically through the Internet to access the services provided by the central entity (Roman et al., 2013). This approach is, for instance, adopted by Amazon Echo that requires a constant active connection to the Internet (e.g., to support Alexa cloud-based voice service), and is broadly used by other commercial companies such as Google and Apple.

One of the challenges with the centralized architecture is that implementing effective privacy measures in this solution is arguably less flexible. For instance,

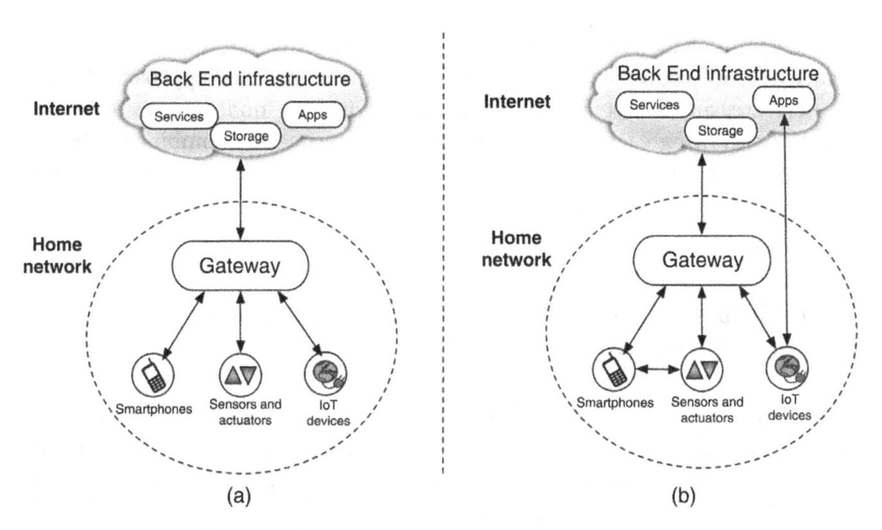

Figure 13.6 (a) Centralized and (b) distributed architectural models.

here since the "intelligence" is positioned at the central entity, the other entities (the edge devices) do not have much control over the data they generate and process. Instead, this is delegated to the central object that can decide what data elements to process, whether to share a particular data stream, and so on (Roman et al., 2013).

13.5.2 Distributed

In the distributed approach, all the devices are self-sufficient capable to retrieve, process, assimilate, and provide information and services to other entities (Roman et al., 2013). In comparison to the centralized approach where information flows through a common node, in the distributed model, the architecture is similar to that of a peer-to-peer system and information is only exchanged when needed. It can be argued that connected devices in this model are considered smart on their own. Thus, in this case the majority of decisions are done locally and the Internet does not have a major role as is the case for a centralized architecture. However, it may still be needed for coordination and analysis (Spender, 2015).

An example of this could be a connected oven that detects the presence of connected light bulbs and flashes them when it finished cooking. Given the market fragmentation, this approach is difficult to have in practice. Distributed architectures may leverage a service-oriented or a mobile agent approach (Badica et al., 2013). One technology that is used to implement this type of architecture is Jini platform. This is used, for instance, by ProSyst Software company to provide products to control household devices (Toschi et al., 2016).

One of the challenges with the distributed model is that applying network security is more complex when compared to the centralized approach. In this architecture, any smart device can connect with any node at any time, not necessary knowing each other beforehand. This, especially combined with the fact that some devices might be heavily constrained, makes key management here a significant challenge (Roman et al., 2013).

13.6 Smart Connected Home Challenges and Research Directions

There are various challenges that need priority attention from the smart connected home developers and researchers to advance the smart home further. Four of the major challenges are elaborated in the subsequent sections (Brush et al., 2011; Stojkoska and Trivodaliev, 2016).

13.6.1 Interoperability

Interoperability is the ability of different devices and services to work together. In the smart home environment, a use case could be when a wearable device, such as a smart watch, detects a householder waking up, and sends a signal to the connected coffee machine to start the brewing process. As discussed earlier, there are diverse integration platforms available in the market. These platforms, for example, Apple HomeKit, help in implementing such a use case, however, if the devices were manufactured by different vendors making them work together might be problematic. Albeit, they may speak IP, disparate protocols, cryptographic algorithms, and proprietary certificate formats interoperability make true interoperability a challenge (Cloud Security Alliance, 2016).

Additionally, an organization that controls different parts of a vertical market (e.g., the smart entertainment) may dominate a market, stifling competition, and creating barriers of entry for competitors. Dissimilar standards, proprietary technologies, and architectures can also lock consumers into one ecosystem of products making it hard to transfer their data from one platform to that of a competing device manufacturer or service provider. For instance, a related case occurred when the company selling the Revolv Hub home automation system[29] shut down its servers making the hardware and related apps useless.

A central comment is that interoperability is key to build generic smart home solutions (Ko et al., 2011), to open markets to competitive solutions in IoT (Misra et al., 2015), and as well to ease integration with the current Internet. Interoperability is a necessary component to evolve the smart connected home to a more intelligent home where devices not only connect together but are able to collaborate together toward achieving a unified goal. In achieving this, possibly devices and applications need to be decoupled. Furthermore, this may require architectures that lean toward the distributed or hybrid approaches combining the benefits of centralized and distributed styles. However, at the moment the device and the applications (apps) tend to be handled by the same stakeholder (typically the device manufacturer). Also, most of the business models adopted by vendors and most commercial applications and research works focus on cloud-based models utilizing centralized architectures.

13.6.2 Security and Privacy

Smart home devices are tightly coupled with the householders, especially in terms of their lifestyles, habits, and sensitive data such as personal photos, videos, and digital diaries. Therefore, all the information gathered, exchanged, and stored within the smart connected home, an environment meant to serve as

29 http://revolv.com/ (accessed September 4, 2017).

a private space, can lead to severe security and privacy risks. While the risks vary across the different smart home systems and services, detailed personal information can be collected by devices, and used for profiling individual behavior, amongst other things.

An example, a smart thermostat may collect device usage statistics and environmental data to learn the householders' behavior. While for beneficial reasons, this data may be shared with legitimate entities such as energy providers for energy efficiency purposes, but may also be eavesdropped by malicious threat agents such as cybercriminals. This information can then be used potentially to steal the house at a time when there is no peak activity being registered on the device implying that the residents may be away. Such a scenario and related ones raise important questions regarding data governance and as to whether collected data are being kept private and processed in a secured manner by all stakeholders across the chain.

Additionally, as the smart connected home integrates more devices, it provides more entry points for malicious software and different threat sources. With heterogeneous devices in the home, the risk impact can range from low to critical. For example, attacking an Internet-connected device may disrupt a service such as preventing a device like Amazon Echo to go online and answer a query. However, attacking an actuator, such as a wireless insulin pump or a network-connected pacemaker, may have life-threatening outcomes. Moreover, the complexity of software APIs, integration middleware, and software stacks create opportunities for different threat agents to exploit. A schematic illustration of malicious threat agents attacking a smart connected home is shown in Figure 13.7.

Thus, a central concern is to put more effort in studying both malicious threat agents and their capabilities. This is an important first step to conduct a privacy and security risk assessment and eventually to develop effective risk management strategies and countermeasures. Especially in the home, which by most people is considered to be the most private and intimate place, there is also an urgent need for more knowledge on the sensitivity of the data influx, as well as, on the means to protect it from harm or misuse. However, work in this area falls behind especially in providing generic data models for smart connected homes.

13.6.3 Reliability

Smart connected homes must neither fail nor behave in unpredictable manners if the things go amiss. However, in comparison to the traditional electrical and electronic devices such as televisions, microwave ovens, and camcorders, IoT devices are of greater complexity and arguably not as reliable. Alongside with fault tolerance, reliability is a principal challenge that must be overcome for the smart home to gain a wider acceptance (Friedewald et al., 2005). Reliability is

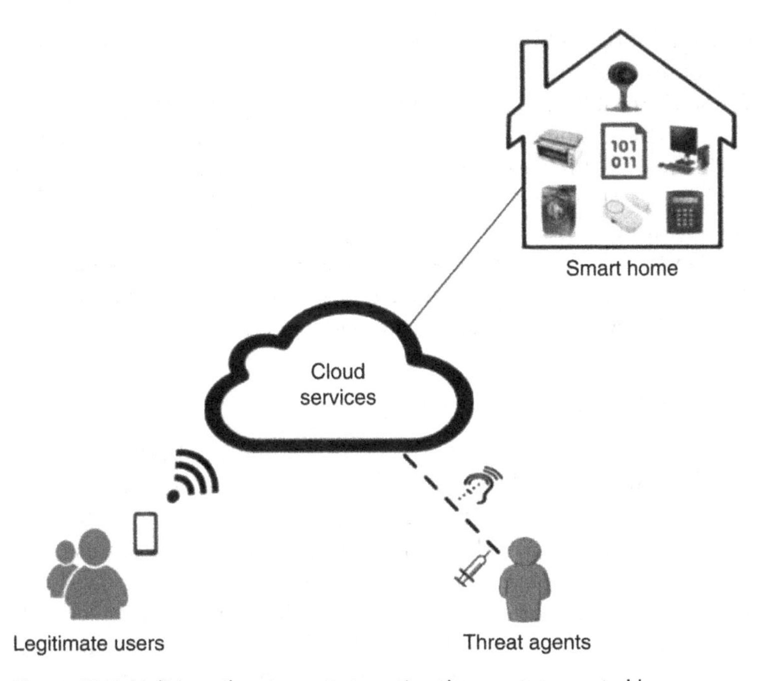

Figure 13.7 Malicious threat agents targeting the smart connected home.

especially important for functions that control the physical environment as not doing so may result in accidents (e.g., people could fall off the stairs when lights are switched off unexpectedly) (Friedewald et al., 2005).

There are also several different facets of the reliability challenge. Specifically, when comparing connected devices to "traditional" computer systems, there are differences in the development culture, technological approaches, expectations of the markets, and regulations, as factors that affect reliability (Edwards and Grinter, 2001). These factors extend beyond the research community to stakeholders that develop, regulate, and consume these services (Edwards and Grinter, 2001).

Another aspect is that the smart connected home being integrated with external service providers transformed the house into a system of systems. This is evident when looking, for instance, at service areas, such as energy. This encompasses devices such as smart meters and smart sockets that integrate the home possibly to a smart grid infrastructure. While the benefits are many, this may also result in the home to behave in unreliable somewhat unpredictable ways. For instance, service outages, software malfunctions, and performance delays can be the direct consequence of problems in external systems. A specific question this scenario raises is how to troubleshoot such problems. This eventually makes it crucial to have resilient architectures that allow for fault-tolerance, standard protocols and interfaces, and regulatory bodies that supervise the development of such systems.

As an observation, it can also be noted that ensuring reliability in smart connected homes is key for assisted living. This is particularly crucial for the health care domain. In this area, medical devices making inaccurate measurements or inferences about the nature of the individual's behavior or health profile could result in severe or life-threatening injuries. Thus, the technology developers have to incorporate means of dealing with errors, particularly in safety critical situations. In solving this challenge, regulations and standards that guarantee a certain level of safety to the householders is key.

13.6.4 Usability

Smart home devices may be used by nonexpert users. These individuals are not familiar with inner workings of technologies. Usability defines the way an end user will interact with various smart devices and system preferences (Holroyd et al., 2010). Fundamentally, it is the quality that provides for ease of use and simplicity from the end user point of view. For instance, a generic user may have difficulty to use strong authentication mechanisms that may require a lengthy text-based password or PIN code. Usability makes the system from the end user perspective characterized as simple, transparent, and unobtrusive, ideally without compromising security and privacy.

Diverse recent advancements have been made to tackle the usability challenge. In particular, voice activation appears to be the leading natural user interface method that facilitates interaction between the householders and the smart home services (Allied Business Intelligence, 2016). Despite this, in order to have voice sensing effective, it must be able to capture voice from anywhere in the house not necessarily limited to rooms that are in close proximity to the listening devices. An alternative to voice sensing is gesture detection. This makes it easier for elderly and disabled people to interact with smart home systems through simple gestures in front of cameras (Bien et al., 2005). However, this method may not be the ideal for people with very limited or impaired body movements. Likewise, certain groups of people may prefer interacting with a smart home through a physical device (e.g. computer, smartphone, or remote control) instead of through modalities such as speech and gesture (Jeong et al., 2009). Moreover, adaptive smart home interfaces that can be tailored to fit the needs of multiple user groups possibly differing in gender, age, and physical capabilities are needed (Jeong et al., 2009).

It can be observed that more research effort is needed to find the right approach to connect end users with the smart connected home system, with the objective for a high degree of usability. Over the recent years, research into Augmented Reality and Brain Computer Interface has been getting a lot of traction to improve ways of interaction with the physical environment. Despite this, their application to commercial smart connected homes has not yet gained momentum. In solving this, standards for usability are critical, and a cross-collaboration between engineering science, behavioral science, and psychological research is needed.

13.7 Conclusions

The smart connected home offers enriched services and accessible information to householders. Such homes incorporate smart devices that enable residents to utilize network-based services, such as energy management, home health care, and smart entertainment. The chapter presented various underlying technologies and different architectural models that are employed in the development of contemporary smart home systems.

While the benefits of IoT technologies in homes are many, the development of effective smart homes is hampered by notable challenges. In particular, these are related to data security and privacy, interoperability support, product and service reliability, and usability. Arguably, security and privacy pose the greatest concern to householders. This is especially because the house, being the natural place where privacy is expected, is being confronted by escalating reports of cybersecurity attacks.

Despite these challenges, more homeowners are adopting smart home technologies, and many commercial stakeholders are intensifying their interest in smart home activities. There are several initiatives and regulations that are currently forming to address the mentioned challenges yet considerable efforts are still needed. Ultimately, addressing these efforts will benefit the householders' comfort, convenience, and piece of mind, and concurrently contribute toward a more energy-efficient, safe, and possibly a healthier society and environment.

Acknowledgments

This work has been carried out within the research profile "Internet of Things and People", funded by the Knowledge Foundation and Malmö University in collaboration with 10 industrial partners.

References

Alam, M.R., Reaz, M.B.I., and Ali, M.A.M. (2012) A review of smart homes: past, present, and future. *IEEE Transactions on Systems, Man, and Cybernetics, Part C: Applications and Reviews*, **42**(6), 1190–1203.

Allied Business Intelligence (2016) ABI Research Deems Voice Control the New Breakout Star in Smart Home Technology, Available at https://www.abiresearch.com/press/abi-research-deems-voice-control-new-breakout-star/ (accessed December 12, 2016).

Amiribesheli, M., Benmansour, A., and Bouchachia, A. (2015) A review of smart homes in healthcare. *Journal of Ambient Intelligence and Humanized Computing*, **6**(4), 495–517.

Badica, C., Brezovan, M., and Badica, A. (2013) An overview of smart home environments: architectures, technologies and applications. BCI (Local), 78.

Balta-Ozkan, N., Davidson, R., Bicket, M., and Whitmarsh, L. (2013) Social barriers to the adoption of smart homes. *Energy Policy*, **63**, 363–374.

Bien, Z.Z., Park, K.H., Jung, J.W., and Do, J.H. (2005) Intention reading is essential in human-friendly interfaces for the elderly and the handicapped. *IEEE Transactions on Industrial Electronics*, **52**(6), 1500–1505.

Brezovan, M. and Badica, C. (2013) A review on vision surveillance techniques in smart home environments. *19th International Conference on Control Systems and Computer Science*, IEEE, pp. 471–478.

Brumitt, B., Meyers, B., Krumm, J., Kern, A., and Shafer, S. (2000) *Easyliving: Technologies for Intelligent Environments. Handheld And Ubiquitous Computing*, Springer, Berlin Heidelberg.

Brush, A.J., Lee, B., Mahajan, R., Agarwal, S., Saroiu, S., and Dixon, C. (2011) Home automation in the wild: challenges and opportunities. *Proceedings of the SIGCHI Conference on Human Factors in Computing Systems*, ACM, pp. 2115–2124.

Chan, M., Esteve, D., Escriba, C., and Campo, E. (2008) A review of smart homes: present state and future challenges. *Computer methods and programs in biomedicine*, **91**(1), 55–81.

Cloud Security Alliance (2016) Future-proofing the Connected World:13 Steps to Developing Secure IoT Products, Available at https://goo.gl/Qtev4f (accessed December 12, 2016)

De Silva, L.C., Morikawa, C., and Petra, I.M. (2012) State of the art of smart homes. *Engineering Applications of Artificial Intelligence*, **25**(7), 1313–1321.

Edwards, W.K. and Grinter, R.E. (2001) At home with ubiquitous computing: seven challenges. *Ubicomp 2001: Ubiquitous Computing*, Springer, Berlin Heidelberg, pp. 256–272.

Friedewald, M., Costa, O.D., Punie, Y., Alahuhta, P., and Heinonen, S. (2005) Perspectives of ambient intelligence in the home environment. *Telematics and Informatics*, **22**(3), 221–238.

Hargreaves, T., Nye, M., and Burgess, J. (2013) Keeping energy visible? Exploring how householders interact with feedback from smart energy monitors in the longer term. *Energy Policy*, **52**, 126–134.

Hassan, Q. (2011) Demystifying cloud computing. Cross Talk: The Journal of Defense Software Engineering, 16–21.

Holroyd, P., Watten, P., and Newbury, P. (2010) Why is my home not smart? *International Conference on Smart Homes and Health Telematics*, Springer, Belin Heidelberg, pp. 53–59.

Jeong, K.A., Proctor, R.W., and Salvendy, G. (2009) A survey of smart home interface preferences for US and Korean users. *Proceedings of the Human Factors and Ergonomics Society Annual Meeting*, **53**(8), 541–545.

Kailas, A., Cecchi, V., and Mukherjee, A. (2012) A survey of communications and networking technologies for energy management in buildings and home automation. *Journal of Computer Networks and Communications*, **2012**. doi: http://dx.doi.org/10.1155/2012/932181.

Ko, J.G., Terzis, A., Dawson-Haggerty, S., Culler, D.E., Hui, J.W., and Levis, P. (2011) Connecting low-power and lossy networks to the internet. *IEEE Communications Magazine*, **49**(4), 96–101.

Misra, P., Rajaraman, V., Dhotrad, K., Warrior, J., and Simmhan, Y. (2015) An Interoperable Realization of Smart Cities with Plug and Play based Device Management. arXiv preprint arXiv:1503.00923.

Ni, Q., García Hernando, A.B., and de la Cruz, I.P. (2015) The elderly's independent living in smart homes: a characterization of activities and sensing infrastructure survey to facilitate services development. *Sensors*, **15**, 11312–11362.

Pedrasa, M.A.A., Spooner, T.D., and MacGill, I.F. (2010) Coordinated scheduling of residential distributed energy resources to optimize smart home energy services. *IEEE Transactions on Smart Grid*, **1**(2), 134–143.

Portet, F., Vacher, M., Golanski, C., Roux, C., and Meillon, B. (2013) Design and evaluation of a smart home voice interface for the elderly: acceptability and objection aspects. *Personal and Ubiquitous Computing*, **17**, 127–144.

Qin, X., Lin, L., Lysecky, S., Roveda, J., Son, Y.-J., and Sprinkle, J. (2014) A modular framework to enable rapid evaluation and exploration of energy management methods in smart home platforms. *Energy Systems*, **7**, 215–235.

Reinisch, C., Kofler, M., Iglesias, F.Ã., and Kastner, W. (2011) Thinkhome energy efficiency in future smart homes. *EURASIP Journal on Embedded Systems*, **2011**, 1–18.

Rivero-Espinosa, J., Iglesias-Perez, A., Gutierrez-Duenas, J.A., and Rafael-Palou, X. (2013) SAAPHO: an AAL architecture to provide accessible and usable active aging services for the elderly. ACM SIGACCESS Accessibility and Computing (107), 17–24.

Roman, R., Zhou, J., and Lopez, J. (2013) On the features and challenges of security and privacy in distributed internet of things. *Computer Networks*, **57**(10), 2266–2279.

Rose, K., Eldridge, S., and Chapin, L. (2015) The Internet of things: An overview understanding the issues and challenges of a more connected world. Available at https://www.internetsociety.org/sites/default/files/ISOC-IoT-Overview-20151014_0.pdf (accessed December 12, 2016)

Spender, A. (2015) Build Your Blueprint for the Internet of Things. Available at http://www.gartner.com/smarterwithgartner/build-your-blueprint-for-the-internet-of-things/ (accessed December 12, 2016).

Stojkoska, B.L.R. and Trivodaliev, K.V. (2016) A review of Internet of Things for smart home: Challenges and solutions. *Journal of Cleaner Production*. **140**, 1454–1464.

Toschi, G.M., Campos, L.B., and Cugnasca, C.E. (2016) Home automation networks: A survey. *Computer Standards & Interfaces*. **50**, 42–54.

Wang, K., Lian, S., and Liu, Z. (2015) An intelligent screen system for context-related scenery viewing in smart home. *IEEE Transactions on Consumer Electronics*, **61**(1), 1–9.

Yu, Y.-C., You, S.D., and Tsai, D.-R. (2012) Magic mirror table for social-emotion alleviation in the smart home. *IEEE Transactions on Consumer Electronics*, **58**(1), 126–131.

Zhang, D., Shah, N., and Papageorgiou, L.G. (2013) Efficient energy consumption and operation management in a smart building with microgrid. *Energy Conversion and Management*, **74**, 209–222.

14

The Emerging "Energy Internet of Things"

Daniel Minoli[1] *and Benedict Occhiogrosso*[2]

[1]*IoT Division, DVI Communications, New York, NY, USA*
[2]*Intellectual Property Division, DVI Communications, New York, NY, USA*

14.1 Introduction

As discussed throughout this book, the Internet of Things (IoT) is finding applications in a wide array of industrial and commercial settings ranging from process control, to e-health, logistics, physical surveillance, crowdsensing, smart cities, smart buildings, and home automation applications. One area of particular interest where IoT can be used effectively is in supporting energy applications. The U.S. Energy Information Administration (EIA), the European Environment Agency (EEA), and the International Energy Agency (IEA), track and publish national and/or global energy data. These data help motivate why advanced management technologies are needed. According to the IEA, in 2013, the aggregate world energy consumption was 12.3 TW and expenditures on energy exceeded $6 trillion, or about 10% of the world total economic output (Energy Information Administration, 2015). In 2012 (the most recent data available), the world's electricity consumption was 18,608 TWh, which equates to a disbursement of approximately $2 trillion yearly. Europe consumes approximately 25% of the world's energy expenditures, North America approximately 20%, Asia Pacific approximately 40%, and the rest of the world (ROW) approximately 15% (Leonardo Energy/European Copper Institute, 2016). By way of electricity generation, coal is around 41% globally, natural gas 21%, hydroelectric 16%, nuclear 11%, oil 4%, and renewables and other sources 7% (Energy Information Administration, 2015). Consumption has been growing in recent years at a rate of 2–3% a year, which over a 30-year period accrues to a 50% cumulative increase (U.S. Energy Information Administration, 2016).

Emerging IoT techniques and enabling technologies can be leveraged to facilitate efficiencies in energy production, management, distribution, and consumption. Even a small efficiency improvement translates into a significant

Internet of Things A to Z: Technologies and Applications, First Edition. Edited by Qusay F. Hassan.
© 2018 by The Institute of Electrical and Electronics Engineers, Inc. Published 2018 by John Wiley & Sons, Inc.

savings (e.g., a 20% saving equates to almost a half-a-trillion dollar saving a year). The term "Energy Internet of Things" (EIoT) is being used by many to describe the application of IoT in the energy sector (Burger, 2018; Feeney, 2016; Haus, 2016a).

These data provide the motivation for the use of IoT in the energy sector, since the possible paybacks, return on investment (ROI), and economic breakeven points are compelling. There are two domains of interest and applicability: the grid that effectively *supplies* the energy resource, and the endpoints that represent the *demand* for the resource. Both domains can benefit from the capabilities afforded by the EIoT; hence, the EIoT term is used here to cover both the grid evolution and the end point evolution in terms of energy generation, transmission, and efficient consumption.

Electric power is distributed over a physical interconnecting *grid*; Figure 14.1 depicts a simplified traditional power grid. We define here an energy element (EE) to be any energy-related element in the power grid that entails any of the following: energy generating, transmitting, transforming, storing, conditioning, measuring, or consuming. Operational technology (OT) is the set of hardware and software tools that is utilized in a process control context, which detects or affects the entities or subentities supporting such process. OT achieves its functionality through monitoring and/or control of physical devices, device status, ecosystem events, or process conditions. IoT approaches enable operations systems (in the universe of energy-related OT) to gather actionable real-time information to facilitate real-time decision-making by the energy producers, distributors, and consumers. The EIoT interconnects the grid-wide energy elements and equipment, appropriate field sensors and actuators, the personnel that handle and manage the energy elements, and the set of OT/operations

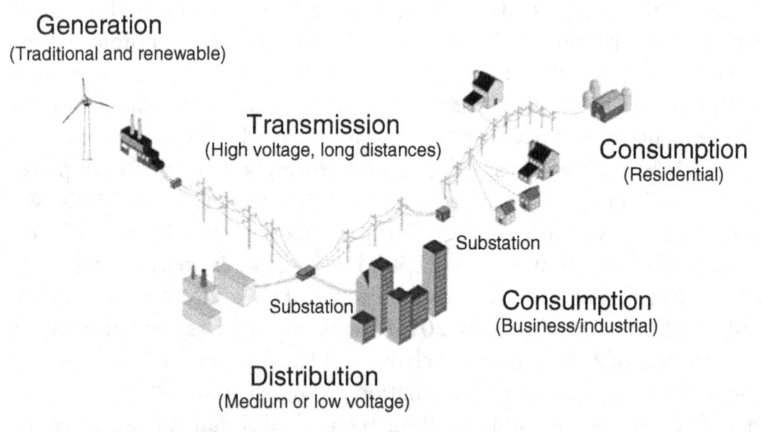

Figure 14.1 A traditional power grid that interconnects EEs supporting the power generation stage, the power transmission stage, the power distribution stage, and the power consumption stage.

support systems (OSS)—enterprise or cloud based—that support the analysis and decision-making of all functional aspects of the energy lifecycle. The EIoT will provide interactive and granular control of the energy supplies, distribution, and consumption.

In the United States (and other advanced countries), stakeholders are in the process of upgrading the nation's electrical power grid into an advanced infrastructure with interactive capabilities for collecting and transmitting information, controlling energy-related equipment, and distributing (electrical) energy (Iorga and Shorter, 2012). The transition aims at transforming the classical grid into a smart grid (SG). A SG is an electrical grid that incorporates a variety of information and communication technology (ICT[1])-enhanced energy resources including energy-efficiency resources, renewable energy resources, smart appliances, and smart meters. Figure 14.2 depicts a contemporary, evolved environment where new users, new sources, and traditional and new providers are depicted. An extensive body of literature and research on the topic of SGs is available (e.g., see Bush (2014) and Mouftah and Erol-Kantarci (2016)). SGs typically entail the following disciplines: advanced distribution network architectures, smart metering, demand response, integration of renewable sources, smart cities, home intelligence, market integration, storage, privacy, and data security. Transmission and distribution systems are currently facing a multiplicity of change drivers, including the emergence of distributed renewable energy generation along with the increasing demand of electrical energy by businesses and consumers. A SG includes effective control of the production and distribution of electricity and power-conditioning capabilities.

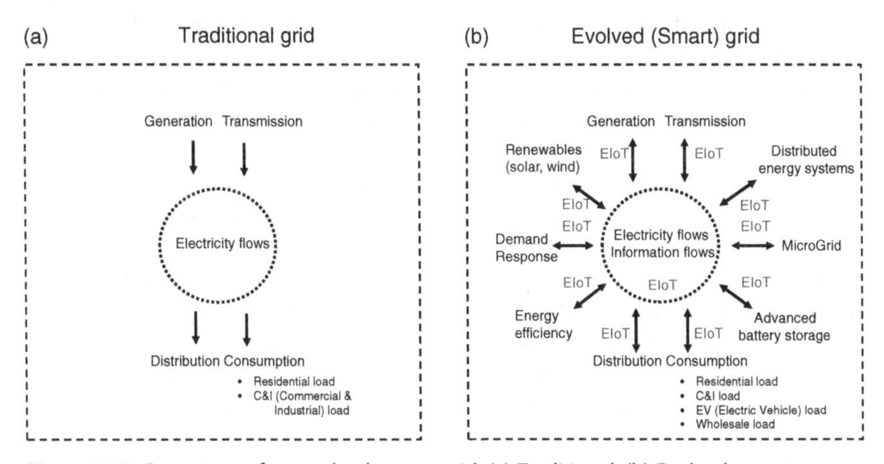

Figure 14.2 Ecosystem of an evolved power grid. (a) Traditional. (b) Evolved.

1 ICT is also generally known as IT (Information Technology).

Comprehensive, real-time management mechanisms are needed to address these and related issues; consequently, the power industry is increasingly seeking to incorporate ICT into its operations, including at the edges. SGs have seen three generations: Smart Grid 1.0 was focused on getting smart meters installed and integrated; Smart Grid 2.0 endeavors to utilize meter data to support functions such as dynamic pricing, prepayment, and line loss analysis. Smart Grid 3.0 entail broad deployment of IoT/EIoT capabilities to support the new energy ecosystem that includes the capabilities just listed, such as distributed generation, renewables, demand response, and smart cities/buildings/homes (Carvallo and Cooper, 2015).

"Decarbonization" and technological development are increasing reliance on renewable technologies, primarily variable weather-dependent resources such as wind and solar. However, the effort of accommodating the relatively less-controllable and less-predictable output of the new technologies is driving changes in the composition and operation of the entire power grid (Sweco Staff, 2015). While the transition to SGs started before the emergence of the EIoT, the IoT is greatly accelerating this process; thus, the IoT is and will be playing a critical role in that transformation. The EIoT is a set of enabling technologies that support the evolved electric grid. EIoT is a data management platform to manage—and transform—the planning, engineering, and operations of the smart grid, including the integration of distributed energy resources (DER) resources. EIoT aims at providing end-to-end solutions. The EIoT is a catalyst to the enhancement of the SG to a fully automated system where every participant on the grid has interactive interfaces to the grid and to grid resources (this, clearly under an appropriate set of authorization, authentication, and access control policies.) SG control requires wide-area network coverage that spans the generation, transmission, distribution, and the consumer portions. For this purpose, cost-optimized communication networks are needed. Additionally, there are IoT-based energy-related initiatives afoot in the smart cities, smart campuses, smart buildings, and home automation contexts to improve the efficiency of energy usage through automation, control, and new endpoint technologies.

As discussed elsewhere in this book, the basic IoT infrastructure is comprised of sensors (and actuators) that can use the IP protocol (IPv4, IPv6, or a dual stack), a variety of personal-, local-, or wide-area networks, and (typically) cloud-resident analytic engines. The EIoT collects operational (e.g., grid, consumption) and nonoperational (e.g., weather, mechanical status of equipment) data that is in turn analyzed in a fine-grain manner by analytics owned by the operator or in the cloud, to assist real-time operations, utility planners, and asset managers. EIoT applications range widely from appliance control to grid element management, from energy efficiency to logistics (Minoli, 2013). As a specific opportunity, utilities are interested in optimizing the integration and use of DERs, including renewables such as solar systems and MicroGrids. DERs are small power generators typically located at users' sites; the energy they generate is principally used at said site. A MicroGrid is a localized power grid that can operate independent of the utility electric grid but still connects to the power grid at a point of common coupling (PCC); in fact, MicroGrid owners can often sell excess power while they are connected to the grid. A MicroGrid supports secure electric power supply mechanisms that are required to operate critical infrastructure within a defined area, such as an industrial park or a campus. MicroGrids may typically include loads with varying criticality; multiple variable and static generation sources and energy storage; and various grid-related sensors, meters, and safety equipment (Staff, 2018).

There has been interest in recent years to reduce power consumption of the grid's end points—the demand points—and increasing overall efficiency, both on a broad scale as well as in specific environments, such as data centers and networks (Bui et al., 2012; Cho et al., 2013; ETSI, 2011, 2018a; Kyriazis et al., 2013; Lai et al., 2012; Serra et al., 2014; Shroufa and Miragliottaa, 2015; Minoli, 2010, 2011). In many localities worldwide, there are government mandates to reduce power consumption and maximize efficiency. In the United States (and elsewhere), there are many State Energy Efficiency Resource Standards requiring efficiency targets to be met. Goals of 20% reduction by 2020 and even more ambitious goals in same parts of the world are well established. However, many cities are behind schedule in these efforts. IoT technologies can be a major driver in achieving these efficiency goals particularly under a smart grid, smart city, and smart building paradigm. To that end, Machine to Machine (M2M) IoT technologies may be directly applicable to energy and grid management (ETSI, 2011, 2018a), especially with the increased interest (and opportunities) in supporting automated consumer interactions with the grid. Furthermore, EIoT sensors for energy management (in offices or homes) are planned to be employed for temperature sensing, CO_2 sensing, automated shading and lighting controls, occupancy sensing, ventilation opening, and task sensing.

Data from the latest Commercial Buildings Energy Consumption Survey (CBECS) indicate that in the United States, buildings (commercial and residential) are responsible for about 40% of total energy consumption (U.S. Energy Information Administration, 2012); data from the U.S. EIA imply that lighting accounts for about 20% of all the electricity used by the commercial sector.[2] These data are expected to be similar for other industrialized countries. In 2012, U.S. commercial buildings utilized 6,963 trillion (British thermal units) Btu of total site energy: 341 trillion Btu of district heat (hot water or steam from a campus central plant or a utility), 134 trillion Btu of fuel oil, 2248 trillion Btu of natural gas, and 4241 trillion Btu of electricity. Natural gas and electricity usage increased by 7% and 19%, respectively, since 2003, the last year for which CBECS results are available. The U.S. CBECS data in Table 14.1 allows one to interpolate that the yearly cost of energy in the commercial sector arena is in the range of

2 http://www.eia.gov/tools/faqs/faq.cfm?id=99&t=3.

Table 14.1 Total U.S. electricity consumption and intensities, 2012.

	All buildings using electricity			Electricity consumption	
	Number of buildings (000)	Total floor space (M ft.)	Total (B kWh)	per building (000 kWh)	per square foot (kWh)
All buildings	5234	84,869	1243	237	14.6

$150 billion in the United States alone (1243 billion kWh at $0.12/kWh), and firms with 25,000 and 100,000 square feet offices spend $43,800 and $175,200 a year, respectively. A 20% saving would result in a bottom line annual saving of $8,760 and $35,040, respectively. Considering the data listed above, EIoT solutions that address energy consumption in the context of heat, ventilation and air conditioning (HVAC), lighting, and computers and office equipment would be very useful: reducing power consumption would mean limiting the need for new power plants, thus also reducing pollution or nuclear waste; also, this would require less mining, drilling, or fracking of the earth's resources; and further limiting the dependence on energy from unstable parts of the world. Specifically, there is a desire to reduce energy consumption and associated expenditures by both commercial and government building owners and their tenants; there is a desire to reduce customer energy consumption by the energy suppliers who are looking to reduce peak-rate consumption and construction of peaking power plants. EIoT solutions can help achieve these goals.

Basic IoT technology and networking are not covered per se in this chapter since they are covered elsewhere in this book; this chapter is *not about how to design IoT systems* but *where* to use IoT systems in the energy context ("how" to use IoT is fairly basic, as covered earlier in this book: deploy sensors all over and utilize the underlying system and communications technologies that are intrinsic to the IoT).

14.2 Power Management Trends and EIoT Support

This section highlights both where the electric energy industry is migrating to and where the EIoT can play a critical role to support such migration. It is expected that in the SG the generation and transmission systems, storage, distributed generators, solar/wind generators, and commercial and residential establishments will all have EIoT-enabled capabilities (Herbst, 2015; Minoli et al., 2017a). Some recent grid-impacting factors include the promulgation of state energy efficiency resource standards (alluded to above), extensive deployments of solar rooftops as well as utility-deployed photovoltaic (PV) power generation, electric energy storage mandates, the deployment of smart meter/

interval meter, the emergence of large number of electric vehicles (EVs) (in the United States alone, 1.5 million by 2025), and the evolution of phasor measurement units (PMUs)[3] and microPMU (Kiliccote, 2015). State regulators and utilities are now advocating for demand-side energy management solutions; these solutions enable improved grid stability over the aging electric infrastructure in conjunction with the influx of renewable energy sources (Sawyer, 2018). Industry proponents make the case that EIoT has the potential to give the renewable energy industry an impetus regarding its effectiveness and ROI; this is due to the reduced amount of infrastructure required for implementation (Haus, 2016a). EIoT affords a bridge between OT and ICT. Areas where the EIoT can play a key role include the following:

- Integration of renewable sources
- Advanced distribution network architectures (under the smart grid paradigm)
- Smart cities/smart building
- Smart metering
- Demand response (DR)
- Office/home intelligence
- Energy storage
- Grid robustness

Some of these items are discussed next.

14.2.1 Integration of Renewable Sources

Integration of renewable sources is becoming increasingly important both at the commercial generation process and also at the home (e.g., when solar panels are used); this integration is facilitated by advanced distribution network architectures, being made available under the smart grid paradigm. Control of the EE that comprises the renewable sources using EIoT mechanisms is critical to achieving the desired integration and efficiencies.

14.2.1.1 Control

In the final analysis, IoT deals with control of physical entities. As depicted in Figure 14.3, the typical control elements of a modern (smart) grid include the following (all of which will soon include EIoT capabilities, also as noted in the figure):

- *SCADA* (*Supervisory Control And Data Acquisition*). It is an industrial automation control system in wide use; SCADA use is nearly ubiquitous

---

3 PMUs measure the electrical waves on the grid utilizing a common time source for time synchronization. PMUs allow synchronized real-time measurements of multiple remote points on the grid—some observers believe that PMUs may eventually become important measuring devices in the grid of the future.

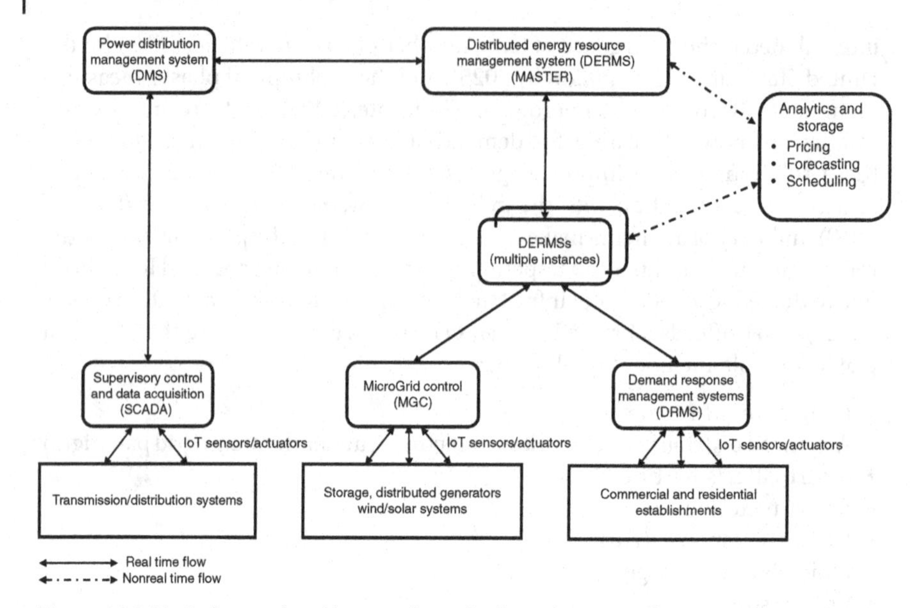

Figure 14.3 Typical control architecture for a distributed smart grid.

in power grid environments, including transmission and distribution systems. A SCADA system encompasses four functions: data acquisition; networked data communication; data presentation; and control. These functions are performed by four kinds of SCADA components: (i) Sensors (either analog or digital) and control relays that interface with the managed system under consideration. (ii) Remote telemetry units (RTUs). These are compact computerized elements deployed in the field at specific locations. RTUs serve two functions: local collection points for gathering reports from sensors, and control points for delivering commands to control relays. (iii) SCADA master units. These are larger computer systems that serve as the central processor for the SCADA system. Master units automatically regulate the managed system in response to sensor inputs and also provide operator's interface to the system. (iv) The communications network that connects the SCADA master unit to the dispersed RTUs. IoT-based systems will likely support and/or interface with SCADA elements.

- *Distribution Management System* (*DMS*). The IEEE provides the following definition "A DMS is a decision support system that . . . assists the distribution system operators, engineers, technicians, managers, and other personnel in monitoring, controlling, and optimizing the performance of the electric distribution system without jeopardizing the safety of the . . . workforce and the . . . public and without jeopardizing the protection of electric distribution assets" (Funabashi, 2016; Normandeau, 2015). A DMS is typically composed of number of collaborating systems—many DMSs in operation actually entail separate solution platforms supplied by several vendors.

- *Distributed Energy Resource Management System* (*DERMS*). It is an integrated system that provides utilities with capabilities to manage and optimize distributed energy resources, automate business processes, and engage with customers. Typical interfaces include meters, distributed generation (DG) systems, loads, grid/SCADA, markets and pricings, system operators, aggregators, and customers. Clearly, such a system will interconnect to IoT elements and benefits from connecting to such elements. See Figure 14.4, which also depicts where the EIoT will play a role. Unmanaged DERs experience issues such as voltage/frequency forced outside of operating tolerances; unplanned reverse power flows; excessive operations of voltage control devices; increased circuit losses and inadequate VAR support; lack of coordination with spinning reserves resulting in inefficient use of resources, among others (Funabashi, 2016). VAR (volt-ampere reactive) is a unit of measurement. It is used to measure the reactive power in alternating current. Voltage/VAR management and control is critical to utilities' ability to deliver power within appropriate voltage limits to consumers so that consumers' equipment operates properly, and simultaneously, to deliver power at an optimal power factor to minimize losses (National Electrical Manufacturers Association, 2013). DERMS mitigates these issues, among others. IoT is expected to play an important role in this ecosystem (Kley, 2016).

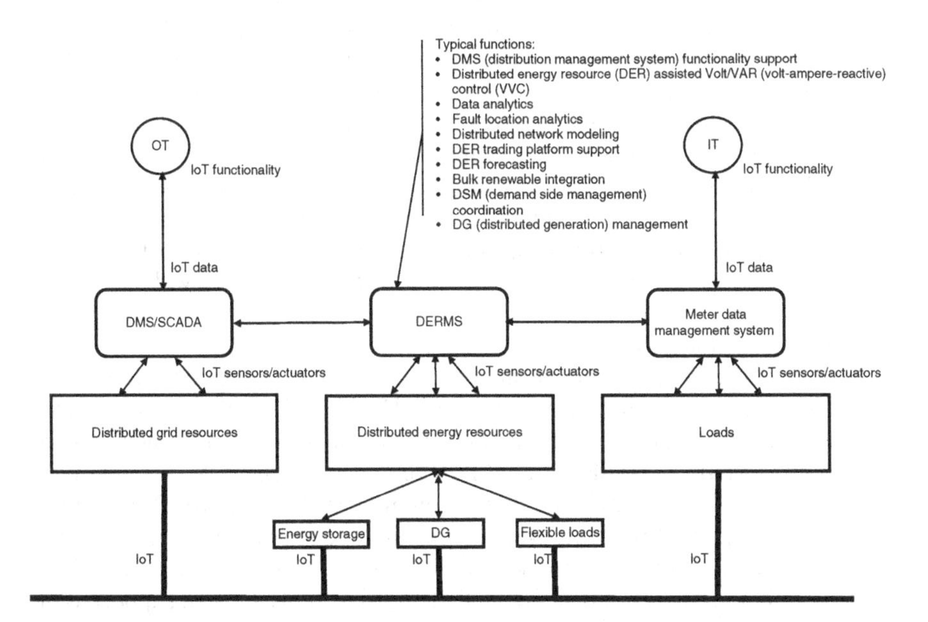

Figure 14.4 DERMS using IoT principles.

- *MicroGrid Control* (*MGC*). It is a management control system for real-time monitoring and control of a MicroGrid, also providing historical reports, analytics, and context-based automation. These functions are needed to achieve electrical reliability for critical loads when the loads are grid connected, islanded, or transitioning between the two. The MGC is integrated with SCADA systems dedicated to individual resources. The MGC system supports the following functions: (i) sensing grid conditions; (ii) directing connection to/disconnection from the grid; (iii) monitoring and controlling affiliated power resources; and (iv) providing building/load controls (Staff, 2018). The MGC monitors both the traditional grid and the MicroGrid. To accomplish this, the MGC issues commands to MicroGrid resources, to energy storage, and to capacitor banks in order to attain unity before closing grid-tie breakers. Other management capabilities include event-driven control logic to initiate programmed functions, autonomous decision-making, and data historization. See Figure 14.5, which also depicts where the EIoT will play a role.
- *Demand Response Management System* (*DRMS*). It is a system (or collection of systems) that enable utilities to manage their DR programs utilizing an integrated system. These systems enable the delivery of resource optimization, distributed energy management, and effective DR programs needed for successful SG deployments. Some of the capabilities are (Staff of ABB Inc, 2015) the development, optimization, and dispatch of virtual power plants that allow a utility to improve distribution efficiency, monitoring, and control; the development, communication, marketing, and implementation of SG programs that incent the energy consumer to make "green choices" and also facilitate the widespread introduction of renewables; the ability to make advanced metering initiatives real by affording commercially actionable information through analysis of consumption data that enable optimization of DR capabilities; and the integration of the consumer into energy markets and grid operations by utilizing SG programs such as green choice programs, voluntary reduction programs, in-home price signaling, and in-home load signaling. See Figure 14.6, which also depicts where the EIoT will play a role.

14.2.1.2 EIoT Roles in Integration of Resources

Balancing of generation and consumption at all times requires flexibility in the system. Traditionally, flexibility was achieved by monitoring the supply side and controlling the generation. Increasing reliance on variable renewable energy sources (VRES) for relatively large fractions of the electricity production in power systems (especially in advanced economies such as in Europe) introduces new challenges to power system planning and operation. Extensive IoT capabilities will be required. Some key desiderata of an SG are listed next (U.S. Department of Energy, 2015); notice that all these items are well supported by IoT sensors, actuators, networks, analytics, and data storage.

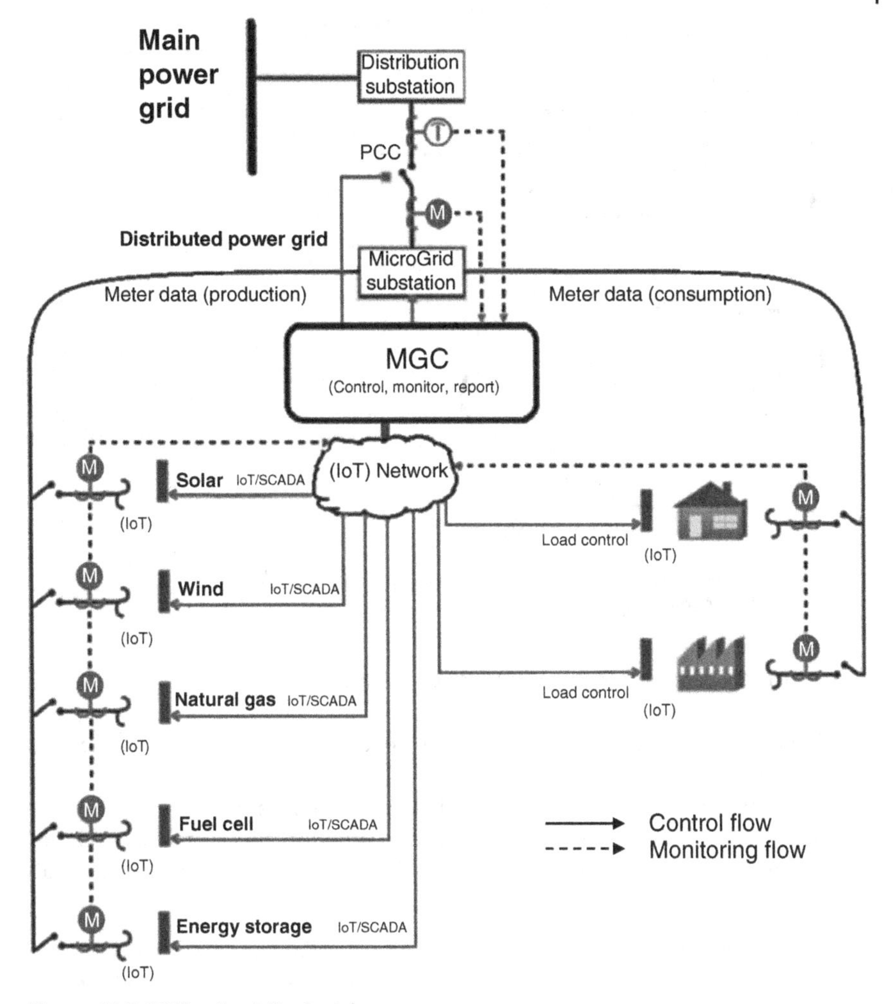

Figure 14.5 MCG using IoT principles.

1) *Interoperability.* The SG must be compatible and technology agnostic, allowing for seamless integration with distribution utility systems.
2) *Dispatchability.* The SG must ensure that power is available, when and where it is needed and at the desired amounts (available on-demand), in a fashion that is better than or at least comparable to conventional power plants.
3) *Communications.* The SG must bring about an infrastructure that is capable of informing, monitoring, and controlling the generation, transmission, distribution, and consumption of solar energy, under broad spatial and temporal scales.

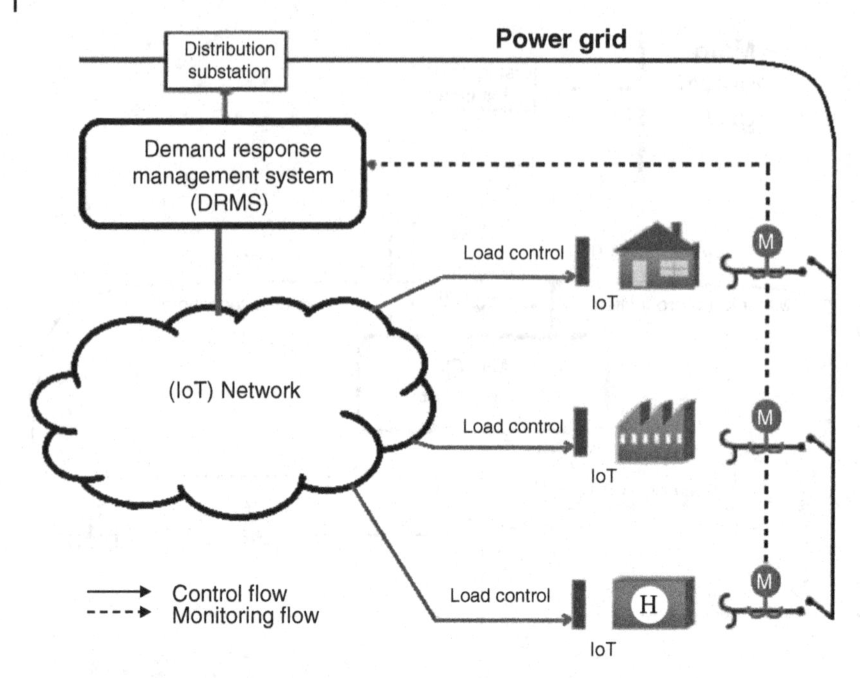

Figure 14.6 Concept arrangement of a DRMS using IoT principles.

4) *Configurability.* The SG must be able to dynamically adapt to changing operating environments.
5) *Visibility.* The SG must be constantly aware of its state.
6) *Intelligent.* The SG must ideally be able to function autonomously without human intervention and have the flexibility to respond to local or central control.

Independent system operators (ISOs), regional reliability organizations, balancing area authorities, and utilities are all developing strategies to better integrate DG in general and variable generation system (e.g., wind systems, solar systems) in particular (Ela et al., 2014). It is expected that DG elements will all have EIoT-enabled capabilities and that the EIoT elements interact with the traditional control systems in order to integrate DG into the SG (as was depicted in Figure 14.3.) The goal of advanced grid integration is to remove barriers to distributed energy grid interconnection and accelerate deployment of DER. DER includes solar systems, microgrids, and storage systems, which are sometimes called "nongrid assets". Some observers perceive DER as a disruptive technology (Sinopoli, 2016).

Utilities have also been deploying wind power systems for electricity generation (Ela et al., 2014). The installation cost of photovoltaic power generation and concentrating solar power have fallen significantly in the recent past, fueling growth and accelerating deployment of solar energy systems (estimated to be in the millions just in the United States (Kiliccote, 2015)). In fact, there are a number of efforts afoot to make solar energy cost competitive with traditional energy sources by 2020. For example, the Systems Integration program of the SunShot Initiative of the U.S. Department of Energy envisions that hundreds of gigawatts of variable solar generation will be interconnected to the grid (U.S. Department of Energy, 2015).

Wind and solar power generation are examples of variable generation. In variable generation, the available power typically changes over time (variability) and cannot be predicted with high accuracy, giving rise to uncertainty. Furthermore, wind power, especially variable-speed wind power (which is the bulk of all wind plant capacity of the world), is not synchronized to the electrical frequency of the power grid. These characteristics—asynchronism, uncertainty, and variability—represent challenges for maintaining a reliable and predictable power system. The classical grid architecture was developed around central power generation with unidirectional power flow to distributed loads. This type of grid, however, represents an inflexible and problematic framework for high penetration scenarios of solar power. The injection of large numbers of solar systems in the proximity of the customer load gives rise to two-way power flows. This invariably creates technical and regulatory challenges that impact the reliability and safety of the power system (issues include voltage and VAR regulation, unintentional islanding, power quality issues, protection coordination, sufficiency of distribution modeling tools, distribution system visibility and control, and creation and adherence to codes and standards (U.S. Department of Energy, 2015)).

EIoT-based technologies (along with other intrinsic technologies) enable DG systems' installations to actively improve the power quality of the electric grid (refer again to Figure 14.4). EIoT principles, new communication technologies, and new control architectures are needed to collect, store, analyze, and interpret real-time operations data (Berganza and Sanchez-Fornie, 2016; Lehpamerm, 2016). A Systems Integration (SI) program seeks to enable the broad deployment of cost effective, reliable, and safe renewable energy on the traditional grid by addressing the intrinsic technical and regulatory challenges. IoT principles have been advocated to address these issues. Specifically, limited visibility or control is available at the distribution level for solar installations, especially when these are behind the meter. To effectively manage a grid with distributed (solar) generation, visibility of the generators is required at multiple spatial and temporal scales. The goal is to deploy a reliable, resilient, integrated cost-effective and secure solution that supports real-time vulnerability assessment and optimization tools for distribution planning and operation. However, the evolved electric grid from a centralized and hierarchical network architecture to a more distributed configuration injects considerable system complexity. DR, energy storage, and improved wind power/solar power forecasting techniques offer potential mitigation strategies to the variability, uncertainty, and asynchronism issues related to DG. All of these techniques benefit from IoT/M2M capabilities built into the grid and its components. Another technique is fast timescale active power control (APC) (Ela et al., 2014). APC enables the adjustment of a resource's active power in various response timeframes to assist in balancing the generation and load; the goal is to improve the reliability of the power system. APC also benefits from IoT/M2M capabilities since APC is clearly based on sampling techniques (by appropriate sensors) and analytical control algorithms implemented somewhere in the ICT cloud. Three common types of APC are automatic generation control (AGC) regulation (also called load frequency control and secondary control), primary frequency control (PFC), and synthetic inertial control. AGC regulation is typically provided by resources with the dispatch of a control signal from a centralized control; PFC is the process following inertial control that modulates the output of generators to balance generation and load during loss-of-supply (or loss-of-load) events. Inertial control is the immediate response to a power disturbance based on a supply–demand imbalance.

There is already extensive vendor support of the energy analytics, whether at the grid level or at the consumer level, whether traditional or in the EIoT environment. For example, in 2015, GE introduced Predix, a cloud-based analytics fabric that processes IoT data, and soon after GE unveiled Current, which applies the Predix data analytics to energy-related programs. Many commercial building management systems (BMS) are available and leading providers are incorporating EIoT technologies in their BMSs. Furthermore, industry trackers also prognosticate that in the not too distant future, companies will create portals where businesses and consumers can trade their energy and their consumption online; these customers will be able to sell surplus energy, for example, from batteries charged by solar panels. One such company, Upside Energy, has been testing its system with plans for a full commercial rollout in 2017 (Haus, 2016a).

Wireless technologies are expected to play a significant role in EIoT, not only in the rural transmission portion of the grid, but also for DG and for some BMSs. In particular, evolving 5G wireless technologies, low-power wide area connectivity technologies (using unlicensed portions of the radio spectrum), and satellite services will be important (Minoli et al., 2017a).

14.2.2 Smart Cities/Smart Buildings

Chapter 12 in this book discusses the concept of smart cities. There are many applications of IoT/M2M in smart city environments (e.g., (ETSI, 2018b)) but energy management is clearly an important application. At this time, building architects, in response to building codes that are becoming more restrictive, are promoting zero-net energy solutions; in parallel, building owners are demanding smart building technologies that can support efficient, healthy, and resilient work environments over the useful life of the building (Minoli et al., 2017b; Sawyer, 2018; Strong and Burrows, 2016). The term *connected buildings* is also used to describe buildings with advanced BMS support, especially when the BMS incorporates EIoT/IoT principles. Smart lighting systems, daylighting, and integrated building envelope can reduce perimeter zone energy use and peak demand by 30–50% compared to ASHRAE 90.1-2013 (Lee, 2015). Weather is the biggest driver of home energy use (50% of the energy use is driven by weather (Ali, 2016)), hence a comprehensive view of the weather conditions in the geography of interest is important (Earth Networks is an example of a large IoT Sensor Network that provides global weather observations and danger alerting sensors.) Studies show, for example, that intelligent device optimization for home use can achieve energy savings in the 16% range (Ali, 2016).

Large-sized commercial buildings have used BMSs for decades to improve occupant comfort and reduce energy costs. But small- and medium-sized commercial buildings—which not only consume half the sector's energy usage, but also account for more than 95% of the commercial building stock—have lacked a cost-effective BMS solution and the savings that come with it; fortunately, newer systems that are optimized for medium-size buildings are appearing (U. S. Department of Energy, 2016). There is interest in bringing to market integrated, interoperable lighting and fenestration systems that enable less complex, reliable, and cost-effective reductions in peak demand and routine energy use and at the perimeter zone in buildings; these benefits are achieved by leveraging the capabilities of low-cost IoT nodal processors to optimize cost control and improve analytics over the life of the building (Lee, 2015).

The manufacturing sector can gain from the adoption of EIoT principles in the context of optimizing factory energy costs for machine-intensive applications, but can also achieve additional IoT-based benefits, where connected factories allow improved productivity, enhanced process reliability, and streamlined inventory planning.

Broad deployment of clean energy technologies requires advanced approaches to electric grid coordination and building equipment integration. Improved energy and business efficiencies can be achieved through co-optimization approaches that reach beyond the electrical meter (Lynn, 2015). Interoperability is defined in this context as the ability to exchange actionable real-time information between two or more systems across vendors, components, or organizational boundaries. The Connected Building is seen as negotiating and transacting energy services across the meter; integrating and coordinating connected equipment (such as loads, generators, storage) for energy efficiency and financial benefits; supporting the scalable integration technologies such as PV and EV chargers; and providing awareness, visibility, and control to serve the preferences of its managers, operators, and occupants. Some value opportunities are hidden while other are untapped (e.g., time-dependent value of energy). Currently, the large majority of buildings are not "connected" in the sense just described. The "connectedness" is limited by existing control and coordination technology. However, interoperability between the BMS and the grid is needed to facilitate the full end-to-end optimization of energy resources (including interactions between and among market service providers, distribution service operations, building service providers, and building operations). Some standards have emerged, but additional work is needed in this area. Some of the standards include the following: for management, supervision, and control (partial list): OpenADR2, BACnet Classic OPC, BACnet WS, IEC 72746, OPC-UA, OBIX, SNMP; for devices (partial list): CC-LINK, CE-Bus, DALI (Digital Addressable Lighting Interface), DSI, Insteon KNX, LonWorks, Modbus, VSCP, xAP, xPL, Z-Wave.

14.2.3 Smart Metering and the Advanced Metering Infrastructure

Advanced metering infrastructure (AMI) is an integrated system of data management systems, communications networks, and smart meters that enables bidirectional communication between a utility and its customers. Customer systems include energy management systems and other customer-side-of-the-meter equipment that enable SG functions in residential, commercial, and industrial facilities. For business customers, BMSs can also be integrated and for residential customers home area networks (HANs) can also be integrated (SmartGrid.gov, 2018). A traditional electrical meter measures the total consumption of electricity in the home or office; smart meters, on the other hand, can communicate with in-home displays to inform the consumer how much energy they are using as a function of time (over the day, week, or month). The consumer is able to use this type of information to monitor usage and manage the energy efficiency and costs (Spikes, 2011). Typically, a smart meter can be interfaced with a home energy management system, that allows the user to view and process the information with a PC-based GUI. The AMI is one example of this modernization effort. The AMI is the ICT mechanism that is put in place between the consumer and the power utility. In the aggregate, it is an integrated assembly of smart meters, networks, and data processing systems that supports bidirectional communication between providers and consumers (Cárdenas and Berthier, 2014; Smart Grid, 2018). AMI is a significant component of the overall approach for implementing the SG and the main component for achieving DR.

Smart meters (also known as advanced meters or digital meters) are a physical element that is added to or replaces a traditional electric meter, while fitting in the same mechanical footprint. Smart meters are part of the physical backbone of SG and must operate in unison with other AMI/SG elements (Doris and Peterson, 2011). A growing number of utility companies are planning to roll out intelligent metering services by incorporating M2M devices in their meters. By

2020, the expectation is that there will be 1 billion smart meters and 1.5 billion utility-managed connected devices in use globally; in Beijing, for example, almost 100% of residential homes were equipped with smart meters by the end of 2015. The cost of rollout, however, is not trivial and is estimated to be in the $100 billion range globally (Haus, 2016a). Smart meters can cost up to $250 per meter, and the cost can approach $500 when smart grid costs are also considered.

Smart meters are invariably computerized; these meters allow for remote data collection through periodic (e.g., 15-min., hourly, daily) interaction with the utility to forward domicile energy use. Utilities can thus gather information on energy consumption and, in turn, reduce peak demand; load reductions allow utilities to avoid or mitigate costs for building new power plants or relying on older, less efficient plants. Smart meters also have the ability to provide output data to customers; these data include real-time energy use, ultimately to allow for behavioral modifications. Grids with smart meters endeavor to predict energy demand; these grids are also able to react to rapid changes in demand and supply to deliver reliable, efficient, and sustainable power. Specifically, smart meters and IoT-ready systems provide previously unavailable real-time and location-specific energy usage data. Many utilities are now able to capture customer-specific real-time or interval-based usage data for a large fraction of their commercial and residential customers. These data can provide new insights about how people make energy decisions. This kind of "behavior analytics" can be useful to a utility. For example, heterogeneous and disaggregated information about actual energy use allows energy efficiency and/or DR program implementers to target specific behavior-based programs to specific households. Additionally, this information enables the evaluation, measurement, and verification (EM&V) of energy efficiency programs; and may provide improved insights into the energy and peak-hour savings that can arise with energy efficiency and DR programs (The State and Local Energy Efficiency Action Network, 2014).

SG devices and smart appliances are increasingly designed to be "demand response enabled." To realize efficiency and load-management goals, utility companies are adopting two-way networking connectivity to enable consumers to monitor and possibly reduce their energy usage (Fouda et al., 2012; Samadi et al., 2012). A centralized entity polls or informs the meter to gather appropriate consumption information. The goal is to improve the efficiency of the distribution of energy by utilizing detailed real-time information about the end user's (consumer or business) consumption, especially in conjunction with incentivized billing plans. Connecting appliances to the network for the purpose of managing such devices goes back to at least 1987 (if not earlier) when Xerox provided remote interactive communications to collect information on copiers and printers. These days the capability of connecting dishwashers, clothes washing machines, refrigerators, thermostats, toasters, garage door openers, and alarm systems is relatively routine.

HANs are also being deployed in support of a smart home. Furthermore, vendors have started to design products that integrate built-in-communication systems that interact with the AMI-enabled meter and with the HAN. The combination of the AMI-enabled meter and the HAN allows consumers to remain aware of their electricity usage and associated costs on a quasi-real-time basis. Codifying information about the cost of power and the consumer's preferences, appliances schedule their operation such as deferring or adjusting the operating parameters to reduce peak energy consumption. Therefore, this method of intelligent management of energy supply potentially reduces expenditures as well as the peak demand. Peak demand reduction helps save money by avoiding the cost of auxiliary power plants' construction that is put in place to handle peak loads.

14.2.4 Demand Response

Demand response is process where customers voluntarily curtail usage, benefiting the grid. Some customer systems that could shed load include lighting space cooling, space heating, water heating, (noncritical) industrial process, and irrigation. DR can provide energy services by either shedding load or shifting its use between different times (U. S. Department of Energy, 2006) (Figure 14.7). DR can improve grid reliability and reduce operational costs. DR resources enhance grid flexibility and can aid in integrating variable generation. However, integration into the grid has encountered some challenges; for example, including DR in grid modeling can be complex due to its variable and

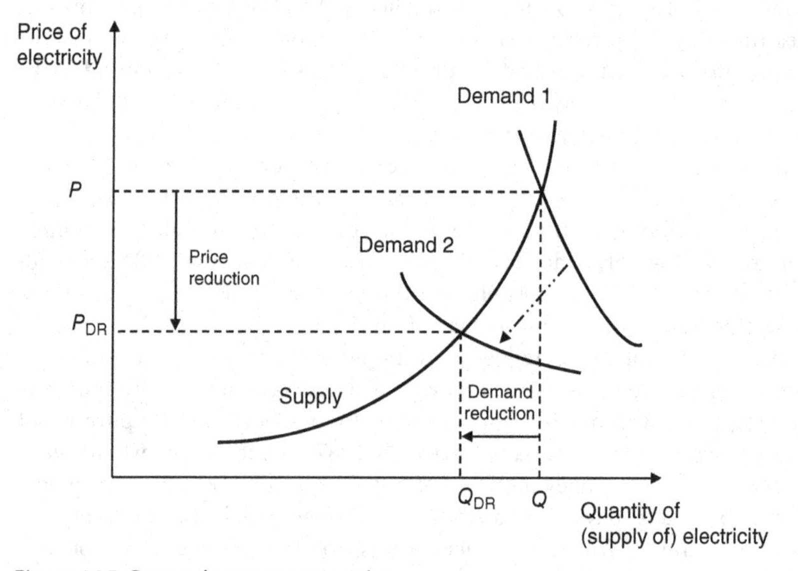

Figure 14.7 Demand response principles.

nontraditional response characteristics, compared to traditional electricity generation. New methods and tools for predicting demand response availability profiles are emerging (Hummon et al., 2013). EIoT-based sensing and connectivity can greatly ameliorate the integration process.

Common types of DR programs include direct load control, where customers are offered incentives for allowing a utility a certain (preestablished) level of control over certain equipment; demand bidding/buyback programs, where customers offer bids to self-reducing load when wholesale market prices are high; emergency reductions, where customers receive incentive payments for load reductions should these be needed to improve reliability; capacity market programs, where customers are offered financial incentives for providing load reductions as substitutes for system capacity; and interruptible control, where customers receive a discounted rate for agreeing to reduce load on request. Pricing options include time of use (TOU) rates, where rates have fixed price blocks based on the time of the day; critical peak pricing (CPP), where rates include a predefined rather high rate that is triggered by the utility for a limited number of hours, as needed to address demand/supply; and real-time pricing (RTP), where rates vary continuously (over certain time granularity, e.g., hourly) in response to wholesale market prices (Sedano, 2010).

As alluded to above, a key element of a SG is a cost-effective communications network (e.g., as supported by the AMI) that enables automated metering capabilities. Specifically, such a network collecting power consumption data allows a utility to automatically control customer loads, especially during peak demand periods. Additionally, it enables the utility to remotely and automatically update the grid configuration, and also control supply of power to certain customers. Among other benefits of smart metering are the ease of site access, power usage information, statement accuracy, and monetary savings associated with DR and management of power demand.

A variant application of the smart meter's DR capabilities can also be utilized to support a prepayment arrangement where a consumer purchases a specified amount of the commodity (electricity, gas, etc.); the data related to the amount purchased is stored on the M2M analytics engine or associated storage, and is then downloaded to the metering device. When the purchased volume has been consumed, the supply is halted (Rehmani et al., 2016). As noted, many grid-related functions have been supported utilizing SCADA-based mechanisms, which have traditionally been wireline oriented. The migration to wireless thus entails supporting the SCADA mechanisms in this new media along with requirement for improved cybersecurity over said wireless media.

14.2.5 Office/Home Intelligence

Solid-state lighting (SSL) is a key technology trend driving performance improvements and cost reductions. SSL is now experiencing wide deployment.

SSL lighting uses semiconductor light-emitting diodes (LEDs), organic light-emitting diodes (OLED), or polymer light-emitting diodes (PLED) as sources of illumination in place of the traditional electrical filaments, plasma (used fluorescent lamps), or gas. In recent years, one has seen major improvements in SSL performance and significant reductions in SSL price. SSL is now considered a viable option for most lighting applications. SSL and connected lighting are seen as offering efficiency and efficacy in application; being diminutive (small form factor); being organic, layered, locally responsive; providing accent and ambient moods to support circadian cycles, having a full dynamic spectrum, all the while being highly reliable (Quinlan, 2016).

LED lighting is more energy efficient and is also less expensive to maintain than traditional commercial lighting methods (such a fluorescent or high-intensity discharge lighting): not only are they brighter per watt, they also last longer than even the preferred florescent bulbs (LED replacement requirements are at approximately 50,000 h, or a 10-fold improvement). Additionally, LED lighting emits less heat, reducing the need to expend energy on cooling a building, which results in lower electricity costs. The typical T12 four-lamp fixture uses 172 W of power between the lamps and ballast while LED equivalents typically use only 50 W, 71% less energy per fixture. On an aggregate building basis, it typically results in a 20% energy savings. Additionally, LED lighting often incorporates lighting controls that have occupancy sensors.

In addition to the simple upgrade to LEDs, there is an area of research into the concept of "connected lighting," namely, where lights are interconnected over an IP (Internet Protocol) network. The connected lighting paradigm makes use of sensor(s); controllable and intelligent SSL sources; wired and/or wireless network (both in-building and into the cloud); interface to cloud storage, computing, analytics as a service, and IoT principles on systems and data. All of these capabilities fall under the rubric of IoT/M2M technologies. Very strong statements such as this made at the department of Energy (DOE) 2016 Connected Lighting Systems Workshop point to the criticality of IoT (U.S. Department of Energy (DOE, 2015)):

> "There is growing awareness that connected solid-state lighting systems may become an IoT backbone, or even *the* IoT backbone. This vision is bolstered by three key factors: SSL's microelectronic nature, which readily facilitates the integration of network interfaces and sensors; the growing integration of SSL devices into the built environment; and the fact that lighting is ubiquitous infrastructure that can be found pretty much everywhere people congregate."

As of 2015, more than 245 patents existed on the topic of LED and/or connected lighting with almost 200 on-market products; estimates were made that over 143 TBtu of energy savings were achieved in 2014, equating to $1.4 billion (Brodrick, 2015). New lighting system are responsive to people, allowing for personal preferences and optimizing their environment. At press time, the industry was wrestling with questions such as the following:

- Why do lighting systems need to become more connected?
- Why should we (and how can we) enable lighting systems to report their own energy consumption?
- Where and when do we need interoperability?

Currently, relatively few buildings have some form of lighting control, but the opportunity is far greater. It has been reported that about 18% of the commercial buildings have light scheduling, 16% have occupancy sensors, 7% have multilevel lighting or dimming, 2% have daylight harvesting, 4% have demand responsive lighting, and 4% have BMSs that support lighting. Given that the penetration of advanced networked lighting controls in commercial buildings is currently around 2%, there is a clear opportunity to use the IoT to accelerate this automation and achieve significant energy savings (Designlight Consortium, 2015). Daylight harvesting is an energy management approach intended to reduce indoor lighting requirements by making use of the ambient light present in a space; this light includes natural light as well as artificial light. Basic techniques entail dimming or switching of luminaries when the space is unoccupied or sufficient ambient light is already available. Figure 14.8 depicts various lighting management strategies that may be controlled by the BMS, while Figure 14.9 provides a logical view of the connected lighting arrangement

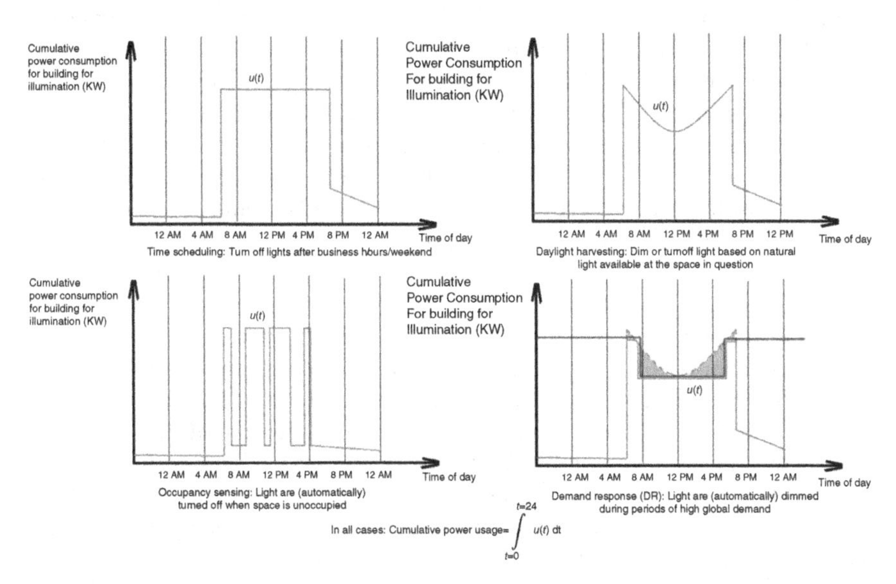

Figure 14.8 Lighting management strategies.

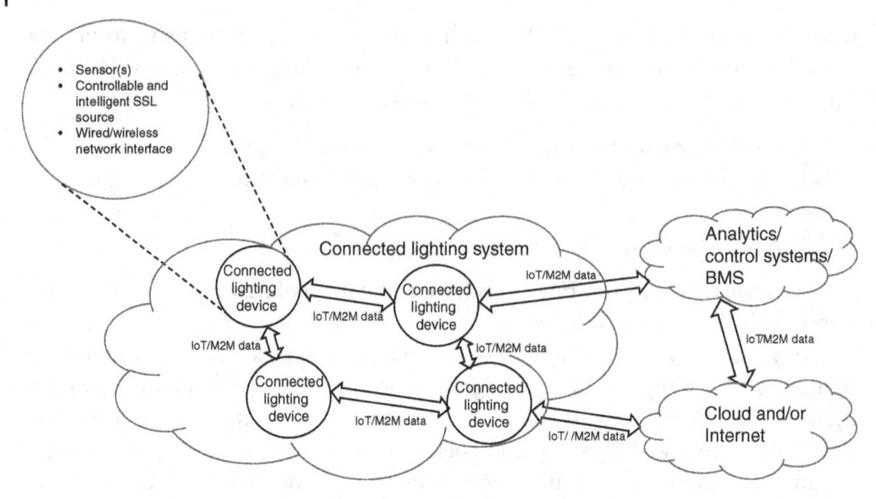

Figure 14.9 Logical view of connected lighting arrangement.

(additional information in Matiko and Beeby (2017); Sinopoli (2016); and Hofbauer et al (2016)). There is a lot of industry interest in this space. To identify just one announced venture, Philips recently entered in a lighting partnership with Cisco, SAP, and Bosch. The technology has broad applicability in homes, buildings, offices, hospitals/healthcare, public spaces, smart cities, industrial sites, and retail sites, to list just a few. It is estimated that there are 4 billion streetlights and 500 billion light sockets, and "Each one is a potential IOT End Node," as advocated in Bhide (2016). In the United States, Legislation (Title 24, ASHRAE 90.1) is driving requirements field task tuning, occupancy sensing, and daylight harvesting. These requirements can be met by implementing sensors in luminaires, but with a(n IoT) network to connect back to the BMS.

Sensors are migrating to minisensors; controllers and drivers are migrating to integrated smart LED drivers; LED assemblies and LED fixtures are migrating to LED and printed circuit board assemblies (PCBA); and the entire system (sensor, driver, controller, LED, and fixture) is expected to migrate to a single PCBA. Scheduling, according to various sources, is expected to result in savings of 10–40%; daylight harvesting a saving of 5–60%; occupancy a saving of 20–60%; task and/or personal tuning a saving of 5–36%; and combining multiple strategies from 38 to 75% (Poplawski, 2015a). Lighting can be seen as in an "advanced state" (e.g., replace traditional luminaries with LED-based luminaries); in a "networked state" where they can be controlled/monitored centrally/remotely, and ultimately in an "integrated state" where lighting is part of a multisystem performance optimization system, such as a BMS that also controls other systems. Typically, lighting has synergy with other systems in smart buildings, including, occupancy sensors for HVAC and security with lights activated by a security system, which affords overlays with digital surveillance

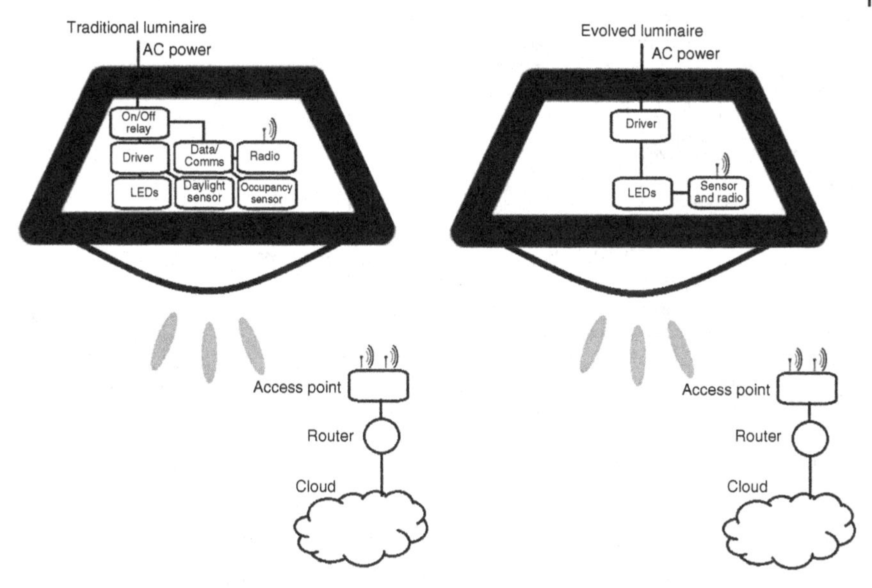

Figure 14.10 Evolved luminaire.

and security. There already is an evolution afoot in connected lighting, which aims to reduce complexity and cost, where every fixture becomes a(n IoT) wireless point (Duine, 2016). Figure 14.10 illustrates the evolution.

The goal is to achieve a state where connected lighting is widely adopted and the technology is "integrated, standardized, interoperable, and interchangeable." This requires that sensors, intelligence, communication are incorporated into every device, into every luminaire, namely, it requires IoT technologies. The introduction of connected lighting entails a paradigm shift from hardware to software, services, and data (via the IoT). Integration and standardization canonically result in lower costs. BMS use in-building networks such as Serial (RS-485), Ethernet (IEEE 802.3), Wi-Fi (IEEE 802.11), ZigBee (IEEE 802.15.4) ZigBee API, OpenADR, Smart Energy Profile (SEP), BACnet, and Modbus to interconnect the various traditional sensors. IoT-based technologies typically make use of these networks plus others. One standard that is being offered as a common open standard for residential connected lighting applications is ZigBee's Light Link. ZigBee's Light Link is part of the more inclusive ZigBee standard depicted in Figure 14.11, and is an easy-to-use global standard for the lighting industry. It allows consumers to have wireless control of light bulbs, LED fixtures, timers, and switches. Consumers can alter lighting features remotely to support current or desired ambiance, natural lighting, tasks, or season (ZigBee Alliance, 2018). Another initiative is the *oneM2M Partnership Project* that seeks to provide the common IoT functions to applications so that

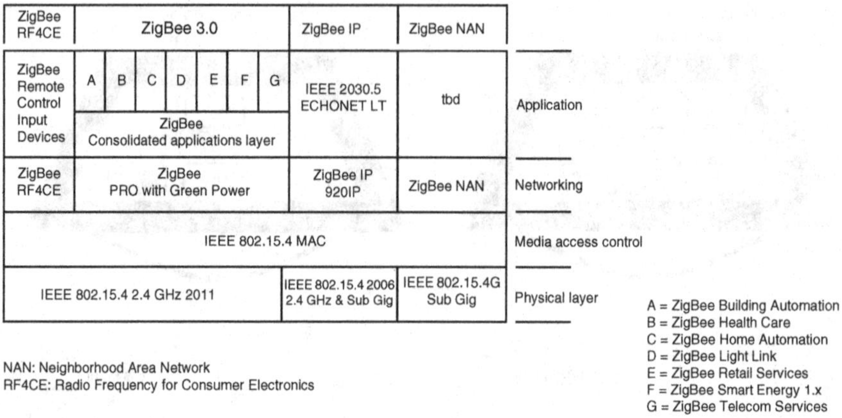

ZigBee RF4CE | ZigBee 3.0 | ZigBee IP | ZigBee NAN

ZigBee Remote Control Input Devices | A B C D E F G | ZigBee Consolidated applications layer | IEEE 2030.5 ECHONET LT | tbd | Application

ZigBee RF4CE | ZigBee PRO with Green Power | ZigBee IP 920IP | ZigBee NAN | Networking

IEEE 802.15.4 MAC | Media access control

IEEE 802.15.4 2.4 GHz 2011 | IEEE 802.15.4 2006 2.4 GHz & Sub Gig | IEEE 802.15.4G Sub Gig | Physical layer

NAN: Neighborhood Area Network
RF4CE: Radio Frequency for Consumer Electronics

A = ZigBee Building Automation
B = ZigBee Health Care
C = ZigBee Home Automation
D = ZigBee Light Link
E = ZigBee Retail Services
F = ZigBee Smart Energy 1.x
G = ZigBee Telecom Services

Figure 14.11 ZigBee stack.

applications can focus on their own application logic (Remy Marcotorchino, 2015). oneM2M aims at bridging systems from different vertical industries (lighting, energy, security, fleet, environment) while allowing each system to use its own semantics. It is designed to work with any communication technology (Cellular, xDSL, Satellite, Wi-Fi, Bluetooth, ZigBee, and other transmission systems).

A related, applicable technology is Power over Ethernet (PoE): PoE is increasingly being seen and used as a step along the continuum of the IoT deployment in smart/green buildings, also for connected lighting (Eisen, 2009; Electrical and Electronics Engineers (IEEE, 2018); Feiden, 2018; Hua et al., 2016; Wesemann et al., 2015). PoE could have a disruptive impact in smart building technologies. The basic concept of PoE is the ability to transmit content (LAN-based transmission) and also deliver device-supporting power over the same cable. A standard was developed as the IEEE 802.3af (2003) that supported power delivery in the 15 W range (at 48 V DC). The capabilities were extended a few years later with the introduction of IEEE 802.3at (2009); power delivery in the 30 W range (at 53 V) was standardized (Eisen, 2009). Extensions are underway to support power in the 50–70 W range (IEEE 802.3bt, Type 3) and possibly even 100 W (IEEE 802.3bt, Type 4). Delivering higher power to endpoint devices enlarges PoE's application scope. With PoE DC power is transmitted on the unshielded twisted pair conductors by applying a voltage to each pair; since Ethernet utilizes differential signaling, the superimposition of a power signal does not interfere with data transmission. Building and operations services facilitated by PoE include digital signage, video distribution, wireless access points, Voice Over IP (VoIP), sensor-rich environments, energy metering, HVAC, physical security/surveillance, access control, lighting, and tenant services such as smart meeting spaces. The PoE business case is rooted in the economics of an integrated infrastructure, and reducing material and installation costs (and also including conduit space). In addition, it simplifies the deployment of endpoint devices (such as digital cameras, Wi-Fi wireless access points, and other sensors), without requiring the installation of a traditional high-voltage AC electric circuit. The expectation is that by 2020 20% of the BMSs, 20% of the lighting systems, 20% of the access systems, 50% of the security cameras, and 80% of the wireless access points will utilize PoE; over 90% of the desks in a commercial building will have PoE (Staff, 2015).

Integrated control of energy (including linking to microgrids) along with integrated control of comfort elements at the perimeter zone of buildings, will not only provide cost savings to the consumer but also foster demand for innovative envelope and lighting component technologies. For example, it would be very beneficial to coordinate a nighttime cooling strategy with the following day's automatic scheduling of motorized shades and dimmable lighting in order to keep afternoon temperatures within comfortable limits without the need for additional cooling. There is work underway at Lawrence Berkeley National Laboratory to advance the science of envelope and lighting integration, among other institutions (Sawyer, 2018). Integrated lighting systems that also encompass dynamic façades are difficult to implement in buildings yet offer demonstrable benefits over conventional unintegrated or fragmented approaches in terms of increased resilience, comfort, and energy efficiency, all at a lower cost. Integrated systems require coordinated control and data exchange between the various elements involved. Often lighting controls have occupancy sensor data, but combining the data with façade and HVAC control systems to implement load reduction or demand-side ventilation strategies has been difficult because the data are stored in different and inconsistent formats, with ill-defined descriptors. IoT/M2M principles can address these issues. See Figure 14.12 (simplified from Sawyer (2018)).

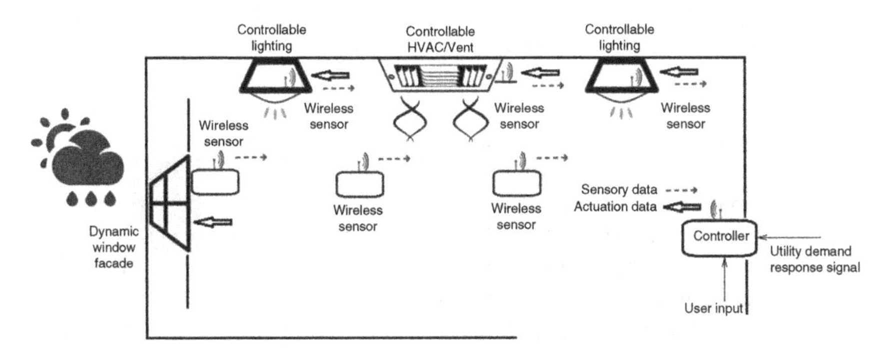

Figure 14.12 Integrated control of energy and comfort at the perimeter zone of buildings, Lawrence Berkeley National Laboratory research model.

14.2.6 Energy Storage

Peak smoothing can be achieved through energy storage. There are some challenges with DG systems such as solar systems (and microgrids): Solar generation does not coincide with residential distribution system peaks and solar output curve does not match typical residential customer daily usage profile (Bialek, 2015). Therefore, the use of residential energy storage can be useful. IoT/M2M-based sensors, networks, and analytics can play a vital role. Storage affords continuous power during nonproducing times, particularly for wind and solar. The research goals are to create a battery with five times the energy density of existing commercial batteries at one-fifth of their present cost. There have been, in fact, a number of technical advancements in recent years that make these arrangements more practical and cost-effective (Plett, 2015). Examples include organic liquid flow batteries (now in prototype) and improved lithium–sulfur batteries (also in prototype); the former is for sedentary applications and the latter for mobile (transportation) applications (Haus, 2016b). Driven by EV requirements, the global lithium-ion EV battery market is poised to grow significantly by 2020 (estimated at $22 billion in 2020). Companies such as Axeon, Bosch/Samsung, Johnson Controls, LG, Panasonic, Tesla, among others are commercially very active in this space.

14.3 Real-Life Power Management Optimization Approaches

As noted, there has been interest in recent years to reduce power consumption and increase overall efficiency in commercial buildings. It was noted above that about 80% of the energy is consumed in HVAC, lighting, and computer/office equipment. Applicable energy and cost saving analysis methodologies are described in American Society of Heating, Refrigerating and Air-Conditioning Engineers (ASHRAE) Standard 90.1-2004, 90.1-2007, 90.1-2010 (Thornton and Rosenberg, 2011) (Figure 14.13). The analysis includes a total assessment of the energy-consuming elements of a building; however, lighting represents low-hanging fruit. The replacement of traditional lighting fixtures with LED-based fixtures that also incorporate occupancy sensing (and perhaps makes use of PoE technologies) may well get an entity close to meeting the 10% reduction mandates that many jurisdictions have issued. Recall, as noted already, that lighting comprises 17–20% of the total consumption.

A useful metric is energy use intensity (EUI). It is a unit of measurement that describes a building's energy use; specifically, EUI represents the energy consumed by a building relative to its size. The optimizing agent will typically focus the analysis first on those buildings that have *large Building EUI* (especially *exceeding the median*); are *large* (by square footage, since these have the highest

ASHRAE Standard 90.1-2010
Large Office Modeling

Form
- Location (Representing 8 Climate Zones)
- Available fuel types
- Building Type (Principal Building Function)
- Total Flooor Area (sq. feet)
- Building shape
- Aspect Ratio
- Number of Floors
- Window Fraction (Window-to-Wall Ratio)
- Window Locations
- Shading Geometry
- Azimuth
- Thermal Zoning
- Floor to floor height (feet)
- Floor to ceiling height (feet)
- Glazing sill height (feet)

Exterior walls
- Construction
- U-factor (Btu / h * ft² * °F) and/or R-value (h * ft² * °F /Btu)
- Dimensions
- Tilts and orientations

Roof
- Construction
- U-factor (Btu / h * ft² * °F) and/or R-value (h * ft² * °F /Btu)
- Dimensions
- Tilts and orientations

Window
- Dimensions
- Glass-Type and frame
- U-factor (Btu / h * ft² * °F)
- SHGC (all)
- Visible transmittance
- Operable area

Skylight
- Dimensions
- Glass-Type and frame
- U-factor (Btu / h * ft² * °F)
- SHGC (all)
- Visible transmittance

Foundation
- Foundation Type
- Construction
- Thermal properties for ground level floor: U-factor (Btu / h* ft²* °F) and/or R-value (h * ft²* °F / Btu)
- Thermal properties for basement walls
- Dimensions

Interior Partitions
- Construction
- Dimensions

Internal Mass

Air Barrier System
- Infiltration

HVAC

System Type
- Heating type
- Cooling type
- Distribution and terminal units

HVAC Sizing
- Air Conditioning
- Heating

HVAC Efficiency
- Air Conditioning
- Heating

HVAC Control
- Thermostat Setpoint
- Thermostat Setback
- Supply air temperature
- Chilled water supply temperatures
- Hot water supply temperatures
- Economizers
- Ventilation
- Demand Control Ventilation
- Energy Recovery

Supply Fan
- Fan schedules
- Supply Fan Total Efficiency (%)
- Supply Fan Pressure Drop

Pump
- Pump Type
- Rated Pump Head
- Pump Power

Cooling Tower
- Cooling Tower Type
- Cooling Tower Power

Service Water Heating
- SWH type
- Fuel type
- Thermal efficiency (%)
- Tank Volume (gal)
- Water temperature setpoint
- Water consumption

Lighting
- Average power density (W/ft²)
- Schedule
- Daylighting Controls
- Occupancy Sensors

Plug load
- Average power density (W/ft²)
- Schedule

Occupancy
- Average people
- Schedule

Elevator
- Quantity
- Motor type
- Peak Motor Power (W/elevator)
- Heat Gain to Building
- Peak Fan/lights Power (W/elevator)
- Motor and fan/lights Schedules

Exterior Lighting
- Peak Power (W)
- Schedule

Figure 14.13 ASHRAE Standard 90.1-2010 Modeling of energy opportunities in an office building.

consumption) and/or have *high usage*; and have the *largest variance in the Building EUI difference from the average*; the other buildings in the same agency will be analyzed thereafter. Moreover, while prioritizing projects for retrofit, equally important is to prioritize facilities where systems are near the end of life and need to be replaced. In its analysis, the optimizing agent will assess the source energy versus site energy metrics (a site with higher source energy will get focused attention). *Site energy* is the amount of heat and electricity consumed by a building as reflected in utility bills. *Source energy* takes into consideration the resources consumed in the distribution, transmission, and generation of electricity, incorporating all transmission, delivery, and production losses. Since source energy represents the total amount of raw fuel that is required to operate the building, it allows one to make a complete assessment of energy efficiency in a building. In a recent initiative focused on energy improvements for state government-owned buildings in New York State, the mean Source EUI (weighted for square foot) for the state's building portfolio is 252 kBtu/sq. ft. However, the median is 183 kBtu/sq. ft., implying that there are a few energy-intensive facilities that move the mean higher than it may, otherwise, be expected.

An energy optimization approach typically follows the methodology depicted in Table 14.2 (and Figure 14.14).

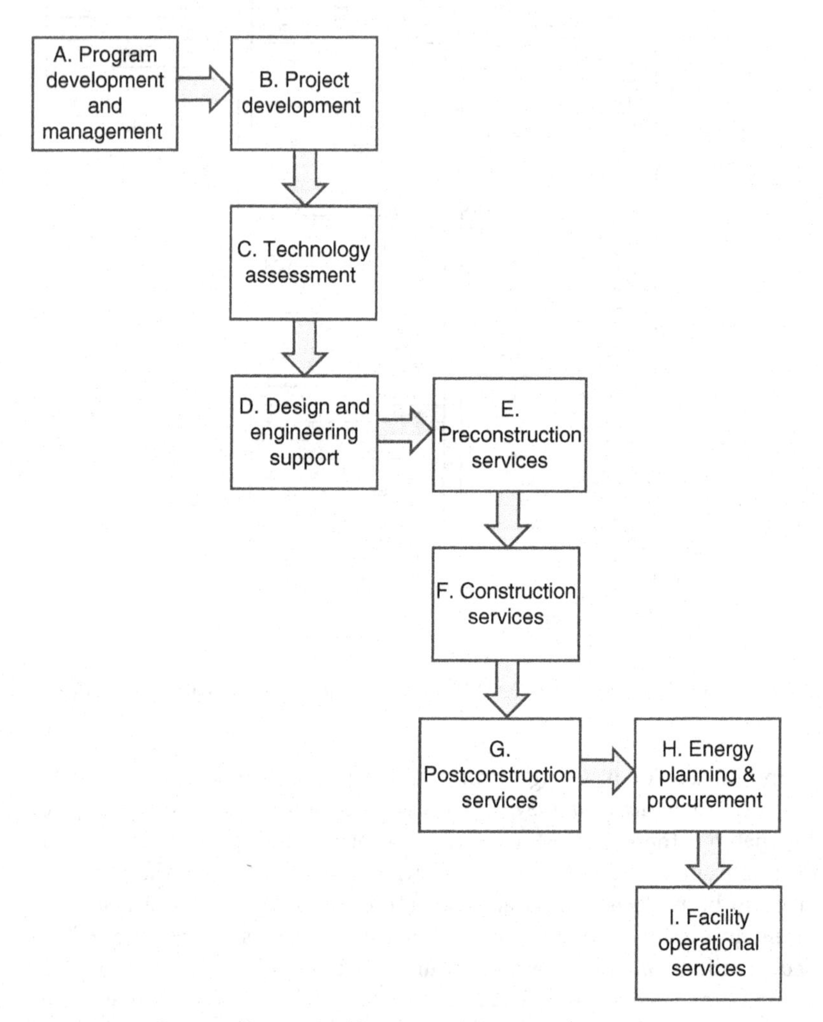

Figure 14.14 Energy efficiency building upgrade process.

Table 14.2 Typical process used in energy optimization studies.

Step	Description of step	Description
1	Energy program development and management	The optimizing agent will typically evaluate the *universe* of existing or new energy management programs of the entity (building, city, agency) in question. This includes, but is not limited to, market research, customer engagement, needs analysis, determination of required resources for delivery, and benefits
2	Project development	The optimizing agent will typically provide the analysis of potential energy efficiency undertakings that can be *practically deployed in the context specific to the entity* in question. Activities normally include: • Quantify the benefits and explore any technical challenges that may arise during implementation. • Analyze customer's current energy use, expenses, projected energy savings and develop payback analyses to determine the cost-effectiveness of recommended measures. • Undertake preliminary project design, including assessment of site conditions and loads, initial equipment identification, sizing and location, system layout and feasibility to cooperate with existing systems. • Determine economic, environmental, resiliency, and operational impacts of various proposed projects.
3	Technology assessment	The optimizing agent will typically provide the entity in question with technology intelligence, including technical and cost trends, market readiness, customer benefits, economics, case studies, operation and maintenance.
4	Design and engineering support	The optimizing agent will typically review engineering analysis, plans, drawings and specifications for energy projects. The optimizing agent will typically evaluate engineering alternatives, including selection of equipment and systems proposed for installation to meet customer's objectives and facility requirements. The optimizing agent will review project designs and review and/or develop project cost and/or savings estimates to meet the intent of the project design. Optimizing agent will identify or confirm the agencies and documents required for permitting the project.
5	Pre-construction services	The optimizing agent will typically support the entity in question in activities prior to commencing construction, such as • Development of RFPs, and appropriate contract documents for energy projects and proposal evaluation criteria, as well as identify potential bidders. • Evaluate contractor proposals and proposed savings guarantees and provide technical support for procurement and contract awards. • Develop and review project schedules to meet the customer's requirements.

(*continued*)

Table 14.2 *(Continued)*

Step	Description of step	Description
6	Construction services	The optimizing agent will typically provide construction services including, but not limited to: • Review contract documents and provide recommendations for improvements. • Provide on-site construction management oversight, review project schedules to ensure scope and milestones can be achieved, and provide critical path analysis and schedule recommendations. • Provide detailed cost estimates. • Review construction change orders for scope, cost, and schedule impact. • Conduct site visits to daily activity reports, punch lists, anticipated scope, cost, or schedule changes, permitting and special inspection requirements, and any other construction documentation or conditions.
7	Post-construction services	The optimizing agent will typically provide services for projects upon completion, such as: • An assessment of contractor performance on projects; equipment and system performance evaluation including energy savings measurement and verification; project closeout process. • Provide closeout documentation inventory and review. • Prepare/review operator training programs as well as operations and maintenance programs and procedures.
8	Energy planning and procurement	The optimizing agent will typically provide energy planning and procurement services such as: • Hedging strategy assessment and design; risk management. • Energy portfolio management plans including energy and demand forecasting and energy budgeting. • Energy procurement including identification of energy suppliers/vendors, development of RFP, management of solicitation process, proposal evaluation, vendor selection, negotiation of contract. • Tariff analysis and energy billing analysis.
9	Facilities operational services	The optimizing agent will typically provide operations and maintenance (O&M) program services, including, but not limited to, planning, developing, procuring, and managing preventative maintenance and staffing programs for the entity in question.

14.4 Challenges and Future Directions

There is a stated interest in the industry to pursue a path as follows, which rationalizes the use of IoT in the energy space (Lee, 2015):

- *Near term:* Initial assessment that enables one to identify EIoT technologies and infrastructure that can be utilized for cost-effective open interoperable solutions.
- *Intermediate term:* Start deploying the broad range of available EIoT devices/sensors/systems that support equipment plug and play and open data exchange.
- *Long term:* Make use of the EIoT platform to achieve more cost-effective, deeper, and reliable savings across entire buildings, campuses, and smart cities in coordination with the SG evolution.

However, as is the case in general, there are a number of open issues and challenges related to the EIoT that need to be addressed by the industry—many of these issues have already been identified in previous chapters. There are challenges slowing down the IoT's ability to scale in the energy and lighting environment, including issues of security, privacy, and compliance; fragmentation of vertical markets; issues related to IT/OT and legacy infrastructure integration; availability or lack thereof of cost-effective ubiquitous connectivity; availability or lack thereof of big data models for data correlation; and dearth of accepted standards and limited interoperability (Curran, 2016; Pal and Purushothaman, 2016). These challenges are slowing down the IoT's ability to scale in the energy and lighting environment, although there is active work in this arena.

With the development of architectures and accepted standards EIoT will, while reducing the cost of incremental enhancement, also reduce or eliminate the risk of device or manufacturer obsolescence, of limited hardware, software, data, and service choices, while facilitating the emergence of open multivendor systems and improved data exchange under a service-based architecture (Poplawski, 2015b). Some of the challenges include but are not limited to the following (Lynn, 2015):

- Business/government policies do not (yet) encourage interoperability; the current administrative/regulatory thrust is more on technology advancement and efficiency savings than on product interoperability among the providers.
- Interoperability can be perceived by some as a commoditization threat: Some suppliers see interoperability as a potential commoditization and competitive threat.
- State of standards-making has not encompassed business processes or aligned business objectives; current standards are targeted at much lower layers of an overall IoT architecture (for example at the physical, data link, or networking

layer, rather than at a vertical application level); thus, interoperability is still focusing on lower layer protocols and not yet entering the informational level.
- Interoperability decisions are often confusing at technology level: A large set of solutions exist, these typically being vendor specific: Currently, there is a broad variety of communication and syntactic choices; furthermore, communication layers are often not cleanly separated from the data model.

One activity addressing interoperability is related to the Open Field Message Bus Protocol (OpenFMB) (Kiliccote, 2015). OpenFMB is a framework for distributed intelligent nodes interacting with each other; it is data centric rather than device centric. It leverages cost-optimized industrial internet technologies applied to SG DER systems, to enable communication via common semantic definitions. In this framework, grid-edge nodes process (some) data locally for control and reporting. OpenFMB supports field-based applications that enable scalable peer-to-peer publish/subscribe architectures.

An important area of development relates to cybersecurity (Minoli et al., 2017c). Threats to the power grid have been well documented in the past. Major advancements are expected to be brought to bear in the future, especially in this critical EIoT segment. There is a recognition that the IoT would benefit from a "security by design" approach, since security infractions are more likely to occur when security is not a consideration throughout the requirement and design process; attempts to "bolt on" security features late in the product development process are both more expensive and more subject to creating product vulnerabilities (The Department of Commerce Internet Policy Task Force & Digital Economy Leadership Team, 2017). Such a bottom-up approach to cybersecurity is, for all intents and purposes, a novel (future) approach. Given the large amounts of data that IoT devices collect, there are privacy concerns raised by IoT, also as related to energy (how much is a certain individual consuming, at what point of the day, occupancy data, and so on); improved privacy protection mechanisms will have to be added to IoT ecosystems.

Some are advocating Named Data Networking (NDN) (also known as Content-Centric Networking (CCN), content-based networking, data-oriented networking, or information-centric networking) for IoT applications, due to the large amount of data generated by the pervasive infrastructure. (Most networks today are device/location/host centric rather than data centric.) Communication in NDN is driven by receivers, the data consumers. This is accomplished through the exchange of Interest and Data packets. Packets carry a name that identifies a piece of data that can be transmitted in one Data packet (Named Data Networking Organization, 2018). A consumer includes the name of a desired piece of data into an Interest packet and transmits it. Network routers utilize the name to forward the Interest toward the data producer(s). Once the Interest packet reaches a node that has the requested data, the node will return a Data packet to the consumer that contains both the name and the content, also including a digital signature by the producer that binds the two elements. This Data packet follows the reverse path in order to get back to the requesting consumer.

The motivation is that a network should allow a user to focus on the data needed (the content), rather than having to identify and reach a specific, physical location where that data is to be retrieved from (Datta and Bonnet, 2016; Named Data Networking, 2018; Zhang et al., 2014). NDN offers benefits such as simpler network devices, which are particularly applicable to the IoT and content caching to reduce congestion and reduce delivery latency. If NDN principles are adopted by the industry for the IoT, this would be a significant paradigm change compared with the present day perspective.

14.5 Conclusion

This chapter assessed the evolution of the power grid to an advanced smart grid. The importance of the EIoT has been discussed.

In the next few years, distribution automation and embedded intelligence will continue to penetrate the grid, and will also be incorporating reliable self-healing. Increasingly, one can expect distributed energy resources to become part of the smart grid and the continued proliferation of these distributed resources will force utility companies to accept changes and make adaptations. While the use of natural gas will continue to grow and even accelerate, the availability of renewable energy will continue to expand given that prices continue to fall. The variability and uncertainty of output from renewable sources create new challenges for the traditional distribution architecture necessitating greater intelligence and more dynamic control of energy generation assets. Additionally, creative demand management along with energy storage will mature into a viable grid-scale resource in the next few years that can serve to better match supply demand perturbations. Distribution Management Systems, both centralized and distributed, will drive system efficiency in conjunction with the Automated Metering Infrastructure, smart homes, smart appliances, smart lighting, and smart home energy management systems (SHEMSs).

The chapter emphasized that emerging IoT techniques and enabling technologies are being leveraged to facilitate efficiencies in energy production, management, distribution, and consumption, and even a small efficiency improvement translates into a significant savings of trillions of dollars worldwide. Greening of the environment is an important governmental and social imperative. The EIoT is critical for the effective integration of renewable sources (both at the commercial generation level as well as the consumption level) and to facilitate the deployment of advanced distribution architectures. EIoT is expected to play a key role in smart cities, smart campus, and smart buildings applications, such as smart metering, demand response implementations, advanced sensor-based connected LED lighting, and in energy storage environments. The level of infrastructure investment in these technologies will be high and beyond the fiscal realities, several daunting problems remain regarding interoperability with the existing infrastructure, the need to maintain different generations of technology as the grid continues to evolve, and the growing importance of cybersecurity in the deployment of these systems to preserve safety and predictability of operation. Despite these challenges, EIoT technology (due to its more distributed and resilient architectures which are also capable of integrating multiple energy sources dynamically—as a function of availability and relative cost-effectiveness) promises to offer a more reliable, cost-effective, and potentially eco-friendly energy creation/distribution/consumption/management system. The future of EIoT has all indications of, indeed, being very "bright." The chapter that follows covers implementation strategies for the EIoT in general and for renewable energy in particular.

References

Ali, A. (2016) DOE Peer Exchange: Better Buildings Residential Network, June 9. U. S. Department of Energy. Available at https://energy.gov/sites/prod/files/2016/07/f33/060916_InteractiveMedia_Summary.pdf.

Berganza, I. and Sanchez-Fornie, M. A. (2016) *Telecommunication Networks for the Smart Grid*, Artech House, Norwood, MA. ISBN 9781630810467.

Bhide, S. (2016) Lighting: A New Ambient Computing IoT Platform based on Intel Architecture, U.S. Department of Energy (DOE) 2016 Connected Lighting Systems Workshop, February 2–4, in Raleigh, NC.

Bialek, T. (2015) Grid Integration of Solar Energy What Have We Learned, Grid Integration of Solar Energy Workshop SDGE, October 29, Office of Energy Efficiency & Renewable Energy, Washington, DC. Available at https://energy.gov/sites/prod/files/2015/12/f27/Tom%20Bialek.pdf.

Brodrick, J. R. (2015) Connected Lighting Systems, DOE 2015 Connected Lighting Systems Workshop, November 16, Portland, OR. U.S. Department of Energy.

Bui, N., Castellani, A. P., Casari, P., and Zorzi, M. (2012) The Internet Of Energy: A Web-Enabled Smart Grid System. *IEEE Network*, **26**(4). doi: 10.1109/MNET.2012.6246751.

Burger, A. (2018) The Emerging "Energy IoT," eniday whitepaper, eniday, Piazzale Mattei,1 00144 Roma. Available at https://www.eniday.com/en/technology_en/the-emerging-energy-iot/.

Bush, S. F. (2014) *Smart Grid: Communication-Enabled Intelligence for the Electric Power Grid*, Wiley-IEEE Press. ISBN 978-1-119-97580-9.

Cárdenas, A. A. and Berthier, R. (2014) A Framework for Evaluating Intrusion Detection Architectures in Advanced Metering Infrastructures. *IEEE Transactions on Smart Grid*, **5**(2), 906–915.

Carvallo, A. and Cooper, J. (2015) *The Advanced Smart Grid: Edge Power Driving Sustainability*, 2nd edn, Artech House, Norwood, MA. ISBN 9781608079636.

Cho, W. T., Lai, Y. X., Lai, C. F., and Huang, Y. M. (2013) Appliance-Aware Activity Recognition Mechanism for IoT Energy Management System. *The Computer Journal*, **56**(8): 1020–1033. doi: https://doi.org/10.1093/comjnl/bxt047.

Curran, J. W. (2016) LEDs and Lighting Controls: Exciting Future and Some Cautions. NFMT2016, National Facilities Management and Technologies, March 22–24, Baltimore.

Datta, S. K. and Bonnet, C. (2016) Integrating Named Data Networking in Internet of Things Architecture. Consumer Electronics-Taiwan (ICCE-TW), 2016 IEEE International Conference on, May 27–29. doi: : 10.1109/ICCE-TW.2016.7520967.

Designlight Consortium. (2015) Why Lighting Systems Need to Evolve, U.S. Department of Energy (DOE) 2015 Connected Lighting Systems Workshop, November 16, Portland, OR.

Doris, E. and Peterson, K. (2011) Government Program Briefing: Smart Metering, NREL, U.S. Department of Energy, Office of Energy Efficiency & Renewable Energy, Technical Report NREL/TP-7A30-52788, September.

Duine, P. (2016) The New World of Connected Light, U.S. Department of Energy (DOE) 2016 Connected Lighting Systems Workshop, February 2–4, in Raleigh, NC.

Eisen, M. (2009) Introduction to PoE and the IEEE802.3af and 802.3at Standards, October 29, Marcum Technology. Available at https://www.ieee.li/pdf/viewgraphs/introduction_to_poe_802.3af_802.3at.pdf.

Ela, E., Gevorgian, V., Fleming, P., Zhang, Y.C., Singh, M., Muljadi, E., and Scholbrook, A. (2014) Active Power Controls from Wind Power: Bridging the Gaps, Technical Report, NREL/TP-5D00-60574, January, National Renewable Energy Laboratory, Golden, CO.

Electrical and Electronics Engineers (IEEE). (2018) IEEE 802.3af and IEEE 802.3at (PoE plus), ieee.org

Energy Information Administration, (2015) Key World Energy Statistics 2015. International Energy Agency, France. Available at www.iea.org.

ETSI, (2011) *ETSI TS 102 690: Machine-to-Machine Communications (M2M); Functional Architecture.* October. Available at http://www.etsi.org/deliver/etsi_ts/102600_102699/102690/01.01.01_60/ts_102690v010101p.pdf.

ETSI, (2018a) ETSI TR 102 691: Machine-to-Machine Communications (M2M); Smart Metering Use Cases. (2010-05). ETSI, France.

ETSI, (2018b) ETSI TR 102 897: Machine to Machine Communications (M2M); Use Cases of M2M Applications for City Automation. (2010-01). ETSI, France.

Feeney, J. (2016) The Energy Internet of Things. Dexmatech Blog. July 25, DEXMA, Spain. Available at http://www.dexmatech.com/energy-internet-of-things-iot/.

Feiden, T. (2018) Power over Ethernet (PoE) Design Considerations. SuperiorEssex Technical Materials, Atlanta, GA.

Fouda, M. M., Fadlullah, Z. M., and Kato, N. (2012) A novel demand control policy for improving quality of power usage in smart grid, IEEE Global Communications Conference (GLOBECOM), December 3–7.

Funabashi, T. (ed.) (2016) *Integration of Distributed Energy Resources in Power Systems – Implementation, Operation and Control*, Elsevier. ISBN 978-0-12-803212-1.

Haus, D. (2016a) Energy trends 2020: the energy Internet Of Things. DestinHaus Webpage, Rancho Santa Margarita, CA, February 10. Available at http://destinhaus.com/2016/02/10/.

Haus, D. (2016b) Energy Storage Innovations. DestinHaus Webpage, Rancho Santa Margarita, CA, December 23. Available at http://destinhaus.com/2016/12/23/

Herbst, T. (2015) Why Lighting Systems Will Become More Connected?, U.S. Department of Energy (DOE) 2015 Connected Lighting Systems Workshop, November 16, Portland, OR.

Hofbauer, J., Wolf, M., and Rudolph M. (2016) Self-sufficient sensors based on energy harvesting validation and evaluation study of tram bearing diagnostics using simulation techniques. 13th International Multi-Conference on Systems, Signals & Devices (SSD), March 21–24, 2016. doi: 10.1109/SSD.2016.7473700.

Hua, R., Gu, Y., He, Z., Yang, J., Cao, T., and Huang, A. (2016) Power over Ethernet Method, Apparatus, Device, and System, US 9268383.

Hummon, M., Palchak, D., Denholm, P., and Jorgenson, J. (2013) Grid Integration of Aggregated Demand Response, Part I: Load Availability Profiles and Constraints for the Western Interconnection. NREL/U.S. Department of Energy Office of Energy Efficiency & Renewable Energy, Technical Report LBNL-6417E, September.

IEEE Power and Energy Society (PES) (2018) Task Force.

Iorga, M. and Shorter, S. (2012) Advanced metering infrastructure smart meter upgradeability test framework. National Institute of Standards and Technology, July (DRAFT NISTIR 7823).

Kiliccote, S. (2015) Integration of Distributed Energy Resources. Grid Integration Solar Energy Technical Workshop October 29, U.S. Department of Energy.

Kley, H. (2016) What is a Distributed Energy Resource Management System? April, Spirae, LLC. Available at https://www.gtai.de/GTAI/Content/EN/Meta/Events/Invest/2016/Reviews/Hannover-messe/smart-grids-forum-2016-presentation-holger-kley.pdf?v=2

Kyriazis, D. Varvarigou, T. White, D. Rossi, A., and Cooper, J. (2013) Sustainable smart city IoT applications: heat and electricity management & eco-conscious cruise control for public transportation. World of Wireless, Mobile and Multimedia Networks (WoWMoM), 2013 IEEE 14th International Symposium and Workshops on a, June 4–7. doi: 10.1109/WoWMoM.2013.6583500.

Lai, C. F., Lai, Y. X., Yang, L. T., and Chao, H. C. (2012) Integration of IoT energy management system with appliance and activity recognition. Green Computing and Communications (GreenCom), 2012 IEEE International Conference on, November 20–23. doi: 10.1109/GreenCom.2012.20.

Lee, E. (2015) High Performance Active Perimeter Building Systems – 2015 Building Technologies Office Peer Review. Lawrence Berkeley National Laboratory and U.S. Department of Energy (DOE). Available at https://energy.gov/sites/prod/files/2015/05/f22/cbi57_Lee_041415.pdf.

Lehpamerm, H. (2016) *Introduction to Power Utility Communications*, Artech House, Norwood, MA. ISBN 9781630810061.

Leonardo Energy/European Copper Institute (2016) World Energy Expenditures, Brussels, Belgium. Available at http://www.leonardo-energy.org

Lynn, K. (2015) Connected Buildings Interoperability Vision, May 20, U.S. Department of Energy. Available at https://energy.gov/sites/prod/files/2015/05/f22/bto_ConnectedBldgInteropVisionWebinar_05202015.pdf.

Matiko, J. W. and Beeby, S. P. (2017) *Applications of Energy Harvesting Technology in Buildings*, Artech House, Norwood, MA. ISBN 978-1-60807-981-0.

Minoli, D. (2010) Designing Green Networks with Reduced Carbon Footprints. *Journal of Telecommunications Management*, **3**(1), 15–35.

Minoli, D. (2011) *Designing Green Networks and Network Operations*, Francis and Taylor.

Minoli, D. (2013) *Building the Internet of Things with IPv6 and MIPv6: The Evolving World of M2M Communications*, John Wiley & Sons, Inc., New Jersey.

Minoli, D., Occhiogrosso, B., and Sohraby, K. (2017a) A Review of wireless and satellite-based M2M services in support of smart grids. 1st EAI International Conference on Smart Grid Assisted Internet of Things (SGIoT 2017), July 11-13, Sault Ste. Marie, Ontario, Canada.

Minoli, D., Occhiogrosso, B., and Sohraby, K. (2017b) IoT considerations, requirements, and architectures for smart buildings – energy optimization and next generation building management systems. *IEEE IoT Journal*, **4**(1), 269–283.

Minoli, D. Sohraby, K. et al. (2017c) IoT Security (IoTSec) Considerations, Requirements, and Architectures" (IEEE CCNC 2017), January, Las Vegas, Nevada.

Mouftah, H. T. and Erol-Kantarci, M. (2016) *Smart Grid - Networking, Data Management, and Business Models*, CRC Press, New York.

Named Data Networking. (2018) Named Data Networking of Things. Available at https://named-data.net/publications/ndn-iotdi-2016/.

Named Data Networking Organization. (2018) NDN Packet Format Specification 0.2-1 documentation. Available at http://named-data.net/doc/ndn-tlv/.

National Electrical Manufacturers Association. (2013) Volt/VAR Optimization Improves Grid Efficiency, August, NEMA, Virginia. Available at https://www

.nema.org/Policy/Energy/Smartgrid/Documents/VoltVAR-Optimazation-Improves%20Grid-Efficiency.pdf

Normandeau, K. (2015) Importance of a Distribution Management System (DMS) in grid modernization, Microgrid Knowledge, October 2. Available at https://microgridknowledge.com/importance-of-distribution-management-system-dms-grid-modernization/

Pal, A. and Purushothaman, B. (2016) *IoT Technical Challenges and Solutions*, Artech House, Norwood, MA. ISBN 978-1-63081-111-2.

Plett, G. (2015) *Battery Management Systems, Volume 1: Battery Modeling*, Artech House, Norwood, MA. ISBN 9781630810238. (G. L. Plett (2015) *Battery Management Systems, Volume II: Equivalent-Circuit Methods*, Artech House, Norwood, MA. ISBN: 9781630810276).

Poplawski, M. (2015a) The Value Of Energy Data, U.S. Department of Energy (DOE) 2015 Connected Lighting Systems Workshop, November 16, Portland, OR.

Poplawski, M. (2015b) DOE Focus Areas, U.S. Department of Energy (DOE) 2015 Connected Lighting Systems Workshop, November 16, Portland, OR.

Quinlan, J. (2016) A New Lighting Paradigm, U.S. Department of Energy (DOE) Connected Lighting Systems Workshop, February 2–4, Raleigh, NC.

Rehmani, M. H., Khan, A. A., and Reisslein, M. (2016) Cognitive radio for smart grids: survey of architectures, spectrum sensing mechanisms, and networking protocols. *IEEE Communications Surveys & Tutorials*, **18**(1), 860–898.

Remy Marcotorchino. (2015) oneM2M Partnership Project, U.S. Department of Energy (DOE) 2015 Connected Lighting Systems Workshop, November 16, Portland, OR.

Samadi, P., Mohsenian-Rad, H., Schober, R., and Wong, V. W. (2012) Advanced demand side management for the future smart grid using mechanism design. *IEEE Transactions on Smart Grid*, **3**(3), 1170–1180.

Sawyer, K. (2018) Active integrated perimeter building systems. Lawrence Berkeley National Laboratory. Available at https://energy.gov/eere/buildings/active-integrated-perimeter-building-systems.

Sedano, R. (2010) Making the most of responsive electricity customers. Mid-America Regulatory Conference, June 8. Available at https://energy.gov/sites/prod/files/oeprod/DocumentsandMedia/MakingtheMostofResponsiveElectricityCustomers.pdf.

Serra, J., Pubill, D., Antonopoulos, A., and Verikoukis, C. (2014) Smart HVAC control in IoT: energy consumption minimization with user comfort constraints. *The Scientific World Journal*, **2014**, 11. doi: http://dx.doi.org/10.1155/2014/161874.

Shroufa, F. and Miragliottaa, G. (2015) Energy management based on Internet Of Things: practices and framework for adoption in production management. *Journal of Cleaner Production*, **100**, 235–246. doi: http://dx.doi.org/10.1016/j.jclepro.2015.03.055.

Sinopoli, J. (2016) *Advanced Technology for Smart Buildings*, Artech House, Norwood, MA. ISBN 978-1-60807-865-3.
Smart Grid. (2018) Advanced Metering Infrastructure and Customer Systems. U.S. Government Smart Grid Website.
SmartGrid.gov. (2018) Advanced Metering Infrastructure and Customer Systems. Available at https://www.smartgrid.gov/recovery_act/deployment_status/sdgp_ami_systems.html.
Spikes, A. (2011) Smart Meters and a Smarter Grid, May 16, Available at https://energy.gov/energysaver/articles/smart-meters-and-smarter-grid.
Staff (2015) BSRIA Study Shows Big Data And Convergence Is Growing. December, BSRIA Ltd., Bracknell, Berkshire. Available at www.bsria.co.uk.
Staff, (2018) MicroGrid Control: a solutions brief. Trimark Associates, Inc. Available at http://trimarkassoc.com/wp-content/uploads/MicroGrid-Solutions-Brief.pdf.
Staff of ABB Inc. (2015) Demand Response Management System, white paper. Available at http://new.abb.com/docs/librariesprovider139/default-document-library/drms-overview_abb.pdf?sfvrsn=2.
Strong, D. and Burrows, V. (2016) *A Whole-System Approach to High-Performance Green Buildings*, Artech House, Norwood, MA. ISBN 978-1-60807-959-9.
Sweco Staff, (2015) Study on the effective integration of Distributed Energy Resources for providing flexibility to the electricity system. Final report to The European Commission, April 20. Proj no: 54697590000, Sweden.
The Department Of Commerce Internet Policy Task Force & Digital Economy Leadership Team (2017) Fostering the advancement of the Internet Of Things, January. Green Report. Available at: https://www.ntia.doc.gov/files/ntia/publications/iot_green_paper_01122017.pdf.
The State and Local Energy Efficiency Action Network. (2014) Insights from Smart Meters: Identifying Specific Actions, Behaviors, and Characteristics That Drive Savings in Behavior-Based Programs. Customer Information and Behavior Working Group, December. Document DOE/EE-1158.
Thornton, B. and Rosenberg, M. (2011) Achieving the 30 percent Goal: Energy and Cost Savings Analysis of ASHRAE Standard 90.1-2010, May. U.S. Department of Energy.
U. S. Department of Energy. (2006) Benefits of Demand Response in Electricity Markets. February.
U. S. Department of Energy. (2016) EERE Success Story—Software Platform Offers Underserved Smaller Buildings Sector an Innovative Solution for Cutting Energy Costs and Waste", August 4. Available at https://energy.gov/eere/success-stories/articles/eere-success-story-software-platform-offers-underserved-smaller.
U.S. Department of Energy. (2015) Grid Integration Of Solar Energy Workshop, October 29, SunShot.
U.S. Department of Energy (DOE). (2015) 2015 Connected Lighting Systems Workshop, November 16, Portland, OR.

U.S. Energy Information Administration (2012) Commercial Buildings Energy Consumption Survey (CBECS), Energy Usage Summary, Edition. Available at http://www.eia.gov/consumption/commercial/reports/2012/preliminary/.

U.S. Energy Information Administration, (2016) International Energy Outlook 2016 (IEO2016), EIA projects 48% increase in world energy consumption by 2040, May 12. Available at http://www.eia.gov/todayinenergy/detail.php?id=26212.

Wesemann, D., Dünnermann, J., Schaller, M., Banick, N., and Witte, S. (2015) Less wires — a novel approach on combined power and ethernet transmission on a single, unshielded twisted pair cable. Factory Communication Systems (WFCS), 2015 IEEE World Conference on, May 27–29. doi: 10.1109/WFCS.2015.7160588.

Zhang, L., Afanasyev, A., Burke, J., and Jacobson, V. (2014) Named data networking. *ACM SIGCOMM Computer Communication Review*, **44**(3), 66–73. doi: 10.1145/2656877.2656887.

ZigBee Alliance (2018) Zigbee Light Link. Available at http://www.zigbee.org/zigbee-for-developers/applicationstandards/zigbee-light-link/

15

Implementing the Internet of Things for Renewable Energy

Lucas Finco[1] *and Daniel Minoli*[2]

[1]*Principal Consultant, Strategain, New York, NY, USA*
[2]*IoT Division, DVI Communications, New York, NY, USA*

15.1 Introduction

The previous chapter described at a broad level the concept of the Energy Internet of Things (EIoT); this chapter takes the discussion in the direction of practical considerations in the process of implementing the EIoT concepts covered in that previous chapter.

Fundamentally, the energy process entails generation, distribution, monitoring, control, and consumption. Each of these areas is currently experiencing innovation. In the generation arena, the promise of renewable sources of electricity, including the emergence of distributed energy resources (DERs) is significant. Renewable generation technologies are clean, abundant, and now widespread. Some of these technologies have low ongoing operational cost, although infrastructure investments are required to build the systems necessary to manage them. At this time, the electricity industry operates on the assumption that humans can control every detail of the production and distribution. However, renewable sources of power generate an amount of electricity that typically fluctuates from moment to moment. Furthermore, they generate electricity at the "wrong times" of day, creating availability when there is low demand and shortage when demand is highest. Thus, while renewable resources have been added to the grid in large quantities, wind, solar and microgrids intrinsically create management challenges due to unpredictable and variable energy outputs.

Distribution can also be a challenge since the solar or wind farms may not be in optimal proximity with either the consumer or the existing distribution infrastructure. There is a need to monitor all the energy-supporting elements of the grid. Additionally, it is desirable to manage usage and/or incentivize conservation; it is desirable to deploy home (and/or commercial establishment) devices,

Internet of Things A to Z: Technologies and Applications, First Edition. Edited by Qusay F. Hassan.
© 2018 by The Institute of Electrical and Electronics Engineers, Inc. Published 2018 by John Wiley & Sons, Inc.

such as programmable thermostats, that manage electricity use while homeowners are absent.

As the power industry searches for a new paradigm of distribution and management that can incorporate renewables into the grid, the use of Internet of Things (IoT) devices can provide solutions by flexibly managing demand. IoT concepts specifically optimized to the energy industry are ideal. EIoT can alleviate the cited challenges by coordinating electric demand with electric supply; this can speed up the adoption of renewable and sustainable electricity technologies. EIoT in the home has produced devices such as programmable thermostats that manage electricity use while homeowners are absent. There is a broad industry support for EIoT technologies; Brundu et al. (2017); Elhebeary et al. (2017); Kaur and Sood (2015); Li et al. (2017); Minoli et al. (2017a, 2017b); Papaioannou et al. (2017); Saber and Khandelwal (2017); and Singh et al. (2017) provide an introductory, if limited, set of recent resources describing various aspects of the applicability and usefulness of IoT in the energy context.

This chapter describes a number of key developments that are required to facilitate the broad-scale deployment of a dynamic Smart Grid (SG), managed under EIoT mechanisms. Issues related to control of renewable generation sources utilizing IoT principles are discussed, as are industry initiatives in the context of required Information and Communication Technologies (ICT)/IoT standardization. Furthermore, the absolutely critical requirements for grid security are highlighted.

15.2 Managing the Impact of Sustainable Energy

The classical power grid relies mainly on electricity generation by stable, dispatchable, controlled sources, such as large centralized fossil fuel plants. *Distributed renewables* are sources of renewable energy produced by local communities and private homeowners and circulated through the common grid. Several examples of distributed renewables are listed in Table 15.1.

The proliferation of new energy technologies presents challenges and opportunities for operators. To examine these issues, it is important to understand the interconnection of energy generation and storage in a grid system. The local generation of energy by many renewable sources presents two challenges: nondispatchability and highly variable output. A nondispatchable energy source cannot be controlled based on the desired energy flow. A variable energy source fluctuates in the amount of energy it produces minute by minute.Therefore, adding large quantities of renewable energy to the power grid also creates large amounts of unpredictability.

Consider the case of distributed solar energy as a variable, nondispatchable renewable energy technology that is undergoing widespread customer adoption

Table 15.1 Distributed renewable generation sources.

Distributed renewable generation sources	Description
Solar	These devices convert sunlight directly into useable energy, such as hot water, steam, and electricity generated by photovoltaics
Wind	Devices that are used to convert wind energy into electricity, or provide angular momentum directly to drive equipment, such as windmills
Hydro	Devices that convert the potential energy of water into useful energy, using dams, or provide angular momentum directly, such as watermills
Fuel cells	Devices that convert chemical potential energy in H_2 and O_2 molecules into H_2O, CO_2, and useful electricity

and needs to be integrated into the energy grid effectively. One example of this can be observed by examining the solar output fluctuations over 1 day in Oahu, Hawaii, as shown in Figure 15.1. The line is the amount of solar electricity production at 3 s intervals through a typical mostly sunny day. Note the extreme volatility of energy available for electricity production at one site. Complicating this variability is the need for energy to be generated and used almost instantaneously. While some believe current electric grids include the capability to store electricity, this is not the case: there is no significant energy storage connected to electric grids that is capable of retaining electricity for a period of time until it is needed. In the absence of this storage capacity, the amount of electricity produced must match the amount of electricity used at any given time of the day or year. If they do not match, there is a risk of poor power quality (brownouts), the shedding of customer loads from the electricity supply (blackouts), or even damage to the grid. To precisely match supply with demand,

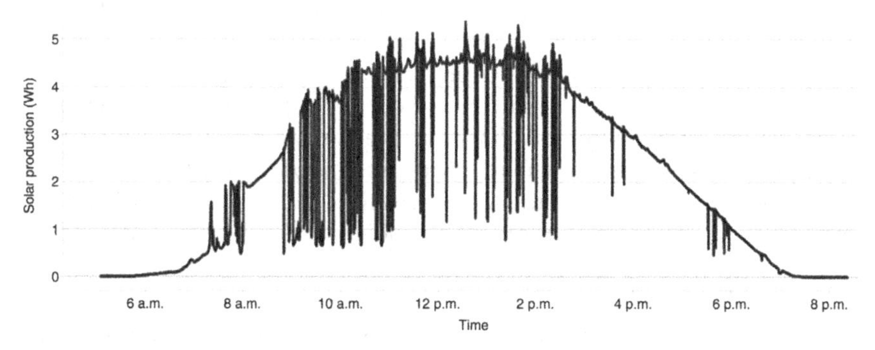

Figure 15.1 Solar output for July 1, 2011 in Oahu, HI.

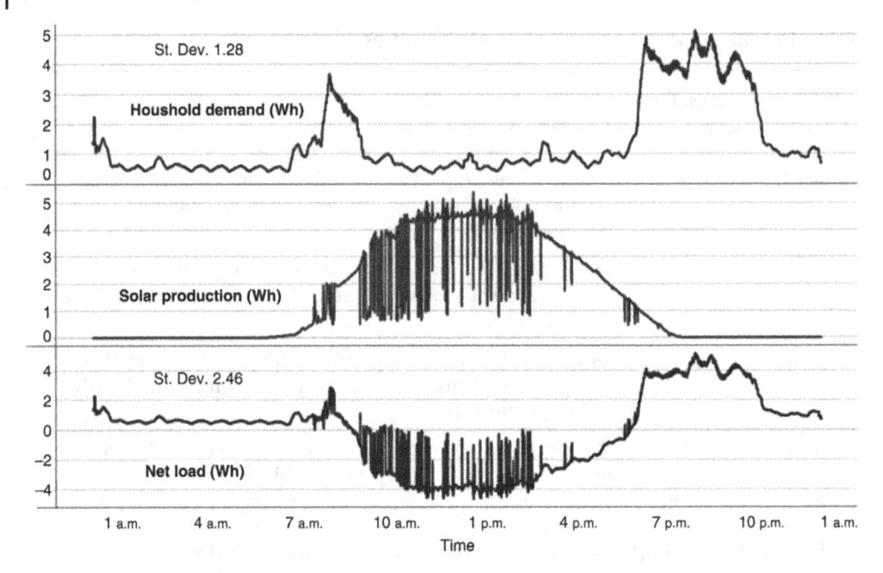

Figure 15.2 Household demand curve, solar output curve, and net load curve.

currently electricity demand is monitored and production controlled by a live system operator, who schedules what electricity is supplied every few minutes, as well as automatic regulation systems, which can adjust the flow of power to the grid in less than a second.

A typical residential demand curve is shown in Figure 15.2. The first curve shows a typical household's electricity consumption through 1 day. The second curve shows a typical day's solar production, scaled such that total solar production equals total household consumption. The final curve is household demand minus solar production, with negative values indicating power flowing back to the electric grid. Note the timing mismatch of consumption with production and also the increased volatility, taking standard deviation from 1.28 to 2.46 Wh. Specifically, this household consumes about 40 kWh/day. The standard deviation, a measure of the variability of the curve, is shown as 1.28 Wh. The variable solar output curve peaks in the middle of the day when sunlight is at its highest, while the household demand curve has peaks in the morning, as residents get ready to go to work, and in the evening, when residents return home to complete daily household activities such as preparing dinner, watching TV, or doing laundry (IEA Energy Conservation, n.d.).[1] Note that when solar production is subtracted from a household's load, the total load drops

1 These data were recorded at 3 s intervals and describe the whole-house electrical usage of a 6-person, single-family residence in Irvine, CA.

Table 15.2 Energy storage technologies.

Energy storage	Description of technology
Chemical batteries	Storage of energy in chemicals in a sealed cell
Hydrogen storage	Use of excess generated electricity to drive electrolysis, creating hydrogen and oxygen that can be stored for later use in a fuel cell to generate electricity
Electric vehicles	Use of plug-connected electric vehicle batteries by charging at off-peak hours and releasing electricity into the grid at peak hours
Hydrostorage	Storage of energy by pumping water uphill using electricity and releasing it back downhill to generate electricity
Uninterruptible power supplies	Energy storage devices for critical applications, such as computers, that require a backup energy supply
Thermal storage	Capturing energy in thermal stores, like ice, that can be released later to provide power intensive services such as air-conditioning

drastically. Excess power flows back to the grid during midday hours (hence, the negative values for net load), and power is demanded from the grid in the morning and evening hours. In this illustration, we matched daily solar output to daily demand, such that the daily net load would total to zero. This means that on net the solar electricity produced by the home is being used by the homeowners. However, notice that the variation of the net load has doubled to 2.46 Wh.

As new renewable generation technologies have matured and become economically competitive, the lack of storage capacity within our electric grids has presented challenges. Examples of the energy storage technologies are given in Table 15.2. Because electricity must be used as soon as it is produced, and until storage capacity can be added to the grid, the *demand curve* (electricity demanded by consumers) must match the *supply curve* (electricity provided by the available power sources). In the case of two prevalent forms of renewable energy, wind and solar, it is challenging to make the supply fit the demand without adding energy storage to the grid. EIoT is one way to do this.

In the residential solar example, if it is possible to store the energy produced, it would be possible even with a simple strategy to smooth out both the rates of production and consumption of energy. This hypothetical but possible scenario can be seen in Figure 15.3, which demonstrates a simulation that incorporates storage. The first curve shows one way to utilize storage over a day by storing solar electricity in the middle of the day and releasing it in the morning and evening hours. By subtracting this storage curve from the *net load* curve from Figure 15.2, we end up with *net load 2*, the resulting load after incorporating solar and storage to a typical household. Note how this *net load 2* stays near zero during the whole day, but is still quite volatile, dropping the standard deviation from 2.46 to 0.73 Wh. Note that the battery charges during

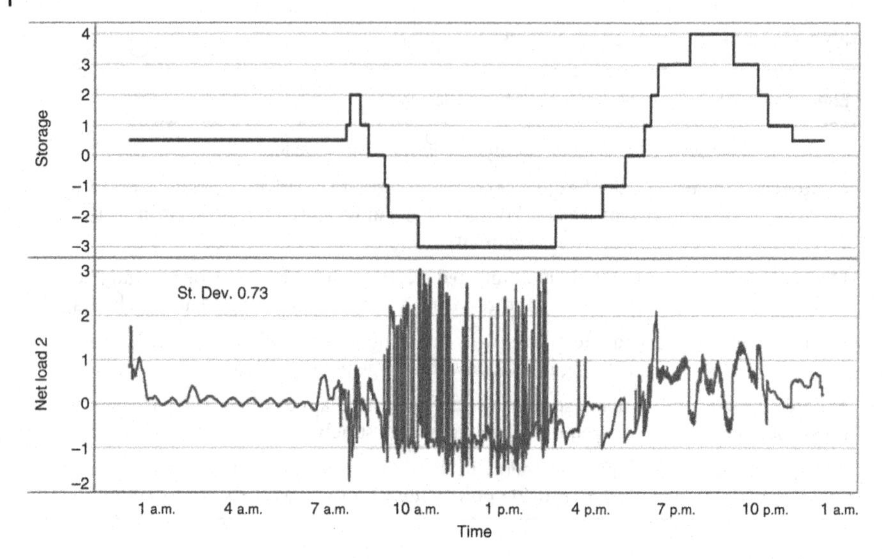

Figure 15.3 Storage simulation.

the day and discharges in the morning and evening hours. In this example, the battery releases 5.9 kWh of energy in the morning, absorbs 24.5 kWh of energy during the day, and releases 17.6 kWh of energy in the evening. The application of this crude storage strategy also drastically reduces variation in the net demand, reducing the standard deviation to 0.73 Wh. Storage is the most effective method to manage the nondispatchability of renewables, while drastically reducing variability of demand.

In the residential solar example, pairing dynamic, short-term reductions from EIoT devices with our storage strategy can provide the best example of EIoT devices managing the complexities of renewable energy. This can be seen in Figure 15.4, which shows a dynamic reduction simulation. The first curve shows how a fast-responding EIoT device could reduce loads quickly in response to solar output fluctuations. The second curve is obtained from subtracting the dynamic reductions from the *net load2* curve from Figure 15.3. Note that the use of dynamically-varying loads drops the standard deviation from 0.73 to 0.61 Wh from Figure 15.3. By addressing the sharp peaks of the load and using devices to manage short-lived peaks, variation in net load drops further (down to 0.61 Wh), while maintaining low net energy flows to and from the grid, and balancing renewable generation and consumption. Fortunately, solar output has a good correlation with midday loads; however, solar does not always produce electricity during peak consumption times. Electricity prices are also high when loads are high. It is the case that the production price of wind and solar energy is very low compared to other energy sources. However, incorporating large amounts of non-dispatchable

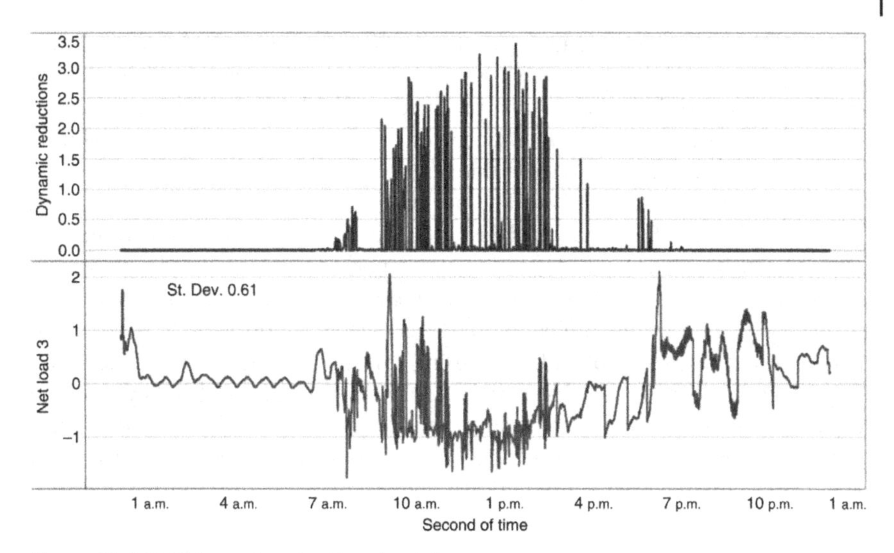

Figure 15.4 EIoT dynamic reduction simulation.

energy sources into the power supply may or may not contribute generation at peak consumption times. This fact can exaggerate the high prices of electricity at those times and will not reduce the capital costs related to generating peak loads.

In the residential solar example, EIoT gives us the ability to shift loads from times when renewables are not available to times that they are. Figure 15.5 shows

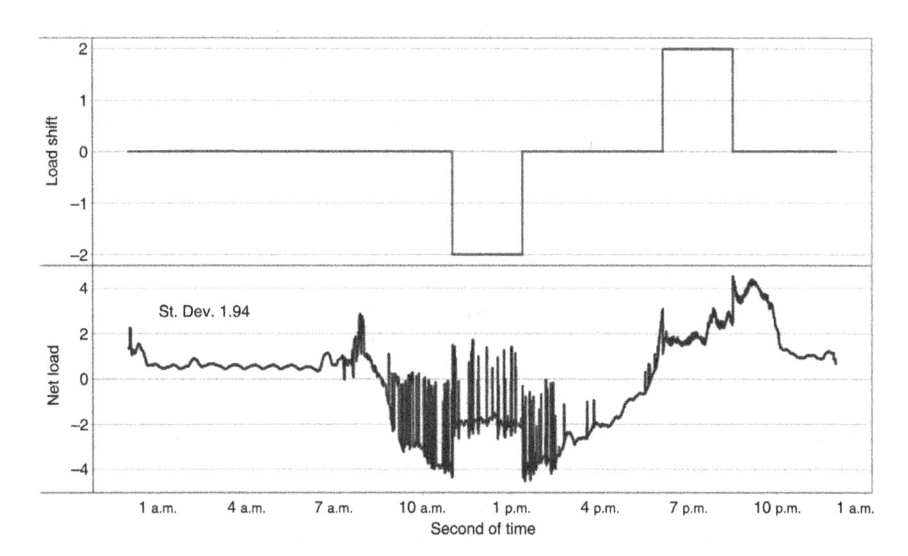

Figure 15.5 Load shift simulation.

an example of shifting a load, by changing the operating schedule of a single 2 kW appliance. The first curve shows the impact of shifting a 2 kW load from evening hours to midday when solar production is high. The second curve shows the load shift added to the *net load* from Figure 15.2. Note that even a very simple strategy like this can move loads in line with solar production and drop standard deviation from 2.46 to 1.94 Wh. The net result is similar to what happens with storage, and also reduces standard deviation, but is not as flexible and granular as storage can be. Still, load shifting is a cheaper alternative to storage, and is another tool to help adapt loads to renewable generation. Although it is difficult to closely match the demand curve to the renewable energy supply curve due to necessities such as use of indoor lighting during evening hours, minimizing disparities between renewable supply and electricity demand is key. Storage capacity that allows electricity to be saved for later use can be combined with EIoT technologies that allow part of the consumer load to be shifted to hours when renewable electricity production is at its highest. This is important because battery storage is currently too expensive to make it economically feasible, requiring technical innovation or mass production of units to bring down its cost in the near future. While the need for cost-effective energy storage is desirable to manage renewable energy sources, the use of EIoT to redistribute consumption loads can reduce the need for both additional energy storage and generation capacity at a low cost today.

15.3 EIoT Deployment

The benefits of using EIoT to manage and incorporate renewables into energy grids and to manage the consumption of energy and environmental impact far outweigh any of the drawbacks of these technologies. In a traditional grid operation model, dispatchable generators are directed to increase or decrease their outputs to ensure that supply matches demand. In an EIoT scenario, demand would be modified to match the supply produced by renewable and conventional energy resources. Internet-connected appliances would use sensors to monitor and alter consumer energy use. This would create dispatchability in consumer demand that would offset nondispatchability in renewable electricity supplies. EIoT devices could measure moment-to-moment variability in electricity generation and adjust their operations accordingly. Overarching patterns of renewable electricity supply could also be considered and accommodated. In combination with energy storage, this process could significantly reduce the need for fossil fuel peakers and allow consumers to avoid paying peak electricity prices.

EIoT devices can be used to enable the integration of renewable energy generation into the electric grid. The variability, unpredictability, and non-dispatchability of renewable energy generation sources such as solar and wind make their integration into the grid a challenge. Much of the existing equipment on the grid for managing variations in voltage, load, or frequency is old, bulky, slow, manually operated, and/or not fast acting. Due to the high costs of implementing a modern smart grid, most utilities have avoided such expenditures. But with the connection of more and more renewables, the need for equipment to balance out such fluctuations is slowly mounting.

EIoT devices offer a streamlined solution to this problem. Instead of implementing utility-scale solutions to manage the grid, EIoT devices can respond to the previously discussed challenges of energy management, including the following:

- *Variability.* Figure 15.1, which shows one day of solar output in Oahu, Hawaii, demonstrates that renewable power sources can have extremely high variability of output, even on one-minute timescales or lower. In a traditional grid operation model, dispatchable generators (typically fossil fuel generators) would be directed to increase or decrease their outputs to ensure that demand was met with the necessary supply. In an EIoT scenario, renewable generation's variability could be directly measured by EIoT devices on the grid, sensors on the grid, or a central control. These EIoT devices could then respond quickly by increasing or lowering electricity loads, mitigating the negative effects of renewable generation variability on the grid.
- *Unpredictability.* The output of solar and wind tends to have some prediction error, especially in the short term. Predicted output is used to schedule other generator assets to meet expected loads on the grid. In a traditional model, when renewables do not provide the generation that was predicted, other dispatchable generation must be used to balance supply and demand. This dispatchable generation tends to be fossil fuel-fired generation, largely defeating the purpose of operating renewables. In an EIoT scenario, any surplus or deficit between renewable power output and forecast can be quickly utilized by EIoT devices. EIoT devices would receive communications that renewable output was not matching forecast and modify their operations appropriately.
- *Nondispatchability.* Renewable generation is generally nondispatchable, meaning that one cannot turn it on and off whenever one wants. Further complicating matters, due to the zero-variable cost of renewables, the grid always accepts renewable energy when it is produced. This causes a plethora of problems, such as the potential for negative prices, required larger generation reserves, and high ramp rates (like the "duck curve"[2]) for

2 When solar electricity production is subtracted from aggregated loads on the grid, it results in a net load curve with a dip in the middle and a sharp ramp up when the sun goes down. Some believe such a curve looks like a duck, hence the name duck curve (Roberts, 2016).

dispatchable generators. EIoT devices provide a unique, readily available, and fast-responding resource that can provide changes in load that respond quickly to changes in supply. This aspect of controlling demand, and not just dispatchable supply, is new to the industry and opens up a world of possibilities to managing the future grid.

The widescale adoption of EIoT requires the deployment of EIoT elements in many, if not all, of the SG components, including generation, transmission, and consumption. This supports the "I" and the "T" of the ICT enhancements required to make EIoT a reality. Additionally, a cost-effective but ubiquitous network infrastructure is needed to tie these elements together. This supports the "C" and the "T" of the ICT enhancements also required to make EIoT a reality. The subsections that follow discuss these two realms.

15.3.1 EIoT Elements

Building a reliable EIoT ecosystem is the best innovation to facilitate the sustainable energy grid of the future. A basic component of the EIoT ecosystem is the EIoT element or device. An EIoT device is a device that controls energy consumption, production, or storage, while having all the characteristics of an IoT device, such as Internet connectivity. Outfitting EIoT devices with sensors, smart controls, processing power, information storage, connectivity, and cloud management creates the capability to solve the intermittency, variability, and unpredictability currently limiting the potential of renewable energy generation (Conti et al., 2017). Use of renewable energy sources and innovation in EIoT can change historic patterns of generation and consumption. This creates new patterns of electricity production and usage that differ from the ways in which electric grid operators manage systems today. Future grid management will involve innovative ways to communicate between all these devices and control their operations in an orchestrated way.

EIoT devices will require outside data to optimize their operations. These data form datasets, based on information from reliable weather prediction services, electric grid prices and loads, and other data relevant to control of EIoT devices. Some of these datasets are listed in Table 15.3. The goals of EIot energy management include the following:

- Minimize energy costs
- Minimize environmental impact
- Increase grid stability
- Increase grid reliability

An important category of EIoT changing the face of grid management is the use of smart controls. These devices are connected both to the Internet and to

Table 15.3 Energy IoT device datasets.

Energy IoT device data	Data examples
Weather data/weather forecast	Outdoor temperature, dew point, wind speed, solar irradiance
Energy consumption	Device energy use, energy consumption patterns, future consumption schedules
Local sensor data	Indoor temperature, line voltage, frequency, and phase angle
Renewables production	Local and regional solar or wind output
Grid data	Local and regional energy loads and prices, peak use indicators, reliability and resiliency indicators

the electric grid. They use machine learning to make intelligent decisions on energy use for the owners of energy-consuming devices. These devices continue to improve in their efficiency and effectiveness while increasing customer comfort. There has been a dramatic increase in smart controls and Wi-Fi-enabled energy consuming devices in recent years. However, these devices still largely lack the coordination necessary to aid in the management of grid challenges.

Ideally, smart controls will be able to make direct measurements of device usage patterns and grid health. EIoT devices will receive information on grid operations and prices. They can then make intelligent decisions using machine learning on when and how to consume or store energy. This analysis will enable them to communicate their energy use and storage schedules to other devices or grid operators and receive instructions to optimize for price, emissions, grid stability, or grid reliability. Examples of potential customer applications of smart controls are listed in Table 15.4.

For these benefits to be realized, all of these new EIoT devices need to be connected through various possible communications technologies to the Internet, where information about device operations can be shared and optimizations performed.

15.3.2 Network Functionality

Another basic component of the EIoT ecosystem is the network. In order for EIoT to be successfully implemented, electronic devices must be able to easily connect to the utility's extranet, to the "cloud," and/or to the Internet (Paek et al., 2017). Specifically, for the SG benefits to be realized, all of the new EIoT devices discussed in the previous subsection need to be connected through various communications technologies to the control and computing infrastructure,

Table 15.4 Smart controls and customer usage.

Smart controls	Customer usage
Smart thermostats	Smart thermostats adjust HVAC usage while customers are out of the home
Smart outlets	Wi-Fi-connected outlets give users internet access to data about their energy usage
Smart chargers	Charging appliances charge user devices according to a schedule or other specific criteria
Wi-Fi-enabled controls	Any energy device can be enabled with smart controls
Smart power strips	Smart outlets help power strip users control their energy usage
Building management systems	Central computers control the energy consumption of all devices in a building
Occupancy sensors	Sensors in rooms adjust energy use in response to detected movement
Smart appliances	Smart appliances offer Wi-Fi connectivity and increased capabilities, such as scheduling and responsiveness to outside signals like price and grid health
Smart inverters	Intelligent inverters convert DC power to AC power, with Wi-Fi connectivity and scheduling options

often realized via the Internet, where information about device operations can be shared and optimizations performed. There are three strategies conceived for managing EIoT coordination: a centrally planned and controlled energy management platform, a web of independent control devices making their own decisions about optimization, or a compromise between the two that utilizes gateway devices. Cost-effective connectivity requires the creation of industry standards that allow devices to interconnect and the adoption of safety measures to protect the grid.

Central Control

Under the central arrangement in Figure 15.6, all EIoT devices would communicate via the Internet with one central energy management platform, likely hosted in the cloud. Examples of EIoT devices include smart thermostats, smart outlets, connected batteries, electrical vehicle chargers, and smart water heaters. This platform would take in sensor measurements, device operation schedules, price signals, and renewable energy generation levels. It would then perform an optimization of device operations to derive the best outcome based on price, carbon emissions, grid stability, and/or reliability. The platform would then send out operation instructions to devices in the field, enabling them to operate according to the platform's instructions.

Figure 15.6 EIoT central control diagram.

Note that this central platform concept does not need to be all-encompassing. It can be used in small cases to manage a single utility feeder, or in large cases to manage price and stability in a Regional Transmission Operator (RTO) or Independent System Operator (ISO). The main point of this concept is that the device sends its information to a central platform that performs the optimization. This helps to focus the computing power in one location, rather than having large amounts of computing power deployed in smart devices, which would sit idle for a large percentage of time. Having one central computing platform could also enable more datasets to be utilized in optimization and more complex optimization methods to be performed, possibly requiring greater computational resources.

Web of EIoT

Under the web arrangement in Figure 15.7, each smart EIoT device communicates over the Internet directly with other EIoT devices. This illustration shows EIoT devices from two different homes and one business communicate with each other, not a central control, to optimize their operations. The solid lines represent communications flows (e.g., parametric data), while the dashed lines represent control flows. A cloud-based data source provides third-party data like weather and electricity prices. Management of grid irregularities are negotiated between EIoT devices as they collect grid data and communicate with each other about the optimal energy use for each device. This arrangement requires no central control to operate, but requires more communications

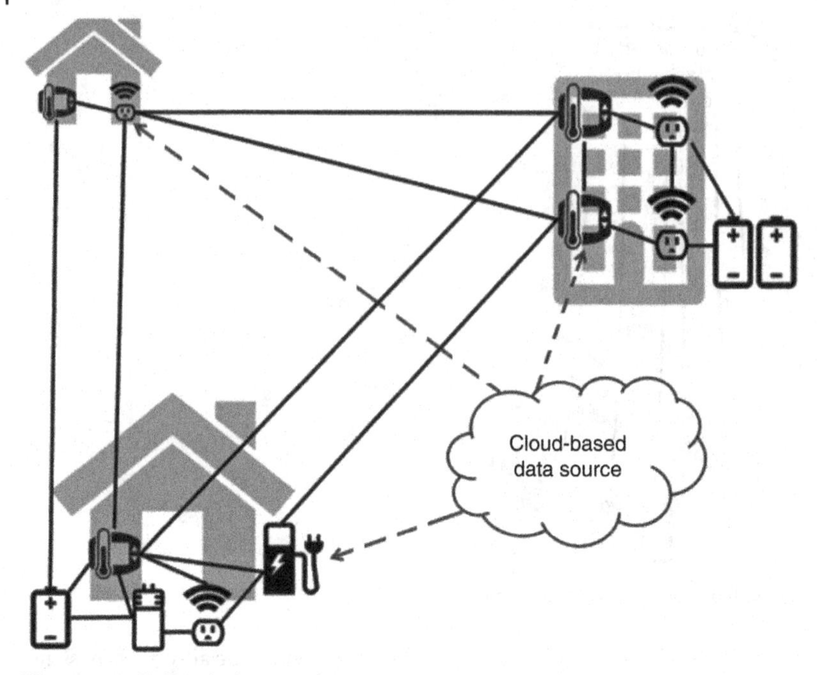

Figure 15.7 EIoT web diagram.

bandwidth between a large number of EIoT devices. It also requires extra data storage and more powerful CPUs, as each device must now collect data and make optimization calculations itself.

Gateway

There is a continuum of options between the two above. Under the gateways arrangement in Figure 15.8, smaller local hubs, or gateways, can act as "localized" central control devices and manage EIoT devices in that building. The illustration shows two homes and one business with EIoT devices that communicate with gateway devices in each building. These gateways optimize EIoT operations, without the complexity of a full web control strategy. This arrangement reduces the need for each EIoT device to manage multiple connections and negotiations, as was the case in the web of devices strategy. Therefore, these devices can do with less powerful CPUs and memory. The gateway can house high power CPUs and extensive memory, and optimize each building's EIoT operations, while negotiating optimal strategies with other buildings. This strategy reduces communications volume, reduces EIoT complexity, but limits flexibility as each EIoT device depends on the gateway for operations scheduling.

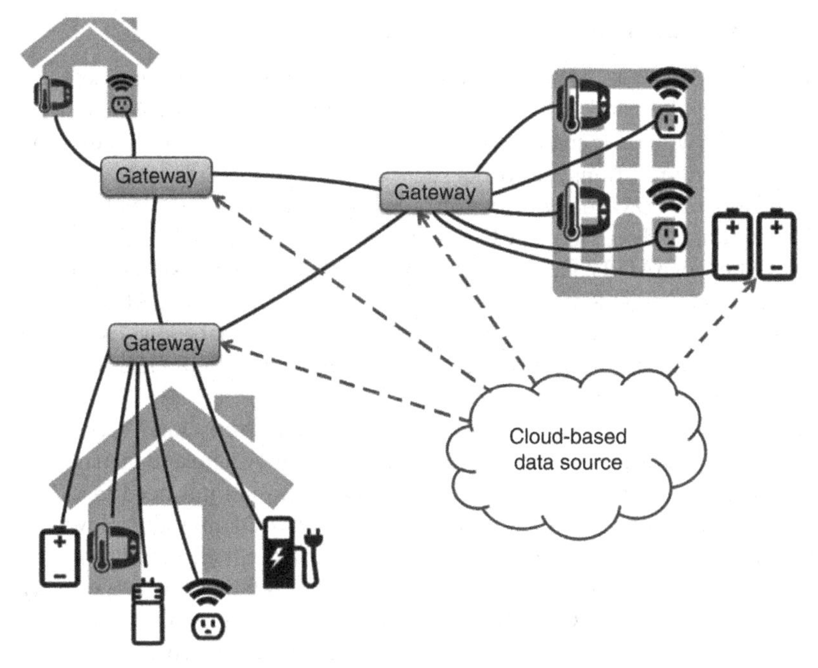

Figure 15.8 EIoT gateways diagram.

15.4 Industry Standards for EIoT

In order to implement EIoT paradigms, it is clear that industry standards will need to be in place. Many different devices, produced by different companies and even different industries, will need to be able to communicate with each other, share information, and respond to control signals. While some generic IoT standards are now emerging, energy-specific standards are also needed (JaeSeung, 2016; Meddeb, 2016). Equipment and/or products from electric utilities, electrical equipment manufacturers, building controls companies, home appliance manufacturers, and indeed any Wi-Fi-enabled device manufacturer will all be interested in adhering to such standards. The reward for ensuring that all IoT devices adhere to standards and interface with the larger EIoT world is that a new stream of value will be unlocked by incorporating these devices into the electric grid. There have been attempts at producing EIoT data standards, which have yielded a range of success. The first and most successful has been the Open Automated Demand Response (OpenADR), a standard for communicating with smart thermostats; others include the Building Energy Data Exchange Specification (BEDES) and IEEE 2030. These are briefly described in the subsections that follow.

15.4.1 Open Automated Demand Response

Created in 2010, OpenADR[3] refers to "an open and standardized way for electricity providers and system operators to communicate DR signals with each other and with their customers using a common language over any existing IP-based communications network, such as the Internet." In this context, *demand response (DR)* refers to methods for reducing consumer load at peak demand times, as discussed in Chapter 14. A group of industry stakeholders launched the OpenADR Alliance, with the intent of developing a system to make DR processes more efficient and, therefore, more cost effective. The Alliance was able to successfully define an energy industry IoT standard that allows electricity providers to seamlessly communicate with and give orders to internet-enabled thermostats in homes around the country.

Many companies that implement OpenADR systems are cloud based and allow a system operator to log on with a browser from anywhere and manage DR events using thousands of thermostats. An operator would not need to install any programs or apps. It is important to note that OpenADR now also allows the implementation of various types of DR programs rather than just direct load control (an industry term for a DR program that allowed the utility to directly control a customer's HVAC unit), including critical peak pricing dynamic rates, ISO/RTO ancillary service, and Electric Vehicle programs. OpenADR listed about 125 approved products at press time.

15.4.2 Building Energy Data Exchange Specification (BEDES)

BEDES[4] is a standard for defining terms, definitions, and formats for data when monitoring building characteristics and energy consumption. First released in October 2014, BEDES was created as a way to standardize the collection of energy data inside buildings, and increase the accessibility and comprehension of data across various products in the EIoT universe. In practicality, BEDES gives building managers the information they need to make better decisions about how to plan, integrate, operate, and maximize renewable energy systems in their buildings. BEDES does this by tracking energy performance and verifying renewable energy production throughout buildings.

As it is relatively new, few products in the market are now using the BEDES standards. However, it is being utilized by government agencies. Currently, the U.S. Department of Energy (DOE) makes use of the BEDES standards in their Commercial Energy Asset Score & Home Energy Score program. The EPA has also embraced the BEDES standards for their Portfolio Manager & Home Energy Yardstick program, which encourages sustainable energy practices (U.S. Department of Energy, 2016).

---

3 www.openadr.org

4 https://bedes.lbl.gov/bedes-online

15.4.3 Institute of Electrical and Electronics Engineers (IEEE) 2030™

IEEE 2030™ is a standard that "provides guidelines in understanding and defining smart grid interoperability of the electric power system with end-use applications and loads. Integration of energy technology and information and communications technology is necessary to achieve seamless operation for electric generation, delivery, and end-use benefits to permit two-way power flow with communication and control" (IEEE, 2011). This standard is designed to empower energy consumers and drive integration of renewable energy, electric vehicles, and EIoT. The IEEE 2030 standard will be a guide to EIoT devices that incorporates grid services and connectivity with the smart grid; it includes three architecture perspectives: power systems, communications technology, and information technology.

15.5 Security Considerations in EIoT and Clean Energy Environments

Security considerations should be a necessity when it comes to managing EIoT applications (e.g., see (Laszka et al., 2017; Liang et al., 2017)). As long as proper grid operation relies on the action of thousands or millions of individual devices working in harmony over an open Internet, there exists a risk that nefarious (or naive actors) can cause harm to the electric grid. Some of the main security concerns are as follows (among others):

- *Cyber Intrusions.* A harmful actor could target EIoT devices in specific areas and disrupt their operations to cause damage to grid equipment or cause a blackout in a specific area. The most well-known examples of harmful hacking activities involve industrial equipment and processes. Those devices tend to be relatively secure compared to EIoT devices, which essentially sit on the Internet exposed. This means that EIoT devices need to prioritize security as if the operation of the entire electric grid relies on them.
- *Software Malfunctions.* Utility planners will design (and have already begun designing) the electric grid to rely on EIoT devices to help manage peak loads. Utility planners are beginning to account for peak load response programs when planning the peak grid loads. This means that utilities are not installing equipment to serve load that is necessary, and instead relying on load-reducing EIoT devices that will respond to peak load signals. If EIoT devices do not deliver on peak load reductions estimated by utility planners, blackouts could result. It is important that EIoT devices act in a repeatable and predictable fashion so that utility planners can act in a prudent manner, reduce capital investments in the grid, and realize the value that EIoT can provide. An interruption at critical peak times can have damaging impacts on the grid.

- *Protocol Flaws.* EIoT devices will also rely on a plethora of communications pathways, technologies, and providers. Wi-Fi, ZigBee, GPS, fiber, cable, and 3G/4G/5G cell services will be utilized by EIoT devices as well as many other technologies. Any disruption in these services and core technologies can create problems for the reliable delivery of electricity to consumers.
- *Operational Complexities.* The control logic of EIoT devices needs to be carefully designed to avoid unintended consequences. Nonlinear effects can emerge in a complex web arrangement of devices. Specifically, adverse feedback loops could cause EIoT devices to operate in unintended ways. Devices might wait until grid capacity is available and inexpensive, increasing their load. This will in turn utilize available capacity, increasing price, and causing devices to decrease their load. The cycle then repeats. If enough devices act this way, unstable oscillations of load can occur on the grid. These oscillations are surely not the way consumers expect their devices to operate.
- *External Dependencies.* All of the EIoT devices installed in the electric grids will rely on many outside data sources, including weather data, weather forecasts, grid data, and more. Any disruption in these data streams could cause unintended operations of EIoT equipment, and possible interruption of operations. These data streams will also need to be safeguarded to ensure their reliability.
- *Unintended Consequences.* It is possible for EIoT devices to have an intentional or unintentional impact on the economics of the grid. If large numbers of the same device become attached to an electric grid, it would be easy for these devices to work in concert to exercise market power on the grid. In this scenario, market power is the ability to influence or control electricity prices.

The electric grid has historically been a very conservative endeavor. Great care is taken to ensure committed resources are available at all hours of the year to meet the demands of electric customers. It carries a critical role in the health of the economy and the well-being of the country's citizens. Ensuring that new technologies do not disrupt this reliable system is of the utmost importance and should not be taken lightly. The safety and security of grid operations should be carefully considered before implementing any EIoT device.

15.6 Conclusion

The increased deployment of green energy has a beneficial effect not only in the context of stewardship of the planet, but also for corporate or national economic reasons. There is strong global interest in deploying renewable energy sources, but also making sure that grid reliability is maintained. However, because green sources often are of the "variable energy source" type, very granular control of the resources is needed. EIoT principles and technologies are well suited to this task. This chapter described a number of key developments that are required to effectuate the deployment of EIoT devices, managed by the SG, managed under EIoT mechanisms.

Because naturally-produced energy can be unpredictable and highly variable, grid operators have two potential responses. They can supplement renewables on the supply side with fossil fuel sources that idle production to offset the variability of renewable energy production. Better yet, they could adjust load on the demand side using EIoT devices that can shift some consumer electricity load to times when renewable production is strong. The first option could result in high emissions, defeating the original purpose of incorporating renewables. Only the second option, the use of EIoT to adjust demand to match renewable supply, ensures that the benefits of renewables are realized. Equipping electrical devices with monitoring sensors, smart controls, and connecting them to the Internet will not allow EIoT appliances to manage their own energy use and be controlled remotely, but will in fact support the goal of energy flow reliability. Thus, EIoT devices will aid in managing the grid and incorporating renewable sources of energy into the electricity supply. They can provide a measure of demand-side control that will help offset the unpredictability of renewable energy sources as these are incorporated into the grid. This would reduce the need for peakers that represent costly infrastructure investments as well as additional sources of carbon dioxide emissions. Ideally, renewables combined with EIoT will transform the current electrical grid into the cleaner and more dynamic grid that we all deserve.

Standardization and security are two key requirements to move the process along. No matter what configuration is chosen, it will be important to standardize EIoT data so that all devices can "talk" to one another successfully. It will also be important to develop appropriate security protocols and apply these to all EIoT devices. There is a risk of harmful actors using EIoT devices to break into the electricity grid and disrupt the electricity supply. If the functionality of the electric grid depends on individual appliances, each of these devices needs to be appropriately secured.

In summary, as it was discussed earlier, injecting large quantities of renewable energy to the power grid is desirable, but it can create global unpredictability unless the process is properly managed: the variability, unpredictability, and nondispatchability of renewable energy generation sources make the integration of these sources into the grid a *prima facie* challenge, but one that is addressable. Related to this predicament it is desirable to manage usage and foster conservation by deploying devices, such as programmable thermostats, that manage electricity use while occupants are absent. Additionally, physical distribution can also be a challenge because the generation elements (solar and/or wind farms) are typically not close to the existing distribution infrastructure or to the consumer. EIoT mechanisms afford management capabilities that can "bridge the gap." Regarding the *variability* challenge, with EIoT mechanisms the renewable generation's variability can be measured by EIoT sensing entities embedded in the grid, also in conjunction with a central control, enabling the EIoT-based devices to respond quickly by increasing or lowering electricity loads, thus mitigating the negative effects of renewable generation variability on the grid. Regarding the *unpredictability* challenge, a surplus or deficit between renewable power output and forecast can be quickly analyzed by EIoT sensing entities and EIoT devices will receive communications that renewable output is not matching forecast and that they should modify their operations appropriately. Regarding the *nondispatchability* challenge, EIoT devices provide fast responding control mechanisms that can provide changes in load to respond rapidly to changes in supply.

The expectation is that the challenges alluded to in the discussion above will be overcome in the next few years and the EIoT-based smart grid will become a practical reality.

References

Brundu, F.G., Patti, E., Osello, A., Del Giudice, M., Rapetti, N., and Krylovskiy, A. (2017) IoT software infrastructure for energy management and simulation in smart cities. *IEEE Transactions on Industrial Informatics*, **13**(2), 832–840. doi: 10.1109/TII.2016.2627479.

Conti, F., Schilling, R., and Schiavone, P.D. (2017) An IoT endpoint system-on-chip for secure and energy-efficient near-sensor analytics. *IEEE Transactions on Circuits and Systems I: Regular Papers*, **64**(9) 2481–2494. doi: 10.1109/TCSI.2017.2698019.

Elhebeary, M., Ibrahim, M., Aboudina, M., and Mohieldin, A. (2017) Dual-source self-start high-efficiency micro-scale smart energy harvesting system for IoT applications. *IEEE Transactions on Industrial Electronics*, **PP**(99). doi: 10.1109/TIE.2017.2714119.

IEA Energy Conservation in Buildings & Community Systems. (2015), Annex 42: The Simulation of Building-Integrated Fuel Cell and Other Cogeneration Systems (COGEN-SIM). Three-Second Daily Demand Data, Part 1: (IEA 3 Second Daily Demand 2_4_05_to_2_14_05 part 1.zip). Three-Second Daily Demand Data, Part 2: (IEA 3 Second Daily Demand 2_4_05_to_2_14_05 part 2.zip). Available at http://www.ecbcs.org/annexes/annex42.htm.

IEEE (2011) Standard 2030-2011 - IEEE Guide for Smart Grid Interoperability of Energy Technology and Information Technology Operation with the Electric Power System (EPS), End-Use Applications, and Loads.

JaeSeung, S. (2016) IoT standards toward its next stage. *IEEE Communications Magazine*, **54**(7), 14–16. doi: 10.1109/MCOM.2016.7514158.

Kaur, N. and Sood, S.K. (2015) An energy-efficient architecture for the Internet of Things (IoT). *IEEE Systems Journal*, **11**(2), 796–805. doi: 10.1109/JSYST.2015.2469676.

Laszka, A., Dubey, A., Walker, M., and Schmidt, D. (2017) providing privacy, safety, and security in IoT-based transactive energy systems using distributed ledgers. *Proceedings of the Seventh International Conference on the Internet of Things*, ACM, Linz, Austria. ISBN 978-1-4503-5318-2/17/10. doi: https://doi.org/10.1145/3131542.3131562.

Li, Y., Orgerie, A.-C., Rodero, I., Parashar, M., and Menaud, J.M. (2017) Leveraging renewable energy in edge clouds for data stream analysis in IoT. *CCGrid '17 Proceedings of the 17th IEEE/ACM International Symposium on Cluster, Cloud and Grid Computing, Madrid, Spain — May 14–17*, IEEE, pp. 186–195. ISBN: 978-1-5090-6610-0 doi: 10.1109/CCGRID.2017.92.

Liang, L., Zheng, K., Sheng, Q., Wang, W., Fu, R., and Huang, X. (2017) A denial of service attack method for IoT system in photovoltaic energy system. *Network and System Security*, International Conference on Network and System Security, NSS Springer, pp 613–622. (Part of the Lecture Notes in Computer Science book series (LNCS, volume 10394).)

Meddeb, A. (2016) Internet of Things standards: who stands out from the crowd? *IEEE Communications Magazine*, **54**(7), 40–47. doi: 10.1109/MCOM.2016.7514162.

Minoli, D., Occhiogrosso, B., and Sohraby, K. (2017a) A review of wireless and satellite-based M2M services in support of smart grids. 1st EAI International Conference on Smart Grid Assisted Internet of Things (SGIoT 2017), July 11–13, Sault Ste. Marie, Ontario, Canada (With B. Occhiogrosso et al).

Minoli, D., Occhiogrosso, B., and Sohraby, K. (2017b) IoT considerations, requirements, and architectures for smart buildings – energy optimization and next generation building management systems. *IEEE Internet of Things Journal*, **4**(1), 269–283. doi: 10.1109/JIOT.2017.2647881.

Paek, J., Gnawali, O., Vieira, M., and Hao, S. (2017) Embedded IoT systems: network, platform, and software. *Mobile Information Systems*, **2017**. Article ID 5921523, doi: https://doi.org/10.1155/2017/5921523.

Papaioannou, T.G., Kotsopoulos, D., Bardaki, C., Lounis, S., and Dimitriou, N. (2017) IoT-enabled gamification for energy conservation in public buildings. Global Internet of Things Summit (GIoTS), Geneva, Switzerland, June 6–9. doi: 10.1109/GIOTS.2017.8016269.

Roberts, D. (2016) Why the "Duck Curve" Created by Solar Power is a Problem for Utilities. Available at https://www.vox.com/2016/2/10/10960848/solar-energy-duck-curve.

Saber, A.Y. and Khandelwal, T. (2017) IoT based online load forecasting. Green Technologies Conference (GreenTech), Ninth Annual IEEE, March 29–31, 2017. doi: 10.1109/GreenTech.2017.34.

Singh, N., Dayama, P., and Randhawa, S. (2017) Photonic energy harvesting: boosting energy yield of commodity solar photovoltaic systems via software defined IoT controls. *e-Energy '17 Proceedings of the Eighth International Conference on Future Energy Systems, Shatin, Hong Kong — May 16-19*, Proceedings Pages, ACM, 56–66. ISBN: 978-1-4503-5036-5. doi: 10.1145/3077839.3077857.

U.S. Department of Energy (2016) Building energy data exchange specification scoping report. Stakeholder & Technology Review, August. Available at https://energy.gov/sites/prod/files/2013/12/f5/bedes_scoping_080113.pdf.

16

The Internet of Things and People in Health Care

Nancy L. Russo and Jeanette Eriksson

Department of Computer Science and Media Technology, Internet of Things and People (IoTaP) Research Center, Malmö University, Malmö, Sweden

16.1 Introduction

Technology has long been viewed as a means of improving health care delivery, ranging from the ability to store and access digital health records to more complex applications such as using artificial intelligence to diagnose disease or using robots to perform surgery. Technology can help overcome limitations in resources (financial and personnel), can reach patients living long distances from health care specialists, and can help communities deal with aging populations. Both the United Nations and the World Health Organization have identified the need and potential for technology to support the delivery of health care services to underserved populations (UN, 2010; WHO, 2010). The ability to capture, store, analyze, and disseminate health information using technology can reduce the human resources needed to manage this process and improve the accuracy of the data. The ability to perform rapid analysis on this health data can ensure that scarce health care resources are allocated where needed most urgently. In remote, rural areas of many parts of the world, health care providers and services are often not available locally, requiring patients to travel long distances to obtain care. Technology, such as mobile device-based videoconferencing, can allow patients to be diagnosed and potentially treated remotely by doctors and other health care professionals. Technology can support communities faced with caring for growing numbers of elderly patients by providing the ability to monitor patients' vital signs, to easily track patients' eating, exercise, and medication schedules and provide reminders, and to strengthen social engagement between elderly patients and other community members. All of these services facilitated by technology can reduce the demands on health care professionals and other caregivers and increase the quality of life for patients.

Internet of Things A to Z: Technologies and Applications, First Edition. Edited by Qusay F. Hassan.
© 2018 by The Institute of Electrical and Electronics Engineers, Inc. Published 2018 by John Wiley & Sons, Inc.

The Internet of Things (IoT) expands the potential for using technology to support health care by connecting not only people, applications, and data, but also sensors and devices that collect biometric and contextual data. These health care ecosystems may include smart devices that are capable of using this data to take actions, such as providing alerts or sending notifications, when a measured value reaches a certain target level, providing recommendations to patients or health care professionals based on analysis of data, and potentially delivering medication based on monitored patient data and medical guidelines.

IoT-based systems can be used to address a broad set of health issues ranging from well-being to sickness, physical to mental health, preventive care to treatment or rehabilitation, and temporary disabilities to chronic disease. Smartphones, watches, and other smart devices as well as additional sensors, devices, and equipment can be connected to IoT networks to provide information about the patient, the patient's activities, and the patient's context or environment. By using IoT technologies to capture physiological measures and kinesiological data as well as other context data, smart health systems can empower people to proactively engage in their health as well as to manage their recovery from illness or injury. In addition, smart health can be used in health care settings to involve patients in their treatment and to share data about things such as treatment effectiveness. The patient, health care providers and caregivers, and other stakeholders, such as the patient's family members and friends, can also be part of the IoT ecosystem. This integration of human and technological agents and the sharing of vast amounts of data made possible by the Internet of Things, along with the applications that provide sophisticated data analysis, support better decisions about the patient's health and ultimately better health outcomes.

The focus of this chapter is IoT applications that support patients' health. These can be part of the broader health care system but are mainly used outside the hospital setting to collect and make use of data during the patient's daily life. Health care systems that are used solely in clinical, surgical, or laboratory settings are not discussed in this chapter, nor are health care systems that are primarily administrative or logistical in nature.

16.2 The Smart Health Care Ecosystem

The model in Figure 16.1 provides an overview of the components that may be included in an IoT-based smart health ecosystem. This is called an ecosystem to represent the complex and interconnected nature of the relationships between the human participants, the autonomous and subordinate devices, and the applications, interfaces, and data. This type of system or ecosystem is different from the early e-health systems that enabled a patient's health care data to be stored digitally and accessed remotely via the Internet and mobile devices (often

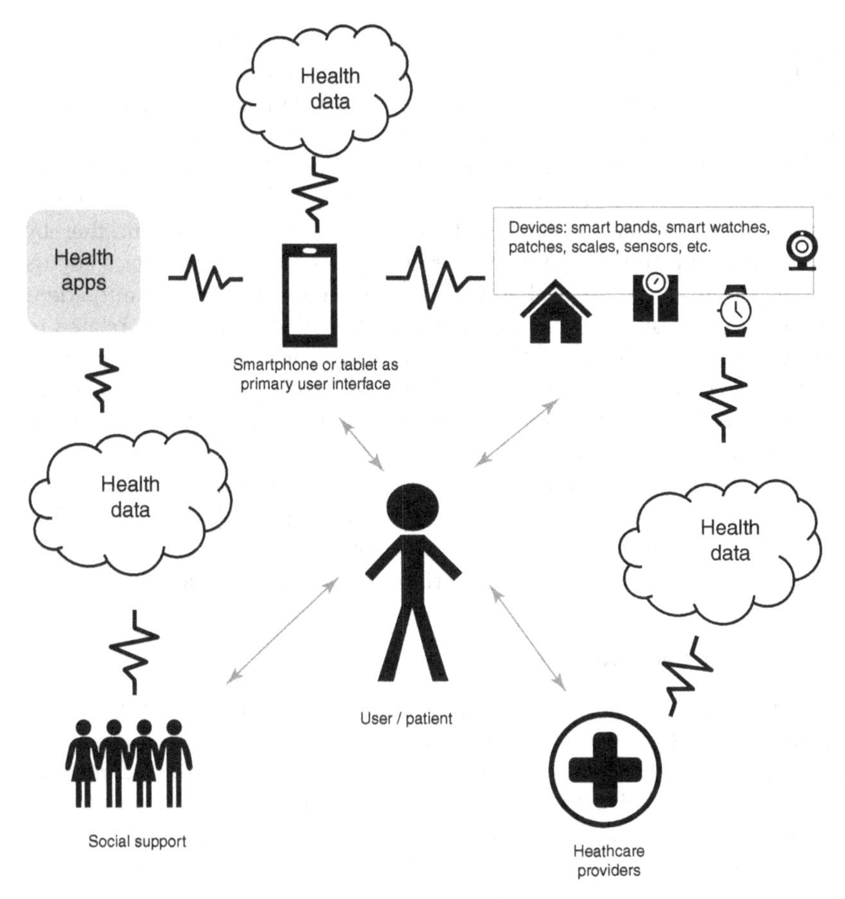

Figure 16.1 Smart health care ecosystem.

called m-health), or the ability to have an online consultation with a physician (telehealth). Now via the IoT, connected devices also store, access, and share data with other components of the ecosystem. Acting alone or in conjunction with other components, some nonhuman members (smart devices or agents) of the ecosystem have the capability to make decisions and take actions; for this reason, the ecosystem is considered to be "smart."

16.2.1 The Patient at the Center

The patient is illustrated at the center of this ecosystem because the primary purpose of a smart health care ecosystem is to support the patient in understanding, monitoring, and managing a health care-related issue. Trends in consumer use of applications and websites show increasing use of apps and websites to obtain information about health, to monitor health, and to more quickly interact with health care providers. This is driven both by the consumers increased health awareness and increasing familiarity in using computer-based products on a daily basis (Glaros and Fotiadis, 2005). People are both asked and enabled to play a more active role in their care. In addition, global economic challenges have led to reductions in resources for health care, which results in health care professionals having less time to spend with each patient, thereby limiting the opportunity to answer questions or discuss treatments. Patients spend less time in the hospital and more time recuperating on an outpatient basis. Therefore, patients have a greater need to obtain information related to their own health and have to take a more active role in managing all aspects of their health, from basic fitness and well-being to recovery from illness or injury as well as managing chronic diseases. Through the IoT, patients can be connected easily to other components of the ecosystem, including health care providers, a social support network, different types of devices and sensors, and applications running both locally and remotely, and can have access to data and services via familiar interfaces on smartphones, tablets, and computers. Each of these components will be described in more detail below.

16.2.2 Health Care Providers

The health care providers interacting with an IoT-based health care system include physicians and other medical specialists, pharmacists, nurses, therapists (such as physical, behavioral, or occupational therapists), laboratory technicians, and other caregivers. Health care providers interact with a smart health care system in a number of different ways, such as monitoring, diagnosing, treating, recommending actions, or communicating with the patient, the patient's family members, or with other health care providers.

A smart health care ecosystem provides the health care providers with access to continuous measures of patients' conditions, possibly on a number of metrics, over time. In addition, sophisticated diagnostic tools (applications and devices) may also be part of the ecosystem. Compared to what is possible with a patient's periodic visits to the physician's office or checkups from a visiting nurse, for example, this volume of data can allow the health care provider to recognize issues more quickly and to respond more rapidly to problems as well as to direct interventions to prevent more serious consequences. Medical professionals can more directly manage a patient's care by changing recommended treatments or medication levels based on analysis of the patent's activities, physiological measures, and other data. In managing patients' health-related behaviors, such as smoking cessation or weight reduction, health care providers may communicate with patients to set goals, monitor adherence, and provide feedback and recommendations to assist the patient in changing behavior.

16.2.3 Devices and Sensors

The number of devices and sensors that can be connected to the IoT is growing rapidly. Some devices and sensors are connected to the patient to measure heart activity (ECG), electric currents in the brain (EEG), retina activity (EOG), emotional stress through galvanic skin response (GSR), blood pressure, blood oxygen levels, and heart rate. Many of these physiological metrics can now be measured via wearable sensors that may be incorporated in a wristwatch-type device or sensors that are built into smartphones. Wearable devices may also track location (via GPS), sleep patterns, and activity levels, typically via number of steps. Gyroscopes and accelerometers in wearable devices measure orientation and speed. Other types of wearables include a vest that measures heart rhythms and can perform defibrillation when abnormal rhythms are detected and booties that monitor babies' heart and breathing rates. Other devices and sensors connected to the IoT include household appliances that can share data, and sensors that detect motion, air quality, temperature, and humidity. Devices such as scales, cameras, blood pressure monitors, eye trackers, glucose monitors, and insulin pumps can be connected. Even specialized clothing and furniture can contain IoT-connected sensors that monitor physiological and kinesiological measures, such as a smart hospital bed that tracks a patient's vital signs. Any device that can produce or receive data can potentially become part of the health care IoT ecosystem.

16.2.4 Applications and Interfaces

To make the data captured by IoT-connected devices useful, applications are needed to process the data and present it to users (whether humans, applications, or other devices) in a meaningful way using well-designed interfaces. Most patients, health care providers, and other stakeholders interact with health care applications via a smartphone, tablet, or computer. Although the most common interfaces are visual (data is displayed on a screen and users interact with the screen directly or through a keyboard and mouse) interactions may also be verbal, motion driven, haptic, or directly from device to device or application to application. Some application interfaces provide a standard set of raw metrics with no interpretation or analysis and no customization features. Other application interfaces provide data that are customized for the particular users, and others allow users to personalize the method of reporting, or even of collecting, data. The application may compare the data to that of a representative population or to medically accepted standards.

Applications allow devices connected to the smart health care ecosystem to have different levels of autonomy. Some applications are passive, in that they record and report data but do not take any actions based on that data. Autonomous applications may be programmed to send alerts when particular thresholds are breached, to take action by itself, or even to direct another device to take some action. Section 16.4 will discuss examples of health care IoT applications.

16.2.5 Other Stakeholders: Social Support

The model of the smart health care ecosystem shows three types of human participants: the patient, the health care providers, and the social support. The individuals included in social support could be family members, friends, co-workers, and members of other social networks to which the patient belongs such as patient support groups. It is also possible that social workers, residence managers, drivers, and other service providers could be included as well.

Many health care situations could benefit from the ability to notify some member of a patient's social support network. For example, if a child's glucose level drops below acceptable levels during the night, an alarm could wake a parent, or if a patient being monitored for depression has not showered in several days (determined by a sensor that measures humidity levels in the bathroom), a friend or family member could be prompted to check on the patient. Patients using a behavior modification application to attempt to lose weight could elect to notify friends and family of their weight loss goal and keep them notified about progress. Some studies have shown that telling others about a goal and receiving the support of friends and family contribute to more success.

16.2.6 Connecting the Components

To enable the sharing of data that underlies the IoT, various devices, sensors, computers, and applications must be connected. In most instances the connection is wireless, but it is not always via the Internet. Connectivity may also be provided by cellular networks (2G–5G or LTE), by Wide Area Networks, or using Bluetooth or NFC (near-field communication) protocols. In some systems where relatively low amounts of data must be transmitted over short ranges, WPAN (wireless personal area networks), such as ZigBee, provide a communication channel that is very efficient in terms of power consumption. An IoT health care ecosystem may utilize multiple communication channels.

16.2.7 Summing up Smart Health Care Ecosystem

As an example that illustrates how all components of the ecosystem interact, assume that a patient is recuperating from surgery to repair a broken hip. The patient wears a device on his wrist that monitors activity (such as walking) and can sense when the patient has fallen, and a sensor in his shoe measures gait and walking speed. Outputs from these devices are stored in the cloud. An application monitors this data and makes it available for a health care provider to check periodically. If the patient has not walked the number of steps "prescribed" by the health care provider, he is sent a reminder via his mobile phone. If the patient's gait and walking speed indicate poor balance, the health care provider is notified so that some intervention may be taken, such as scheduling an appointment with a physical therapist or a checkup with a visiting nurse. If a fall is detected, the patient's family member or other local caregiver is sent an urgent notification via SMS and the health care provider is also notified. It is also possible that emergency services could be contacted. All of the data on these events would be available to the health care provider, along with other data such as medication history, to use when doing periodic assessment of patient treatment plans.

Not every smart health system involves all of the components. All, however, use the IoT as the foundation for sharing data. Although the ecosystem identifies the basic building blocks of IoT systems for health care, it is necessary to look deeper at the applications to explore the many different ways the IoT can be used in patient-centered health care contexts. The following section will describe IoT-supported smart health care applications by evaluating the problem and solution space covered by the applications in terms of seven different characteristics called dimensions.

16.3 Dimensions of Internet of Things Applications in Health Care

As already discussed, a breadth of contexts are currently addressed by smart health care applications that focus on the patient. These health care applications can be described by how they address aspects of the health care situation:

1) *The Type of Health Care Issue.* General maintenance of health or well-being or a particular illness or injury or other health care event.
2) *The Locus of the Health Care Issue.* Physical (of the body) or mental (related to functioning of the mind).
3) *The Time Dimension.* One-time or temporary versus on-going or chronic.
4) *The Goal of the Application.* To prevent a condition, or to prevent a condition from getting worse, or to treat a condition with the goal of curing it.
5) *The Degree of Action Provided by the Application.* Monitoring data about the situation, or providing data to allow users to monitor the data, versus taking action or supporting the user or other stakeholder in taking action to manage the situation.
6) *The Source of Data.* Internal measures related to the condition of the patient (generated by the patient) or measures external to the patient but relevant to the health care situation.
7) *The Primary User of the Application.* Used on an individual level by the patient for self-care, or by a patient's family member(s) for monitoring the

patient versus applications used by health care providers or other caregivers to monitor or assist the patient or to evaluate treatments. It is assumed that informal social support could be used with any application and thus it is not shown as a separate dimension.

To make a visual representation of a smart health care application, the dimensions can be combined into one diagram (Figure 16.2). On this diagram, which resembles a snowflake, it is possible to indicate how strongly the individual dimensions are represented in the application. Each dimension represents a range or continuum between one end-point value and the other. Particular applications could fall anywhere along the range, but because this positioning is not an absolute value there are simply two states shown: some or all. Therefore, a dimension that is partially or somewhat addressed in an application would be marked up to the first 'X' on the dimension on the snowflake. A dimension that is identified as the focus of an application, and thus more fully addressed by the application, would be marked to the 'X' at the end of the dimension range. This identification of the degree to which applications address the dimensions allows categorization and comparisons of different applications on the various dimensions. Each of these seven dimensions is described in more detail below.

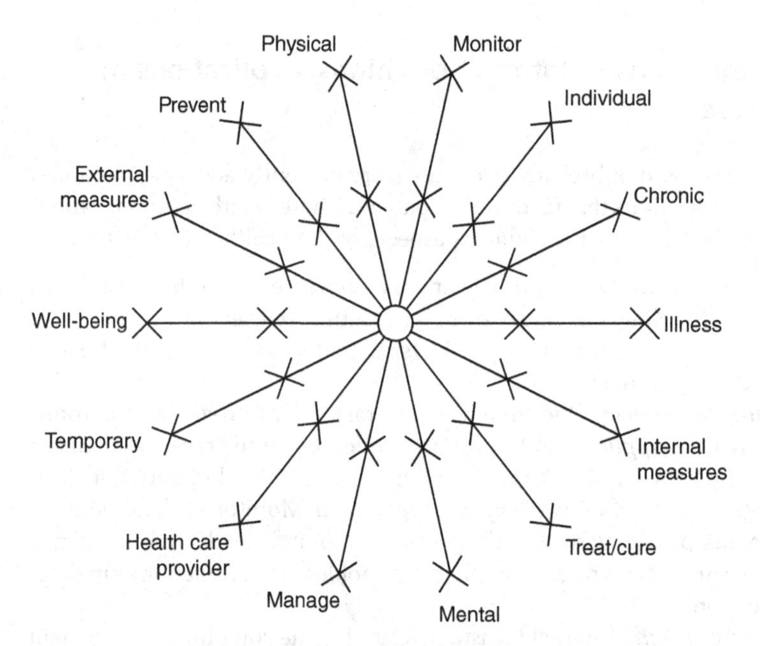

Figure 16.2 A snowflake model representing the seven dimensions of smart health care applications. (*Note:* If the dimension is only partially addressed by an application, the appropriate end of the dimension line will be marked with a bold line to the first "X" (from the center); if the dimension is a major aspect of the application, the entire line relating to that end of the dimension will be marked with a bold line.)

To relate this model to the smart health ecosystem model in Figure 16.1, it should be noted that the focus of this chapter has been patient-centered systems, so all applications of the model involve data about the patient. Several of the dimensions describe aspects of the data: its source, its primary user, and the type of health care issues to which it is applied. The social support component of the ecosystem is not shown as a separate dimension because the output of any application could potentially be shared with others via social media (and at times directly from the application itself), and the individuals providing social support, while important, are secondary to the patient and the health care providers in terms of managing the patient's health.

16.3.1 Well-being—Illness Dimension

The quantified self movement represents an extreme manifestation of self-tracking for the purpose of well-being or health maintenance. It is common today for smartphone applications and fitness tracking bands to automatically track metrics such as steps, sleep, and heart rate. Some applications can also measure stress level and other types of exercise. Many applications allow users to enter information on food consumption, emotional status, weight, and any other metrics the user might want to track. Because these applications are connected to the IoT, they can also exchange information with other devices such as smart scales. These types of applications would for the most part be considered to address the user's general health and fitness. Performance improvement applications, such as those used by runners to track their speed and endurance, would also fall into this category.

At the other end of this dimension, there are applications that address a specific illness or condition such as strep throat, diabetes, asthma, or COPD (chronic obstructive pulmonary disease), injury such as a broken bone or dislocated kneecap, or other health incidents such as stroke. Applications that address a particular health issue may use specialized equipment as well as many of the same sensors used for wellness applications, but in conjunction with specialized programs to analyze the metrics and determine the appropriate response.

16.3.2 Physical–Mental Dimension

Health care can be separated into the two broad categories of physical health and mental health. Physical health includes issues related to the body. Physical health issues could be related to functioning of the various systems (respiratory, vascular, etc.) or to organs or to communications between them; to bones, muscles, and connective tissue and their functioning; and to the ability of the individual to perform tasks related to living, work, and leisure. Mental health encompasses intellectual, cognitive, and emotional health. Conditions related to mental illness, such as depression, schizophrenia, and obsessive-compulsive disorder, as well as addiction, anger management issues, and cognitive impairments, such as memory loss, would be classified as mental health issues.

16.3.3 Temporary–Chronic Dimension

Some health-related issues are temporary in nature—a broken bone has to heal or medication must be taken for a certain number of days to vanquish an infection. Other health-related issues are chronic because they exist over a long period of time, perhaps a lifetime. While it can be assumed that temporary health-related issues are those that will be dealt with within a fixed amount of time, it is of course possible that the particular amount of time may not be easily determined. Certain issues, such as losing a particular amount of body weight, could be considered temporary yet may require lifelong monitoring to maintain. Therefore, weight loss applications would be categorized as temporary while weight maintenance applications would be categorized as addressing a chronic condition.

16.3.4 Prevent–Cure Dimension

Smart health applications that allow users to monitor their weight, food intake, and activity levels can be viewed as applications that could be used with the goal of preventing the onset of a disease such as type 2 diabetes. Applications may also be used to aid in the prevention of serious complications from an illness, such as asthma attacks or uncontrolled bleeding due to lack of coagulants in the blood. Other IoT-based health systems can be used in curing diseases or supporting recovery from an injury, illness, or other health event. Applications that monitor adherence to prescribed medication regimes can increase the likelihood of successful treatment of malaria, tuberculosis, or a bladder infection. IoT systems can aid in the rehabilitation of individuals who are recovering from surgery or have suffered a stroke or a broken bone, for example, and need to retrain muscles or to regain strength and endurance through recommended exercises. Applications can provide information on what movements or exercises should be performed (and those that should be avoided) and can track whether or not they have been performed either autonomously via sensors or via a user's self-reported data. In addition to prompting the patient to perform the exercises or take the prescribed medication, these applications may also be able to inform health care providers and members of the patient's social support network if the patient is not adhering to the care plan.

16.3.5 Monitor–Manage Dimension

IoT systems may be used to both monitor and actively manage health-related conditions. Monitoring systems provide access to data for the patient, health care professionals, and/or other caregivers. While some systems are strictly for personal monitoring, others track data and compare it to a larger population or to previous results for the individual (or both) and allow the sharing of data with others. Personal monitoring systems assist patients in understanding the impact of different activities or situations on their health. Some monitoring systems allow health care professionals and other caregivers to monitor particular metrics remotely. This may be to enable health care professionals to assess treatment effectiveness over many different patients over time, or to monitor whether a patient is progressing in their rehabilitation. In other applications, the intention may be to allow faster interventions by the health care professional if a patient's measures go outside desired ranges.

Applications that manage health care issues are more likely to include interaction with health care professionals, although this is not always the case. Applications that help patients manage depression or anxiety may include active participation by therapists or psychiatrists, particularly if the illness is severe and medication is used. As an extreme example of patients directly managing their own health care situation, some type 1 diabetes patients frustrated with the slow pace of the medical establishment's introduction of diabetes management systems that both track glucose levels and administer insulin have circumvented the medical professionals and regulatory bodies and built their own systems. These Artificial Pancreas Systems link a patient's continuous glucose monitoring device with an insulin pump via the IoT using an inexpensive computer (the Raspberry Pi). Instructions for programming the application are provided online[1]. While the system still requires patients to monitor their glucose levels and make the final determination about insulin dosage, the system can administer the dosage once approved by the patient.

16.3.6 Internal–External Measures Dimension

Some health-related systems rely exclusively on data that come from measures of the individual, often generated by sensors, in smartphones or wearables, or other devices that measure functions of the individual user. This data may be used directly in its raw form or it may be used as input to more complex calculations. Examples of physiological measures that come directly from measurements of the patient include heart rate, number of steps, body temperature, weight, pupil size, brain activity (EEG), and glucose level. Other data relevant to a health care-related application may come from the context of the individual user's environment, such as data about location, humidity, temperature, and altitude or about events on the user's calendar, other applications they have used, social media access, and phone usage. Other data commonly used in health care applications include patient's self-reported data such as food

1 https://openaps.org/

consumption, sleep quality, energy level, mood, stress level, and other information specifically related to the application focus. To determine whether to categorize data as internal or external, the locus of the generation of the data is considered. If it is generated by the patient's body, it is considered internal. If the data is generated outside of the patient's body, it is considered external.

16.3.7 Health Care Provider–Individual Dimension

This final dimension addresses the primary user of the health-related application (particularly the data). Many applications are used primarily by the patient to monitor or manage health-related issues. At this end of the spectrum, patients would provide data to the system via sensors, devices, or self-reporting, and patients would also be the primary recipients of the data to use in their own self-care. At the other end of the spectrum would be applications that are used primarily by health care providers (doctors, nurses, and other specialists) to monitor patients or to acquire information to use in evaluating treatment plans. While it is possible that some health-care-provider-focused applications that collect data from the patients via sensors, devices, or self-reporting would provide no data to the patient, it is more likely that the patient will be able to see the data, although the patient is not the targeted recipient. Most smart health care applications provide data to both the patient and to the health care provider. When data is shared with family members, friends, or other social support network members, this would be considered an extension of the individual dimension.

16.4 Examples of IoT-Related Health Care Applications and Their Dimensions

It would be impossible to identify all of the smart health care applications under development or available today, so this section will describe a sample of these applications. The focus is on applications that focus on the patient, rather than on applications that support administrative or logistical functions such as tracking the locations of medical equipment within a hospital. There is sometimes a misperception that all patient-focused IoT applications are smartphone apps that use the Internet to share data. Although these are quite common, they are not the only types of patient-focused IoT health care applications. Some systems use specialized output to other devices, such as an infusion pump to deliver medication. Data may be shared over Bluetooth or RFID networks, for example. In this section, some of the diversity found in IoT health care applications will be illustrated through examples of several products that are currently in use or in the late stages of testing in preparation for deployment. These examples are provided for educational purposes, and the discussion is not

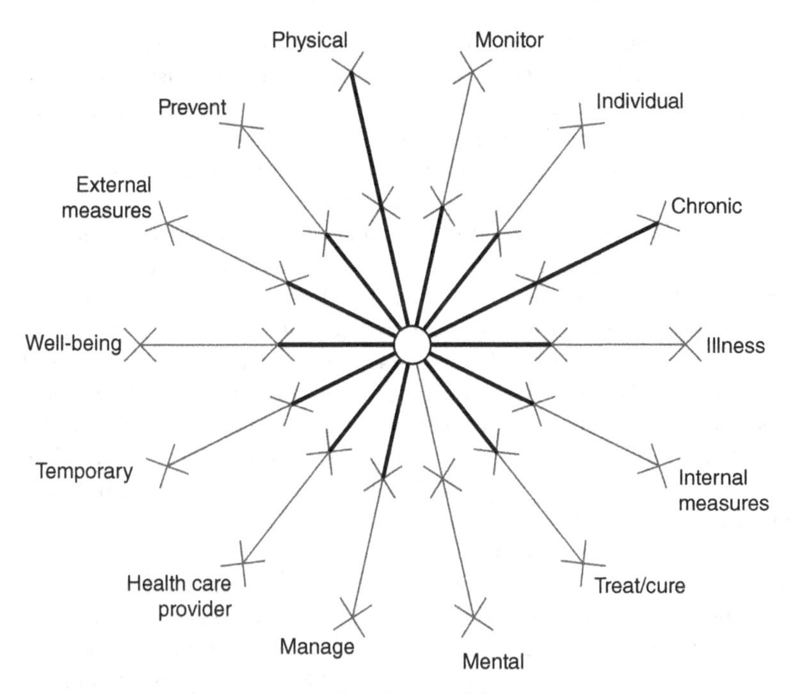

Figure 16.3 EpiWatch represented in the snowflake model.

intended to serve as endorsement for the products. A website (active at the time of publication) is provided for each of these applications. Each application will be described using the snowflake model of health care application dimensions that was introduced in the previous section. The extent to which the application addresses a particular dimension is indicated using a bold line: a bold line going halfway indicates that the application only partially addresses that aspect, and a bold line fully covering one end of the dimension indicates that the particular aspect is a central part of the application. In the text description of the applications, *italics* are used to highlight the particular aspects of each dimension addressed by the application.

EpiWatch[2] is an Apple Watch application that enables users who have epilepsy to track seizures, daily medication use, and possible medication interactions, and to share this information with medical researchers. This data, in conjunction with an accelerometer and a heart rate monitor, may ultimately be able to predict seizures. In Figure 16.3, the dimensions of the EpiWatch are shown in the snowflake model.

The EpiWatch is designed for a *physical chronic* disease, epilepsy. Epilepsy may be caused by a brain injury or stroke, or may be a genetic disorder. In either

2 www.hopkinsmedicine.org/epiwatch

case, the result is abnormal functioning of brain cell activity that leads to seizures. If a person gets epilepsy before the age of ten the disease is often *temporary*, as a child's brain may heal and return to normal functioning, but many people have to live with the disease throughout their life. The EpiWatch app makes it possible for individual patients to *monitor* their seizures and the monitoring makes it possible for researchers and *health care providers* to better understand epilepsy for better *treatment*, and also in the future to be able to detect and *prevent* seizures. Epilepsy is a kind of *illness* in the brain, but the objective of the EpiWatch is to make it possible for people to live a good life without worrying about their next seizure, thus for their *well-being* rather than solely for treating the illness. The sensors in the watch measure *internal measurements*, that is, data automatically generated by the user's body, as it measures heart rate and the user's movements while having a seizure. The app also collects *external data* as the users can input data about their quality of life.

The Owlet *Smart Sock 2*[3] is a baby bootie that uses pulse oximetry to measure an infant's heart rate and oxygen levels while they sleep. The sock, or bootie, is worn on the infant's foot and is held in place by a Velcro strap. The sensor is embedded into the sock. The heart rate and oxygen level data captured by the sensor can be viewed on a smartphone via the app. If heart rate or oxygen levels go outside of predetermined bounds, an alarm uses light and sound to notify a parent. In Figure 16.4 the dimensions of the Smart Sock 2 are presented in the snowflake model.

The Smart Sock is designed for babies and their parents. It is designed for *individuals* and health care providers are not involved at all. The socks are entirely designed for the baby's and the parents' *well-being* by *preventing* the *physical* affliction, early infant death (also called SIDS or sudden infant death syndrome), by alerting parents so they can take action. The Smart Sock is not curing or treating the affliction. By using Smart Sock the parents can monitor their newborn baby and feel safe that the sock will send an alarm if the baby's breathing deviates from the normal. The monitoring is just *temporarily* needed as the risk for early infant death will decrease over time and eventually the risk will fade. The sensors in the sock collect *internal* data like heart rate and oxygen in the blood.

Propeller[4] is an inhaler used by patients with asthma or chronic obstructive pulmonary disease (COPD). For both of these, patients often use an inhaler to get medication into the lungs to facilitate breathing. With Propeller, the user connects a sensor to his or her inhaler and uses the medication as usual. The sensor records when the inhaler was used. Other data captured by the application (such as location, time of day, air quality) allow the user to gain insight into triggers for breathing difficulty. The application can provide results to family and

3 https://www.owletcare.com/
4 www.propellerhealth.com

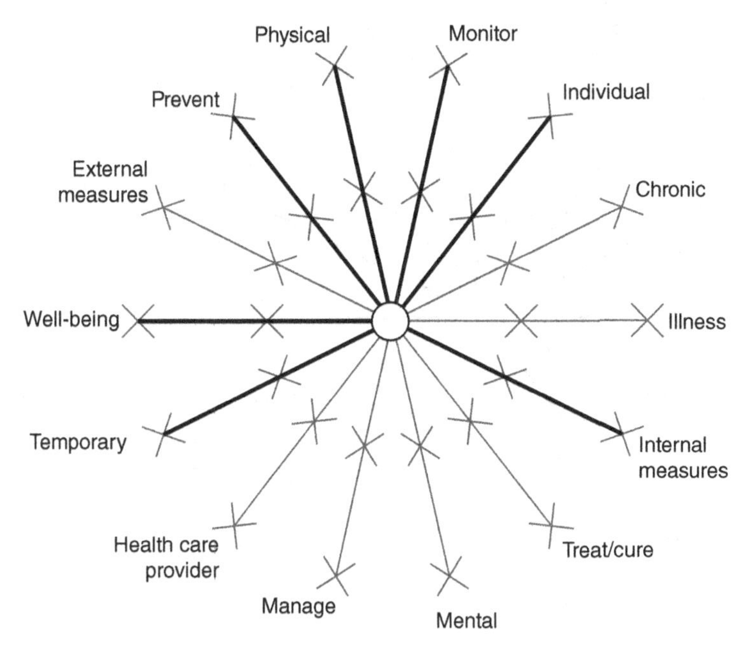

Figure 16.4 Smart Sock 2 represented in the snowflake model.

health care providers. In Figure 16.5 the dimensions of the Propeller device are mapped out in the snowflake model.

Asthma and COPD are *physical illnesses.* Most of the people suffering from asthma have a less severe form, but some patients have *chronic* asthma with seizures. COPD is also something you have to live with throughout your life. The device is only collecting *external* data such as air pollution and context data of the inhaler use. The application can inform the user of the air quality and the user can thereby *prevent* a seizure by taking medicine or avoiding the polluted area. In this way, it is possible for the user to not only *monitor* the environment and the medicine intake, but also makes it possible for the user to *manage* the disease by learning what triggers a seizure and in this way have a freer and better quality of life, thus increasing their *well-being.* The device, of course, benefits the user, but the data is also available for the health care providers that can improve the relationship between the *individual* and the *health care provider* as they both share the knowledge of how the disease impacts the patient's daily life.

Ginger.io[5] is an application for mobile phones combined with a subscription service to manage depression, stress, and anxiety with the support of licensed therapists and psychiatrists. The application allows users to track their moods and collects data from sensors regarding activity levels, use of social networking

5 www.ginger.io

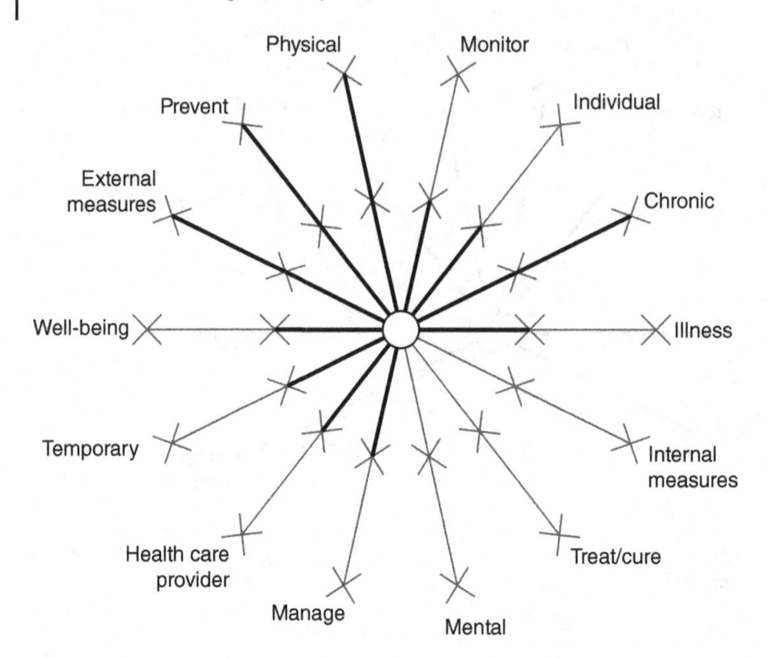

Figure 16.5 Propeller represented in the snowflake model.

applications, and phone use. Personalized counseling is provided via text and interactive video. In Figure 16.6, the dimensions of the Ginger.io are shown in the snowflake model.

Ginger.io is an app for *mental illness*. Depression, anxiety, and stress are *temporary*, but they have a tendency to come back if the person does not *manage* their *well-being* even when they are symptom free. The app in combination with support of the *health care providers* helps the *individuals* to *manage* and *cure* their stress, anxiety, and depression. The sensors are automatically collecting activity data generated by the person. Accordingly, the app collects *internal* measures, but it also collects *external* data like self-reported mood, phone calls, and social media activities. Also, through the app the users can communicate with health care providers about the data that has been recorded.

GPS *SmartSole*[6] allows caregivers to use their smartphones or computers to track the movements of patients who are prone to wandering or getting lost due to Alzheimer's, dementia, brain injury, or other cognitive issues. The SmartSole looks like an ordinary shoe insert (such as those commonly used to provide additional cushioning or arch support) that can be placed in the patient's shoe. Sensors embedded into the insert allow for GPS tracking. In Figure 16.7, the dimensions of the SmartSole are shown in the snowflake model.

---

6 www.gpssmartsole.com

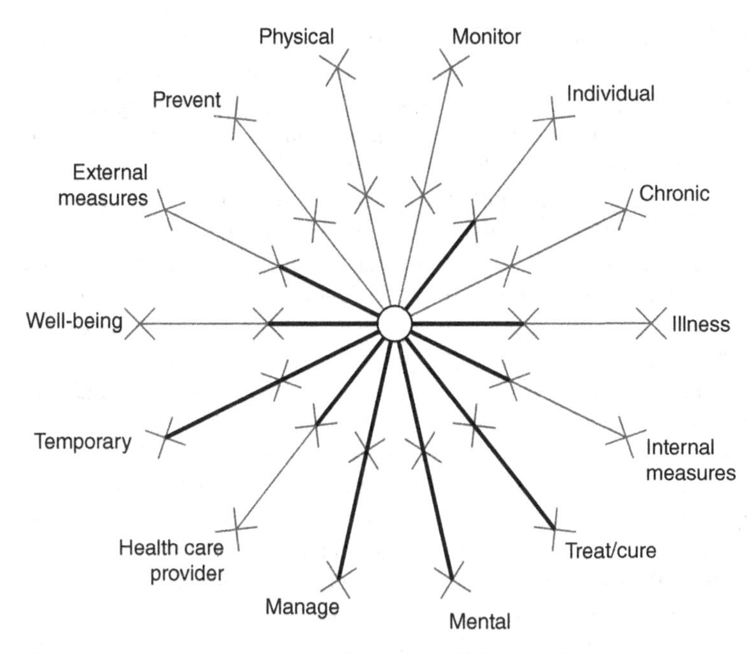

Figure 16.6 Ginger.io represented in the snowflake model.

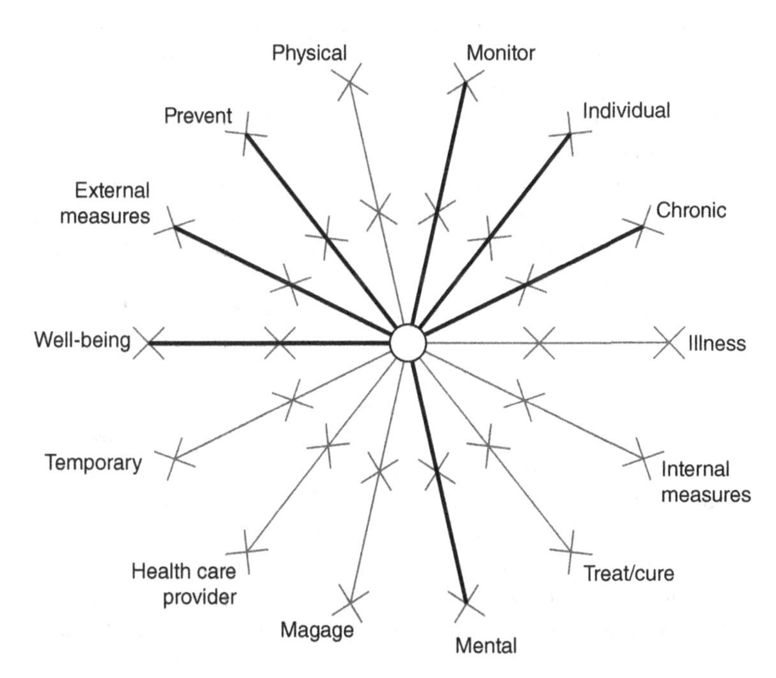

Figure 16.7 SmartSole represented in the snowflake model.

GPS SmartSole is designed to empower people with cognitive impairments by giving them more independence because they can be given more freedom of movement, while at the same time keeping them safer because caregivers can track their location. Cognitive impairment is a *chronic mental* condition and GPS SmartSole supports those with such conditions to *prevent* them from the mental anguish of being lost and potentially from coming to harm. People with cognitive issues are often very dependent on significant others and GPS SmartSole makes it possible for the significant others to *monitor* and track the person's movements from a distance and make sure everything is as it should be. They can make sure the user has not moved outside the "safe zone," for example. GPS SafeSole can in this way promote *well-being* for both the person with cognitive impairments and their caretakers. The sole collects external data as it is equipped with a GPS sensor.

Philips *Automated Medication Dispenser* service[7] uses a smart pill container that is programmed according to the patient's medication schedule and loaded with the appropriate medication by the caregiver. An alarm alerts patients that it is time to take their medication. At this time, the container will unlock and make the appropriate pills available. If pills are not taken on schedule, they will be stored inside the dispenser. If pills are missed repeatedly, the caregiver receives an alert. In Figure 16.8, the dimensions of the automated medical dispenser are shown in the snowflake model.

The automated medicine dispenser is designed for people who might forget to take their medicine. While the medicines dispensed could *treat* both *physical* and *mental illness,* its primary purpose is to serve as a memory aid, thus addressing a mental issue. The dispenser aims to help both the *individual* patient and the *health care provider* to *manage* the medicine intake. The device can be viewed both as *prevent*ive and as providing *treatment* as the reminders prevent the patients from forgetting to take the medicine that is essential for their treatment. The device also prevents the patient from taking more medicine than prescribed. It might be that the patient does not remember that she has taken the medicine and, therefore, takes it twice or more, with potentially serious consequences. To avoid this, the dispenser locks the compartments until it is time to take the new dose. The dispenser can be used by people with both *temporary* and *chronic* diseases. Forgetfulness itself can be temporary or permanent, although it is likely that the patient has an on-going problem with forgetfulness if the dispenser is prescribed by the health care provider. The automated medical dispenser is not only beneficial for the *individual,* but also for the *health care provider* as the nurse handling the medicine knows if the medicine is taken or not and it gives valuable information about further treatment. The medical dispenser is a stand-alone device, not connected to the user at all, meaning that it only involves external measurement. It does,

7 https://www.lifeline.philips.com/health-solutions/health-mdp.html

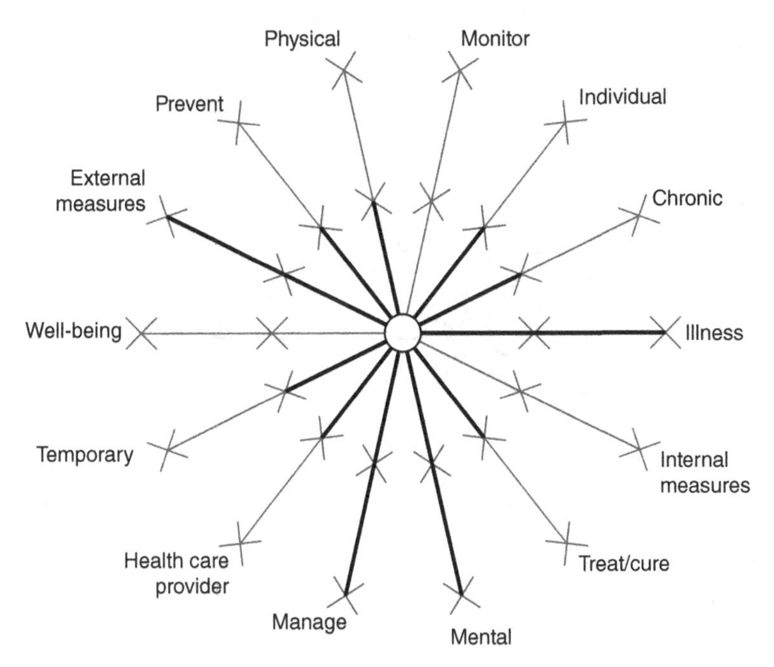

Figure 16.8 The automated medicine dispenser represented in the snowflake model.

however, communicate with other components of the IoT health care ecosystem. It alerts the patient according to a schedule when it is time to take a pill, and alerts the health care provider when the patient has missed taking a medication dose.

CoaguChek®[8] monitors blood coagulation levels for patients who take blood-thinning drugs, such as warfarin, and sends the results to health care providers. The device is fed by a drop of blood and the display shows the INR-value (International Normalized Ratio), which indicates the blood's tendency to clot. The application also collects user comments regarding, for example, diet change, illness, missed medication, and travel, and patients can set reminders to check their blood. The data received by patients indicate trends and whether the blood coagulation levels are within a target range. In Figure 16.9, the dimensions of the CoaguCheck are shown in the snowflake model.

The CoaguCheck is designed for the *individuals* with an increased risk of blood clots and who take warfarin to keep the blood viscosity within a desired range. The accurate dose is established by measuring how long it takes for the blood to clot. The device's main objective is to make it possible for the patients to *monitor* the INR-value on their own without going to the hospital to take a blood sample. Instead of wasting time to travel to the hospital, the patient can call the

8 www.coaguchek.com

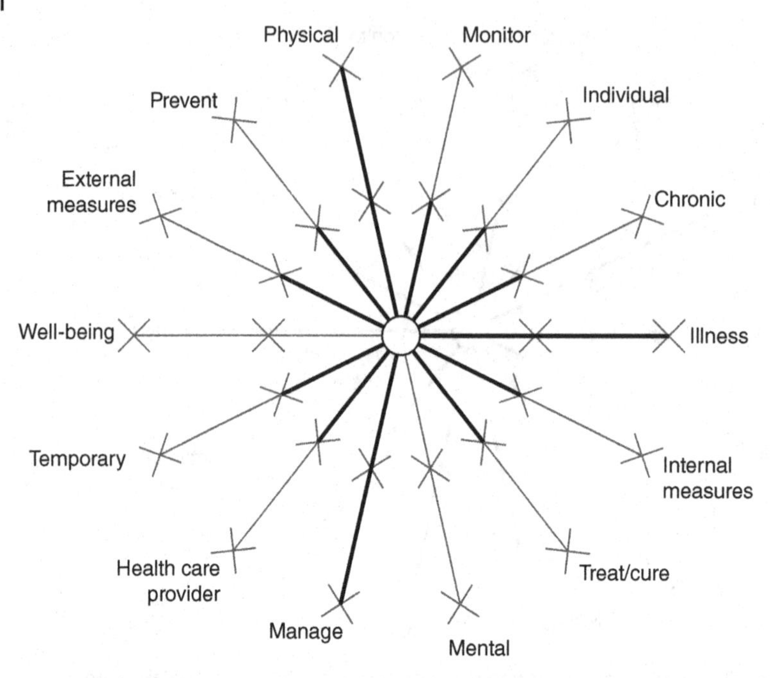

Figure 16.9 CoaguChek represented in the snowflake model.

health care provider and communicate the IRN-value and the health care provider can then change the dose if necessary. The self-measured INR-values can also help the patient to *manage* their value as they can learn how food and their lifestyle affect the INR-value. Measuring the INR-value often makes it possible to better adjust the *treatment* and also to *prevent* the values from being too low or too high because the dose can be changed more frequently. The reasons for taking warfarin can differ. Warfarin can be taken because the patient has had a blood clot. Only when and if the patient is determined not to have an increased risk for having more blood clots does he or she stop taking the medication. In this case, the warfarin is taken only *temporarily,* and accordingly, the CoaguCheck would be used only temporarily. But patients that have medical heart valve replacements need to be medicated with blood thinners like warfarin for the rest of their lives. The CoaguCheck does not only collect *internal* measurements such as the INR, but also *external* values the users report about, such as their diet.

An *activity tracker for oncology treatment* (a collaboration between Medidata[9] and Memorial Sloan Kettering Cancer Center[10]) makes use of ordinary activity

9 https://www.mdsol.com/en/
10 https://www.mskcc.org/

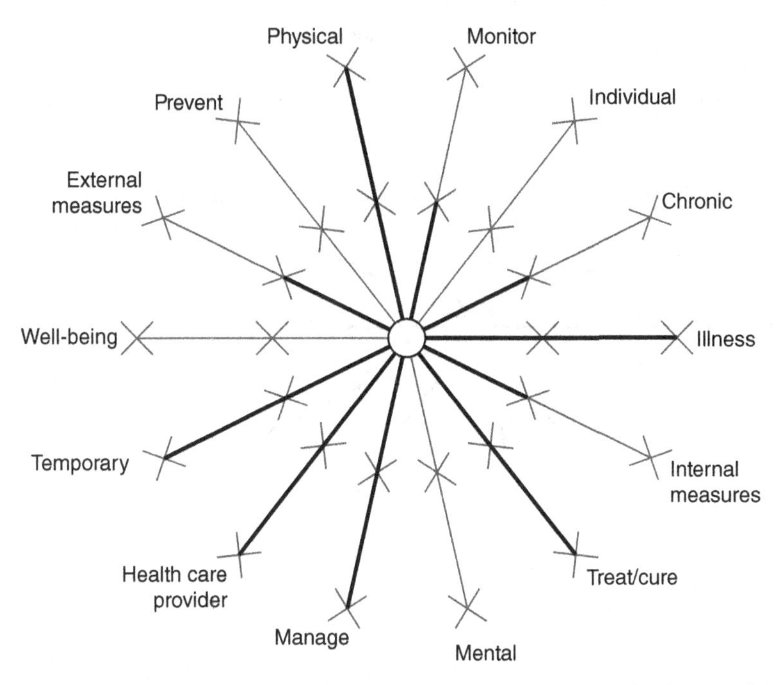

Figure 16.10 The activity tracker for oncology treatment represented in the snowflake model.

trackers on the market to allow health care professionals to measure the efficacy of drug treatment for multiple myeloma (a blood cancer). Wearable activity trackers and patient-reported data on health status enable medical researchers to evaluate the impact of drug treatments on patients. In Figure 16.10, the dimensions of the activity tracker for oncology treatment are shown in the snowflake model.

Today, multiple myeloma is not curable, but the survival rate has increased during the past decade due to improved treatment. The disease is *physical* and the patient has to live with it for the rest of their lives, but the illness is not classified as chronic and for that sake, it is shown as primarily *temporary* in the snowflake model. The activity tracker is used by the *health care provider* to *manage* the *treatment* and *monitor* the quality of the treatment as well as the patient's quality of life to better customize the treatment. As with the ordinary use of activity trackers, *internal* measurements like activity level and pulse rate are captured. In addition, the patients report their health status, so *external* values are also recorded.

Eversense continuous glucose monitoring[11] systems enable diabetics to continuously measure their blood sugar level. A small cylinder-shaped sensor

11 https://eversensediabetes.com/cgm/

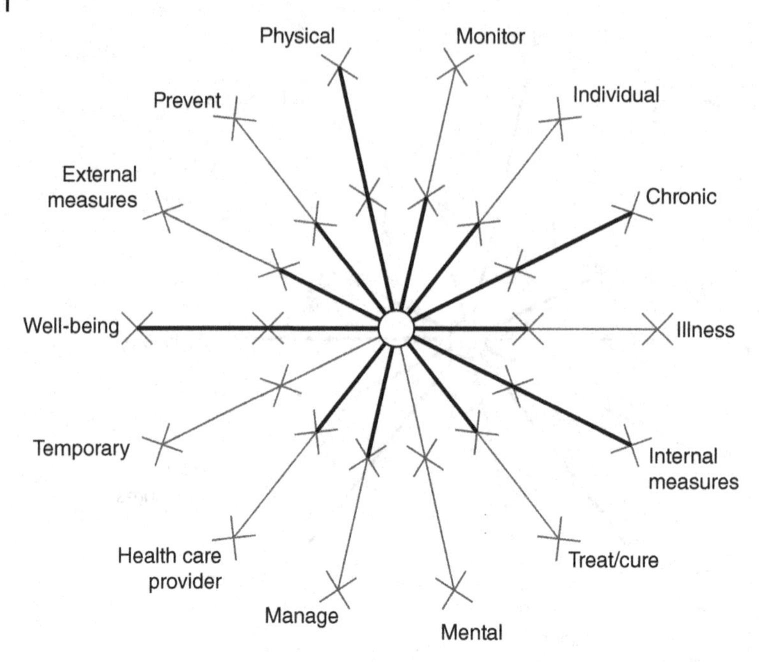

Figure 16.11 The continuous glucose monitoring meter represented in the snowflake model.

embedded under the skin measures levels of blood sugar in fluids in the tissues. A connected transmitter powers the sensor and sends readings wirelessly to a smartphone. The sensor does glucose readings every 5 minutes and the app can send predictive alerts before reaching a too low or high-glucose level. The app also allows the user to manually track events like meals, workouts, and insulin injections. Results can be shared with a health care provider. In Figure 16.11, the dimensions of the continuous glucose monitoring are shown in the snowflake model.

The continuous glucose monitoring system is designed for a *chronic* disease, diabetes. This is a *physical* disease the person has to live with throughout his or her life. For people with diabetes, it is important to both *monitor* and *manage* their blood glucose value. The patient has to monitor the glucose value to make sure the value is not too low or too high. If the values are not in a desirable range, the patient has to manage the blood glucose level in different ways. If the value is too low, the person has to eat something and if it is too high he or she needs to inject some insulin. To do that, the patient has to understand the monitored values and also know the appropriate action to take. The blood glucose monitoring system addresses the *illness,* but it is mainly used to make it possible for people to live a good life, for their *well-being.* While diabetes cannot yet be cured, those people with type 2 diabetes still have some insulin production of their own left and by monitoring and managing their blood glucose values and taking additional action they can actually *treat* their diabetes and *prevent* it from becoming worse. People with type 1 diabetes can try to prevent serious complications later on in life (damage to other organs and eyesight, for example) by keeping their glucose level in a normal range. The sensor itself only measures *internal* data—the blood glucose value— but the user can manually collect *external* data like meals and insulin injections. The device mainly targets the *individual* but is also beneficial for the *health care providers* because the measurements and the trends in the measurements can help them support the patient by recommending insulin doses and actions to manage the blood glucose levels.

In summary, the types of health care applications using the IoT span a broad spectrum. The aforementioned applications focus on the patient, although the metrics and other data collected by some applications are more directly used by health care providers, family members, or other caregivers to monitor the patient. Some applications are quite comprehensive, in that they collect both biometric data from the patient's body as well as contextual data about the patient's environment and activities and feelings, while others focus on a very small set of metrics. Applications may help with the management or treatment of an illness or condition, or may simply allow monitoring of a patient's status on particular metrics. There are applications that support well-being and those that help to prevent illness or complications. By exploring these examples, the great potential for technology, particularly the IoT, to improve the quality of life for many patients becomes clear.

16.5 Challenges

Smart health care ecosystems are multidimensional systems of people, hardware, software, and networks operating in a complex legal and regulatory environment with potentially life or death consequences. Complicating this context are data and platform heterogeneity issues, a lack of comprehensive standards, and concerns with privacy, security, and accuracy of data. These systems have the potential to change the relationship between patients and their health care providers, as they impact the way health care providers perform their duties and the amount of responsibility patients may take in managing their own health. Each of these issues is discussed further.

16.5.1 Lack of Standards

The use of personal devices, such as smartphones and smart watches, sensors, and applications that provide data that can be used by both individual patients and health care providers, are relatively new developments in the technology and health care realms. It is not always entirely clear which regulatory bodies and agencies have jurisdiction over these applications, and to date (as is common), regulations have not kept up with technological developments (Vincent et al., 2015). While the International Standards Organization (ISO) has published international standards for medical equipment, these address safety and usability. Standards for connected medical devices and connected health software are currently under development.

Standards to ensure data compatibility and universal regulations regarding how data can be stored, transmitted, and used are also lacking. These are currently addressed regionally. In the United States, the Health Information Technology Standards Panel has been given this responsibility. A similar body, the Medicine and Health Care Products Regulatory Agency (MHRA) exists in the United Kingdom. However, for smart health care IoT applications to gain widespread use, global standards are needed. Without standards for platforms, functionality, communication interfaces, and data it is difficult to ensure the compatibility of different devices, sensors, and applications, particularly in a global context. This can slow both innovation as developers wait for standards to be set and adoption because certain applications may not be functional or legal in particular locations. In addition, it may not be possible to integrate the data generated by these smart health applications with existing digital medical records or to use them as input to hospital or clinical systems.

16.5.2 Data Issues

There are additional data issues beyond the lack of standards that present challenges in the health care environment. The types of data used by smart health systems may include personally identifiable information (PII) and health-related data that some patients might view as quite sensitive. As discussed in the previous chapters, IoT systems in general are known to have many security vulnerabilities, and these apply to the smart health care IoT ecosystem as well. Medical devices and consumer technologies often have lower levels of protection than do typical computer network devices. There have been many examples of hacking of medical and consumer devices connected to the Internet (e.g., Reel and Robertson, 2015), and these issues must be addressed to safeguard the personal and medical data saved and communicated via the IoT.

Another issue with data is the level of accuracy of the metrics produced by smartphones and wearable sensors. Comparisons of consumer devices such as step trackers have shown quite a bit of variation in the data captured by different devices (Guo et al., 2013). Data that is not quite accurate may not have significant impact for individuals who are tracking their steps, but if the inaccurate sensor is measuring blood glucose levels, for example, incorrect results could lead to serious consequences, including death of the patient. To account for this, most blood glucose monitoring devices require frequent (up to multiple times per day) calibration. The lack of accuracy causes some health care professionals to resist the use of smart health applications, although there have been reported examples of data from personal fitness trackers being used to support medical diagnosis. In one example, emergency room physicians used data from a patient's Fitbit to determine the appropriate treatment for a heart arrhythmia (Rudner et al., 2016).

Providing data to patients in a usable manner is another challenge. Patients have different levels of technological and medical literacy, and their ability to understand the meaning of health metrics will differ. The way in which data is filtered, summarized, and displayed can make a difference in how easily it is understood. Patients may have different preferences regarding how much data they receive, or how often they receive it. Therefore, smart health applications should take these individual differences into account and should enable some level of customization of the system.

16.5.3 Changing the Health Care Provider–Patient Roles

Smart health systems should not be viewed solely from a technological or medical perspective, but from a broader sociotechnical perspective that includes the impact of these systems on roles, relationships, organizations, and society in general. Building these applications requires not just technological competence, but also medical knowledge related to the particular specialty area addressed by the application. An understanding of human behavior, psychology, and motivation is needed to create systems that can assist patients in changing their behavior.

While smart health systems offer patients the opportunity to have more control of their health care, this comes with added responsibility. As patients take a more active role in managing their health, they have more duties in terms of monitoring various metrics, learning what the numbers and other visualizations mean and what actions to take, and possibly doing self-recording of data required by the application. This may be a welcome change for some patients, but for others this may be an unwanted burden.

These smart health systems may also impact the health care professional's role. Physicians and nurses may spend less time meeting with patients face-to-face, and more time monitoring output from patient's devices and sensors. While this should enable health care professionals to provide better care to more patients, it may require more effort and offer less personal satisfaction for some practitioners. Just as the Internet has led to patient self-diagnosis (and often misdiagnosis), the volumes of data available to patients may also lead to some patients using more health care professionals' time as they request assistance in interpreting the data or seek treatment for nonserious issues.

16.6 Conclusion

Both advances in technology and increasing pressures for more effective and efficient methods of addressing health care issues have contributed to the growth in the number of applications of IoT in health care. Through the IoT, data captured by various sensors and devices can be shared with applications used by patients and health care providers to monitor physiological metrics and other data that can be used in the management, treatment, or prevention of various illnesses or other medical conditions. Both physical and mental conditions are addressed by these systems. Some systems focus on well-being and quality of life issues as well. This chapter presented a people-focused view on IoT by providing an overview of the components that may be included in an IoT-based smart health ecosystem and introduced a set of dimensions to consider in smart health applications. In summary, there are many challenges facing the widespread adoption of smart IoT health care applications. However, for many of these applications, the benefits outweigh the risks, and therefore this area is expected to continue to grow at a rapid pace.

Future developments will include the integration of artificial intelligence components to supplement or take the place of the health care professional or other caregiver or the social support agents. Miniaturization (nanotechnology) will enable injectable sensors to become common. Injectable and implanted devices—embedded chips and other sensors to monitor physiological status—are beginning to be used, as are electronic patches and electronic tattoos that serve as sensors (Motti and Caine, 2015). Augmented reality will also be used in smart health care applications both to aid health care professionals in integrating data on patients, conditions, and treatments and to assist patients in using the data collected by these systems to manage their health. Innovations in power consumption (increasing battery life and generating power through the user's movements, for example) and increased bandwidth will allow wearable devices to last longer and to transmit more data more quickly. In addition to advances in sensor technology, user interfaces, and computational power, we can also expect to see more integration across platforms, so that patients can monitor multiple aspects of their health simultaneously thus allowing a more holistic perspective not just addressing one condition or symptom, but instead addressing the broader context of patients dealing with multiple health-related issues. As smart health IoT systems become more accurate, powerful, and reliable and legal and regulatory systems are updated to include these systems, they will become integrated into the health care infrastructure.

Acknowledgments

This work has been carried out within the research profile "Internet of Things and People", funded by the Knowledge Foundation and Malmö University in collaboration with 10 industrial partners.

References

Glaros, C. and Fotiadis, D.I. (2005) Wearable devices in healthcare. *Studies in Fuzziness and Soft Computing*, **184**, 237–264.

Guo, F., Li, Y., Kankanhalli, M.S., and Brown, M.S. (2013) An evaluation of wearable activity monitoring devices. Proceedings of the 1st ACM International Workshop on Personal Data Meets Distributed Multimedia (PDM'13), October 22, 2013, Barcelona, Spain. pp. 31–34.

Motti, V.G. and Caine, K. (2015) An Overview of Wearable Applications for Healthcare: Requirements and Challenges, UBICOMP/ISWC '15 Adjunct, September 07–11, 2015, Osaka, Japan, pp. 635–641.

Reel, M. and Robertson, J. (2015) Hospital gear could save your life or hack your identity. Bloomberg Businessweek, November 2015. Available at https://www.bloomberg.com/features/2015-hospital-hack/ on 05/01/17.

Rudner, J., McDougall, C., Sailam, V., Smith, M., and Sacchetti, A. (2016) Interrogation of patient smartphone activity tracker to assist arrhythmia management. *Annals of Emergency Medicine*, **68**(3), 292–294.

United Nations (2010) The Millennium Development Goals Report 2010. Available at: http://www.un.org/millenniumgoals/pdf/MDG%20Report%202010%20En%20r15%20-low%20res%2020100615%20-.pdf (accessed March 03, 2017).

Vincent, C.J., Niezen, G., O'Kane, A.A., and Stawarz, K. (2015) Can standards and regulations keep up with health technology? *JMIR mhealth uhealth*, **3**(2), e64.

WHO (2010) *Health Systems Financing: The Path to Universal Coverage*, World Health Organization (WHO).

17

Internet of Things in Smart Ambulance and Emergency Medicine

Bernard Fong,[1] *A. C. M. Fong,*[2] *and C. K. Li*[3]

[1]*School of Public Health, Auckland University of Technology, New Zealand*
[2]*Department of Computer Science, Western Michigan University, Kalamazoo MI, USA*
[3]*Add-Care Ltd., Hong Kong*

17.1 Introduction

Recent advances of wearable health devices such as smart watches and non-invasive health monitors have generated interests in expanding IoT health applications to be an increasingly important part of public health. Various data collected by biosensors with varying volumes can be further analyzed for diagnosis and prognosis of chronic diseases. As health care service providers become increasingly reliant on intelligent and interconnected devices in every aspect of health support, critical reliability, data integrity, and interoperability are important considerations that need to be thoroughly addressed. Data analytics and syndromic surveillance for providing effective treatment in remote rescue entail careful consideration from data acquisition, selection, transmission, mining, analysis all the way to manipulation and storage to update electronic patient records (EPR) as well as disease database maintenance. To this end, challenges related to supporting on-scene paramedics by providing them with all necessary information without affecting the way they carry out their rescue mission must be overcome.

Effective on-scene treatment helps minimize the risk of developing medical complications, there are certain cases like in the case of treating asthma patients, it is in fact possible to eliminate the need for transporting the patient to the hospital after providing necessary relief, thereby reducing the demand on hospital accident and emergency (A&E) personnel (Campbell et al., 1995). One of the key challenges of designing a smart ambulance is the confined space limitation of the vehicle itself, very limited space is available for additional equipment. Furthermore, constraints on ergonomics, electromagnetic compatibility (EMC), and ease of cleaning must be considered for such additional

Internet of Things A to Z: Technologies and Applications, First Edition. Edited by Qusay F. Hassan.
© 2018 by The Institute of Electrical and Electronics Engineers, Inc. Published 2018 by John Wiley & Sons, Inc.

equipment being installed in an ambulance. Currently, there is no standard on ambulance specifications such that theoretically any smart and assistive technology can be incorporated in an ambulance platform. Taking the United States as an example, there is an urgent need for IoT-enabled smart ambulances to support mass scale recovery operations. As observed in the recent hurricanes in the south-east to intense fires across large parts of the west, the huge volume of human traffic between states resulting from evacuations has posed immense challenges to emergency rescue services across large parts of America. An important lesson learned from hurricane Katrina on the increasing risk of pandemic across Texas is that there is an urgent need for timely outbreak detection and effective disease-spread simulation analysis to enable health resource management under pandemic outbreaks. Some surveillance systems are currently available such as ESSENCE (Lombardo et al., 2003), a system used by the U.S. Department of Defense, detects infectious disease outbreaks at military treatment facilities. Bio-Sense and EARS were developed to detect and monitor bioterrorism. To safeguard the health of emergency rescue personnel in the aftermath of a major disaster, a surveillance system like influenza-like illness (ILI) and virologic data are needed for a vast area often across multiple states. Using an example of an influenza outbreak, such system tracks ILI and laboratory-confirmed influenza in hospitals throughout the affected region. However, in spite of these surveillance system implementations, the algorithmic capability to accurately detect infectious disease outbreaks and pandemic must be constantly updated for the entire region.

Unlike most A&E staff based in the hospital, paramedics often travel long distances and cope with a large number of patients across vast areas in response to an emergency evacuation. The need for connected smart ambulances that provide paramedics with all necessary real-time information to minimize the risk to emergency support staff is staggering. Part of the challenge is the complexity of disease incident data, which is likely to be heterogeneous, multivariate, intercorrelated, of multiple data types, and often exhibits seasonal patterns over time. There are clearly substantial algorithmic obstacles to interrogating disparate datasets with such complexity or diverse conditioned datasets. IoT plays an important role in this regard, with different connected devices working together to support paramedics cover across states.

This chapter takes an in-depth look into various aspects of technical challenges surrounding the connected smart ambulance environment and to highlight how advances in IoT change the way paramedics carry out their duties under different situations. This chapter commences by discussing the key technologies behind the bridge that links emergency support personnel to their patients and the hospital as they carry out the first line of rescue, investigate data analytics technology for accurate on-scene diagnosis and prognosis in order to provide the best possible treatment in an efficient manner.

17.2 IoT in Emergency Medicine

In realm of emergency medicine, connected devices play an important role within and around a smart ambulance. Instrument in the ambulance are connected to both the hospital via a telemedicine link (Fong et al., 2005) and wearable devices on paramedics with short range communication networks with choices ranging from WLAN to Bluetooth or Zigbee, and more likely a combination of these given that different devices can connect to the same network in a different way. Figure 17.1 shows the basic smart ambulance architecture where the key feature is the three major networks where IoT plays an important role in each of them. There are basically two separate networks that are interconnected together where the smart ambulance serves as a hub. This link supports two-way vehicular communications between the hospital database and the ambulance. Information about the patient's medical history and remote consultation can be provided from the hospital whereas data related to the patient's current state can be sent back to the hospital for both EPR update and advance preparation by A&E personnel. On the other side of the ambulance, there is a local IoT network that connects ambulance instrument with the paramedics that in turn retrieves patient activity log from wearable consumer devices.

The telemedicine link connects the ambulance to the hospital. This is a duplex link that provides on-scene information collected by the paramedic about the patient's current condition so that hospital staff can make advance preparations prior to the patient's arrival. The operational reliability of this link can be greatly affected by a number of environmental factors especially under the influence of heavy rain (Fong, 2003a). Generally, the information exchange entails retrieval and update of EPR. Additionally, paramedics can seek advice on prognosis and treatment in cases such as treating chronic obstructive pulmonary disease (COPD) or asthma where on-site diagnosis can be challenging, as this chapter will take a close look with a case study. The technologies surrounding diagnosis

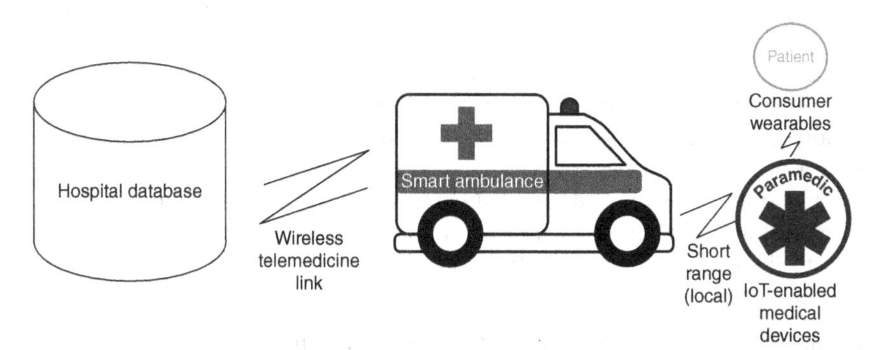

Figure 17.1 Smart ambulance architecture.

and prognosis of any form of COPD, asthma, or asthma COPD overlap syndrome (ACOS) (Postma and Rabe, 2015) as a specific case study will be discussed in more details later in this chapter. What paramedics need in providing first aid treatment is any known history of allergy inherent to the patient (Dennehy, 1996); this is a vital piece of information to minimize any risk of developing further medical complications.

IoT between the ambulance and the patient form a more complex relationship than the above discussion between the ambulance and the hospital as it entails more connected devices of different types. First, a range of diagnosis and supporting tools are available for the paramedics that must remain connected as they are moved around the scene. In addition, useful information may be acquired from the patient's wearable health devices that may include health information related to the emergency event, these could include various vital signs, medications, and activities taken. All these could make a significant contribution to accurate diagnosis of the patient.

17.2.1 Point-of-Care Environment

The effect of the ambient environment at the Point-of-Care (PoC) can have a significant impact on the patient being treated. This involves the study of physical condition as well as chemical and biological contaminants on patient's health while being treated. Advances in IoT for health care allow environmental sensing systems to be incorporated as part of the smart ambulance (Kelly et al., 2013). Through integrating ambient sensing in the smart ambulance as illustrated in Figure 17.2, it is possible to assess the impact of varying environments on changes in the physical condition of patient's state of health as well as the treatment being provided at the PoC in order to determine the environmental factors that are most critical to enable better patient treatment in the ambulance setting. Ambient environmental effects could impair the recovery of patients that have certain respiratory conditions (Roche et al., 2008). Monitoring the environment the patient is in and ensuring it is optimal for specific patient conditions would improve recovery. In the center of the IoT–PoC environment is the smart ambulance that serves as a connected support console. Additionally, it can initiate an automatic request for a rescue helicopter in one of the two conditions: difficult access to patient, road condition either not allowing rapid return or number of patients exceed available road resources; or the patient exhibits time sensitive condition that entails clinically important time saving in reaching hospital. Clinically significant information about the patient can also be sent to the helicopter support staff prior to dispatch.

Other forms of basic paramedic support include the following:

- Vital signs if a patient that may indicate the mortality of a patient such that no treatment remains meaningful.

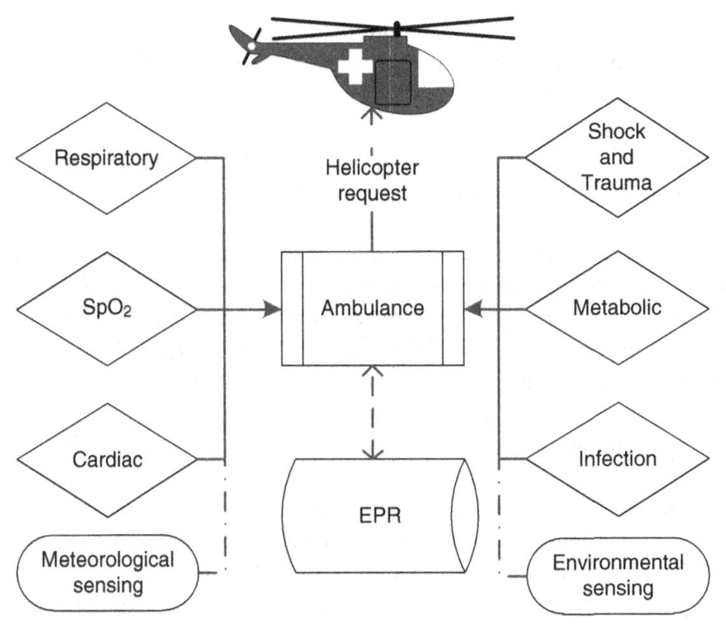

Figure 17.2 Integrated IoT sensing at the point-of-care (PoC).

- Oxygen administration with SpO2 measurement below 94% for identification of high risk patients.
- Respiratory determination of type and severity.
- Cardiac conditions with ECG in the diagnosis of anterior ST elevation myocardial infarction (STEMI) as well as other conditions such as cardiac arrest and atrial fibrillation or flutter.
- Shock and trauma entails image processing engines and determining causes of hypovolemia. The former allows fast and accurate diagnosis of burns severity, spinal cord injury, joint dislocation, and fracture misalignment; whereas the latter may involve mechanism for fluid loss detection or infrared sensing for hyperthermia patients.
- Metabolic analysis involves protection of frontline paramedics' personal safety; this concerns agitated delirium in case a patient exhibits sign of anxiety with physical aggression. In addition, biosensors play an important role in the measurement of altered consciousness.

Perhaps the most important aspect of safety concern is commutable infectious disease. Similar to minimizing the risk of hospital acquired infection (HAI) (Dancer, 2009), keeping the ambulance clean is also a great challenge in the avoidance of infection. An infection occurring in a patient prior to arriving at the hospital as well as occupational infections among supporting staff of the ambulance requires close monitoring of the ambulance environment itself.

However, the current medical settings and guideline-based standard operating procedures (SOP) in providing on-scene treatment does not specify a technology or technique to reduce HAI (Liu et al., 2011). The existing practices are more related to discipline in paramedicine without any associated assistive technology. In this regard, technological advancements are beginning to change the way infection prevention scientists tackle HAIs by adopting IoT and assistive technologies in a smart ambulance. Earlier work has described various technologies and the way it is facilitating measurement and compliance from hand hygiene to environmental cleaning and instrument decontamination (Pyrek, 2014). Any effort in tracking would entail the deployment of wireless system to instrument hygiene events. IoT enables improvement in the analysis of on-scene treatment and prognostics due to the increase of collected data about environmental effects that could affect patient health. Such IoT system would enable determining environmental factors that are most critical to enable better treatment.

Patient monitoring and managing remote rescue operations can deliver significant value in reducing recovery time and operating costs. Before the introduction of connected ambulance when paramedics were inadequately networked, they were effectively serving in a closed system unable to communicate with the hospital network. Capturing, aggregating, filtering, and sharing data from health systems seamlessly into emergency medicine is both a challenge and promise of the Medical IoT, where IoT serves as the point where patient data are directly linked to the national health care system via telemedicine. IoT allows monitoring in remote and demanding locations through the smart ambulance as a hub. Data filtering and aggregation in the ambulance gateway facilitates full duplex data transfer and storage to the hospital.

Central to the PoC environment featured in Figure 17.2 are intelligent wireless networks that consists of both short-range local coverage and long-range links, wearable devices, and low-power sensing arrays coupled with medical data analytics combined to form the basic building block of medical IoT in emergency medicine. In the IoT context, this combination of connected technologies enables a multitude of biomedical and environmental sensors to be temporarily placed in any location not restricted to where network coverage and power source are available. All information paramedics need about the patient and details on treatment options is assembled according to the condition of the patient. The concept of instrumenting "things" such as medical devices, health monitors, consumer wearables with biosensors are all proliferated in the ambulance setting. Even before IoT became popular in recent years, emergency medicine systems have operated as separate networks like one between paramedics and the hospital via traditional cellular phone networks; this is linked to another completely isolated wireless local area network (WLAN) within or surrounding the ambulance. Maintaining adequate network reliability and security for emergency medicine cannot be simply met with traditional network management strategies (Ansari et al., 2006). Moreover, the way these networks are implemented determines whether the patient data manually collected at the scene can be reliably and securely transmitted to the hospital under harsh environments typical of causing accidents under heavy rain (Fong et al., 2003b).

IoT sensing networks facilitate the study of development of hazard response and any methodology can be derived for a sensor network within an ambulance setting. This will complement the biosensing network for monitoring the patient's physiological and vital signs that is already in place. With current focus of biosensor research on acquiring patient data, integrating environmental sensing into the ambulance setting toward human prognostics based on environmental factors would improve on-scene treatment, which expands on the field of biosensing networks to also monitor environmental changes in order to better accommodate the recovery of patients. This would complement the use of traditional wireless sensor networks for patient monitoring and increasing efficiency of the emergency support system itself.

17.2.2 Biosensing Network

Biosensors form the basic building blocks of IoT in emergency medicine that drives the development of biocompatible composite materials and the design of reliable biosensing networks using novel composite materials (Guiseppi-Elie, 2010). In order to maximize biosensors' reliability, the sensor head should be encapsulated in a cavity to avoid interference from external sources as shown in Figure 17.3. The primary consideration here is optimizing cavity creation and the method of actuating the sensor head inside the cavity from the interaction between encapsulation and the biological parameters to be measured. In the IoT context, the wireless biosensing system consists of a network of environmental biosensors that have minimal space requirements. The embedded microelectromechanical system (MEMS) is mounted with an adequate degree of freedom

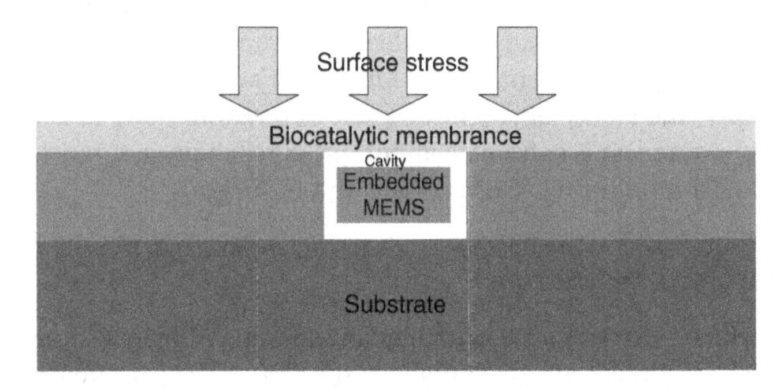

Figure 17.3 Biosensor designed to withstand harsh operating environment.

Table 17.1 Molar heat output of enzyme-catalyzed reactions in biosensors for emergency medicine.

Parameter	Enzyme	Heat (Kcal/mol)
Benzylpenicillin	Penicillinase	16
Cholesterol	Cholesterol oxidase	13
Glucose	Glucose oxidase	19
Starch	Amylase	2
Sucrose	Invertase	5
Urea	Urease	14

to oscillate during use since it may be subject to surface stress and analyte interaction.

Some biosensors employ a biocatalyst that maps a substrate into a corresponding electrical signal for subsequent data processing. This output signal is often weak and degraded by additive noise. Wearable biosensors can effectively double in size when another adjacent sensor, with biocatalytic membrane omitted, taking a reference baseline reading for noise compensation is added to the sensor head. Given that the ambient environment can have a significant impact on biosensors' reliability (Kress-Rogers, 1996), the biosensors in a PoC environment, where many sensors operate under exothermic enzyme-catalyzed reactions, are often deployed for both patient and environmental sensing (Yakovleva, 2013). Heat sensing biosensors commonly found in emergency medicine are summarized in Table 17.1. For improved reliability, thermistors can be used to monitor any changes in ambient temperatures while measurement is taken. It is worth noting that these types of heat sensitive biosensors as a worst case scenario as other types are less prone to variations in the ambient environment.

In the ambulance setting, it is virtually impossible to control the environment where a rescue takes place, attending to an accident scene under snow will be very different from attending to a patient indoor. One major advantage of using IoT in the ambulance is the ability of supporting self-calibration (Song et al., 2014). Environmental awareness is extremely important in reliability assurance through condition-based monitoring (CBM) (Fong and Li, 2012a).

17.2.3 Hierarchical Cloud Architecture

Integrating the biosensing network into an ambulance setting, a hierarchical cloud platform model for context-aware emergency support services would be useful to control both resources and scheduling by supporting a range of context-aware IoT services in the cloud control layer (CCL) (Carvalho et al., 2017). The main advantage is to control each context-aware service for the patient and the paramedic in the user control layer (UCL). This supports interoperability irrespective of communication protocols and operating systems, the important issue of interoperability will be discussed in the next section.

Controlling access in an IoT environment is an important part in providing context-aware services (Perera et al., 2014). When a paramedic accesses a context-aware service upon arriving at the scene, an association between service and the paramedic needs to be generated, this is referred to as a service binding (Chen et al., 2016). The context-aware service binding is controlled by the control process shown in Figure 17.4 that manages access, service, resource, context, and performing binding adaptation management. The main modules of this control system are a cloud control layer (CCL) and a user control layer (UCL).

CCL is responsible for resource and service management, which consists of three main modules, namely, a cloud optimization controller COC) that controls cloud resources and service scheduling, a cloud service controller (CSC) that controls service configuration, and a cloud network controller (CNC) that controls cloud-related network information. UCL is responsible for context management and real-time service binding adaptation, which is made up of four main modules, namely, patient context controller (PCC), binding controller (BC), user network controller (UNC), and environmental device controller (ECC). The PCC controls various services for patient health context whereas

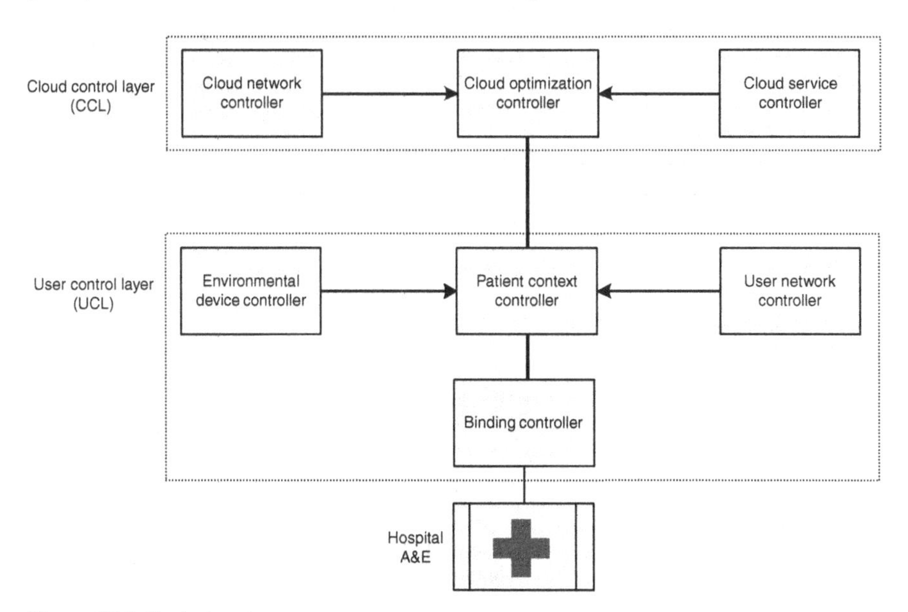

Figure 17.4 Control system.

the BC controls service binding adaptation according to the change of the patient health context that links directly to the hospital via the telemedicine network.

Service binding entails both application and transmission binding (Curbera et al., 2002). Application binding module represents an association among application-related objects, these cover the user interface (UI), programming language, and application protocols; transmission binding module represents an association among data transmission related objects, this is related to interconnection of devices that cover compressor, filter, modulator, queuing, caching, and transmission protocols. Service binding operates upon a binding identifier (BID) such that each object in the binding is also assigned with a system-object identifier (SID). These identifiers are used for the platform-level control as shown in Figure 17.5. Upon detecting a change of service context such as in the case of accessing EPR medical history, BC controls the service binding in conjunction with CCM, ECC, and CNC. The BC needs to initiate real-time service binding adaptation through its control agents through the application binding adaptor (A-adaptor) and transmission binding adaptor (T-adaptor). In this system, both the A and T adaptors update the system objects or protocols inside the application and transmission bindings.

In situations where high demands of network resources are needed, such as sending video images showing the extent of injury suffered by the patient and

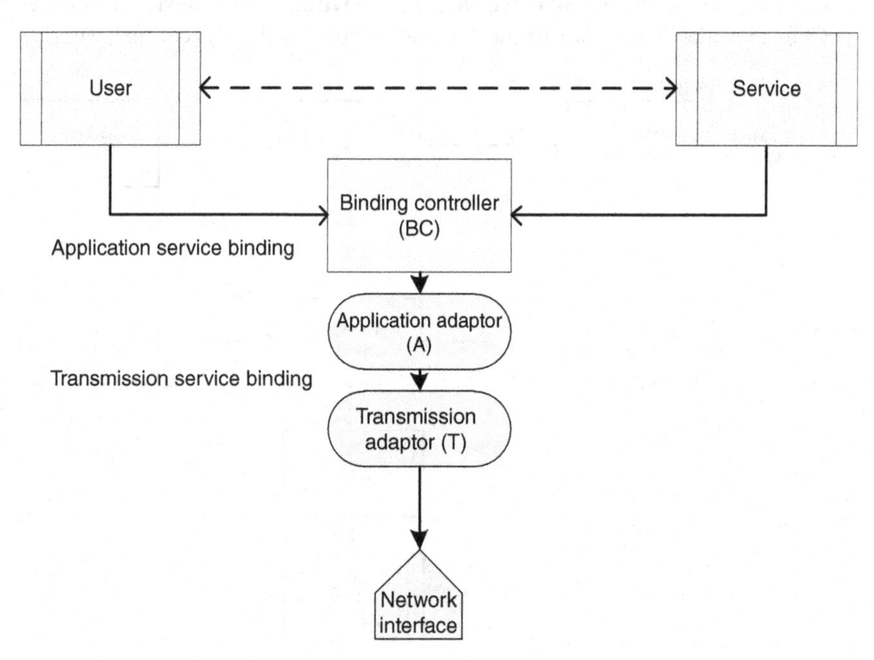

Figure 17.5 Service binding.

information about an accident scene, the paramedic uses video service context such that video streaming is served with adequate resolution. While it is not necessary to support near real-time video playback, the video resolution must be adequately high to show key features needed by hospital staff so that advance preparation can be made prior to the patient's arrival. The A-adaptor and T-adaptor configure image processing objects in the application binding and high-throughput transmission objects in the transmission binding. The network status has to be continuously monitored so that in the event that the network conditions degrade, the T-adaptor will attempt to maintain the connection while performing data compression through local-binding adaptation. In case the network status further deteriorates, the T-adaptor will attempt to maintain the quality of service (QoS) corresponding to the video context by searching for alternative routing path (Zhang and Ansari, 2011). Greatest challenge is posed if the T-adaptor is unable to establish an alternative route and, therefore, unable to maintain the necessary QoS for context delivery, the T-adaptor should initiate a QoS violation to the BCM, so that the CCM determines its new service context, for example, through a store-and-forward approach. The A-adaptor will store the video captured by the paramedic continuously as transmission objects in the transmission binding.

17.2.4 Weather Observation for Remote Rescue

While California battles significant widespread fires, emergency rescue operations also face challenges associated with frequent earthquakes across California and Nevada as reported by the California Earthquake Authority (CEA). Many parts of the Central America face tornadoes that often occur with rescue resource virtually having no advance warning. On the Eastern side, hurricane season in the Atlantic can stretch emergency rescue resources to their limits with millions of peoples affected across multiple states. Throughout most of the United States, ambulances must be able to obtain accurate real-time meteorological information both locally and across adjacent states for efficient and safe rescue operations.

Weather observation plays an important role in remote emergency rescue where accidents are often directly caused by adverse and abrupt change in weather conditions. IoT allows analysis of meteorological information surrounding the ambulance while attending to a remote accident scene (Mendonca et al., 2001). In the context of IoT for meteorological parameters acquisition, IEEE 802.15.4 ZigBee or Bluetooth are commonly used for low-power short-range wireless data communication between connected devices (Tung et al., 2013). Earlier work in wireless sensor networks (WSN) deployment that measure primitive parameters, such as wind speed and direction, can be measured by using a system on chip (SoC) implementation (Du, 2011) such that a microwide velocity sensor and a digital compass are embedded. IoT

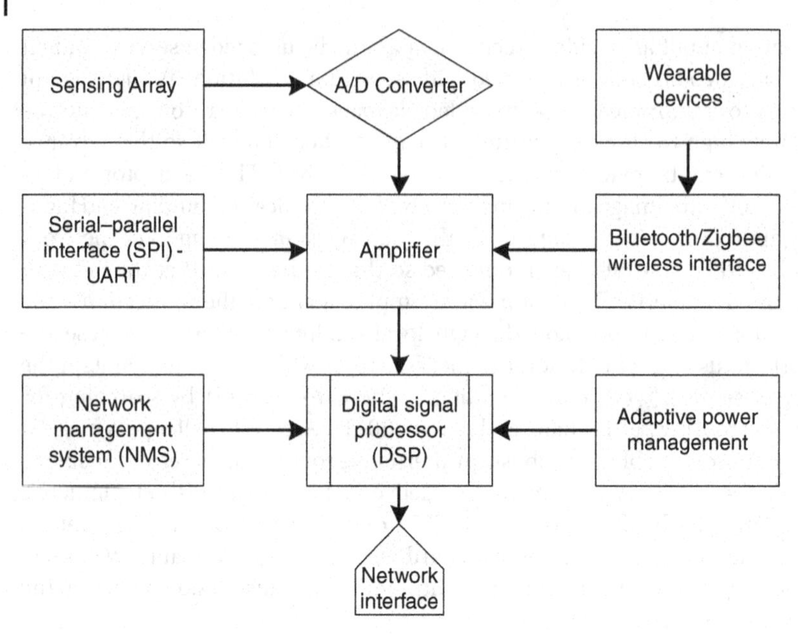

Figure 17.6 IoT-based local area weather observation.

combines different communication networks from stationary meteorological sensors mounted on the ambulance to wearable biosensors that can be moved around by paramedics or patients deployed around an accident scene. While high-precision instruments are needed for the accurate observation of any sudden changes in weather condition that may affect the rescue operation, specific communication interfaces may be necessary for the avoidance of any interference to medical devices (Tung et al., 2014). The basic schematic is shown in Figure 17.6.

In this implementation, a pattern recognition algorithm is implemented inside the universal asynchronous receiver/transmitter (UART) module. The UART is driven by one of the internal timers, which analyzes different types of samples from either internal (for diagnostics and power management) or external (meteorological parameters such as temperature, humidity, and rainfall) sensors. All these sensors are simultaneously interconnected so that any risk that could impact a rescue operation can be detected.

17.3 Integration and Compatibility

IoT is extending the connection to different types of sensors making them part of a comprehensive emergency rescue system, thereby requiring high-computational capabilities for biosignal processing, management, and update to a broader EPR system. Accurate on-scene diagnosis entails acquisition of data from both devices worn by paramedics and patient-centric consumer health care wearables covering a wide geographic area particularly in rural rescue. This poses substantial challenges to being able to maintain reliable short range and long-range communications. While cloud-based telemedicine platforms have become an emerging technology for supporting remote rescue operations, the integration of different devices without any common standards can be particularly challenging. Many consumer wearable devices, such as smart watches and health trackers, require very low latency; this is due to their high degree of mobility inherent nature. These stringent requirements need to be thoroughly considered in the connected ambulance environment particularly because there are many uncontrollable parameters surrounding the scene. Based on all available information to the paramedic, a swift decision must be made on what treatment should be given to the patient. An ambulance-based edge computing platform is needed to support IoT services. This could be either paramedic or patient-based devices such that the smart ambulance would have to act as a road side machine-to-machine (M2M) gateway that routes data from both nearby devices and the hospital.

17.3.1 Operational Consistency and Reliability Assurance

Connected devices can be made self-cognizant by maintaining their own database of certain system health parameters. An individual device can apply statistical modeling to predict the failure of its electronic components and subsystems so that preventive maintenance can be scheduled prior to an anticipated failure. Given that the smart ambulance environment involves many types of devices and sensors with varying life expectancies, accurate prediction of their behavior under different operating conditions would ensure continuing operability.

Operational reliability is perhaps the most important aspect of any service system as its major objective is to safeguard both paramedics and patients by providing timely and accurate information at all times. While calibration guarantees measurement accuracy for a certain period thereafter, further assessments are necessary to deduce the deviation from expected precision. In addition, any impact on measurement due to changes in environmental parameters, such as ambient temperature, humidity, shock, skin condition where sensors are placed on a patient, have to be continually assessed. Calibration is a vital process to ensure long-term operational worthiness. Various aspects of prognostics and system health management will provide solutions for determining an optimal service interval for calibration. It will also help to determine a statistical model that can be used for self-calibration, thereby eliminating the annoyance of having to send the device in for periodic calibration.

Calibration regression analysis with linear interpolation of a reference measurement with the field measurement can therefore provide a system for customized calibration (Maruo et al., 2006). Linear interpolation is used to assign laboratory values to the field measurement, between adjacent pair of the reference measurements performed on a device. Compensation for calibration errors resulting from changes in parameters, such as the user's skin temperature and time of reading taken, should also be taken into consideration. The former can be accomplished by temperature regulation in the confined area during the calibration process, while the latter can be dealt with by statistically modeling the variation pattern caused by use conditions. The known characteristics of an individual device can therefore be deduced. The accuracy will primarily be a function of the number of reference samples taken such that the number measurements is much more than the number of regression terms in the regression algorithm.

While calibration plays an important role in ensuring the trustworthiness of a medical device, continuing reliability assurance entails the development of a comprehensive set of reliability engineering modeling and analysis techniques (Fong et al., 2013). The framework is to facilitate data management and reporting for reliability centered maintenance (RCM) analysis. It will provide support for the different industry sectors and extensive customization options to fit a particular analysis approach.

An important aspect of lifetime analysis is to find a lifetime distribution that can adequately describe the ageing behavior of the device concerned. Most of the lifetimes are continuous in nature and hence many continuous life distributions have been proposed in the literature. IoT allows discrete failure data to be analyzed for the following:

- Reports on field failures are collected continually, and the observations are the number of failures, without specification of the failure times.
- A device operates in cycles and the experimenter observes the number of cycles completed successfully prior to failure. A frequently referred example is a copier whose life length would be the total number of copies it produces. Another example is the number of on/off cycles of a switch before failure occurs.
- An experimenter often discretizes or groups continuous data while attending to patient to construct a model over time.

IoT facilitates analyzing reliability or survival data set to determine which ageing class an individual device belongs to. Thus, tests of stochastic ageing play an important role in such reliability study. Connected devices continually capture data to generalize these concepts to multivariate lifetimes because a complex system usually consists of several components that are working under same environment and hence their lifetimes are generally dependent. Indeed, many such bivariate and multivariate aging concepts have already appeared in the literature for a long time. The concept of dependence permeates throughout our daily life. There are many examples of interdependence in the medicine, economic structures, and reliability engineering, to name just a few. A typical example in engineering is that all outputs from a device will depend on the inputs in a broader sense that include material, equipment, environment, and others. Moreover, the dependence is not deterministic but of stochastic nature.

Prognostics is the process of predicting the future reliability of a product by assessing the extent of deviation or degradation of a product from its expected normal operating conditions (Lau and Fong, 2011). Device or system health monitoring, just like health monitoring of a patient, is a process of measuring and recording the extent of deviation and degradation from a normal operating condition. On-going operational reliability will be assessed by computational algorithms and data collection techniques, condition-based maintenance, prognostics, and system health management for the application of *in situ* diagnostics and prognostics.

At the hospital end, a computer will be running the host-side software, which not only maintains an electronic patient record of each patient but also maintains record of medical devices. When the computer has connection established with the ambulance, the software will also update the data on a central medical device database similar to the case of saving individual patient's medical records for any actions to be taken such as replacing batteries or other consumables. Under normal conditions, the response center does not need to be alerted as data archival and responsive alert generations are all performed automatically. Manual intervention is only necessary for unscheduled system maintenance and update. Prognostics technology will allow the administrator to be warned in advance of any system problem prior to a complete failure so that corrective actions can be taken as appropriate.

The IoT environment enables each device to obtain information about the ambient environment that in turn facilitates the establishment of a maintenance database that is constructed using data on actual use condition of each individual device. One of the important aspects of IoT is the ability to connect devices of different types. Their operational reliability can therefore be closely monitored and at the same time the operating environment can also be monitored. A wide range of information needs to be assessed for ensuring operational reliability; these include disease surveillance-related data, community archive information in social demographic structures, Red Cross blood indicators, Internet search queries, health care facilities operation, and flow data. Data gathered from these sources need to be studied by close linkages for the purpose of compiling available census and research resources. The accuracy, reliability, and robustness of the surveillance algorithms for efficient disease monitoring will be verified by employing principled meta-analysis methods to handle multiple data sources. Another important aspect is to minimize false alarms and missing signals in these surveillance systems.

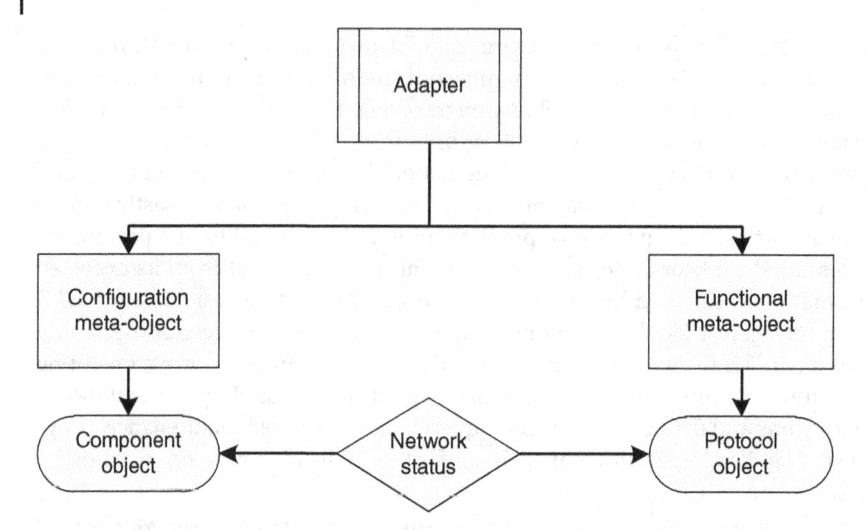

Figure 17.7 Binding meta-objects.

In order to support real-time device monitoring irrespective of platform, it is necessary to deploy a meta-object-based binding adaptation scheme to detect any change to a device's internal system health (Fong et al., 2012b). Meta-class defines a set of generic functions for bindings control in object-oriented (OO) classes (Costanza and Hirschfeld, 2007). A meta-object refers to the instance of the OO classes of which the methods are defined. Referring to Figures 17.5 and 17.6 that illustrate a meta-object space of both A-adaptor and T-adaptor that are utilized for their respective application and transmission binding management, the binding meta-object space can be shown as illustrated in Figure 17.7. The meta-object space contains the main features of configuration and functional meta-objects. The former detects and processes the configurations of the methods, system objects, and the connection topology inside the bindings. It manages connection from point-to-point connection to a point-to-multipoint connection of a specific binding adaptively based on the current network status. The latter detects and manages the binding's protocols and QoS of the bindings that adaptively changes the queuing policy from first input, first output (FIFO) to class-based low latency queue (CB-LLQ) based on current network status.

Before rolling out in an ambulance, it is necessary to develop a series of design study methods specific to a specific country's regulations and validate the patterns of disease models in both controlled laboratory and field experiments inside an ambulance cabin. The laboratory experiment will mainly be conducted in the controlled test chamber facility in a hospital setting that entails a class-100 clean room chamber, a controlled environmental test chamber, and a twin-room test chamber. This should then be verified in an ambulance before being taken out for rescue operations.

The disease prevalence parameters will then be aggregated into the stochastic simulation models to define a set of baseline. The calibration and validation of these models will need to be carried out prior to developing the statistical models.

17.3.2 Electronic Patient Record Retrieval in Multihop Communication

Multihop (ad hoc) communication makes use of multiple wireless hops to relay patient information from the hospital to on-scene paramedics or vice versa. Generally, multihop makes use of either multihop cellular networks or mobile ad hoc networks (MANETs) (Wang, 2008). The former is an important configuration for an ambulance traveling on rural roads. As illustrated in Figure 17.8, vehicle-to-vehicle (V2V) communication can convey data from the nearest base station to other nearby vehicles so that data communication is still possible even when the ambulance is outside cell coverage range. In this example, only Vehicle A is within the boundary of cellular coverage, data packets are relayed from one vehicle to the next with short range V2V direct communication so that the ambulance can still maintain network connection even while traveling outside cellular coverage range.

MANET is also an important topic in supporting smart ambulance in an IoT environment where a MANET on site consists of a group of mobile nodes that communicate without the presence of a wireless link to the hospital. In this arrangement, each node within the proximity of the ambulance is interconnected in a peer-to-peer (P2P) mode instead of the kind of master–slave relationship between nodes. The key feature is that communication between nodes is accomplished by either direct connection or through multiple hop relays.

EPR has been progressively replacing paper scribbled by physicians and other health professionals. In the case of retrieving EPR resources at the ambulance, it usually involves a multihop communication environment. At the same time, it should also allow simultaneous updating of the patient's health while being

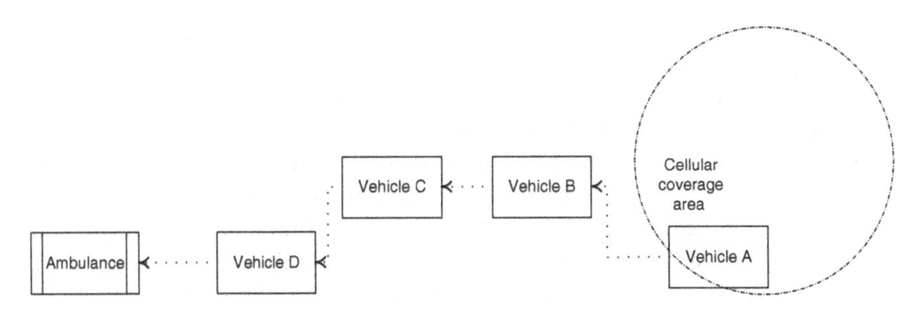

Figure 17.8 Range extension through multihop cellular network.

treated by paramedics. With the interconnection of different sensors and devices discussed earlier in this chapter, the proliferation of these IoT devices form a collection of nodes with unique IDs across different platforms and support data sharing between patient and paramedic devices. In an IoT configuration where different devices are interconnected with cross-platform compatibility, power optimization is necessary to receive and transmit data packets (Han and Ansari, 2014). Nodes in the network are often required to relaying packets destined for other nodes.

Position identification and localization of all IoT devices surrounding the ambulance can provide an efficient method of relaying information to paramedics and becomes possible to extend the coverage range through multihop communication (Abdulla et al., 2012). When a paramedic leaves the ambulance, it can be difficult to continuously track the position through both the distance and angle of the ambulance when using a GPS-enabled device alone. The paramedic cannot be detected if entered a sheltered location or carrying out underwater rescue. While acoustic positioning can provide location tracking (Mohl et al., 2001), there is still a potential issue with the tracking signal having a lower signal-to-noise ratio (SNR) than the ambient noise of the rescue area. Given that offshore rescue is beyond the scope of this text, we shall concentrate our discussion on the technology aspects of providing a reliable link for paramedics to access patient information or any additional information needed for on-scene diagnosis.

17.4 Case Study: Chronic Obstructive Pulmonary Disease

This section takes a closer look at the advances in IoT technology for smart ambulance deployment by using an example of treating COPD. This is an important area in technological advances in emergency medicine for saving lives as an earlier study has reported that paramedics face challenges in providing first aid to patients when distinguishing an asthma strike from other complications due to COPD (Halbert et al., 2006). The key component to the implementation of such a strategy is that paramedics should be able to accurately identify the COPD incidence. A misdiagnosed COPD has the potential to delay a treatment that may alter patient outcome and mortality (Brodie et al., 2006). There is an urgent need for fast and accurate detection as well as effective patient data analysis to facilitate efficient health resource management in treating COPD (Bosse et al., 2011). Current health diagnosis systems lack the ability to interrogate disparate data with diverse conditioned datasets from sources such as the Internet and hospital databases. Access to the amount of information and the patient's medical history may be very limited to the paramedic attending to an emergency, making fast and accurate diagnosis of COPD extremely difficult (Williams et al., 2015).

In view of the increasing risk of developing complications due to delay in providing appropriate treatment, there is an urgent need for timely COPD diagnosis and effective disease simulation analysis to enable health resource management under conditions ranging from patients' homes to polluted workplaces such as dusty environments and mineral mines.

17.4.1 On-scene Diagnosis and Prognosis

When a paramedic arrives at the scene of an accident, it is likely that an initial judgment will be made from observed syndrome since most incidents are reported without accurate medical description suffered by the patient. Commonly observable syndrome such as cough and respiration anomaly can be linked to COPD or infectious diseases such as influenza (Benfield et al., 2008). As observed in the outbreaks of SARS and swine flu, a patient can exhibit multiple syndromes due to a combination of chronic and infectious diseases during pandemic outbreaks (Fong, 2011). Currently, on-scene diagnosis methodologies lack the ability to interrogate disparate data with diverse conditioned datasets from sources such as the national health system and hospital databases. In this respect, an approach that facilitates on-scene disease data collection to enable fast and reliable data-oriented disease prognosis, disease propagation analysis, and risk analysis is needed. By considering a case study involving the on-scene treatment of COPD patient, we have the following developments:

- Syndromic surveillance algorithms for analyzing public health data together with correlated indicators such as patient record data to provide accurate prognosis of respiratory diseases.
- Quantitative methods for modeling disease transmission behavior in highly infectious risk areas and stochastic simulation methods for mimicking infectious disease spreading under complex community contact structures and settings. This is a vitally important topic in ensuring the health and safety of paramedics when arriving at the scene with little knowledge of the patient.
- Aggregate the community-based disease transmission simulation models with physical disease transmission patterns to enable risk modeling, health economic analysis, and performance evaluation of disease surveillance methods and mitigation strategies under a variety of outbreak scenarios.
- Validate infectious disease models through epidemiological knowledge and experience, carefully designed clinical and field experiments, and medical records and data collected during previous pandemic and infectious disease periods.

Advances in IoT enhance algorithms and methodologies that provide quantitative disease modeling solutions to enable the preemptive detection, identification, and comprehension of respiratory diseases as well as scientific justifications for mitigation strategies. Currently, most disease outbreak surveillance methods apply standard monitoring methods used in industrial applications, such as Shewhart, CUSUM, and EWMA charts (Sparks et al., 2010), and disease propagation studies are direct extensions of over-simplified SIR models that were developed over 20 years ago and are often poor indicators of developing disease trends. Part of the challenge is the complexity of disease incident data, which is likely to be heterogeneous, multivariate, intercorrelated, of multiple data types, and often exhibits seasonal patterns over time. There are clearly substantial algorithmic obstacles to interrogating disparate datasets with such complexity or diverse conditioned datasets. Based on the current understanding of surveillance algorithms and modeling methods, an effective, reliable, and credible public health and health care disease monitoring and mitigation framework has not yet been achieved. IoT allows data acquisition from multiple sources such that analysis and modeling methods can enable reliable and data-oriented disease forecasting, disease propagation analysis, and risk analysis. This entails data collection and surveillance algorithms for monitoring multiple streams of data, including disease symptoms, correlated indicators, and incidents under various contagious disease assumptions.

17.4.2 Data Acquisition and Analytics

The smart ambulance entails a wide range of data types from multiple sources as shown in the overall IoT smart ambulance architecture summarized in Figure 17.9.

Automated time series modeling and forecasting methodologies, such as regressions, autoregressive integrated moving average (ARIMA) (Williamson and Hudson, 1999), and Holt–Winter exponential smoothing methods (Burkom et al., 2007), are common syndromic surveillance algorithms used to predict the occurrence of future health events for prognosis. For the purpose of COPD diagnosis and prognosis, a weighted CUSUM chart for detecting patterned mean shifts resulting from forecasting or feedback control can be developed (Shu et al., 2008). Optimizing the effectiveness of a multivariate control chart may relate to the correlation of variables as well as the direction and magnitude of the process shift (Han et al., 2010). Spatial scan statistics have also become popular for the evaluation of geographical disease clusters in a wide range of application areas making it particularly suited for respiratory diseases analysis (Kulldorff, 1997), thereby distinguishing from infectious diseases (Washington et al., 2004). Scan statistic-based prospective spatiotemporal surveillance methods are useful in detecting increase in incident rates in clusters of regions (Woodall et al., 2008). The main objective is to carry out a generalized likelihood ratio test that uses estimated parameters through the maximum likelihood principle. A generic framework based on likelihood ratio statistics for both spatial surveillance and spatiotemporal surveillance under independent or correlated regions is developed for this purpose, as shown in Figure 17.10.

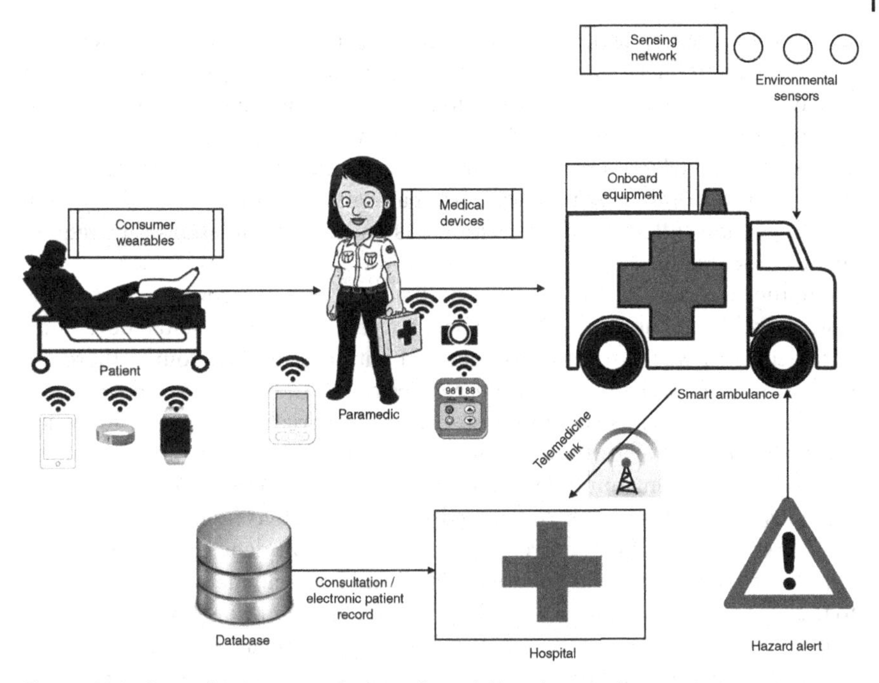

Figure 17.9 Generalized smart ambulance in an IoT environment.

17.4.3 Decision and Selection Process

Most basic methods such as statistical process control (SPC), regression, time series, and forecast-based methods were originally developed as temporal approaches. On the other hand, health surveillance methods such as scan statistics were originally developed as spatial approaches and later extended as temporal and spatiotemporal processes. Most spatial surveillance techniques rely on statistical clustering methods (Sonesson and Bock, 2003). The objectives in temporal surveillance and spatial clustering methods are generally different such that the former involves detection delay whereas the latter is for correct

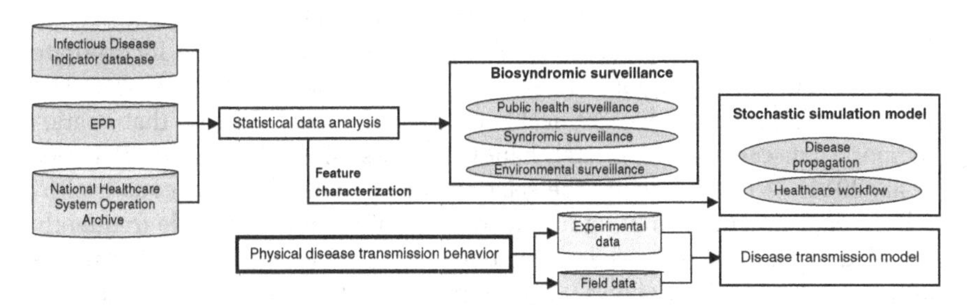

Figure 17.10 Disease modeling using spatio-temporal surveillance.

identification. When deployed in the selection process, this can be conflicting in some situations. It is, therefore, challenging to develop effective methods for timely detection with high identification rates even for standard problems with homogeneous populations and independent distributions of occurrences among individual patients. The disease incident rates are heterogeneous, correlated, and often exhibit seasonal patterns over time particularly with COPD patients. Effective selection methods under these situations thus require further discussion.

Another concern in syndromic surveillance is that it encounters a multiple testing problem. There are multiple data sources from the hospital, devices on paramedics as well as on the patient, and within each data source there are usually multiple series. Many of these series are further divided into subseries. Currently, most series are monitored in a univariate manner, after which multiple detection algorithms are applied to each series. Each of the common methods for handling multiple data streams has its limitations. Commonly used methods like Bonferroni is considered overconservative (Reiner et al., 2003), false discovery rate (FDR) corrections exhibit inherent deficiency with too few hypotheses (Benjamini and Yekutieli, 2001), and Bayesian methods are sensitive to the choice of a prior and it is unclear how to choose a prior (Robert, 2007). In helping with selection process on emergency treatment, faster detection by monitoring disease symptoms and correlated indicators in addition to monitoring observed symptoms is needed.

To eliminate symptoms due to infection, the selection process needs to quantify infection probabilities for microscale or society-scale models of infectious disease spread. Previous studies on pathogen-laden expiratory aerosols do not investigate infectious source strengths under the effects of particle size-dependent dynamics on the removal and dispersion of expiratory aerosols in an actual clinical environment. By integrating fluid dynamic modeling into infection exposure assessment, the selection process entail an accurate prediction model for infection probabilities in contained environments within the PoC to better understand the transmission mechanisms of respiratory diseases.

Most retrospective analyses of health surveillance focus on studies of clusters based on static covariates and exponential smoothing forecasting methods. In order to model discrete count data in health surveillance, it is necessary to construct a dynamic generalized linear model (DGLM) that can incorporate covariates such as subregions, days of the week, holidays, and seasonality. The DGLM accommodates hierarchical modeling of spatial data so that spatio-temporal data can be appropriately handled in a systematic framework (Banerjee et al., 2014). When modeling the relationships among heterogeneous health-related data streams, sampling frequencies are often different while frequently sampled streams may have leading power to predict the key performance indicators. The DGLM allows accurate disaggregation of space–time data so that different levels of disease-related information can be integrated and synchronized in combination with the Bayesian network model (Moauro and Savio, 2005). DGLM can include dynamic covariates for identifying dynamic clusters and their relationships during disease prognosis using data mining such as FDR and variable selection to identify clusters based on dynamic covariates. An algorithm based on a Bayesian framework to calibrate and update model parameters can be used. Predicted values will be compared to observed values, and any significant differences will be indicated in the application for detecting health anomalies.

17.4.4 Patient and the Ambient Environment

The first piece of information to gather upon arriving to the scene to carry out syndromic surveillance is to detect disease outbreaks (Chen et al., 2010) in addition to prior received information from conventional methods through reporting of the case. This can be accomplished by monitoring data that are related to the symptoms, such as influenza-like illness (ILI) symptoms or rapid analysis of the cough sound (Shin et al., 2009), recent over-the-counter (OTC) drug sales, or hospital telephone hotline calls made by the patient. By monitoring various disease-related indicators, the type of disease suffered by the patient can be diagnosed so that countermeasures will be implemented effectively and proactively, possibly even saving a trip to hospital admission given proper treatment at the scene. This entails collection of health care registration data in real time from participating hospitals. To diagnose a COPD patient through elimination of other causes, data acquisition and analytics play a vital role in supporting on-scene diagnosis and prognosis.

The sensing network that consists of paramedics and patient devices will be created as a collection of sensor nodes that sense changes in the environment. In the context of IoT, the physical parameters of the biosensor are environment dependent, as described earlier in this chapter. Deploying wireless sensor networks with specific biomaterial at the sensor head require a paramedic to identify locations that are not suitable for a particular patient affected with a common type of disorder such as in the case of asthma. The biosensor, when networked to a central server, will be vital for PoC environment assessment such that the paramedic can retrieve patient medical history as well as any allergic information in relation to the type of biomaterial to be used. The biosensor module comprises of biomaterial, a processor with on-board memory, analog-to-digital converter (ADC), control circuits, and a wireless transceiver section. IoT will enable real-time monitoring of PoC sites through the architecture in Figure 17.11.

This PoC sensing system will capture, store, and transmit the reading of the biomaterial sensor. For this system, the greatest concern in this implementation is the cost of custom-made components that may end up being expensive as they are likely to be disposable. It is, therefore, more economically desirable to use

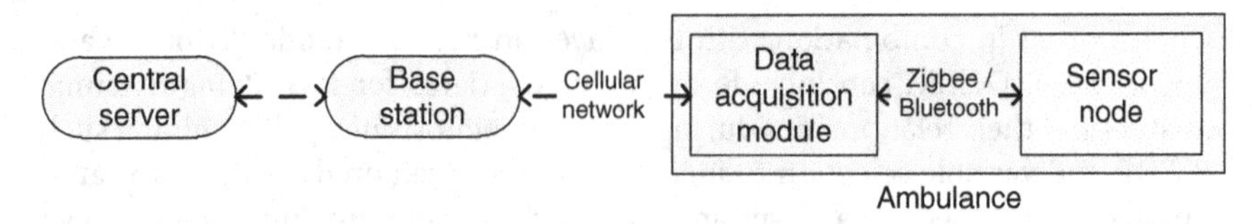

Figure 17.11 High-tier architecture of wireless PoC sensing system.

less powerful hardware on the biosensor module and have more processing done at the receiver station. Recent advances of IoT would support high-speed data transfer and computational processes to be carried out at the server end.

Simulation studies play a unique and significant role in supporting pandemic scenario prediction and enabling understanding of disease spread, which are paramount for mitigation and containment of pandemics. Some disease-spread simulation models aim to understand the effects of changes in citizen behavior or government policies. Others study disease outbreak parameters such as community, demographic, physiological, behavioral, epidemiological, and mitigation-strategy-related features (Cauchemez et al., 2008). PoC sensing provides paramedics with vital information to assess the effectiveness of a treatment plan and nonpharmaceutical interventions to protect themselves from infection while reducing the risk of infection in a hazardous environment (Nuno et al., 2008). This can be accomplished using a Susceptible–Infected–Removed (SIR) model to quantify states of human infection. The Susceptible–Exposed–Infected–Removed (SEIR) model considers the exposed state between the susceptible and infected states, and the compartmental model separates an asymptomatic state from an infectious state by a latent state.

17.5 Smart Ambulance Challenges

User acceptance plays a vital role in the successful rollout of any smart technology. In a smart ambulance setting, this involves end users like paramedics and patients. There are also other supporting personnel such as staff members in the hospital A&E department, emergency response center, and technical support that are all directly involved. To this end, operational reliability is an essential element for all these people concerned since there are both legal and regulatory implications in ensuring that everything incorporated in the ambulance are fit for supporting critical life rescue missions irrespective of operating environment. A number of major challenges must be thoroughly considered for the successful deployment of smart ambulance:

17.5.1 Reliability

Current methods for reliability assessment of electronics-rich systems within the smart ambulance have fundamental flaws due to their inability to keep pace with new technologies. This is due to the fact that any test conducted in a controlled environment during the design phase is unable to account for complex and unpredictable usage profiles. Furthermore, it is equally important to address soft and intermittent faults that are common failure modes in many smart systems. This is especially problematic given the fact that these systems can fail at any time. Vehicle crashes caused by failed electronics, LED cabin lighting systems that cannot survive several months, a host of automotive engine controller failures, life supporting apparatus shutdowns caused by failed sensors, and patients' medical history cannot be retrieved due to failed servers. The impact on safety, availability, and cost pose tremendous challenges to the reliability of smart ambulance.

IoT utilizes connected sensors and monitors that continuously track system performance. These include system degradation assessment, fault diagnostics, and the real-time prediction of reliability through prognostics of subsystems within the smart ambulance. System health information can be gathered while the ambulance is in operation. Such information can be sent to a backend server for analysis and automated scheduling of preventive maintenance based on actual state of health (SoH) of the system.

17.5.2 Standards

The fact that there is currently no standards to ensure operational compatibility across different connected devices mean that data exchange between the paramedics and the patients may not go as smoothly as what standardized devices can offer. It would be virtually impossible to define one single set of standard for all connected devices in a smart ambulance setting given the diverse range of devices involved. There are medical devices that comply with FDA regulatory requirements, legacy instrument, consumer wearables, wireless nodes, and integrated smart home consumer electronics such as assistive elderly care systems and smart drug dispensary systems. Standardizing all these devices and systems can be extremely challenging.

17.5.3 Staff Training and Operating Procedures

Over decades, paramedics are professionally trained to strictly adhere to a set of standard operating procedures optimized for a specific country. When the ambulance goes smart, paramedics too need to be trained to adopt new technologies. It is natural to assume that these health care professionals are not trained with in-depth technical skills.

User interface (UI) must therefore be made to be suitable for the unique environment of the ambulance setting. For example, special gloves may be needed for interaction with touchscreens while these gloves must be suitable for paramedics carrying out their regular duties. Furthermore, user interaction must be minimized in order to ensure that the paramedic can concentrate on the emergency task when supported by smart technologies.

During the process of revolutionizing the way paramedics are supported, paramedics need to be trained with the necessary technical skills without affecting their regular roster for carrying out emergency support. Furthermore, it will also be necessary to aggregate the standard operating procedures with actual operating patterns to enable risk modeling, health economic analysis, and performance evaluation of surveillance methods and mitigation strategies under a variety of rescue scenarios.

17.5.4 Security and Privacy

The successful deployment of smart ambulance relies on sharing of information about the patient such as recent activities, drug consumption, and health indicators. Given that it is highly unlikely that all the necessary information for diagnosis and prognosis are stored in a stand-alone dedicated device, that is, such information is almost certainly stored in an ordinary consumer electronic device that is multipurpose by nature. There is an inherent challenge with having all the necessary health information being made available to on-scene paramedics without any consent being explicitly given by the patient (this is extremely important since the patient may not even be conscious when being attended by the paramedic). At the same time, absolutely no other kind of information can be made accessible to the paramedic or anyone at the scene.

The fact that the smart ambulance environment has so many different types of connected devices with varying connection methods pose a security challenge in ensuring that many possible points of attack within the entire network are all safeguarded.

17.6 Conclusions

The smart ambulance consists of a network of connected medical devices, sensors, and wearable assistive devices worn by paramedics and extends to consumer health devices worn by patients. An IoT platform serves as a bridge that links the paramedics to both patients and the hospital network that provide support from patient medical history retrieval to receiving remote support for providing on-scene treatment. Wearable and noninvasive sensors are vital elements in an IoT environment that supports continuous monitoring of the patient's state of health, thereby providing the paramedics a much better picture about the patient than what traditional methods of emergency medicine practices can offer.

One of the main challenges of interconnecting patients' devices to the health care system is that consumer health devices made by different manufacturers do not always connect using the same protocols so that issues related to cross-platform compatibility among paramedics' devices and consumer health care gadgets without any standards have to be addressed in addition to the problems associated with privacy and security.

Advances in wearable technology and increase in demand for reliable life-critical rescue services aid in the tremendous growth in supporting paramedics with IoT in emergency medicine. Utilizing IoT in a smart ambulance environment, data from a diverse range of devices and biosensors provide vital real-time information for offering the best initial treatment. This chapter described the use of IoT surrounding a telemedicine backbone for emergency medicine with connected devices and sensors gather information about the patient through multistream data analysis. With a direct link established between the hospital and the ambulance, IoT not only supports paramedics carrying out rescue operation but also ensures continuing operational reliability of the smart ambulance through updating self-diagnostic data on a central medical device database similar to the case of saving individual patient's medical records for any actions to be taken such as replacing batteries or other consumables. Under normal conditions, the response center does not need to be alerted as data archival and responsive alert generations are all performed automatically. Manual intervention is only necessary for unscheduled system maintenance and update. Prognostics technology will allow the administrator be warned in advance of any system problem prior to a complete failure so that corrective actions can be taken as appropriate.

Future research and development in IoT for emergency medicine will entail the integration of mobile learning and operating procedures monitoring in providing treatment. Such enhancements will facilitate both continuing training to improve paramedics' skills as well as to reduce the risk of human errors. Furthermore, integration of video processing technology will allow hazard detection to provide preemptive alerts to paramedics of any dangers surrounding the scene.

Enhancements will include the integration of artificial intelligence components to supplement or take the place of the health care professional or other caregiver or the social support agents. Miniaturization (nanotechnology) will enable injectable sensors to become common. Injectable and implanted devices—embedded chips and other sensors to monitor physiological status—are beginning to be used, as are electronic patches and electronic tattoos that serve as sensors (Motti and Caine, 2015). Augmented reality will also be used in smart health care applications, both to aid health care professionals in integrating data on patients, conditions, and treatments and to assist patients in using the data collected by these systems to manage their health. Innovations in power consumption (increasing battery life and generating power through the user's movements, for example) and increased bandwidth will allow wearable devices to last longer and to transmit more data quickly. In addition to advances in sensor technology, user interfaces, and computational power more integration across platforms will also be developed so that patients can monitor multiple aspects of their health simultaneously, thus allowing a more holistic perspective not just addressing one condition or symptom, but instead addressing the broader context of patients dealing with multiple health-related issues. As IoT systems become versatile with increasing types of connected devices and sensors emerging into the health care industry, sophisticated ambulance systems will support emergency rescue more efficiently, maximizing the chance of survival for a greater number of patients in the foreseeable future.

References

Abdulla, A. E., Nishiyama, H., Yang, J., Ansari, N., and Kato, N. (2012) Hymn: a novel hybrid multi-hop routing algorithm to improve the longevity of WSNs. *IEEE Transactions on Wireless Communications*, **11**(7), 2531–2541.

Ansari, N., Fong, B., and Zhang, Y. T. (2006) Wireless technology advances and challenges for telemedicine. *IEEE Communications Magazine*, **44**(4), 39–40.

Banerjee, S., Carlin, B. P., and Gelfand, A. E. (2014) *Hierarchical Modeling and Analysis for Spatial Data*. CRC Press.

Benfield, T., Lange, P., and Vestbo, J. (2008) COPD stage and risk of hospitalization for infectious disease. *Chest*, **134**(1), 46–53.

Benjamini, Y., and Yekutieli, D. (2001) The control of the false discovery rate in multiple testing under dependency. *Annals of statistics*, **29**, 1165–1188.

Bosse, G., Schmidbauer, W., Spies, C. D. et al. (2011) Adherence to guideline-based standard operating procedures in pre-hospital emergency patients with chronic obstructive pulmonary disease. *Journal of International Medical Research*, **39**(1), 267–276.

Brodie, B. R., Hansen, C., Stuckey, T. D. et al. (2006) Door-to-balloon time with primary percutaneous coronary intervention for acute myocardial infarction impacts late cardiac mortality in high-risk patients and patients presenting early after the onset of symptoms. *Journal of the American College of Cardiology*, **47**(2), 289–295.

Burkom, H. S., Murphy, S. P., and Shmueli, G. (2007) Automated time series forecasting for biosurveillance. *Statistics in Medicine*, **26**(22), 4202–4218.

Campbell, I. A., Colman, S. B., Mao, J. H., Prescott, R. J., and Weston, C. F. (1995) An open, prospective comparison of beta 2 agonists given via nebuliser, nebuhaler, or pressurised inhaler by ambulance crew as emergency treatment. *Thorax*, **50**(1), 79–80.

Carvalho, G. H., Woungang, I., Anpalagan, A., Jaseemuddin, M., and Hossain, E. (2017) Intercloud and HetNet for mobile cloud computing in 5G systems: design issues, challenges, and optimization. *IEEE Network*, **31**(3), 80–89.

Cauchemez, S., Valleron, A. J., Boëlle, P. Y., Flahault, A., and Ferguson, N. M. (2008) Estimating the impact of school closure on influenza transmission from Sentinel data. *Nature,* **452**(7188), 750.

Chen, H., Zeng, D., and Yan, P. (2010) *Infectious Disease Informatics: Syndromic Surveillance for Public Health and Bio-defense.* Springer Science & Business Media.

Chen, R., Bao, F., and Guo, J. (2016) Trust-based service management for social internet of things systems. *IEEE Transactions on Dependable and Secure Computing,* **13**(6), 684.

Costanza, P., and Hirschfeld, R. (2007) Reflective layer activation in ContextL. Proceedings of the 2007 ACM symposium on Applied computing, pp. 1280–1285.

Curbera, F., Duftler, M., Khalaf, R., Nagy, W., Mukhi, N., and Weerawarana, S. (2002) Unraveling the web services web: an introduction to SOAP, WSDL, and UDDI. *IEEE Internet computing,* **6**(2), 86–93.

Dancer, S. J. (2009) The role of environmental cleaning in the control of hospital-acquired infection. *Journal of Hospital Infection,* **73**(4), 378–385.

Dennehy, C. E., Kishi, D. T., and Louie, C. (1996) Drug-related illness in emergency department patients. *American Journal of Health-System Pharmacy,* **53**(12), 1422–1426.

Du, L., Zhao, Z., Fang, Z., Xu, J., Xiao, L., Geng, D., and Zhao, J. (2011) Wireless networks wind sensor based on micro wind velocity chip and electronic compass. 2011 IEEE International Conference on Nano/Micro Engineered and Molecular Systems (NEMS), pp. 859–862.

Fong, B., and Li, C. K. (2012a). Methods for assessing product reliability: looking for enhancements by adopting condition-based monitoring. *IEEE Consumer Electronics Magazine,* **1**(1), 43–48.

Fong, B., Rapajic, P. B., Hong, G. Y., and Fong, A. C. M. (2003a) Factors causing uncertainties in outdoor wireless wearable communications. *IEEE Pervasive Computing,* **2**(2), 16–19.

Fong, B., Rapajic, P. B., Fong, A. C. M., and Hong, G. Y. (2003b). Polarization of received signals for wideband wireless communications in a heavy rainfall region. *IEEE Communications Letters,* **7**(1), 13–14.

Fong, B., Fong, A. C. M., and Hong, G. Y. (2005) On the performance of telemedicine system using 17-GHz orthogonally polarized microwave links under the influence of heavy rainfall. *IEEE Transactions on Information Technology in Biomedicine,* **9**(3), 424–429.

Fong, B., Ansari, N., and Fong, A. C. M. (2012b). Prognostics and health management for wireless telemedicine networks. *IEEE Wireless Communications,* **19**(5). doi: 10.1109/MWC.2012.6339476.

Fong, B., and Fong A. C. M. (2011) A prognostics approach for cardiac abnormalities detection from ECG signals. The Lancet/JACC Summit.

Fong, B., Fong, A. C. M., Li, C. K., Lee, W. C., and Tsang, K. F. (2013) A study on the reliability optimization of LED-lit backlight units in mobile devices. *Journal of Display Technology,* **9**(3), 131–138.

Guiseppi-Elie, A. (2010) Electroconductive hydrogels: synthesis, characterization and biomedical applications. *Biomaterials*, **31**(10), 2701–2716.

Halbert, R. J., Natoli, J. L., Gano, A., Badamgarav, E., Buist, A. S., and Mannino, D. M. (2006) Global burden of COPD: systematic review and meta-analysis. *European Respiratory Journal*, **28**(3), 523–532.

Han, T., and Ansari, N. (2014) Powering mobile networks with green energy. *IEEE Wireless Communications*, **21**(1), 90–96.

Han, S. W., Tsui, K. L., Ariyajunya, B., and Kim, S. B. (2010) A comparison of CUSUM, EWMA, and temporal scan statistics for detection of increases in Poisson rates. *Quality and Reliability Engineering International*, **26**(3), 279–289.

Kelly, S. D. T., Suryadevara, N. K., and Mukhopadhyay, S. C. (2013) Towards the implementation of IoT for environmental condition monitoring in homes. *IEEE Sensors Journal*, **13**(10), 3846–3853.

Kress-Rogers, E. (1996) *Handbook of Biosensors and Electronic Noses: Medicine, Food, and the Environment*. CRC Press.

Kulldorff, M. (1997) A spatial scan statistic. *Communications in Statistics: Theory and methods*, **26**(6), 1481–1496.

Lau, D. K. and Fong, B. (2011) Prognostics and health management. *Microelectronics Reliability*, **51**(2), 251–253.

Liu, C. C., Lai, M. C., Liu, C. M., Chiu, Y. N., Hsieh, M. H., Hwang, T. J., et al. (2011) Follow-up of subjects with suspected pre-psychotic state in Taiwan. *Schizophrenia Research*, **126**(1), 65–70.

Lombardo, J. S., Burkom, H. S., Elbert, Y. A., Magruder, S. F., and Lewis, S. H. (2003) A systems overview of the electronic surveillance system for the early notification of community-based epidemics (ESSENCE II). *Journal of Urban Health*, **80**(2), 32–42.

Maruo, K., Oota, T., Tsurugi, M., Nakagawa, T., Arimoto, H., Tamura, M., et al. (2006) New methodology to obtain a calibration model for noninvasive near-infrared blood glucose monitoring. *Applied Spectroscopy*, **60**(4), 441–449.

Mendonca, D., Beroggi, G. E., and Wallace, W. A. (2001) Decision support for improvisation during emergency response operations. *International journal of emergency management*, **1**(1), 30–38.

Moauro, F., and Savio, G. (2005) Temporal disaggregation using multivariate structural time series models. *The Econometrics Journal*, **8**(2), 214–234.

Mohl, B., Wahlberg, M., and Heerfordt, A. (2001) A large-aperture array of nonlinked receivers for acoustic positioning of biological sound sources. *The Journal of the Acoustical Society of America*, **109**(1), 434–437.

Motti, V. G., and Caine, K. (2015) Users' privacy concerns about wearables. *International Conference on Financial Cryptography and Data Security*, Springer, Berlin, Heidelberg, pp. 231–244.

Nuno, M., Reichert, T. A., Chowell, G., and Gumel, A. B. (2008) Protecting residential care facilities from pandemic influenza. *Proceedings of the National Academy of Sciences*, **105**(30), 10625–10630.

Perera, C., Zaslavsky, A., Christen, P., and Georgakopoulos, D. (2014) Context aware computing for the Internet of Things: a survey. *IEEE Communications Surveys & Tutorials*, **16**(1), 414–454.

Postma, D. S., and Rabe, K. F. (2015) The asthma–COPD overlap syndrome. *New England Journal of Medicine*, **373**(13), 1241–1249.

Pyrek, K. M. (2014) The Economics of Hand Hygiene Compliance Monitoring.

Reiner, A., Yekutieli, D., and Benjamini, Y. (2003) Identifying differentially expressed genes using false discovery rate controlling procedures. *Bioinformatics*, **19**(3), 368–375.

Robert, C. (2007) *The Bayesian Choice: From Decision-Theoretic Foundations to Computational Implementation*. Springer Science & Business Media.

Roche, N., Zureik, M., Soussan, D., Neukirch, F., and Perrotin, D. (2008) Predictors of outcomes in COPD exacerbation cases presenting to the emergency department. *European Respiratory Journal*, **32**(4), 953–961.

Shin, S. H., Hashimoto, T., and Hatano, S. (2009) Automatic detection system for cough sounds as a symptom of abnormal health condition. *IEEE Transactions on Information Technology in Biomedicine*, **13**(4), 486–493.

Shu, L., Jiang, W., and Tsui, K. L. (2008) A weighted CUSUM chart for detecting patterned mean shifts. *Journal of Quality Technology*, **40**(2), 194.

Sonesson, C., and Bock, D. (2003) A review and discussion of prospective statistical surveillance in public health. *Journal of the Royal Statistical Society: Series A*, **166**(1), 5–21.

Song, Z., Lazarescu, M. T., Tomasi, R., Lavagno, L., and Spirito, M. A. (2014) High-level Internet of Things applications development using wireless sensor networks. *Internet of Things* Springer International Publishing, pp. 75–109.

Sparks, R., Carter, C., Graham, P. et al. (2010) Understanding sources of variation in syndromic surveillance for early warning of natural or intentional disease outbreaks. *IIE Transactions*, **42**(9), 613–631.

Tung, H. Y., Tsang, K. F., Tung, H. C., Chui, K. T., and Chi, H. R. (2013) The design of dual radio ZigBee homecare gateway for remote patient monitoring. *IEEE Transactions on Consumer Electronics*, **59**(4), 756–764.

Tung, H. C., Tsang, K. F., Lam, K. L. et al. (2014) A mobility enabled inpatient monitoring system using a ZigBee medical sensor network. *Sensors*, **14**(2), 2397–2416.

Wang, Z., Liu, L., Zhou, M., and Ansari, N. (2008) A position-based clustering technique for ad hoc intervehicle communication. *IEEE Transactions on Systems, Man, and Cybernetics: Part C (Applications and Reviews)*, **38**(2), 201–208.

Washington, C. H., Radday, J., Streit, T. G. et al. (2004) Spatial clustering of filarial transmission before and after a mass drug administration in a setting of low infection prevalence. *Filaria Journal*, **3**(1), 3.

Williams, T. A., Finn, J., Fatovich, D., Perkins, G. D., Summers, Q., and Jacobs, I. (2015) Paramedic differentiation of asthma and COPD in the prehospital setting is difficult. *Prehospital Emergency Care*, **19**(4), 535–543.

Williamson, G. D., and Hudson, G. W. (1999) A monitoring system for detecting aberrations in public health surveillance reports. *Statistics in Medicine*, **18**(23), 3283–3298.

Woodall, W. H., Brooke Marshall, J., Joner Jr, M. D., Fraker, S. E., and Abdel-Salam, A. S. G. (2008) On the use and evaluation of prospective scan methods for health-related surveillance. *Journal of the Royal Statistical Society: Series A (Statistics in Society)*, **171**(1), 223–237.

Yakovleva, M., Bhand, S., and Danielsson, B. (2013) The enzyme thermistor—a realistic biosensor concept. A critical review. *Analytica Chimica Acta*, **766**, 1–12.

Zhang, J., and Ansari, N. (2011) *On Assuring End-To-End QoE in Next Generation Networks: Challenges and a Possible Solution*. IEEE. Communications Magazine, **49**(7), 185–191.

18

Internet of Things Applications for Agriculture

Lei Zhang, Ibibia K. Dabipi, and Willie L. Brown Jr.

Department of Engineering and Aviation Sciences, University of Maryland Eastern Shore, Princess Anne, MD, USA

18.1 Introduction

The history of agriculture starts at least 22,000 years ago when mankind learned to collect wild grains as food. Various crops have been cultivated as earlier as 9500 BC in Levant according to archaeological discoveries (Hillman, 1996; Walsh, 2009). Over tens of thousands of years since then, significant innovations have been made from time to time to increase the agricultural yield and reduce the heavy human labor needed. However, the demand for more foods from the increasing population will never get satisfied. It is predicted that the world's population will reach 9.7 billion by 2050, which is about 33% more than today (un.org, 2015). Consequently, to keep pace with such population growth, the global production of food has to increase at least 70% to feed the world.

Meanwhile, only a small portion of earth's surface is available for agriculture uses, due to various limitations, including temperature, climate, topography, soil quality, and technologies. Agricultural land use is also shaped by political and economic factors, such as land tenure patterns, environmental regulations, and population density (learner.org, 2016). In fact, the total agricultural land used to produce food has been decreasing for the last few decades. In 2013, total agricultural land used to produce food was around 18.6 million square miles, which covers 37.73% of the world's land area. In comparison, in 1991, these numbers were 19.5 million and 39.47%. So, humanity is facing a daunting challenge of how to feed more people with less land, as shown in Figures 18.1 and 18.2 (WorldBank, 2016).

The answer to the critical issue lays in a new technology "precision agriculture" (PA) that will have a profound effect on the lives of billions of people. The precision agriculture techniques and technologies aim to improve

Internet of Things A to Z: Technologies and Applications, First Edition. Edited by Qusay F. Hassan.
© 2018 by The Institute of Electrical and Electronics Engineers, Inc. Published 2018 by John Wiley & Sons, Inc.

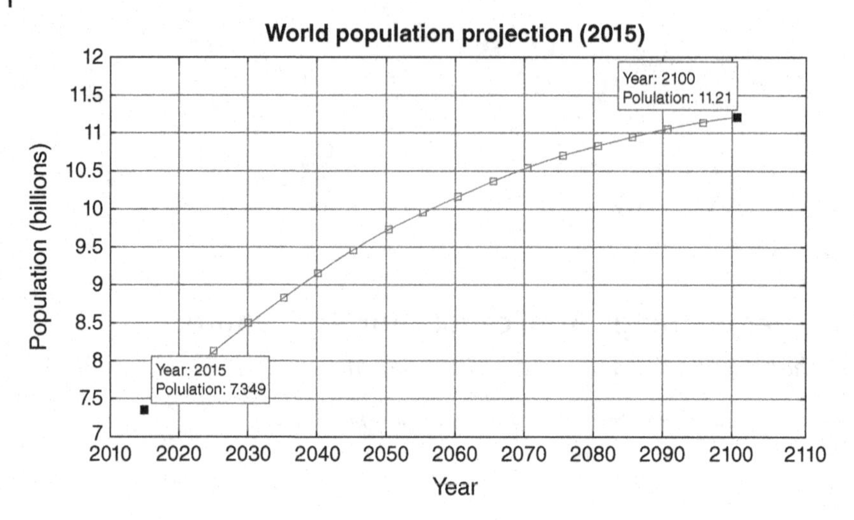

Figure 18.1 World population, 2015–2100.

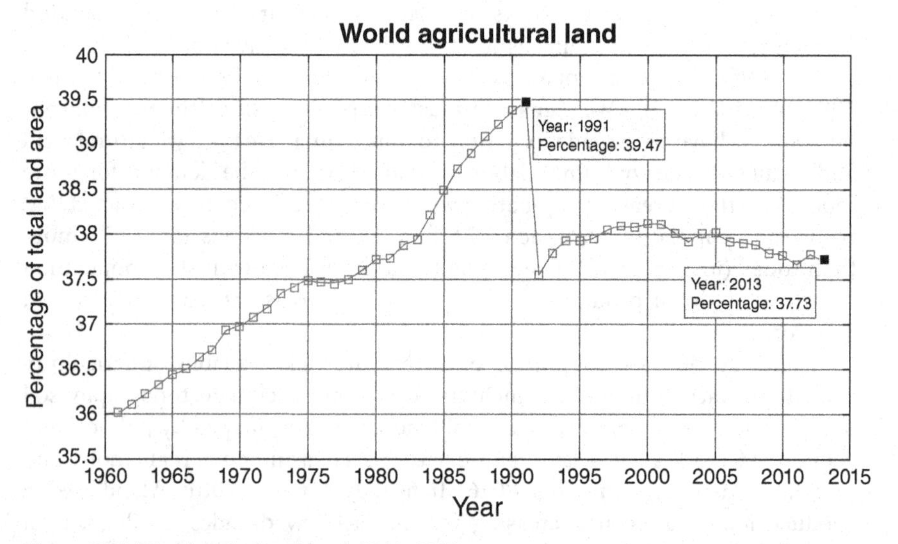

Figure 18.2 World agricultural land, 1961–2013.

the efficiency of agriculture to maximize food production, minimize environmental impact, and reduce cost. Basically, PA, or site-specific agriculture (SA), is integrated information and production-based farming system that can collect precise data on every site in the field and accordingly customize the cultivation of each site independently. In the traditional way, agricultural operations such as planting or harvesting are performed by following a predetermined schedule. However, the effectiveness of practice and schedule could be greatly improved with smarter decisions based on real-time data and predictive analytics on weather, soil quality, crop maturity, equipment, labor costs, and availability. The PA is a farming management system based on observing, measuring, and responding to agricultural variabilities in different aspects. In PA, the large fields are managed as a group of small fields and each small one will be treated precisely and independently with reduced misapplication of water, seed, and nutrients in order to increase crops and farm efficiency.

As an emerging paradigm, the IoT is considered to be *the next big thing* that can have a significant influence on the future of the world. By applying latest IoT technologies in agriculture practice, traditional ways of farming can be fundamentally changed on every aspect, to pave the way to a new agriculture pattern of PA. Briefly, the implementation of PA relies on three stages: (i) the real-time data acquisition, (ii) data analysis and decision-making, and (iii) corresponding precise treatments. All these three stages can be greatly facilitated with the advancing of the IoT technologies in recent years. First of all, IoT provides the fundamental network infrastructure through which enormous smart objects, spanning from microsensors to heavy agricultural vehicles can easily interconnect to each other and to the Internet. This enables the simplest solution of data collection, gathering, exchange, and transmission. Second, backboned by the Internet, IoT provides a revolutionary way in data processing and intelligent decision-making. Solutions for all kinds of services are available online offered by providers all over the world, from industry giants to start-up companies in remote countries. Most of these services, from satellite imaging and processing to chicken health and welfare monitoring, can be cloud based. By this way, they can process data to make an intelligent decision in real-time 24×7 automatically with no human intervention needed. Finally, agricultural treatment decisions and procedures generated from the cloud will be transferred back to the farm. Within the IoT environment, automated agricultural devices, machines, and vehicles will operate accordingly to cultivate crop and livestock in an optimal way (Burrus, 2016). Meanwhile, users can enjoy an unprecedented convenient way of Internet-based data and information access, visualization, and presentation.

Consequently, the seamless integration of IoT technologies into PA raises the agriculture to a new level that was unimaginable before, by which the whole agriculture industry will be reshaped with enormous profits. In general, within the scope of PA, IoT can improve or solve critical issues such as drought response, crop yield optimization, land management, and pest control.

This chapter introduces typical IoT applications in PA, including basic concepts, related technologies, system organization, and the implementation with available products.

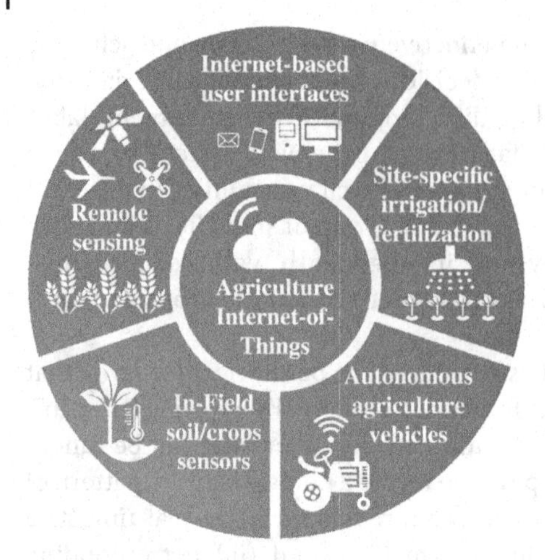

Figure 18.3 IoT-backboned PA.

18.2 Internet of Things-Based Precision Agriculture

Emerging IoT technologies can be integrated into PA in various aspects to improve the agriculture efficiency and productivity. As an overview, Figure 18.3 illustrates typical functions and features provided by this integration.

The IoT technologies can be classified into three categories: data collection, cloud-side data analysis and decision-making, and IoT-assisted agricultural operation.

18.2.1 Data Collection

The goal of PA data collection is to collect soil parameters and crop status/yield data on each site for guiding later operation such as planting, fertilizing, and irrigation. The data collection can be achieved majorly in two ways. The first is through multifunctional imaginary devices equipped with remote sensing platforms, including satellites, agriculture airplanes, balloons, and unmanned aerial vehicles (UAVs). The second is from various sensors installed in different sites across the farmland. Diversified sensors have been developed for the measurement of humidity, temperature, nitrate levels, and so on, to meet requirements of different PA schemes. All data must be tagged with the precise location information, which normally is generated from GPS devices, to support the site-specific treatment later on.

In the next step, all data will feed into a geographic information system (GIS) to generate a crop or soil index map. The GIS can process the data to visualize agricultural environments and status for managing cultivation. In PA scheme, GIS technologies can be used to examine farm conditions, measure and monitor the effects of farm management practices, including crop yield estimates, soil amendment analyses, and erosion identification/remediation. Furthermore, with the help from GIS, reduction in farm input costs such as fertilizer, fuel, seed, labor, and transportation can be achieved (esri.com, 2016).

18.2.2 Site-Specific Operation

Different to the ancient farmer, most agriculture works in modern large-scale agriculture have taken over by equipment such as tractors and harvesters. More specifically, within the scenario of PA, agriculture vehicles will be equipped with GPS and GIS systems, and can operate precisely, site-specifically, and autonomously in dealing with various tasks, including seeding, fertilizing, and harvesting. Another major operation necessity in agriculture is irrigation and the understanding of natural precipitation (e.g., satisfied or unsatisfied water needs of plants). In PA, the irrigation is precisely managed to cover the deficit between crops' optimal water needs and natural supplies on each site independently (CEMA, 2016).

18.2.3 IoT Application in PA

The IoT technologies can play key roles in the PA implementation. IoT provides not only the communication infrastructure to interconnect every smart object from sensor, vehicle, to user mobile device through the Internet, but also functions including local/remote data acquisition, in-cloud intelligent information analysis and decision-making, data access, visualization, user interfacing, and agriculture operation automation.

In general, IoT has two perspectives, to be either Internet centric or smart device centric (Gubbi et al., 2013). Within the PA scenario, the Internet-centric IoT systems have better functionality, flexibility, and extendibility. Systems in this category can take advantage of various Internet services and will have more powerful computing capability from the cloud side. A conceptual schematic architecture of such IoT system is shown in Figure 18.4.

As can be seen in Figure 18.4, the agriculture IoT model has three basic layers. The bottom is the data acquisition layer, in which environmental/crop data are collected through either sensors or remote sensing devices such as UAV then uploaded to the cloud storage through an Internet gateway.

The second layer is the cloud computing function layer (Hassan et al., 2012). Diversified cloud computing functions and services are integrated and provided in this layer. For any required function, there are always multiple solutions available from different providers. In the general process, at first, raw data will be filtered and processed by data analysis tools to abstract information. Then purified information will be converted to essential knowledge by data mining and machine learning tools. With the consideration of

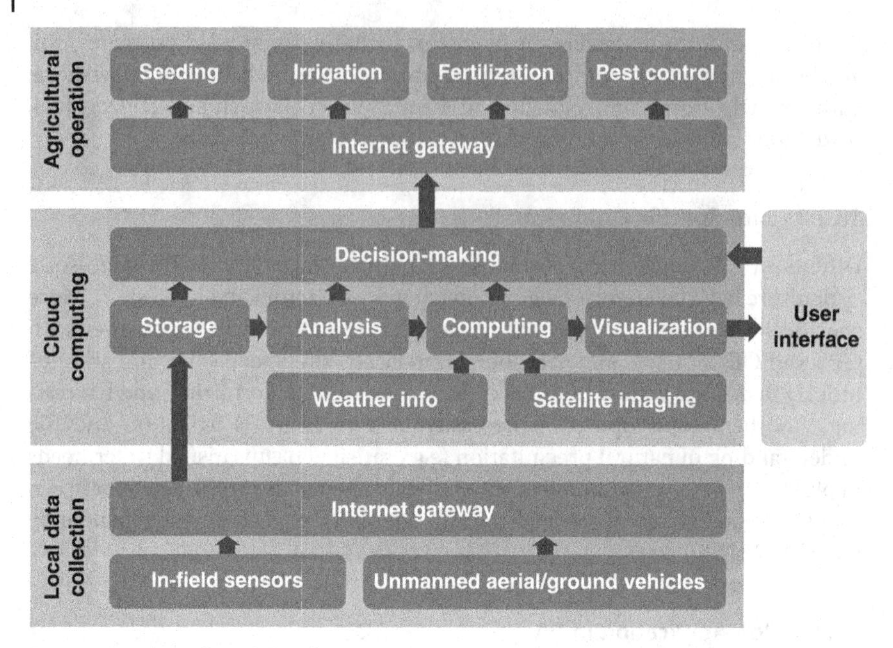

Figure 18.4 Agriculture IoT architecture.

specific factors, such as soil condition, fertilizing pattern, crop state, weather and environment situation and so forth, knowledge derived in the last step will be used for decision-making or the generation of field index maps for different purposes. Such maps will guide agriculture operations in the next stage. Meanwhile, this layer also provides data visualization and presentation services for user access.

Ultimately, the top layer is the control of all agricultural operations, including seeding, irrigation, fertilizing, and harvesting. Corresponding operations are implemented with agricultural devices, machines, vehicles, and irrigation system based on decisions or index maps generated on the cloud side during the last stage. Related control commands will be transmitted through the Internet gateway to agricultural systems. Assisted with GIS, these agriculture systems will treat each atomic plot precisely, by which optimal efficiency and productivity can be expected.

18.3 IoT Application in Agriculture Irrigation

Here are some facts about water. About 71% of the Earth's surface is covered with water, of which about 96.5% of all Earth's water is salt water held by oceans (USGS, 2016). Only the rest 3% is fresh water. Hence, over two thirds is frozen water possessed by glaciers and polar ice caps. Furthermore, most of the remaining unfrozen freshwater is underground, with only a small fraction (0.5%) staying above ground or in the air (GreenFacts, 2016). Shortly, humanity relies on this 0.5% for all of man's and ecosystem's fresh water needs. Furthermore, about 70% of all the accessible freshwater is consumed by agriculture. In comparison, industry consumes about 23% and dwarfing municipal consumes about 8%, which altogether count less than a half of agriculture (UN-Water, 2016).

The traditional ways of water consumption are facing severe challenges on both scarcity and environmental impacts. For example, water has become increasingly scarce in the western U.S. states during the last few decades. In the next 10 years, 40 over 50 states will experience water shortage at different levels (Sunding et al., 2016). And the situation will get worse with the growing demands from cities and industries. Meanwhile, enough fresh water must be kept in rivers and lakes to sustain ecosystems. Consequently, improving water use efficiency or enhancing agricultural water productivity is a critical response to growing water scarcity.

In general, the agricultural water productivity is defined as the ratio of the net benefits obtained from all agriculture sectors, including crop, livestock, fishery, forestry, and so on, to the amount of water consumed during the process that those benefits are produced (Sharma et al., 2015). The key to improving agricultural water productivity is to incorporate crop demand-dependent irrigation schedule that can save water without affecting crop yields. However, practically it is quite challenging to accurately estimate the water demand of crops, which involves many factors such as crop type, irrigation method, soil type, precipitation, crop needs, and soil moisture retention. In fact, even in the year of 2013, crops visual inspection were still playing the key role as a method for irrigation decision-making in nearly 80% of farms with irrigated land in the United States (USDA, 2016).

The current situation is expected to turn around in the next decade with the adoption of emerging IoT technologies. By adopting new IoT-based approaches such as crop water stress index (CWSI)-based irrigation management, the traditional crop watering scheme will be revolutionized to significantly improve the agricultural water productivity.

18.3.1 Crop Water Stress Index

Optimal irrigation relies on the precise measurement of crops' water demand. Since early 1970s, researchers have attempted to monitor crop's water demand by measuring crop's surface temperature. In recent years, this approach has attracted more and more attention, especially with the thriving of remote sensing technologies.

Shortly, crop's water demand of a specific site can be characterized as a CWSI. The CWSI value can be derived in different ways. One popular method to calculate CWSI is based on the following equation (USDA, 2016):

$$\mathrm{CWSI} = \frac{\mathrm{d}T - \mathrm{d}T_{\mathrm{l}}}{\mathrm{d}T_{\mathrm{u}} - \mathrm{d}T_{\mathrm{l}}} \tag{18.1}$$

In Equation 18.1, $\mathrm{d}T$ is the difference between the crop canopy and the air temperature measurements; $\mathrm{d}T_{\mathrm{u}}$ is the difference between the canopy upper limit and the air temperature (nontranspiring crop); and $\mathrm{d}T_{\mathrm{l}}$ is the difference between canopy lower limit and the air temperature (well-watered crop). A CWSI value of 0 indicates that the crops have no water stress, and the CWSI value of 1 represents that crops are suffering the maximum water stress. In conclusion, the crop–water stress signals the site-specific crops' need for irrigation. Moreover, in irrigation management in dealing with crops' water stress, many other factors should be considered, such as yield response to water stress, probable crop price, and water cost (USDA, 2016).

In CWSI-based irrigation management, the first step is to acquire crops' water demand of each site in the entire farmland, in which modern remote sensing technologies can contribute a lot. Then on each site irrigation outlets will be controlled independently based on the crops' water demand; by this way, optimal irrigation efficiency can be achieved.

Various methods have been developed (Idso et al., 1981; Idso, 1982; Maes and Steppe, 2012; Taghvaeian et al., 2012) to determine canopy limits $\mathrm{d}T_{\mathrm{l}}$ and $\mathrm{d}T_{\mathrm{u}}$ in Equation 18.1 in advance. Practically, periodical acquisition of crop canopy and air temperature measurements is needed to calculate the $\mathrm{d}T$ for CWSI derivation.

18.3.2 Data Acquisition

In deriving the value of $\mathrm{d}T$, normally the air temperature can be easily acquired, for example, by the use of a temperature sensor installed meters above ground (Taghvaeian et al., 2012). The difficulty stands on how to measure the crop canopy on each site through the entire farmland with the minimum cost, time delay, and labor needed with an acceptable precision. Different solutions have been developed for this purpose based on different platforms such as satellites, airborne platforms, ground station, and unmanned vehicle.

Thermal infrared imaging data can be acquired via the remote sensing of satellites, for example, the MODIS (Moderate Resolution Imaging Spectroradiometer) system on the Terra and Aqua satellites launched by NASA (NASA, 2016). There are many successful cases of CWSI irrigation management by the use of MODIS data (Xiaoning et al., 2007; Tingting et al., 2008).

Crop canopies may also be obtained by mounting a thermal camera on a balloon or in an aircraft (Jackson, 1982; Hatfield and Pinter, 1993; Jones et al.,

2009). Satellite and other airborne platforms offer the highest efficiency on data acquisition. However, none of them are convenient for an individual farmer to access.

Another way to collect crop canopy values is through in-field ground stations. Commercialized devices are available on the market for this temperature measurement. For example, the infrared radiometer products can acquire high accuracy, noncontact surface temperature measurement (ApogeeInstruments, 2016). Coupled with other options, the installation cost is expensive and linear to the area of farmland and could be problematic for the operations of agricultural vehicles due to scattered ground stations. In contrast, the benefit of this approach is that the data collection is local, direct, and 24×7 in real time allowing continuous monitoring.

Technology advancing and cost reduction of unmanned vehicles (UV) especially in recent years have greatly stimulated their applications in many fields, including the agriculture. Equipped with a thermal camera, small-scale UAVs can provide inexpensive solutions for convenient crop canopy measurement (Pipia et al., 2012; Park et al., 2015). Comparing to all the other methods, UAV solution is low cost, user-friendly, flexible, and customizable, with no installation needed.

18.3.3 IoT Irrigation System

The system diagram of a typical IoT irrigation system is shown in Figure 18.5. The system deploys wireless sensor network (WSN) to connect all in-field sensors for crop canopy and air temperature measurements, and then feed all data to the network gateway. Among the many solutions available for this purpose, ZigBee is a popular technology that is easy to setup and configure (Baronti et al., 2007). Due to the low-data transmission bandwidth requirement in this application, the gateway can access the Internet wirelessly by the use of technologies, such as 4G LTE mobile communication network, with

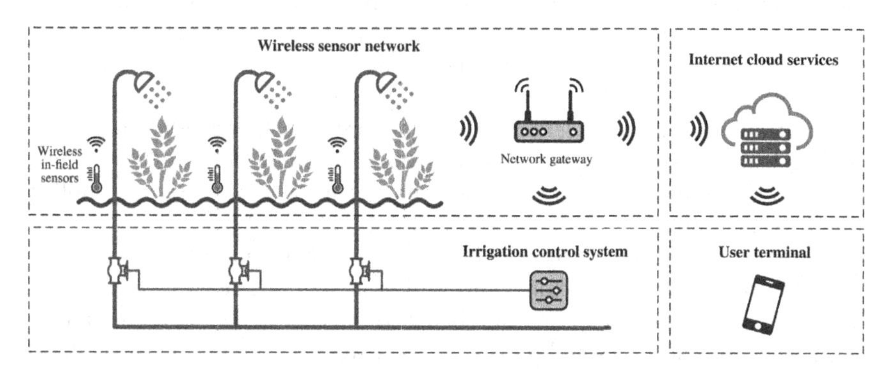

Figure 18.5 IoT irrigation system diagram.

comparatively low cost. The data transmitted over mobile network will be received by subscribed web services on the cloud. Corresponding intelligent software applications will analyze data from farmland, with the consideration of information from other sources, such as weather service and satellite imaging, to apply CWSI models for water need assessment, and finally, produce irrigation index value for each site. These results will be transmitted back to the network gateway, and then forwarded to a controller to manage irrigation. In addition, farmers can access all data and results, and make adjustments with terminals such as smart phones through specifically developed web applications (Zhang, 2011; Guo and Zhong, 2015; Harun et al., 2015; Khelifa et al., 2015).

18.4 IoT Application in Agriculture Fertilization

Thousands of years ago, ancient Egyptians, Romans, and Babylonians have learned to use fertilizers, such as minerals and manure, to enhance the productivity of farms. In general, fertilizer refers to natural or synthetic materials that can supply essential nutrients for plants to grow. The three main macronutrients required by plants are nitrogen (N) for leaf growth; phosphorus (P) for root, flowers, seeds, and fruit development; and potassium (K) for stem growth, water movement, and promotion of flowering and fruiting (Kiiski et al., 2009).

Plants need nutrients from fertilizers to maintain the healthy life. However, applying nutrients improperly can be detrimental or even lead to the death of plants. More important, excess fertilizer is harmful to the environment by depleting the soil quality, poisoning ground water, and even contributing to the climate changes across the globe (Environment.co.za, 2015).

The key to minimizing the negative effects of agricultural fertilizing to the environment stands on precisely applying the required dose of crops, which is known as PA fertilization. The implementation of PA fertilization requires site-specific soil nutrient level measurement. However, practically, the PA fertilization is far more complicated than precise irrigation schemes introduced in previous sections, since the determination of the amount of fertilizer needed for soil patches is a very complicated problem. The fertilizer amount needed relates to various factors, including crop type, soil type and capability of soil absorption, product yield, fertility type, fertilizer utilization rate, weather condition, climate, and agricultural technological factors. The soil nutrient level measurement is expensive, inconvenient, and time consuming, normally involves the sampling and investigation of soil samples at each location.

The advancing of technologies in remote sensing has tremendously strengthened the capability in data collection. With that support, new IoT approaches have been developed for estimating spatial patterns of fertilizing requirements with acceptable accuracy and minimum labor work required. The Normalized Difference Vegetation Index (NDVI) is a powerful method for crop nutrient status monitoring using aerial/satellite photographic images. In brief, NDVI value represents the reflection of visible and near-infrared light from vegetation. NDVI can be used for relative estimation of crops health, vegetation vigor, and density, and can also contribute to the soil nutrient level assessment (Rouse et al., 1974; Rahman et al., 1994; Schumann, 2006).

NDVI can be calculated from individual measurements according to the following equation:

$$\mathrm{NDVI} = \frac{\mathrm{NIR} - \mathrm{VIS}}{\mathrm{NIR} + \mathrm{VIS}} \tag{18.2}$$

In Equation 18.2, VIS and NIR represent the measurement of spectral reflectance acquired in the visible and near-infrared regions, respectively. NDVI value is in the range of $(-1, +1)$ (Weier and Herring, 2011).

The basic procedure of PA fertilization is similar to that shown in Figure 18.5. The first step is to acquire NDVI value by the use of satellite, airborne platforms, or IoT ground stations for generating the site-specific fertilization index map. Then, the application of fertilizers is implemented by automated agriculture machines and vehicles according to the index map. Shortly, such precise fertilization can significantly improve the application efficiency of fertilizers and reduce the side effects to the environment, which has been proved by many successful cases reported (Van Alphen and Stoorvogel, 2000; Tianhong et al., 2003; Schumann, 2006).

There are quite a few new enabling technologies contributing to the IoT-based PA fertilization (CEMA, 2016):

- *High Accuracy Global Positioning System (GPS).* Currently, there are four different satellite-based global positioning systems: GPS (US), GLONASS (Russian), Galileo (EU), and BeiDou (China). Until now, the accuracy of GPS signal is 4 m root mean square (rms) (7.8 m at a 95% confidence level) (DoD, 2001). This is far from meeting the requirements of PA fertilization. The high-precision GPS is a key technology needed for agriculture vehicle to obtain its position in the field based on geographic coordinates (latitude and longitude). Real Time Kinematic (RTK) is a new technique developed in recent years that can significantly improve the GPS precision. Its measurements rely not only on the signal content but also on the phase of the carrier wave of the signal. Moreover, RTK technology uses the signal from reference station or interpolated virtual station for real-time coordinate corrections, to reach up to centimeter-level accuracy (Keller et al., 2001). Currently, there are various RTK GPS products available (Topcon, 2016), with price ranges from a few hundred to tens of thousand dollars.
- *Autonomous Driving System.* The autonomous driving systems for agriculture vehicles are much simpler than those on the road, since its working space is clear and known. Autonomously driving agriculture vehicles can design path,

autosteer, make turns, and follow edges and rows to cover the entire field while applying fertilizers and other matters.

- *Geo Mapping.* Geo mapping tools are required to produce site-specific index maps for input/output data, including soil type, nutrients levels, irrigation, and fertilization (understanding layers for PA data analysis and operation guidance of agriculture vehicles).
- *IoT Communication Infrastructure.* IoT-based communication mechanism is needed to interconnect every object and the commander in the cloud for real-time operation.
- *Variable Rate Technology (VRT).* It denotes the technology that enables the variable rate application of materials, which is required for most PA equipment operations. Simply, VRT is a variable rate (VR) control system on application equipment that can precisely control time and/or location rates of different matters to achieve site-specific application. Through this way, for instance, seed or fertilizer can be applied according to the index map to meet exact variations in crop growth, or soil nutrient needs (CEMA, 2016).

In addition, most water-soluble matters, including fertilizers, soil amendments, and pesticides, can be applied by injecting into irrigation system. This technology is denoted as fertigation (chemigation), which has been widely applied for several years (Threadgill, 1991). Fertigation is recognized as the best management practice to improve effectiveness of various agriculture matters (Wright et al., 2002). Obviously, fertigation system can be seamlessly integrated into the IoT-based PA infrastructure as introduced.

18.5 IoT Application in Crop Disease and Pest Management

Plant diseases and pest have caused severe losses to humans from the birth of agriculture. In history, a crop disease known as "potato blight" had resulted in a significant potato yield reduction in Ireland around 1850, and it led to a famine well known as the "Great Hunger," in which around one million Irish people died (O'Neill, 2010). Even nowadays, every year corn growers in the United States are still experiencing the economic loss of up to one billion dollars due to the crop disease "southern corn leaf blight" (Maloy, 2005). Roughly, direct yield losses caused by animal pests and pathogens are responsible for losses ranging between 16 and 18% of global agricultural productivity (Oerke, 2006; Savary et al., 2012).

The rapid advancement of IoT provides a solid foundation for the development of effective approaches in dealing with crop disease and pest (Iotworm.com, 2016). Compared to the traditional disease/pest control that is calendar or prescription based, the IoT-based modern disease/pest management allows for disease forecasting, modeling, or real-time monitoring, and hence to be more proactive (Maloy, 2005).

The reliability of a crop disease and pest management system depends on three aspects: sensing, evaluation, and treatment (supported by IoT technologies). In an IoT disease/pest management system, the first stage is to collect real-time crop physiopathology-related data. The main approach for disease/pest recognition is image processing, in which raw image can be acquired through either in-field sensors or remote sensing devices on satellite/aircraft (Blakeman, 1990). In general, remote sensing imagery has higher efficiency and less cost, yet higher threshold as well. Meanwhile, in-field sensors can offer more functions in collecting data, information, and even samples of the environment, plant health, and pest situation anytime in every corner. For example, IoT automated traps (Semios, 2016; Spensa, 2016) can capture, count, and characterize insects and upload data to the Cloud for later analysis, which is beyond the capability of remote sensing.

On the second stage, all data collected remotely or locally will be forwarded to the management center sitting in the IoT cloud. The managing center is a combination of sophisticated models and algorithms to process and analyze the raw image and data to provide a batch of related functions such as disease/pest identification, pest behavior prediction, and expert system. Models and algorithms have been developed for dealing with different diseases/pests of different crops. The whole system keeps monitoring the field 24×7, to automatically provide early warning, disease/pest problem report, and even suitable intervention suggestion on an hourly base. This information can be presented to the farmer by the managing center wirelessly in various ways such as text messages or emails to ensure that he/she is notified at the first moment.

The last step of crop disease/pest management is the precise application of corresponding matters required. In disease treatment and pesticide application, similar approaches in PA fertilization can be used, such as autonomous vehicle precise spray or automatic VRT chemigation. Moreover, the advancing of robotic technology provides another solution. Equipped with multispectral sensing devices and precision-spraying effector, an agricultural robot is capable of locating and dealing with crop problems under the manipulation of remote IoT disease/pest management system (Oberti et al., 2016).

18.6 IoT Application in Precision Livestock Farming

A new production pattern denoted as precision livestock farming (PLF) has been practiced in animal husbandry for years, to revolutionize the traditional labor intensive process of animal production (Corkery et al., 2013). Relies on real-time data collection and analyses, PLF offers fully automatic animal health and welfare monitoring and product yields improvement with less environmental impacts. All functions PLF require can find a perfect match in the emerging IoT paradigm that offers various functions, including animal and environmental sensing, in-cloud data analysis and decision-making, and equipment automation. The embedding of IoT into PLF enables optimal animal feeding and nutrient utilization by which higher production efficiency, environmental protection, and high-quality products can be achieved. Related innovative applications such as *smart chicken farm, smart cow farm,* and *IoT aquaculture* have been developed all over the globe.

18.6.1 Smart Chicken Farm

Chickens have been domesticated for both eggs and meat for thousands of years (Clauer, 2016). In the twenty-first century, poultry production industries are involved in highly advanced technology (e.g., sophisticated rearing operations). Actually, modern poultry facilities are biological reactor vessels in which numerous inputs, including feeding materials, water, ventilation air, heat, and lighting, will be converted to the output of either meat or eggs (McNulty and Grace, 2009).

In general, the success of a modern poultry production system relies on the knowledge of nutritional requirements, the capability to match these requirements by tuning inputs in real time, and stable the comfort of environmental conditions such as air, temperature, humidity, and lighting. By taking advantages of versatile capabilities offered by IoT technologies, *smart chicken farms* can manage these key factors effectively and efficiently (Corkery et al., 2013).

In daily operation of a poultry facility, first IoT sensors collect live chickens' data through modern methods such as multispectral imagery using internal environmental information that includes air, temperature, humidity, and light. Then, the in-cloud central control system processes collected data to monitor and evaluate chickens' real-time condition, such as distress level, thermal comfort, live weight, behavior, avian influenza evaluation, and so forth. Next, the intelligent algorithms are applied in optimal decision-making to operate automated equipment accordingly in the farm, including mechanized feeding systems, internal environment controllers, mechanical handling and disposal of poultry waste, automation of lighting systems, crate washing and sanitization, and mechanical egg collection and quality control systems (McNulty and Grace, 2009). Comparing with the traditional operation, IoT-based smart farms have much lower cost and higher reliability and extendibility in sensing, better flexibility, updatability and intelligence in data analysis and decision-making, higher capability in machine operation automation, and unprecedented convenience to the user in live/history data access and system remote control. Overall, these advantages will lead to improvements in chicken health and welfare, product quality assurance, and reduction of labor involvement, operation cost, and running risk. Currently, commercialized IoT solutions have been well developed for farmers to build *smart chicken farms* (Iotreecloud, 2016).

18.6.2 Smart Cow Farm

Another application of IoT technology in livestock farming is in cow farm. With IoT sensors attached to cows, farmers can easily locate cows in the field through smartphone or tablet app, and also check more important animal welfare issues such as rumination levels and lameness (Miles, 2016). More and more farmers are adopting IoT technologies to convert their animals into the so-called *Internet of Cows, Connected Cows,* or *Smart Cows.*

In general, the IoT paradigm can help livestock farming or ranching to improve the productivity of water, energy, food, and other resources use, while maintaining the well-being of animals. It also helps farmers to make lists, prepare reports, sort cows by category, and track each animal's overall lifetime.

Difficult problems that have been bothering farmers for centuries now found their solutions in IoT scope; for instance, the identification of cow estrus which is critical to milk yields. Normally, a farmer needs to spend 20–30 min each time, four to five times a day in the stables to check if a cow is in heat, which is a sign of estrus. However, over 60% of estrus cases happen at night when the farmer is asleep (Anderson, 2016). Another example is the cattle lameness that has a large impact on a cow's performance in terms of yield, fertility, and longevity (Van Metre et al., 2005).

With the advancing of IoT technologies, these kinds of issues now find their solutions (Ilapakurti and Vuppalapati, 2015; SCR, 2016). Miniaturized IoT sensor tags can be attached to ear, neck, or leg of cows to monitor activity and well-being of each of them, 24 h every day (Miles, 2016). Then sophisticated analytical/empirical models will be applied to analyze the data to identify if a cow has gone into an estrus state. As the direct results, the detection rate of IoT solution can reach up to 95% (Fujitsu, 2016), whereas the detection rate of traditional method is around 55%. In addition, aiming to dynamic lameness detection, fully integrated IoT health monitoring platforms are also developed with available products on the market (Adityan, 2015).

Mostly, in *smart cow farm,* diversified data will be collected from both on-animal sensors and other in-premise sensors. The in-cloud intelligent management system will process data to monitor animal status, control all automated equipment and devices, and provide professional suggestions for user decision-making. With the continuing miniaturization, cost decreasing, and measuring capability increasing of diversified sensors, more and better IoT solutions will be developed to benefit cow farmers further.

18.6.3 IoT Aquaculture

The IoT technology is also reshaping the aquaculture by offering new functions and integrating them into an intelligent autonomous system. Large-scale distributed wireless sensor network comprised of diversified sensors are equipped to collect all kinds of data, including fish behavior, water quality, and equipment status. In a higher level in the IoT system, optimization models and intelligent algorithms are deployed to deal with different jobs, including the following (Zhang et al., 2013; Chen et al., 2015; ShaoHua, 2015):

- Aquaculture water quality monitoring and maintenance
- Fish status monitoring and precise feeding
- Fish behavior analysis, early warning, and disease diagnosis, control, and prevention
- Facility management and fault diagnosis
- Automatic equipment operation
- Information management, storage, visualization, and user access
- Logistics and fishery products quality traceability

The IoT can also help to the better understanding of how industrial fishing impacts important species in the blue water. Advanced aquatic sensor tags have been developed by adopting the latest IoT technologies in RFID Tag and pop-up satellite archival tag (PSAT). These sensor tags can be used to collect data regarding depth, temperature, and moving speed (microwavetelemetry.com, 2016). A sensor tag will be attached to an aquatic animal and held for days, weeks, or months, then released automatically by itself at the end of the measurement period. Then it will float (pop up) to the water's surface and transmit stored data to researchers through the satellite link. Collected data can reveal many facts, including how the fishing operation will change the life of marine animals. The PSAT has been deployed in various marine research projects all around the world (Musyl et al., 2011). Primarily, IoT technologies are penetrating many aspects of livestock farming. Successful applications are reported that span from *Internet of Pigs* (farmingfuture.org, 2016), *Internet of Goats* (Huawei, 2016) to *Smart Oyster Farm* (Microsoft, 2016).

18.7 Conclusion

Agriculture is the basis of human society. Moreover, as Masanobu Fukuoka said, "The ultimate goal of agriculture is not just growing crops, but the cultivation and perfection of human beings" (Fukuoka, 2009). Fortunately, advancing of technologies, especially the combination of the Internet of things and precise agriculture, is paving the road for reaching the goal. Human now is on the cusp of the second green revolution, which is largely built on the IoT and related technologies. IoT-based PA promises to make the farm of the future more productive and efficient with less labor work needed. It is grounded on the use of data to form more efficient and effective farming practices and drive associated environmental and social benefits. The chapter has presented many related technologies developed to serve this purpose, for example, CWSI, NDVI, RTK approaches and advanced sensors. These technologies and associated applications enable farmers to treat crops and animals more precisely. Future implications of data collected through these technologies also allow farmers to make more strategic and effective decisions to increase productivity with fewer environmental impacts. In conclusion, IoT technologies will take center stage on the farm of the future (Sarni et al., 2016). It is predicted that 75 million IoT devices will be used for agricultural purposes by 2020, and the smart agriculture market is expected to reach \$18.45 billion in 2022 (Mittal, 2016).

References

Adityan, V. (2015) Smartbell: The Internet of Cows™. Available at http://startupcompete.co/startup-idea/internet-life-sciences-green/smartbell-the-internet-of-cows/51003 (accessed November 17, 2016).

Anderson, M. (2016) Saving the World With The Internet of Cows. Available at https://blog.equinix.com/blog/2016/03/01/saving-the-world-with-the-internet-of-cows/ (accessed November 17, 2016).

ApogeeInstruments. (2016) http://www.apogeeinstruments.com/infraredradiometer/ (accessed November 17, 2016).

Baronti, P., Pillai, P., Chook, V. W., Chessa, S., Gotta, A., and Hu, Y. F. (2007) Wireless sensor networks: a survey on the state of the art and the 802.15. 4 and ZigBee standards. *Computer communications*, **30**(7), 1655–1695.

Blakeman, R. (1990) The identification of crop disease and stress by aerial photography. Applications of Remote Sensing in Agriculture, 229–254.

Burrus, D. (2016) The Internet of Things Is Far Bigger Than Anyone Realizes. Available at http://www.wired.com/insights/2014/11/the-internet-of-things-bigger/ (accessed November 17, 2016).

CEMA. (2016) http://cema-agri.org/ (accessed November 17, 2016).

Chen, J. H., Sung, W. T., and Lin, G. Y. (2015) Automated Monitoring System for the Fish Farm Aquaculture Environment. 2015 IEEE International Conference on Systems, Man, and Cybernetics (SMC).

Clauer, P. (2016) History of the Chicken. Available at http://extension.psu.edu/animals/poultry/topics/general-educational-material/the-chicken/history-of-the-chicken (accessed November 17, 2016).

Corkery, G., Ward, S., Kenny, C., and Hemmingway, P. (2013) Incorporating smart sensing technologies into the poultry industry. *Journal of World's Poultry Research*, **3**(4), 106–128.

U.S. DoD (2001) Global positioning system standard positioning service performance standard. Assistant secretary of defense for command, control, communications, and intelligence.

Environment.co.za. (2015) How Do Fertilizers Affect the Environment. Available at https://www.environment.co.za/environmental-issues/how-do-fertilizers-affect-the-environment.html (accessed November 17, 2016).

esri.com. (2016) Farming the Future. Available at http://www.esri.com/library/ebooks/farming-the-future.pdf (accessed November 17, 2016).

farmingfuture.org. (2016) The Internet of Pigs. Available at http://www.farmingfutures.org.uk/blog/internet-pigs?page=3 (accessed November 17, 2016).

Fujitsu. (2016) Akisai Food and Agriculture Cloud GYUHO SaaS (Cattle breeding support service). Available at http://www.fujitsu.com/jp/group/kyushu/en/solutions/industry/agriculture/gyuho/ (accessed November 17, 2016).

Fukuoka, M. (2009). *The One-Straw Revolution: An Introduction to Natural Farming*, New York Review of Books.

GreenFacts (2016) http://www.greenfacts.org/ (accessed November 17, 2016).

Gubbi, J., Buyya, R., Marusic, S., and Palaniswami, M. (2013) Internet of Things (IoT): a vision, architectural elements, and future directions. *Future Generation Computer Systems*, **29**(7), 1645–1660.

Guo, T. and Zhong, W. (2015) Design and implementation of the span greenhouse agriculture Internet of Things system. 2015 International Conference on Fluid Power and Mechatronics (FPM).

Harun, A. N., Kassim, M. R. M., Mat, I., and Ramli, S. S. (2015) Precision irrigation using Wireless Sensor Network. 2015 International Conference on Smart Sensors and Application (ICSSA).

Hassan, Q. F., Riad, A. M., and Hassan, A. E. (2012) Understanding cloud computing. *Software Reuse in the Emerging Cloud Computing Era*, Information Science Reference, pp. 204–227.

Hatfield, P. and Pinter, P. (1993) Remote sensing for crop protection. *Crop Protection*, **12**(6), 403–413.

Hillman, G. (1996) Late Pleistocene changes in wild plant-foods available to hunter-gatherers of the northern Fertile Crescent: possible preludes to cereal cultivation. *The Origins and Spread of Agriculture and Pastoralism in Eurasia*, UCL Press, 159–203.

Huawei. (2016) Huawei Helps Modernize Stock Breeding Using the Internet of Things. Available at http://eblog.huawei.com/chinas-first-internet-goats (accessed November 17, 2016).

Idso, S. B. (1982) Non-water-stressed baselines: a key to measuring and interpreting plant water stress. *Agricultural Meteorology*, **27**(1), 59–70.

Idso, S., Jackson, R., Pinter, P., Reginato R., and Hatfield, J. (1981) Normalizing the stress-degree-day parameter for environmental variability. *Agricultural Meteorology*, **24**, 45–55.

Ilapakurti, A. and Vuppalapati, C. (2015) Building an IoT framework for connected dairy. 2015 IEEE First International Conference on Big Data Computing Service and Applications (BigDataService), IEEE.

Iotreecloud. (2016) Cloud Based Monitoring and Automation. Available at https://iotreecloud.com/ (accessed November 17, 2016).

Iotworm.com. (2016) Agriculture Internet of Things (IoT) Technology/ Applications. Available at http://iotworm.com/agriculture-internet-of-things-iot-technology-applications/ (accessed November 17, 2016).

Jackson, R. D. (1982) Canopy temperature and crop water stress. *Advances in irrigation*, **1**, 43–85.

Jones, H. G., Serraj, R., Loveys, B. R., Xiong, L., Wheaton, A., and Price. A. H. (2009) Thermal infrared imaging of crop canopies for the remote diagnosis and quantification of plant responses to water stress in the field. *Functional Plant Biology*, **36**(11), 978–989.

Keller, R. J., Nichols, M. E., and Lange, A. F. (2001) Methods and Apparatus for Precision Agriculture Operations Utilizing Real Time Kinematic Global Positioning System Systems. U.S. Patent 6,199,000 B1.

Khelifa, B., Amel, D., Amel, B., Mohamed, C., and Tarek, B. (2015) Smart irrigation using Internet of Things. 2015 Fourth International Conference on Future Generation Communication Technology (FGCT).

Kiiski, H., Dittmar, H., Drach, M., Vosskamp, R., Trenkel, M. E., Gutser, R., and Steffens, G. (2009) Fertilizers, 2. types. Ullmann's Encyclopedia of Industrial Chemistry.

learner.org. (2016) https://www.learner.org/ (accessed November 17, 2016).

Maes, W. and Steppe, K. (2012) Estimating evapotranspiration and drought stress with ground-based thermal remote sensing in agriculture: a review. *Journal of Experimental Botany*, **63**(13), 4671–4712.

Maloy, O. C. (2005) Plant disease management. *The Plant Health Instructor*, **10**. doi: 10.1094/PHI-I-2005-0202-01.

McNulty, P. B. and Grace, P. M. (2009) Agricultural Mechanization and Automation, Report from Agricultural and Food Engineering Department, National University of Dublin, Ireland.

Microsoft. (2016) Internet of Oysters: The Yield Delivers Sunnier Results for Australian Oyster Farmers. Available at http://news.microsoft.com/en-au/features/internet-of-oysters-the-yield-delivers-sunnier-results-for-australian-oyster-farmers/#sm.0001wirg1i1dvlfb0qt82zsbf9m01 (accessed November 17, 2016).

microwavetelemetry.com. (2016) http://www.microwavetelemetry.com/fish/ (accessed November 17, 2016).

Miles, S. (2016) Internet of Cows is now a thing as UK start-up creates cow tracking app. Available at http://www.pocket-lint.com/news/136825-internet-of-cows-is-now-a-thing-as-uk-start-up-creates-cow-tracking-app (accessed November 17, 2016).

Mittal, S. (2016) IoT ecosystem: How the IoT market will explode by 2020. Available at http://blog.beaconstac.com/2016/03/iot-ecosystem-iot-business-opportunities-and-forecasts-for-the-iot-market/ (accessed November 17, 2016).

Musyl, M., Domeier, M., Nasby-Lucas, N., Brill, R., McNaughton, L., Swimmer, J., Lutcavage, M., Wilson, S. G., Galuardi, B., and Liddle, J. (2011) Performance of pop-up satellite archival tags. *Marine Ecology Progress Series*, **433**, 1–28.

NASA. (2016) http://modis.gsfc.nasa.gov/ (accessed November 17, 2016).

O'Neill, J. R. (2010) *Irish Potato Famine*, ABDO.

Oberti, R., Marchi, M., Tirelli, P., Calcante, A., Iriti, M., Tona, E., Hočevar, M., Baur, J., Pfaff, J., and Schütz, C. (2016) Selective spraying of grapevines for disease control using a modular agricultural robot. *Biosystems Engineering*, **146**, 203–215.

Oerke, E.-C. (2006) Crop losses to pests. *The Journal of Agricultural Science*, **144**(01), 31–43.

Park, S., Nolan, A., Ryu, D., Fuentes, S., Hernandez, E., Chung, H., and O'Connell, M. (2015) Estimation of crop water stress in a nectarine orchard using high-resolution imagery from unmanned aerial vehicle (UAV). International Congress on Modelling and Simulation (MODSIM) (Tony Weber and Malcolm McPhee 29 November–04 December 2015), Modelling and Simulation Society of Australia and New Zealand.

Pipia, L., Pérez, F., Tardà, A., Martínez, L., and Arbiol, R. (2012) Simultaneous usage of optic and thermal hyperspectral sensors for crop water stress characterization. 2012 IEEE International Geoscience and Remote Sensing Symposium.

Rahman, S., Vance, G. F., and Munn, L. C. (1994) Detecting salinity and soil nutrient deficiencies using spot satellite data. *Soil science*, **158**(1), 31–39.

Rouse Jr, J. W., Haas, R., Schell, J., and Deering, D. (1974) Monitoring vegetation systems in the Great Plains with ERTS. *NASA special publication*, **351**, 309.

Sarni, W., Mariani, J., and Kaji, J. (2016) From dirt to data: the second green revolution and the Internet of Things. Deloitte Review.

Savary, S., Ficke, A., Aubertot, J.-N., and Hollier, C. (2012) Crop losses due to diseases and their implications for global food production losses and food security. *Food Security*, **4**(4), 519–537.

Schumann, A. W. (2006) Nutrient management zones for citrus based on variation in soil properties and tree performance. *Precision Agriculture*, **7**(1), 45–63.

SCR. (2016) Cow intelligence. Available at http://www.scrdairy.com/cow-intelligence/cow-intelligence-overview.html (accessed November 17, 2016).

Semios. (2016) Integrated pest management. http://semios.com/ipm/ (accessed November 17, 2016).

ShaoHua, H. (2015) Dynamic monitoring based on wireless sensor networks of IoT. 2015 International Conference on Logistics, Informatics and Service Sciences (LISS).

Sharma, B., Molden, D., and Cook, S. (2015) Water use efficiency in agriculture: measurement, current situation and trends. *Managing Water and Fertilizer for Sustainable Agricultural Intensification*, International Fertilizer Industry Association, Paris, p. 39.

Spensa. (2016) Z-Trap Available at http://spensatech.com/ (accessed November 17, 2016).

Sunding, D., Rogers, M., and Bazelon, C. (2016) The Farmer and the Data: How Wireless Technology is Transforming Water Use in Agriculture.

Taghvaeian, S., Chávez, J. L., and Hansen, N. C. (2012) Infrared thermometry to estimate crop water stress index and water use of irrigated maize in Northeastern Colorado. *Remote Sensing*, **4**(11), 3619–3637.

Threadgill, E. (1991) Advances in irrigation, fertigation and chemigation. Expert Consultation on Fertigation/Chemigation, Cairo (Egypt), September 8–11, 1991.

Tianhong, L., Yanxin, S., and An, X. (2003) Integration of large scale fertilizing models with GIS using minimum unit. *Environmental Modelling & Software*, **18**(3), 221–229.

Tingting, D., Miao, J., Fengkui, Q., and Zengxiang, Z. (2008) Researching on extracting irrigated land in northern China based on MODIS data. EORSA 2008. International Workshop on Earth Observation and Remote Sensing Applications, 2008.

Topcon. (2016) http://www.topcon.co.jp/ (accessed November 17, 2016).

un.org. (2015) Probabilistic population projections based on the World Population Prospects: the 2015 revision. http://esa.un.org/unpd/ppp/ (accessed November 17, 2016).

UN-Water (2016) http://www.unwater.org/ (accessed November 17, 2016).

USDA. (2016) http://www.usda.gov (accessed November 17, 2016).

USGS. (2016) water.usgs.gov (accessed November 17, 2016).

Van Alphen, B. J. and Stoorvogel, J. J. (2000) A methodology for precision nitrogen fertilization in high-input farming systems. *Precision Agriculture*, **2**(4), 319–332.

Van Metre, D. C., Wenz, J., and Garry, F. (2005) Lameness in cattle: rules of thumb. *Proceedings of the American Association of Bovine Practitioners*, **38**, 40–43.

Walsh, F. J. (2009) Rupert Gerritsen: Australia and the origins of agriculture. *GeoJournal*, **74**(5), 499–501.

Weier, J. and Herring, D. (2011) Measuring vegetation (ndvi & evi). NASA online publication.

WorldBank. (2016) http://data.worldbank.org/indicator/AG.LND.AGRI.ZS (accessed November 17, 2016).

Wright, J., Bergsrud, F., Rehm, G., Malzer, G., and Montgomery, B. (2002) Nitrogen application with irrigation water–chemigation. College of Agriculture, Food, and Environmental Sciences, University of Minnesota.

Xiaoning, S., Shifeng, H., Liu, Q., and Li, X. (2007) Vegetation water inversion using MODIS satellite data. 2007 IEEE International Geoscience and Remote Sensing Symposium.

Zhang, F. (2011) Research on water-saving irrigation automatic control system based on internet of things. 2011 International Conference on Electric Information and Control Engineering (ICEICE).

Zhang, Y., Hua, J., and Wang Y. B. (2013) Application effect of aquaculture IOT system. *Applied Mechanics and Materials*, **303–306**, 1395–1401.

19

The Internet of Flying Things

Daniel Fernando Pigatto,[1] *Mariana Rodrigues,*[2] *João Vitor de Carvalho Fontes,*[3] *Alex Sandro Roschildt Pinto,*[4] *James Smith,*[5] *and Kalinka Regina Lucas Jaquie Castelo Branco*[2]

[1]*Graduate Program in Electrical and Computer Engineering (CPGEI), Federal University of Technology Paraná (UTFPR), Curitiba, Paraná, Brazil*
[2]*Institute of Mathematics and Computer Sciences (ICMC), University of São Paulo (USP), São Carlos, São Paulo, Brazil*
[3]*São Carlos School of Engineering (EESC), University of São Paulo (USP), São Carlos, São Paulo, Brazil*
[4]*Federal University of Santa Catarina (UFSC), Blumenau, Santa Catarina, Brazil*
[5]*Computer Science and Creative Technologies (FET), University of the West of England (UWE), Bristol, England, United Kingdom*

19.1 Introduction

Unmanned aircraft systems have received a lot of attention lately, especially due to their flexibility and reduced acquisition costs. However, in many regions, legislation issues have emerged that curtail their operation in critical environments. In response to well-reported instances, it seems likely that in many countries "no-fly" zones will be established around critical areas, such as airports (where accidental "drone-strikes" could pose a threat to jet planes similar to "bird-strikes"), prisons (where cases have been reported of drones being used to transport contraband goods in to prisoners), and military/confidential areas (where government is combating drones with trained eagles). Security threats from terrorist groups also pose a risk to key infrastructure. In the future, it seems likely that international consensus may arise around certain areas (e.g., commercial airports) but the picture is likely to remain fluid for some time. Meanwhile, significant research efforts are exploring the current capabilities of UAVs, and their potential for autonomous action beyond the line of sight of a dedicated operator, which is likely to fuel further debate and legislation.

The relatively recent concept of Internet of Things (IoT), which consists of a new form of connecting and sharing resources among devices, has been

Internet of Things A to Z: Technologies and Applications, First Edition. Edited by Qusay F. Hassan.
© 2018 by The Institute of Electrical and Electronics Engineers, Inc. Published 2018 by John Wiley & Sons, Inc.

considered as a candidate for potential integration with unmanned aircraft. Such collaboration may provide a new degree of freedom for old applications and a completely new spectrum of applications.

This chapter reviews the main characteristics of the Internet of Flying Things and how the term is related to unmanned aircraft systems and the Internet of Things. The chapter describes how this new concept solves known issues, but also introduces different challenges to the design of systems.

19.2 Flying Things

The popularity and flexibility of embedded systems have introduced new applications in vehicular segments such as cars, drones, and maritime underwater or surface vehicles in recent decades. This section outlines the main concepts of the aerial segment under a new name, flying thing, encompassing not only a limited range of drones, but any type and/or classification of unmanned aircraft system.

19.2.1 Unmanned Aircraft Systems

Unmanned Aerial vehicles (UAVs), popularly known as *drones*, are considered enablers of a completely new way of performing tasks that were previously either unreachable or high cost, so fulfilling gaps in many modern applications (Marshall et al., 2015). As shown in Figure 19.1, these aircraft can be of many different sizes and shapes, and missions can be accomplished by either a single or multiple UAVs. This richness of form allows them to be used in diverse applications such as search and rescue, surveillance missions, and goods delivery.

UAVs are usually considered as part of a bigger unmanned aircraft system (UAS) that includes all the needed elements to accomplish a mission (Fahlstrom and Gleason, 2012). The components of the UAS may vary according to UAV type, size, and mission, but typically include the ground control station (GCS), the communications subsystem, and the safety and recovery mechanisms.

Creating a UAS with multiple cooperating UAVs brings additional, highly demanding connectivity requirements and in order to meet them, a new network approach has emerged, the Flying Ad Hoc Network.

19.2.2 Flying Ad Hoc Networks

Communication is a crucial aspect of, and one of the biggest challenges in, the design of multiple vehicle systems (Bouachir et al., 2014; Chung et al., 2011). In a UAS, there are three main types of communications: (a) internal machine communications (IMC), which encompasses any communication between UAVs' internal modules or devices, such as the automatic pilot or cameras;

Figure 19.1 Examples of UAVs. (a) A drone. (b) A remotely piloted aircraft. (c) A balloon.

(b) machine-to-machine communications (M2M), which encompasses communications among UAVs; and (c) machine-to-infrastructure communications (M2I), which encompasses the communication between UAVs and the network infrastructure (which can be, for instance, a ground control station or a satellite (Frew and Brown, 2009), or even a combination of both).

In the simplest scenario, all vehicles are directly connected to a common infrastructure, and this can act as an intermediary for all communications between them. However, this strategy has several problems. First, each vehicle must be equipped with expensive and complex hardware in order to perform the long-distance communication with the control station or satellite. Second, many factors may compromise communication reliability such as changing environmental conditions, the high mobility of vehicles, different terrain topologies, or obstacles. Finally, the typical use of a ground control station (GCS) to provide the communication infrastructure limits the mission target locations to the GCS coverage area, since beyond that vehicles disconnect from the network, and become unreachable.

The implementation of an ad hoc network connecting all vehicles is one of the most feasible alternatives to infrastructure-based communication. An ad hoc network is composed of nodes that also function as routers, forming a temporary network with no fixed topology or centralized administration (Sarkar et al., 2008). This approach increases the mission target area, since communications between vehicles and the GCS can be routed through other vehicles in a series of hops. Also, even if there is no connection to a GCS, the nodes can form an ad hoc network to share information or work in cooperation.

Ad hoc networks are classified according to their implementation, utilization, communication, and mission objectives. If the nodes that compose an ad hoc network are mobile, the network is classified as MANET (Mobile Ad hoc NETwork). For vehicle-specific applications, MANETs are subdivided into UANET (Underwater Ad hoc NETwork) for aquatic vehicles, VANET (Vehicular Ad hoc NETwork) for terrestrial vehicles, or FANET (Flying Ad hoc NETwork) for aerial vehicles (Bekmezci et al., 2013; Sahingoz, 2013), as illustrated in Figure 19.2.

Each type of vehicular network faces different, unique challenges: For instance, a UANET must deal with an underwater transmission medium and VANETs often encounter unexpected road obstacles. However, it has been recognized that FANETs have to address more challenging issues than other ad hoc networks (Bekmezci et al., 2013; Sahingoz, 2013), because of the following specific characteristics:

- *Higher Node Mobility.* FANET nodes typically have higher mobility than those in other types of MANET. As a result, a FANET's network topology can

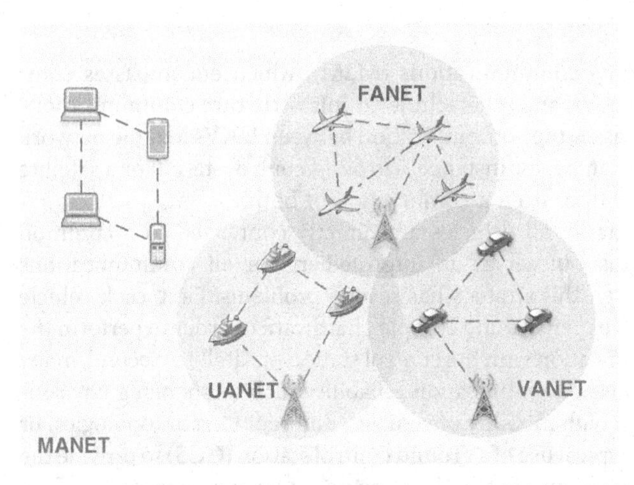

Figure 19.2 Relationships between different types of Mobile Ad Hoc Networks (MANET): Underwater Ad Hoc Networks (UANET), Vehicular Ad Hoc Networks (VANET), and Flying Ad Hoc Networks (FANET). (Figure adapted from Pigatto et al. (2014).)

change more frequently, which increases the overhead caused by connecting and routing operations.

- *Multiple Connections.* In many applications, the nodes in FANETs collect environmental data and then retransmit it to the control station, similarly to wireless sensor networks (WSN) (Rieke et al., 2011). Therefore, FANETs have to manage multiple communications between UAVs and monitoring stations, as well as providing support to peer-to-peer connections among UAVs.
- *Very Low Node Density.* Typical distances among nodes in FANETs are usually longer than in MANETs and VANETs (Clapper et al., 2007); thus, the communication range in FANETs must also be greater than in other networks. This imposes more demanding requirements for radio links and other hardware elements.
- *Heterogeneity.* UAV systems may include heterogeneous sensors, and each of them may require different strategies for data distribution.
- *Obstacles.* Due to the higher node mobility, obstacles may randomly block links among UAVs, which must be addressed in order to provide different temporary communication paths, avoiding the disconnection of nodes.

19.2.3 Flying Things: Unmanned Aerial Vehicles and More

Common to the many different visions for the Internet of Things is the ubiquitous presence of everyday objects equipped with identifying, sensing, networking, and processing capabilities that communicate with each other to achieve a common goal (Atzori et al., 2010; Whitmore et al., 2015). In the context of emerging network models such as IoT, a new name for UAVs starts to take place. As a *Flying Thing,* UAVs (and any other elements able to fly, like autonomous or nonautonomous air vehicles) can be integrated into a network of physical interactive objects that are able to communicate with other internet-enabled devices and systems. This integration has been called the *Internet of Flying Things* and has the potential of addressing some of the issues in unmanned systems, as well as introducing new, powerful applications.

19.3 The Internet of Flying Things

Following the tendency of our increasingly connected world, UAVs are likely to be integrated with other elements and systems to perform missions with higher complexity. The new Internet of Flying Things (IoFT) gives a new degree of freedom to unmanned aircraft systems, broadening the limits of their missions and enabling new applications, by increasing their connectivity, improving collaboration and cooperation between systems, and also enabling up-to-date data provision.

Although the new Internet of Flying Things is applicable in several different segments, one particular highlight is the affordance it offers in environments that are currently badly served by existing (wired) infrastructures, such as rural areas or poor communities. These examples are big motivators for the creation of solutions merging the flexible adaptive IoT with fast, cheap UAVs, which together are likely to provide high quality services for end users in remote areas. However, an important concept takes place while considering hard-to-reach areas—fog computing. Close relations can be established among fog computing, cloud computing, and Internet-based models such as the Internet of Flying Things. These concepts will be presented in the next subsection.

19.3.1 Fog and Cloud Computing

There are two main trends shaping our networking today: cloud computing and the proliferation of mobile computing. Around 90% of global Internet users now rely on cloud-based service, and that is mainly due to the fact that smartphones worldwide shipment overtook that of PCs in 2011 (Luan et al., 2015). An emerging wave of Internet deployments, such as the notable IoT, requires not only mobility support and geodistribution, but also location awareness and low latency.

Cloud computing is a key concept for the provision of IoFT. A cloud is an on-demand computing model composed of autonomous, networked IT (hardware and/or software) resources (Hassan et al., 2012). As most of the communication managed by IoFT is transmitted via IoT infrastructure, quality of service (QoS) becomes an important aspect, which meets the same requirements of cloud computing. Service providers offer cloud services with predefined QoS terms based on the Internet as a set of managed scalable, easy-to-use, and inexpensive services to gather clients on a subscription basis (Hassan et al., 2012). Therefore, it is essential for the IoFT infrastructure to meet QoS requirements on the provision of services from and to the flying thing, since the criticality of its operations might be higher than ordinary fixed IoT elements. Moreover, some IoFT applications need real-time response, which is done by delegating computational tasks to the cloud, due to limited capacity of mobile devices (Luan et al., 2015). Such fact amplifies the necessity of assertive services provision.

On the other hand, in case of remote areas with limited or no internet connectivity, it would not be always possible to arbitrarily access cloud services with acceptable QoS response times and real-time requirements. However, some applications could still benefit from IoFT by temporarily working with local copies of data and services, imitating the structure of a cloud computing, and then allowing the provision of services that do not rely on frequent updates. This is the case where fog computing takes place as a key element for IoFT networks. Fog computing consists of a highly virtualized platform to compute, store, and provide networking services between end devices and traditional cloud computing data centers, which are typically, but not exclusively located at the edge of the network (Bonomi et al., 2012). On the IoFT perspective, for instance, a smart farm with limited access to the Internet could benefit from a fog infrastructure processing most of the data and services needed for the usual tasks, uploading and downloading only relevant and/or essential information to the Cloud.

As discussed in Chapter 4, cloud/fog computing characteristics have important intersections with IoT scenarios. The main requirements of the IoFT are met by fog computing due to following characteristics: (a) low latency and location awareness allowing the IoFT network to operate at least within a limited range and time, providing services and performing tasks and missions; (b) mobility, a main necessity of IoFT applications; (c) very large number of nodes, which meets the requirements of both IoT and IoFT networks; (d) predominant role of wireless access; (f) strong presence of streaming and real-time applications; and (g) heterogeneity (Bonomi et al., 2012).

A flying thing might play two different roles in an IoFT scenario supported by fog computing, as seen in Figure 19.3. The first is as processing units in a fog layer, serving as providers for the Internet of Flying Things. To give two examples, in remote or disaster areas where connectivity is hard to be achieved, a flexible mobile structure that could provide or extend fog services would be an important enabler. Equally, UAV's could have a role as fog providers in highly connected environments, where some applications might work well being offline for short or even long periods of time, reducing data usage, and creating batches of information to be uploaded to the cloud all-in-once. In a different role, UAVs could function as edge nodes (end users) benefiting from a fog infrastructure.

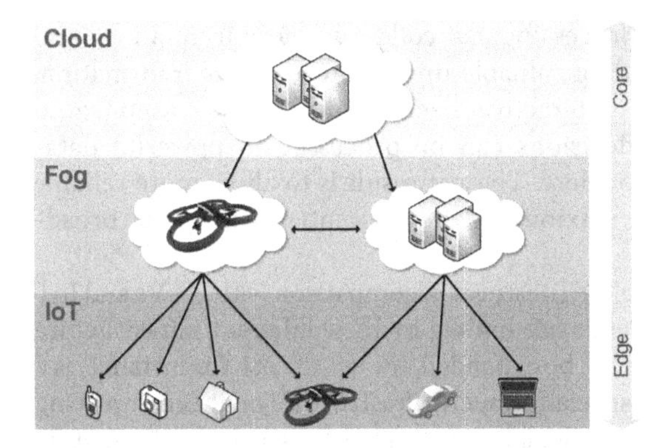

Figure 19.3 The UAV can be placed either in fog or IoT layers when it comes to remote areas applications.

19.3.2 Characteristics of the Internet of Flying Things

The IoFT is indeed flexible. This characteristic is important for the provision of almost every feature in such model. It helps increase overall cooperation and collaboration, is ready for real-time operations, is usually up-to-date due to the highly connected environment and easy access to the Internet, and is assisted by a powerful remote cloud and/or local fog structure. In terms of cost, since the IoFT merges the benefits of two well-known paradigms, namely, IoT and UAVs, which may vary from cheap to expensive commercial off-the-shelf products, there will be an affordable and adaptive solution for most needs.

Cooperation and *collaboration* are desired features for most of modern computing systems. Many modern applications distribute tasks and share information in real time, providing better results quickly. In particular, an IoT-ready environment is usually designed to be equipped with more than one way of acquiring data, interacting, and automating specific tasks (Ungurean et al., 2014; Andreev et al., 2015). Although IoT is a scalable model, its expansion can mean high costs for relatively small returns. If an environment does not expand easily, it might have its flexibility compromised, resulting in limited cooperation and collaboration. IoFT addresses this tension in different ways—such as by setting up UAVs in strategic areas serving as gateways, fog, or cloud data link providers, and also by being capable of replacing sensors and actuators in more active and inexpensive ways (for instance, if a traffic light fails, a flying thing might be used to temporarily replace its task). Moreover, real-time operations are also a priority of the model, since it can be reconfigured to meet requirements and provide the best connection to servers and services available locally or through the Internet.

The IoFT integration with IoT infrastructures achieved by strategically positioning UAVs helps the model to meet some key features. Linking to internet-based information processing can collate services from all around the world in real time, providing valuable up-to-date accurate information. This in turn can facilitate interactive decision-making in response to dynamic situations. Those decisions can be processed in powerful data-centers available as cloud providers. The net result is to allow more reliable and more adaptive missions, maximizing their potential benefits, or broadening their applicability.

Table 19.1 summarizes these IoFT features in comparison with UAVs and IoT paradigm. IoT and UAS segments are limited by their inherent infrastructure characteristics. Although they can be expanded, the setup cost, for instance, is a con that must be considered, especially if such infrastructure might end up being underused. In such case, the use of flexible flying things for sensing and actuating is an advantage.

Table 19.1 A comparison of available features of unmanned aircraft systems, Internet of Things, and the new Internet of Flying Things.

Features	Internet of Things (IoT)	Unmanned aircraft systems (UAS)	Internet of Flying Things (IoFT)
Cooperation	Limited by IoT infrastructure	Limited by FANET infrastructure	Includes all the IoT and FANET infrastructure capabilities
Collaboration	Limited by IoT infrastructure	Limited by FANET infrastructure	Includes all the IoT and FANET infrastructure capabilities
Real-time operations	Limited to the network coverage	Limited to the actuation areas	Reduced limitations due to increased connectivity
Connectivity	Internet connected	Locally connected by a FANET	Highly connected—not just to the Internet, but also locally connected
Up-to-date data/ services	Available	Weakly available	Available
Internet-based information processing	Available	Weakly available	Available
Interactive decision-making	Available	Available	Available with higher flexibility
Mission-assistive multisource information providers	Available	Weakly available	Available with higher variety of sources

19.3.3 General Modern Applications of the Internet of Flying Things

UAVs and IoT are popular subjects that have gained attention due to their flexibility and low-cost achievements. The benefits of the Internet of Flying Things go beyond the traditional applications being seen on media press lately. The following paragraphs will present some examples of how applications could be taken to a new level by exploiting the flexibility of IoFT.

19.3.3.1 Applications in Emergency Situations

- *Search and Rescue.* In an emergency scenario with victims, flying things can detect and report the position of the victims in real time. This information can be coordinated with that from weather/traffic sensors providing an efficient rescue.

- *First Aid and Supplies.* Once the positions of the victims have been identified, flying things which carry first aid and supplies can be moved toward to the specific positions in order to help the victims, and this can be coordinated using information about the availability of useful resources/supplies in nearby locations. Moreover, if the victims were moving for any reason, the location information can be updated in real time. This update can be performed by a flying thing that detects the position and sends this information to the first aid supplier flying thing.

19.3.3.2 Applications in Smart Cities

- *Surveillance.* Flying things can be used to add additional capacity to existing systems for monitoring crowds or responding to emergency signals. For instance, in a smart city, people can be connected to the city's services via wearable technology; in a dangerous situation, a person may notify the infrastructure in several different ways, and trigger an appropriate action such as assigning a surveillance flying thing to monitor the area (see A in Figure 19.4).
- *Traffic Monitoring.* Although smart cities have devices and sensors to monitor the traffic, these devices are costly and might be unreliable, so coverage is not always complete. Flying things can be used in two ways when planning a smart city infrastructure: to provide flexible additional coverage that might be planned (for example, to monitor the occasional, but predictable high-traffic density around a sports arena) or unplanned (for example, following a traffic incident in an area that is not, or poorly covered by fixed sensors). Thus, flying things can be used to monitor these areas and help the car traffic management (see B in Figure 19.4).

Figure 19.4 Smart city applications taken to the next level with IoFT. A: Surveillance, B: traffic monitoring, and C: commercial package delivery.

- *Commercial Package Delivery.* Flying things provide a flexible resource for delivery services. They could be used both to augment vehicles for meeting existing demand types and also to generate new revenue streams for high-speed delivery in hard-to-access or congested areas (see C in Figure 19.4).

19.3.3.3 Applications in smart farms

- *Surveillance.* Considering that farms are private areas, they may be vulnerable to invasions. The use of flying things to monitor the farms' border might help the surveillance by reporting potential problems in real time using the IoT infrastructure of the smart farm. Moreover, the IoFT serves as a backend infrastructure to IoT-based applications (see A in Figure 19.5).
- *Service Integration.* Different tasks are required in smart farms, and this is expensive if each is implemented separately. A flexible flying thing can monitor crops, animals, spread products, load supplies, and other tasks. In particular, it can also be connected to other flying things through the Internet even with no line of sight. This feature can improve services in a smart farm by overcoming limitations of traditional communication media (see B in Figure 19.5).
- *Identification of Fire or Other Issues.* A crop can be damaged by several factors, such as wild animals, fire, and frost. Either through their own sensors or by collected data from fixed sensors, flying things may detect problems in crops (see C in Figure 19.5).

19.3.3.4 Government Official Missions

- *Border Surveillance.* Considering the increasing globalization, countries are tending to protect their borders by different ways. Focusing on the improvement of automated surveillance, flying things can identify border crossings and follow illegal immigrants while sending location information in real time

Figure 19.5 Smart farm applications taken to the next level with IoFT. A: Surveillance, B: services integration, and C: identification of fire or other issues.

via Internet. Such information associated with sensors spread on the ground should provide more accurate results.

- *Forest Fire Detection and Illegal Logging.* Concerns about global warming are growing stronger every day. One of the causes is the reduction of forest areas around the planet due to fire and illegal logging. Manned aircraft are often used but expensive. Flying things may become a solution to increasing coverage and hence the possibility of identifying problems in real time, so possibly preventing major fires or helping apprehend illegal loggers.

19.3.4 Novel Applications of the Internet of Flying Things

FANETs are enablers of inter-UAV connectivity that simplify the cooperation and collaboration among flying things. However, they are not necessarily connected to the Internet since they operate over swarms of UAVs, which cooperate sharing resources and data in a local offline network. Taking the flying things to a new paradigm, such as IoFT, provides "wings" to the concept of FANETs and makes them more powerful. Imagine flying things geographically distant with no possible line of sight (LoS), but still sharing data, network resources, and services just like if they were all in the same local network. That becomes possible with the introduction of IoFT-oriented networks.

The Wide FANETs (WFANETs) merge the concept of FANETs with IoT, and take wide area networks (WAN) as an inspiration to benefit from all paradigms. A visual representation is seen in Figure 19.6, which illustrates a real scenario with WFANETs.

The main characteristic of ad hoc networks is the existence of at least one path to reach every single node in the network. It is usually possible if nodes are within a limited range and if there are strategic LoS links, guaranteeing that the main links will not be interrupted. The new concept of WFANETs benefits from the connection to an IoT infrastructure already available due to smart environments nearby (e.g., smart cities, smart roads, smart homes, smart farms, and even smart cars). Once connected to an infrastructure that can provide broadband Internet access to the local FANET, a tunnel can be created for information exchange and cooperation with remote FANETs. The act of connecting flying things from different geographical locations through a transparent cloud-based tunneling, providing the same features as a local FANET, creates a WFANET.

WFANETs allow for different approaches and applications. A good example is big events, for example, football matches, Olympic games, and concerts, which require high security in order to manage large crowds, monitor suspicious activities, manage ongoing events, and so on.

Another example of application improved by WFANETs is the pavement scanning for distress, helping the provision of better quality roads. A flying thing scanning roads may be helpful for spreading warnings to intelligent transportation systems, reducing car accident rates, improving security and safety services,

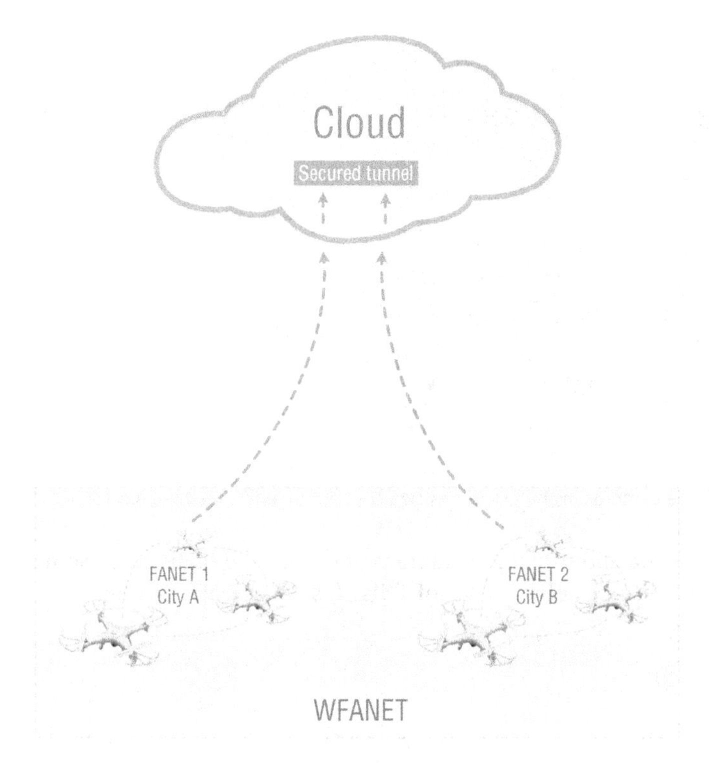

Figure 19.6 A Wide-FANET.

and also providing information about possible trip delays and/or personal agenda-based notifications. Additional information such as weather conditions, for example, could also be integrated into the scenario, improving services provision.

By logical extension, one can envisage that the sensors carried on UAVs could also be considered as flying things in their own right, that happen to be colocated with a set of others on board. If these services become advertised through an automatic broker, many other agencies might choose to pay to access them. For example, a food retailer might choose to pick video streams from any UAV-mounted cameras that happened to be in the area if they could be used to predict and respond to incoming customer flows (see Figure 19.7).

Furthermore, the integration of flying things into smart environments allows improvements to the UAV's operation. For instance, GPS (global positioning system) location is a critical information that is essential for a flying thing. With the existence of an infrastructure to which the flying thing is connected, high accurate location information may be available, helping the UAV operate more precisely. Other sensors such as weather stations mounted on buildings might

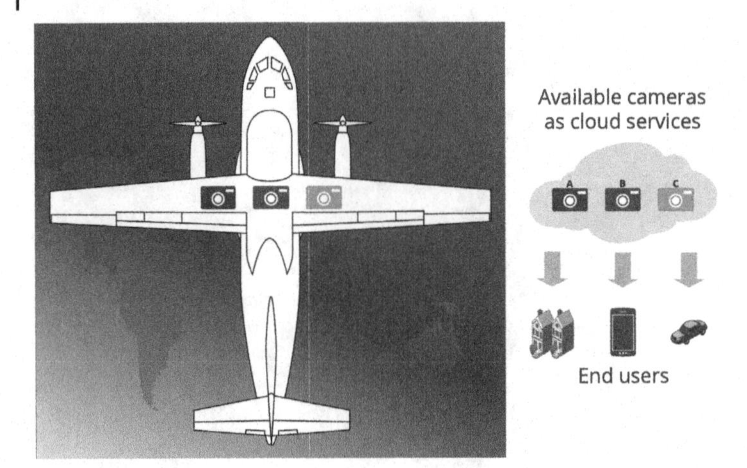

Figure 19.7 Available cameras in the cloud can be accessed by end users as paid services.

provide real-time data about flying conditions ahead on the flight path. Some security issues can also be solved with IoFT itself, such as GPS spoofing.

19.4 Challenges

We have already outlined some of the benefits to UAV systems from coupling to an IoT network of sensors, and even more benefits that could arise from adding flexible airborne sensing and networking to an IoT application. However, some security and safety concerns come from opening up the "closed-world" of a UAV system, which is discussed here. This section also highlights how the Internet of Flying Things can help solving old issues as well.

19.4.1 General Issues

The new Internet of Flying Things model combines the best of IoT and UAVs into one solution that inherits characteristics, but it also introduces new challenges that must be overcome. In short, these fall into three categories.

The first set of issues relates to public safety, and to ethical concerns around the collection and distribution of data. Other obvious concerns would be around snooping—either by physically capturing images or by moving into the proximity of local secure wireless networks. It seems likely that many countries will enact legislation requiring certification and authorization to fly UAVs and benefit from the possibilities of IoFT.

The second set of issues arises from a desire to avoid the past mistakes of the computing industry and lower the thresholds to engagement for key stakeholders, so that the entire population can reap the benefits of this technology. These issues are related to standardization, and must involve both the industry and governments. From a technical point of view, standardization is necessary to make all things actually able to “talk” to one another, instead of creating a “Babel Tower Effect” in which devices become split into disjoint subsets (for instance, all devices from the same manufacturer) that can only talk to others from the same subset. Hardware and software design must take into account limitations of memory, storage, processing capabilities, and power source. This is likely to lead to different solutions for flying objects being developed by different manufacturers, using various hardware architectures, platforms, and communication protocols. Thus, there is a pressing need for the emerging standards for communication and data transfer between devices to be implementable in highly resource-efficient algorithms. From a social and economic point of view, standardization will favor the entrance of small and medium companies in the market, stimulating entrepreneurship and competition, benefiting the final customer, and spreading the use of the technology.

The third set of issues that arises is mainly related to security and Big Data. The heterogeneity will also extend to the data being collected, since it is expected that IoT—and consequently IoFT—will bring a large amount of new, most likely nonstructured data. How and where (locally or remotely) this data will be stored and recovered following real time and security requirements will involve the use of Big Data techniques and technologies.

These considerations apply to a single IoFT device, but in many scenarios more than one device may be deployed. As the number of connected devices grows, addressing and managing all of them without prejudicing quality of service become a critical issue. This is compounded by the fact that new devices, offering or requesting new services can join the ad hoc network at any time. Hence, applications must be designed from the ground up to enable extensible services and operations. Moreover, IoFT and WFANETs have a common issue to deal with, which is the possible service unavailability regarding an Internet connection. This raises questions such as what measures should be taken if the LTE/3G/4G connection is lost without prejudicing the mission and data privacy?

19.4.2 Security Issues at Different Internet of Flying Things Conceptual Layers

In general, security challenges are big concerns of computing systems. There are several security issues that could exist in IoFT-based applications, which are mostly inherited from the underlying networks and technologies (e.g., UAS, IoT). The main security issues that could be imposed by the different network layers are listed as follows:

- *Physical Layer.* Both *jamming* and *tampering* attacks are known issues for this layer. Jamming is a well-known attack that causes interferences to the radio frequencies that network's nodes are using. It can interrupt the network if a single frequency is used throughout the network: at worst, interrupting communications with flying things; and at best causing excessive energy consumption. On the other hand, if an attacker can physically tamper nodes, a tampering attack takes place which damages, replaces, and electronically "interrogates" the nodes to acquire information (Sen, 2012). *GPS spoofing* attacks, which happen in this layer, consist in the use of a signal that is stronger than and mimics the attributes of a genuine GPS signal to take over a GPS receiver, have become more frequent. Such attacks can cause the aircraft to completely lose control, which is a very critical issue.

- *Data Link Layer. Collision, exhaustion,* and *unfairness* are the most likely attacks at this level. A collision happens when two nodes simultaneously attempt to transmit on the same frequency, resulting in either partial or complete packet disruption, which will cause an erroneous data transmission through a communication channel. In an IoFT, it is a big issue because the possibility of having an intermittent network condition is likely to cause chaos in a FANET. Therefore, each flying thing can be affected and could crash with each other due to the loss of information on the fly. If repetitive collisions take place, the system may suffer from resource depletion, that is, an exhaustion attack. Finally, rather than blocking access to a service outright, an attacker can degrade it to gain an advantage such as causing other nodes in a real-time MAC protocol to miss their transmission deadline, characterizing an unfairness attack (Sen, 2012).
- *Network Layer.* There are several attacks found on this layer. In a *selective forwarding* attack, malicious nodes attempt to stop the packets in the network by refusing to forward or drop messages passing through them (Khan et al., 2012), which could compromise a FANET that relies on strategically placed flying things to reach all the destinations within the network. If an attacker makes the compromised flying thing look more attractive to surrounding ones, which is considerably an easy task since IoFT applications usually take place in open environments, the selective forwarding attack becomes very simple. Then, through the affected flying thing, a data transfer situation may be started leading to a *sinkhole* attack (Dener, 2014). Another attack on network layer is the *Sybil* attack, in which a flying thing exhibits multiple identities to other flying things in the network (Dener, 2014). On the other hand, a *wormhole* is an out-of-band connection between two nodes using wired or wireless links, which can facilitate forwarding packets faster than via normal paths (Wallgren et al., 2013). According to Wallgren et al. (2013), a *HELLO flood* attack refers to overriding "HELLO" messages by broadcasting a stronger signal, allowing an attacker to introduce himself as a neighbor to many nodes, possibly the entire network, which is very likely to be seen in scenarios with flying things. In operation, the communications systems must balance the desire to maximize throughput (the amount of data transferred

per unit time—which suggests large packets) and minimize latency (maximum delay accepted for the data transmitted and received—which suggests small packets). The final issue when multiple devices are deployed is that nodes in FANETs can be fixed or not and the distance among nodes can vary while in motion. To assure the route among nodes, FANETs rely on broadcast techniques, which becomes a problem that can be interpreted as an attack to the communication system: the growing amount of data traffic between flying things and IoFT-infrastructure elements; the increasing number of manned aircraft sharing the same airspace; and the progressive forthcoming of flying things to dispute nonsegregate airspace. To solve the problem of broadcast packets to all nodes and avoid packet redundancy and its associated problems (i.e., Broadcast Storm Problem - BSP), techniques to mitigate BSP must be taken into account (de Melo Pires et al., 2016).

- *Transport Layer. Flooding* attack causes immense traffic of useless messages on the network. It may result in congestion, and eventually lead to nodes exhaustion (Sastry et al., 2013). The *desynchronization* attack is made when the adversary repetitively pushes messages, which convey sequence numbers to one or both of the endpoints (Dener, 2014). The GPS spoofing may help attackers to *hijack* flying things, which is another issue that is strongly related to situations where an attacker secretly relays and possibly alters the communication between two parties, also known as *man-in-the-middle attacks*. This kind of attack allows an attacker to land the flying thing in an unauthorized place, and taking advantage of its legitimate network access. However, such problem can be neutralized by the high connectivity in IoFT environments: In most cases, a solution lies in techniques that check the accuracy of GPS signals by comparing to the ones provided by access points and other fixed known infrastructures.

From another point of view, some researchers have organized IoT in a new security architecture that also applies to IoFT (Jing et al., 2014). Such architecture is mainly divided in three layers: perception, transportation, and application. This organization with three layers only implements most of the features of Open Systems Interconnection model (OSI model), despite the fact it is not as well separated. Figure 19.8 shows a comparison between the IoT model proposed by Jing et al. (2014) and OSI model. Each layer of IoT model faces specific challenges, which will be discussed further.

- *Perception Layer.* This layer is mainly about information collection, object perception, and object control. In the perception layer within the Internet of Flying Things, tasks related to security of RFID (radio-frequency identification), WSN, RSN (RFID Sensor Network), GPS technology, and so on will be performed (Wu et al., 2010). The heterogeneity of flying things and ordinary IoT/IoFT-ready devices is one of the main problems that may arise in this layer, which can lead to compatibility problems. Another issue is the

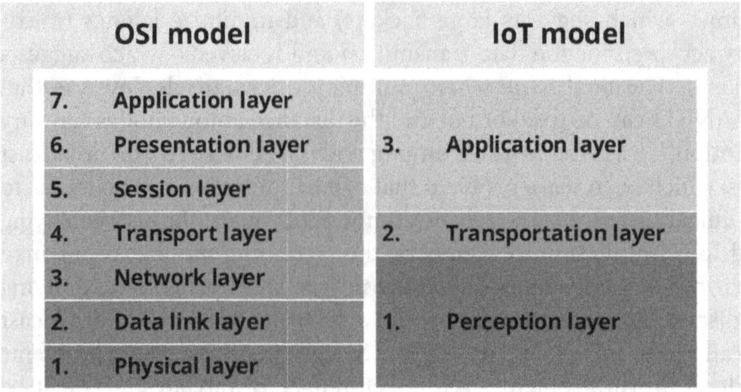

OSI model	IoT model
7. Application layer	3. Application layer
6. Presentation layer	
5. Session layer	
4. Transport layer	2. Transportation layer
3. Network layer	1. Perception layer
2. Data link layer	
1. Physical layer	

Figure 19.8 A comparison between OSI model and IoT. (Model proposed by Jing et al. (2014).)

limitation of power, computing ability, and storage capacity, especially in flying things, that make them more vulnerable to attacks, allowing for physically stealing information and also functioning modification (Jing et al., 2014).

- *Transportation Layer.* Also referred to as network layer, the transportation layer's main function is transmitting information obtained from the perception layer (Wu et al., 2010). This layer encompasses the Wi-Fi, establishing and maintaining MANET/FANET and 3G/4G/5G networks, leading to a heterogeneity problem for the exchange of information among different networks, which is even more challenging when it comes to the IoFT and its inherent integration among different networks (IoT, FANETs). Moreover, it also leads to new vulnerabilities on the relevant segments for the implementation of IoFT networks. For instance, the main issues related to Wi-Fi are *phishing access attacks, malicious AP, DDoS/DoS attacks,* and so on. On MANET/FANET side, the security problems that may be faced are *data security, network routing,* and *DDos/DoS* issues. Finally, related to the 3G/4G/5G networks, *data security* and *unlawful attacks* are the main concerns (Jing et al., 2014).
- *Application Layer.* This layer supports all sorts of business services and realizes intelligent computation and resources allocation in screening, selecting, producing, and processing data. The security issues it faces cannot be solved in other layers of the IoT model, such as *privacy* protection issue, which can become a real demand in certain special contexts. Thus, in operation, the application support layer must be able to recognize untrusted data (e.g., spam data and malicious data), and filter them in real time. The application layer can be organized in different ways according to different services, and usually includes middleware, M2M, cloud computing platform, and service support

platform (Jing et al., 2014). The *privacy* concerns in flying things have been recently discussed as a big threat as storage systems get more sophisticated. There is a tendency to store as much as possible into the flying thing memory, so guaranteeing that the needed information is always available (Hartmann and Steup 2013). However, such approach ends up being a very critical security threat (Jing et al., 2014). As the flying thing is inserted in a highly connected environment, it can be physically stolen or its control can be taken. Thus, the attacker may still use it as a gateway to obtain information from the network, since the flying thing is authorized and will be able to access private and confidential information. This is a consequence of the high connectivity of things and the increased contact surface that generates more possible threats to be explored by malicious entities.

19.4.3 Safety Issues of the Internet of Flying Things

The growing popularization of UAVs has increased the research in this field and is fostering the use of such technology in many applications. There are roadmaps published periodically by military and civil organizations—for example, United States Army (US Army), American Department of Defense (DoD), European RPAS Steering Group (ERSG), and Federal Aviation Administration (FAA)—that outline the expected advances for UAVs (US Army, 2010; Yearbook, 2011; UK Civil Aviation Authority, 2012; DoD, 2013). However, there are not enough studies on safety for the specific integration of UAVs and IoT, which is one of the most important topics to be discussed and an open opportunity for researches in safety-critical systems.

There are five challenges of UAS integration, as stated by Dr. Wilson Felder, the Director of the William J. Hughes Technical Center of the FAA, reported by Stark et al. (2013): procedural, technical, aircraft safety, crew credentials, and public acceptance. Sense and Avoid Systems remain as one of the largest obstacles for the safe integration of UAS into airspace. Any person or computer-based system that meets the three mandatory activities to operate an aircraft (flight, navigation, and communication) should assume the command of an aircraft, be it manned or unmanned (Baraldi Sesso et al., 2016). Regulations regarding collision detection and resolution efforts must be met by any UAS designed for nonsegregated airspace.

In short, FAA requirements demand that UAS meet safety levels equivalent to those of manned aircraft. It includes the frequency of collision of a UAS being operated in a FAA-controlled airspace, which is currently 1×10^{-7} events per hour of operation for manned aircraft (Asmat et al., 2006). As for IoFT applications, there will be merged approaches that may include both manned and unmanned aircraft at the same time, leading to the necessity of using the nonsegregated airspace. This fact increases the need for meeting safety requirements and the first step is to point them out and discuss possible solutions.

Despite the security applied to each layer, there is also the need to find joint approaches that ensure safety individually and collectively. Reducing the breaches throughout the layers will consequently reduce the overall chances of attacks to the network. Such characteristics regarding security and safety make it more difficult to ensure the system safety, composing one of the main threats in the area, which is finding approaches that deal with both concepts at once.

The concept of safety has a long tradition for vehicles (ISO, 2011). This is a mature area, and several standards exist for creating safe systems such as RTCA/DO-178C for UAV software (RTCA Inc., 2011) and RTCA/DO-254 for UAV hardware (RTCA Inc., 2000). Safety deals with minimizing the frequency of accidents or failures in a system, mainly when related with loss of life, high-value assets, and it is related with incautious actions or events.

Providing a safe wireless communication means ensuring that the information transmitted is received without any transmission error and loss of the information. Due to noise, interference, and fading effects, wireless network cannot have zero transmission error, since there is no system with zero risk. For wireless network, transmission error and loss of information cannot be avoided, but they can be overcome by reducing or by detecting them. In order to guarantee communications with safety, Pendli (2014) listed a number of requirements that must be satisfied. Communication links should be reliable and immune against noise, jamming signal, interference, and fading effects in order to provide a link without errors and losses. Since flying things are safety-critical systems, communication channels must be continuously available and provide timely delivery information without failure. Ensuring real-time performance means that the technology used must take into consideration the delay during information transmission and retransmission and be able to cope with burst errors. Device mobility, and the changing external environment requires communication links to be robust even under adverse conditions against channel fading, low SNR (signal-to-noise ratio) conditions, and channel losses.

For flying vehicles in general, there are regulation and legislation requirements, which demand that any factors affecting safety must be taken into account, and incorporated into a risk model meeting certain minimum standards before certification is granted. For IoFT devices, the possible failure rates of solutions to all the summarized communications issues must be incorporated. Indeed, in some applications where the flying thing operates above cities or farms, which are critical fields, the need for certification is even higher. Although there is some way to go, it is hoped that the robustness of the IoT infrastructure may help achieving the low failure rates needed.

In short, developers and users of IoFT systems must take safety as one of the main concerns. However, the combination of high mobility and wireless communications highly increases the exposure of these systems to malicious threats and to faults deriving from uncertain connectivity or communication timeliness. Nonfunctional requirements like security have thus become harder to fulfil, creating new challenges to such safety-critical embedded systems (Bloomfield and Lala, 2013). In fact, further research on the development and assurance of both safety and security must be performed, addressing the needs of multidisciplinary approaches like integrated control systems, communication, security mechanisms, artificial intelligence, neural networks, safety assets, and other technological concerns. A key challenge is that the architectural solutions to ensure safety may open further weaknesses from the point of view of security. On the other side, security weaknesses may lead, if exploited by attackers, to safety violations and the implementation of a given security mechanism may impact safety. In the few cases where security is taken into account, the only problem that is addressed is the open network communication, such as wireless systems. Security at large is not handled (or handled in a very general way) without a full support to identify and mitigate security threats. Therefore, the relationship between safety and security seems to be still an open issue in the community.

19.5 Case Studies

Three case studies will be carried out in this section exemplifying real-world applications of the Internet of Flying Things. The first is on smart farms highlighting WFANETs application. The second is on the provision of Internet access and IoT-based services on remote areas, especially on rural zones and peripheries of smart cities. Finally, the third case study is on the management of big events and provision of targeted services.

19.5.1 Case Study 1: WFANETs for Surveillance Tasks in Smart Farms

Farms of hundreds of hectares with varied topology can be formed and subjected to different climatic conditions along their area. They can perform many different tasks in agriculture like raising livestock, becoming specialized units (e.g., vegetable or fruit farm, dairy, pig and poultry farm or even used for the production of natural fibers, biofuel, and other commodities). All of them also have a large infrastructure that may include plantations, pasture, feedlots, orchards, greenhouses, silos, barns and other buildings, and the farmhouse. In modern times, such term has been extended to include industrial operations as wind farms and fish farms, both of which can operate on land or sea. For more information about the various applications of IoT in agriculture, please refer to Chapter 18.

The concept of smart farms has been used to denote a ubiquitous IoT-ready farm environment for increasing general productivity and optimization of everyday tasks. Internet connectivity can be achieved in such environments mostly on strategic administrative buildings, but also in some dedicated fields with specialized technologies for sensing and monitoring. This case study analyzes a scenario from the IoFT perspective through the application of Wide-FANETs.

The concept of Wide-FANETs is an ad hoc network that benefits from the connection to an IoT infrastructure already available due to smart environments nearby. Such connection can provide broadband Internet access to local FANETs and, thereafter, a tunnel can be created for information exchange and cooperation among remote FANETs. More than one flying ad hoc network from different geographical locations connected through a transparent cloud-based tunneling consists of a WFANET.

19.5.1.1 The Problem

Although smart farms are already benefiting from IoT, they still have limited connectivity to the Internet, which usually leads to big areas with no connectivity and no surveillance. Thus, network structures that do not depend on Internet connections have been applied as temporary alternatives for monitoring animals, crop fields, forests, schools of fish, and so on. It is common to find WSN and wireless body area networks (WBAN) spread in many fields. However, the feedback from these networks and sensors is frequently limited, delayed, and poorly updated, which leads to less dynamic operations and limited overall farm management.

The problem addressed in this case study is related to general surveillance of a smart farm, including farm borders, controlled cattle, or suspect activities within the farm field. Such problems get bigger and more complex according to the farm size and also due to the number of different tasks that demand monitoring for safety and security reasons. They can also be considered critical since high-value assets are put at risk.

19.5.1.2 Proposed Solutions

Following topics will discuss the hypothetical application of FANETs and WFANETs to three farm scenarios: border control, cattle monitoring, and crop monitoring. Figure 19.9 illustrates the scenarios discussed further.

The surveillance of farm borders to identify potential threats, such as intruders (humans or not), and trigger appropriate procedures is one of the most important tasks for safety purposes. In a typical scenario, the task would be performed by human patrol and/or surveillance cameras. This solution, besides costly, is not likely to be effective, since the risk of human failure is always present.

A WFANET would provide border control at a lower cost. The UAVs would gain from the available IoT infrastructure to provide surveillance images or data in real time to a control central (e.g., for security guards, a specialized security company or a police station). The use of UAVs would reduce the necessity of

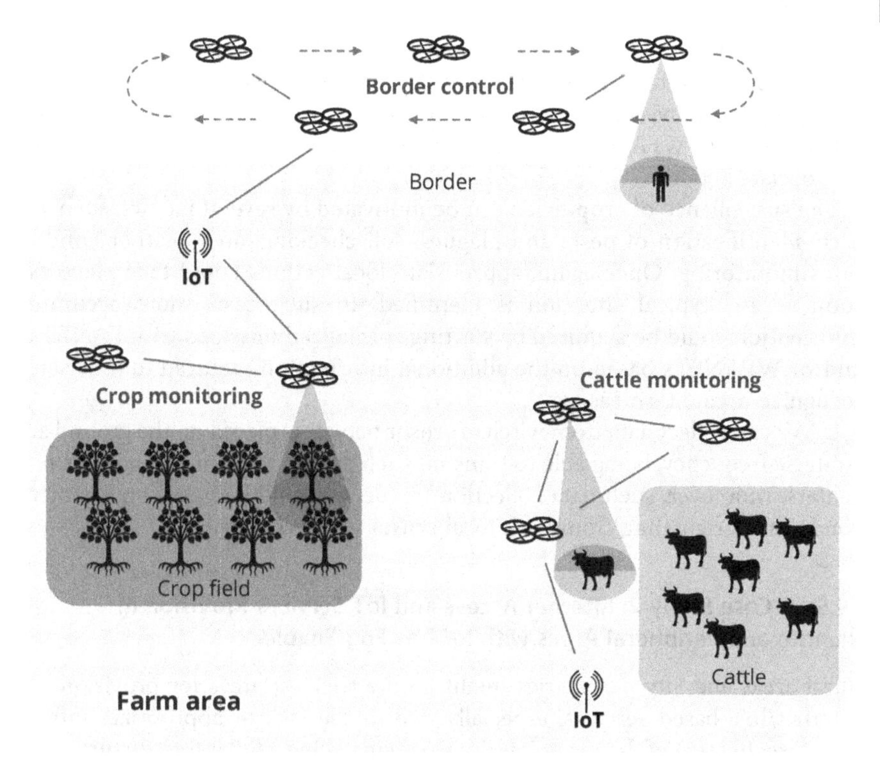

Figure 19.9 Surveillance situations for the study of WFANETs application in smart farms.

surveillance cameras installed over the property border and its required infrastructure (such as batteries or solar cells for power), again reducing the task cost[1]. The advantages would be even bigger in bigger farms, since the use of WFANETs would allow the exchange of relevant information among more than one local FANET, in a cooperative live border monitoring. Additionally, data from official governmental web services may be used to identify fugitives.

Another situation is related to cattle monitoring. In big farms, it is normal to specify strategic land areas intended for cattle raising. Such areas must be constantly monitored in order to avoid escapes or unauthorized human intervention. Once again, if the solution relies on human resources, the coverage of monitoring might be subject to failures or be too slow to take appropriate actions with little damage to the cattle or other farm fields.

An IoT infrastructure would provide means of cattle tracking by installing cheap sensors in each single animal. Moreover, by surveying the area with UAVs, data acquisition would flow fast from the cattle raising area to monitoring

1 *Guidebook to the Tools, Software and Applications for Internet Enabled Drones.* Available at http://www.postscapes.com/iot-mashup-003-wearable-drone-control/.

centers, providing means of taking timely actions. As an alternative, automatic actions could also take place as soon as an unwanted behavior is identified, for example, closing backup gates in case of cattle escape. For that, FANETs (in small areas) or WFANETs (in bigger areas) would identify and trigger the appropriate action.

The surveillance of crop fields can be motivated by several factors, such as early identification of pests and plagues, soil checking, and weather conditions monitoring. Once again, appropriate local actions could take place as soon as an atypical situation is identified. In such case, more accurate information could be acquired by starting specialized missions with FANETs and/or WFANETs based on the additional information required to precisely recognize a countermeasure[2,3].

UAVs can collect data from wireless sensor networks placed on the ground at strategic frequency, being able to transmit such data in real time to monitoring centers. Moreover, such data collection frequency could be based on weather conditions, originating from both local sensor and web services.

19.5.2 Case Study 2: Internet Access and IoT Services Provision in Remote and Peripheral Areas with IoFT as Fog Enabler

Rural areas and city peripheries might be the trickiest areas for provision of Internet/IoT-based services, especially due to the lack of appropriate infrastructure. In most of the cases, it is not worth installing a full infrastructure that will be rarely used in remote areas. Moreover, as the demand is usually low, there will be no interest by government and business companies to update such infrastructure in order to match the newest technologies requirements.

On the other hand, there are situations when the support of Internet/IoT-based services would positively improve tasks related to search and rescue, life quality levels, special events, people/objects tracking, and so on. This case study will investigate the benefits of using the Internet of Flying Things to amplify the coverage of common smart city services. By doing so, there is a potential opportunity of providing Internet connectivity and IoT-based services for the temporary improvement of special tasks in an inexpensive manner.

19.5.2.1 The Problem

The downtown of a smart city is likely to be the geographic region to first experience novel efforts and updated technologies, while the peripheries will usually be the last ones to face a full integration and also to get relevant

---

2 Agricultural crop surveillance. Available at http://www.precisiondrone.com/drones-for-agriculture.html.

3 *IoT for Agriculture – Drones/UAV.* Available at http://www.slideshare.net/ChristopheRaix/iot-for-agriculture-drones-uav-presentation.

investments. That is the natural process given a business model full of potential opportunities which focuses on highly populated areas to be profitable.

In a different context, rural areas might not need Internet connectivity at all times, but they do need to update/synchronize data at some stage. For this, data mules can be used, for example, in vehicles that physically carry computers with dedicated storage servers allowing a slow, limited, once-a-day synchronization. Although this approach can be considered inexpensive and efficient in many situations, it is poorly flexible and does not provide the benefits of a fully connected infrastructure.

The same issue is experienced in emergency situations, for example, search and rescue. A local network infrastructure does not provide the real-time support that such an operation would need to properly and efficiently do the search and also the rescue tasks. In some cases, the inexistence of cellular network coverage limits even more the connectivity, which leads to the necessity of a flexible, inexpensive, easy-to-set-up approach.

Pursuant to such issues, the IoFT paradigm might be a relevant alternative of either temporarily or permanently minimizing the already discussed problems. Next subsection will discuss proposed solutions for recurrent cases demonstrating how this paradigm would solve practical cases.

19.5.2.2 Proposed Solutions

Here, three main applications of this case study will be highlighted. The first investigates how an IoFT network would be important for smart cities peripheries. The second addresses the environmental monitoring in rural areas helping the environmental police to identify illegal actions and take appropriate countermeasures. Last, natural phenomena and emergency operations support are discussed considering that in such situations connectivity becomes an issue due to the loss of infrastructure nearby. Figure 19.10 illustrates the scenarios that will be discussed further.

By strategically moving drones to the edges of smart cities, a powerful connection to the Internet will be available for restricted areas helping with the provision of connectivity to city peripheries. From this connection, a complete range of IoFT services will be available for end users nearby for a specific period, allowing the execution of relevant tasks[4].

For instance, the smart city's electric power company might automate the reading process of residential energy consumption. Such task is usually performed by a person taking note about the consumption in each residence/building, which takes more time and is susceptible to misreading. Applying appropriate identifiers to each residence (such as RFID tags), which in turn will be recognizable by the IoFT infrastructure, would provide means of reading the energy consumption of a big area in several minutes. That is possible due to the

4 UAV as a Service. Available at http://www.aerialiot.com/.

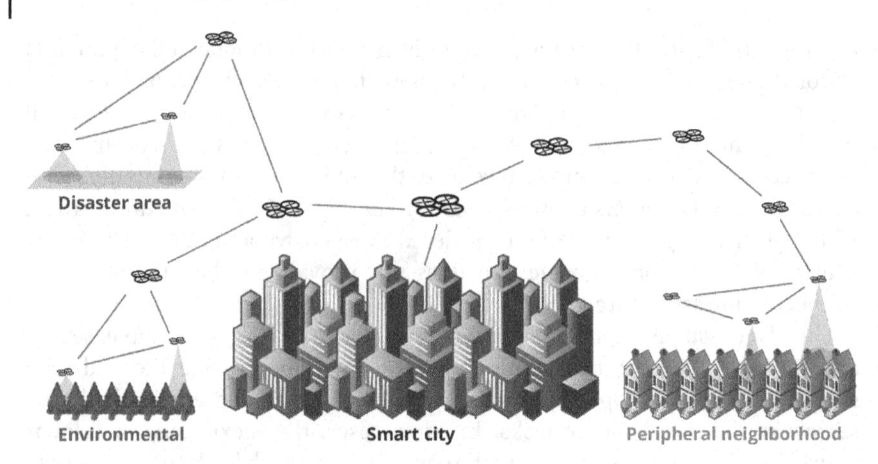

Figure 19.10 Applications of IoFT in smart cities.

existence of a FANET/WFANET flying over the area, providing such class of services and being able to provide real-time information to both the customer (e.g., bad debt warnings) and the electric power company (e.g., reading issues in specific residences that would require a technical visit).

Moreover, the existence of a single flying thing or a FANET/WFANET over a neighborhood that is not yet equipped with the smart city infrastructure can be used as a traffic support to drivers in that regions. Smart cars would benefit from IoFT warnings about the traffic. Regular cars could also get important updates about the traffic via smartphones connected to the IoFT infrastructure. Actions like these would contribute for a safer and more accurate use of streets, not to mention the emergency situations that could take place, and fully benefit from IoFT infrastructures.

Rural areas usually do not demand Internet connectivity at all times. In some cases, data mules can be used, allowing the transfer of data between remote locations to effectively create a data communication link. Similarly, an IoFT network can be created to transport data to a desired location, but with an intermittent link provided by a mobile ad hoc network connecting the city and the remote area.

The concept of fog computing is clear in this situation, but the IoFT network emerges as an enabler of the model. Section 19.3.1 discussed the application of a UAV as a fog computing infrastructure using the IoFT elements to provide an updated, flexible solution. Such application fits well for rural areas that need better services delivery for more advanced tasks performing.

This model also meets the requirements of environmental monitoring by governmental official agencies, which is a trend due to the global issues being reported lately. The real-time monitoring by flying things can be remotely analyzed by specialists in central offices that will be able to use updated images to identify suspicious activities. If an environmental illegal practice is taking place, the environmental police can promptly move to the region being supported by the IoFT infrastructure at all times, allowing efficient red-handed operation.

A situation that is also frequent in remote areas is related to natural phenomena and emergency operations support. Natural disasters and catastrophic situations can be predictable but usually not controllable. Their destruction might lead to the unavailability of essential services from local infrastructure, such as cellular networks, causing chaos situations.

The application of IoFT to provide means of connectivity from the heart of the city throughout a FANET that reaches the disaster area can be an efficient temporary action to support rescue agents to do their job as qualified as possible[5]. For instance, if a disaster has led to deaths, one might consider the real-time recognition of bodies by relatives. However, it is not always safe or appropriate to transport people to the disaster area. Also, it is not possible in some situations to remove all the bodies right away, resulting in panic and people flooding agencies with calls demanding information.

IoFT in emergency situations will provide connectivity for basic services provision, governmental central offices, and news agencies to be updated with recent discoveries, and the complete support for ambulances being moved to the target area.

19.5.3 Case Study 3: Targeted Services Delivery on Big Events with IoFT

Big events take place throughout the year basically everywhere. They can be of many types such as local exhibitions, big theatre presentations, and hundreds of simultaneous concerts and presentations in several different venues in a music festival. These big events can be crowded by people coming from multiple places with different accessibility demands.

General safety is an issue that can be supported by an IoFT infrastructure, especially when an event takes place in a remote location (e.g., gigantic music festivals). It is similar to the situation described in Case Study 1, when surveillance tasks may be demanded. However, this case study will focus on targeted services delivery for big events provided by IoFT networks.

19.5.3.1 The Problem

In big events one of the most alarming problem is to lose people, especially children. Sometimes a cellular network signal will not be available and finding someone in a crowded area is almost an impossible mission for one person.

5 Medical drones poised to take off. Available at http://www.mayoclinic.org/medical-professionals/clinical-updates/trauma/medical-drones-poised-to-take-off.

Moreover, due to the high number of people, it is hard to move from one location to another to check if your favorite band started playing or if a boring presentation is about to end. These are desired information that are usually unavailable or partially available in paper guides, which can become outdated in minutes.

19.5.3.2 Proposed Solutions

The applicability of IoFT in the cited situations might be a simple and efficient solution. The following subsections will discuss two cases specifically. First, the application of finding people will be discussed. Second, real-time information about multiple venues in music festivals will be addressed. Figure 19.11 presents the scenario of this case study.

It is common to suddenly find yourself separated from your friends or family in crowded places. If a cellular network is not available for any reason or your phone ran out of battery, then you might have a problem. Plus, the situation gets even worse if a kid is lost by any reason and the parents are desperately looking for him/her.

The identification of people using fixed cameras coupled to the infrastructure of venues is efficient most of the times. However, in crowded places the angle of the camera might be a problem. In case of lost children, it is even more problematic, since the height would negatively influence the camera coverage.

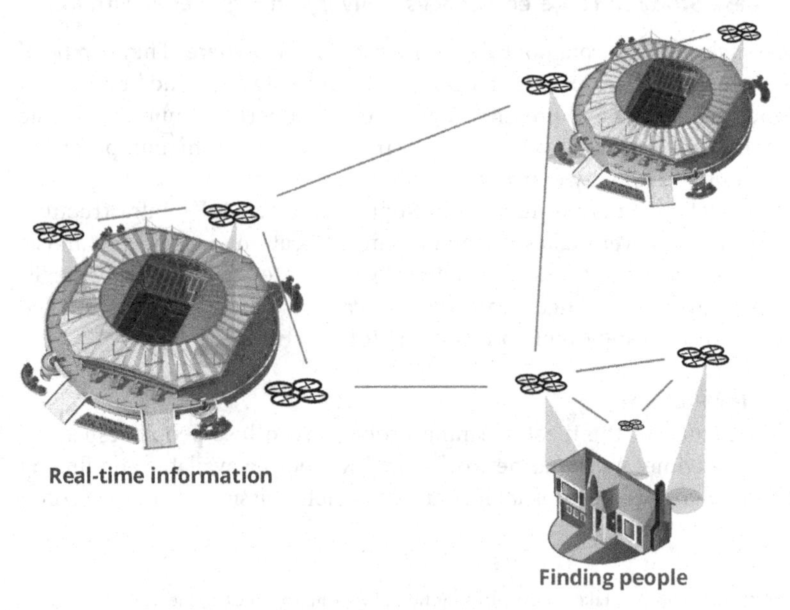

Figure 19.11 Applications of IoFT in big events.

For cases like this a reasonable solution would be a flying thing connected to a FANET/WFANET for accurate, fast search on site.

The real-time streaming can be monitored by trained people and also relatives. Such effort would significantly reduce the chances of any hurt to an unattended child and also provide effective means to supporting families.

Another relevant topic in big events, especially the ones with simultaneous attractions in multiple venues, is the possibility of getting access to information in real time. One can wonder whether a concert meant to be played at a specific time has started or is delayed. Moreover, why not have access to the audio or a video streaming to check how is the performance going or how crowded it is?

An IoFT network plus fixed cameras and sensors could provide a set of information about multiple venues in real time to a central server that would redistribute such information all over the network. Personal smartphones or special stations could get access to such information, improving the delivered services during the days of the event.

19.6 Conclusions

The Internet of Flying Things integrates flying things to the Internet of Things paradigm. This new concept increases the cooperation and collaboration among UAVs, amplifying the network coverage, and creating new possibilities for UAS applications, such as interactive decision-making. In the context of IoT, being aware of flying things' basic characteristics, such as capabilities, targets, and limitations, is a key factor to determine how eligible they are to perform sophisticated missions and how likely they are to be integrated onto the IoT. Despite the fact that flying things could be treated as generic nodes on the IoFT paradigm, the development of specific approaches might help reaching more optimized, safe, and secure results.

There are understandable concerns about the threat that networked UAVs could pose to privacy and safety. Legislation is likely to address the design and usage of IoFT-based systems, and to help with public confidence the largest open challenge facing the field is the development and adoption of robust standards for security and safety—of both devices and the data they carry and transmit. Considering the many physical forms that a UAV might take, security policies and algorithms must be devised that are resource efficient, work on many different types of hardware (from data storage devices right through to different aircraft chassis) and software (across the layers of network protocols). When one extends the considerations to multi-UAV swarms, or even disparate groups connected via a WFANET, policies must then take into account the greater "vulnerable surface" of an ad hoc network, and redundancy of information (encryption keys, etc.) that will arise as mobile devices become unavailable for periods of time as they move around.

Despite all the challenges, the Internet of Flying Thing is a promising paradigm with high chances of applicability. For designers of airborne missions, connecting UAVs to smart environments will allow the acquisition and provision of updated information, increasing the accuracy of tasks they perform. Coupling UAVs to the Internet means they can outsource processing to cloud servers, so reducing the need for sophisticated hardware and making them more flexible and easier to use for new scenarios. Conversely, for infrastructure and IoT designers, Flying Ad Hoc Networks offer a flexible new way of connecting environments that might have been separated due to their remoteness, or for temporal reasons such as natural disasters or infrastructure breakdowns. These improvements are a very important innovation in the field, bringing possibilities of new applications, especially hard real-time ones, for example, search and rescue, critical surveillance tasks, and sensitive fields monitoring.

Acknowledgments

Research was sponsored by FAPESP under Process Number 2012/16171-6. Research was also sponsored by the Army Research Office and was accomplished under Grant Number W911NF-18-1-0012. The views and conclusions contained in this document are those of the authors and should not be interpreted as representing the official policies, either expressed or implied, of the Army Research Office or the U.S. Government. The U.S. Government is authorized to reproduce and distribute reprints for government purposes notwithstanding any copyright notation herein.

References

Andreev, S., Galinina, O., Pyattaev, A., Gerasimenko, M., Tirronen, T., Torsner, J., Sachs, J., Dohler, M. and Koucheryavy Y. (2015) Understanding the IoT connectivity landscape: a contemporary M2M radio technology roadmap. *IEEE Communications Magazine*, **53**(9), 32–40. doi: 10.1109/MCOM.2015.7263370.

Asmat, J., Rhodes, B., Umansky, J., Villavicencio, C., Yunas, A., Donohue, G., and Lacher, A. (2006) UAS safety: unmanned aerial collision avoidance system (UCAS). Systems and Information Engineering Design Symposium, 2006 IEEE, 43–49. doi: 10.1109/SIEDS.2006.278711.

Atzori, L., Iera, A., and Morabito, G. (2010) The Internet of Things: a survey. *Computer Networks*, **54**(15), 2787–2805. doi: 10.1016/j.comnet.2010.05.010.

Baraldi Sesso, D., Vismari, L. F., Vieira Da Silva Neto, A., Cugnasca, P. S., and Camargo, J. B. Jr. (2016) An approach to assess the safety of ADS-B-based unmanned aerial systems: data integrity as a safety issue. *Journal of Intelligent and Robotic Systems*, **84** (1–4), 621–38. doi: 10.1007/s10846-015-0321-0.

Bekmezci, İ., Sahingoz, O. K. and Temel, Ş. (2013) Flying ad-hoc networks (FANETs): a survey. *Ad Hoc Networks*, **11**(3), 1254–70. doi: 10.1016/j.adhoc.2012.12.004.

Bloomfield, R., and Lala, J. (2013) Safety-critical systems: the next generation. *IEEE Security & Privacy*, **11**(4), 11–13. doi: 10.1109/MSP.2013.95.

Bonomi, F., Milito, R., Zhu, J., and Addepalli, S. (2012) Fog computing and its role in the internet of things. Proceedings of the First Edition of the MCC Workshop on Mobile Cloud Computing, 13–16. doi: 10.1145/2342509.2342513.

Bouachir, O., Abrassart, A., Garcia, F. and Larrieu, N. (2014) A mobility model for UAV ad hoc network. 2014 International Conference on Unmanned Aircraft Systems (ICUAS), 383–88.

Chung, H., Oh, S., Shim, D. H., and Sastry, S S. (2011) Toward robotic sensor webs: algorithms, systems, and experiments. *Proceedings of the IEEE*, **99**(9), 1562–86.

Clapper, J., Young, J., Cartwright, J., and Grimes, J. (2007) Unmanned Systems Roadmap 2007–2032. Office of the Secretary of Defense, p. 188.

de Melo Pires, R., Arnosti, S. Z., Pinto, A. S. R., and Branco, K. R. L. J. C. (2016) Experimenting broadcast storm mitigation techniques in FANETs. 2016 49th Hawaii International Conference on System Sciences (HICSS), 5868–77. doi: 10.1109/HICSS.2016.726. IEEE.

Dener, M. (2014) Security Analysis in Wireless Sensor Networks. *International Journal of Distributed Sensor Networks*, **10**. doi: https://doi.org/10.1155/2014/303501.

U.S. DoD (2013) Unmanned Systems Integrated Roadmap: 2013–2038. Washington, DC, USA.

Fahlstrom, P. G., and Gleason, T. J. (2012) *Introduction to UAV Systems*. 4th edn, John Wiley & Sons, Ltd. Chichester.

Frew, E. W. and Brown, T. X. (2009) Networking issues for small unmanned aircraft systems. *Journal of Intelligent and Robotic Systems*, **54** (1–3), 21–37.

Hartmann, K. and Steup, C. (2013) The vulnerability of UAVs to cyber attacks: an approach to the risk assessment. 2013 5th International Conference on Cyber Conflict (CyCon), 1–23.

Hassan, Q. F., Riad, A. M., and Hassan, A. E. (2012) Understanding cloud computing. *Software Reuse in the Emerging Cloud Computing Era*, IGI Global, pp. 204–227.

ISO. (2011) ISO 26262-1:2011 - Road Vehicles. Available at http://www.iso.org/iso/catalogue_detail?csnumber=43464.

Jing, Q., Vasilakos, A. V., Wan, J., Lu J., and Qiu, D. (2014) Security of the Internet of Things: perspectives and challenges. *Wireless Networks*, **20**(8), 2481–2501. doi: 10.1007/s11276-014-0761-7.

Khan, W. Z., Xiang, Y., Aalsalem, M. Y., and Arshad, Q. (2012) The selective forwarding attack in sensor networks: detections and countermeasures.

International Journal of Wireless and Microwave Technologies (IJWMT), **2**(2), 33.

Luan, T. H., Gao, L., Li, Z., Xiang, Y., and Sun, L. (2015) Fog computing: focusing on mobile users at the edge. Eprint arXiv:1502.01815, 1–11. doi: 10.1016/j.jnca.2015.02.002.

Marshall, D. M., Barnhart, R. K., Shappee E., and Most, M. T. (2015) *Introduction to Unmanned Aircraft Systems*, CRC Press.

Pendli, P. K. (2014) *Contribution of Modelling and Analysis of Wireless Communication for Safety Related Systems with Bluetooth Technology*. Kassel university press GmbH.

Pigatto, D. F., Goncalves, L., Pinto, A. S. R., Roberto, G. F., Fernando Rodrigues Filho, J. and Branco, K. R. L. J. C. (2014) HAMSTER: healthy, mobility and security-based data communication architecture for unmanned aircraft systems. 2014 International Conference on Unmanned Aircraft Systems (ICUAS), 52–63. IEEE. doi: 10.1109/ICUAS.2014.6842238.

Rieke, M., Foerster, T., and Broering, A. (2011) Unmanned aerial vehicles as mobile multi-sensor platforms. Proceedings of the 14th AGILE International Conference on Geographic Information Science, Utrecht, NL, USA, 18–21.

RTCA Inc. (2000) RTCA/DO-254, Design Assurance Guidance for Airborne Electronic Hardware. Available at http://www.do254.com/.

RTCA Inc. (2011) DO-178C Software Considerations in Airborne Systems and Equipment Certification. Available at http://www.rtca.org/store_product.asp?prodid=803.

Sahingoz, O. K. (2013) Mobile networking with UAVs: opportunities and challenges. 2013 International Conference on Unmanned Aircraft Systems (ICUAS), 933–41. doi: 10.1109/ICUAS.2013.6564779. IEEE.

Sarkar, S. K., Basavaraju, T. G., and Puttamadappa, C. (2008) *Ad Hoc Mobile Wireless Networks: Principles, Protocols and Applications*. 1st edn, Auerbach Publications, Boca Raton, FL.

Sastry, A. S., Sulthana, S., and Vagdevi, S. (2013) Security threats in wireless sensor networks in each layer. *International Journal of Advanced Networking and Applications*, **4**(4), 1657.

Sen, J. (2012) Security in wireless sensor networks. *Wireless Sensor Networks: Current Status and Future Trends* CRC Press, Boca Raton, FL, p. 407.

Stark, B., Stevenson, B., and Chen, Y.Q. (2013) ADS-B for small unmanned aerial systems: case study and regulatory practices. 2013 International Conference on Unmanned Aircraft Systems (ICUAS), 152–59. doi: 10.1109/ICUAS.2013.6564684.

UK Civil Aviation Authority. (2012) Unmanned Aircraft System Operations in UK Airspace -- Guidance (CAP722). UK.

Ungurean, I., Gaitan, N.-C., and Gaitan, V. G. (2014) An IoT architecture for things from industrial environment. 2014 10th International Conference on Communications (COMM), 1–4. doi: 10.1109/ICComm.2014.6866713. IEEE.

US Army. (2010) Unmanned Aircraft Systems Roadmap 2010–2035. Alabama, USA.

Wallgren, L., Raza, S.. and Voigt, T. (2013) Routing attacks and countermeasures in the RPL-based Internet of Things. *International Journal of Distributed Sensor Networks*, **9**. doi: https://doi.org/10.1155/2013/794326.

Whitmore, A., Agarwal, A., and Da Xu, L. (2015) The Internet of Things—a survey of topics and trends. *Information Systems Frontiers*, **17**(2), 261–74. doi: 10.1007/s10796-014-9489-2.

Wu, M., Lu, T. J., Ling, F. Y., Sun, J., and Du, H. Y. (2010) Research on the architecture of Internet of Things. ICACTE 2010, 2010 3rd International Conference on Advanced Computer Theory and Engineering, Proceedings, 5, 484–87. doi: 10.1109/ICACTE.2010.5579493.

UAS Yearbook (2011) *Unmanned Aircraft Systems: The Global Perspective 2011/ 2012*. Blyenburg & Co, pp. 1709–1967.

Part V

Relevant Sample Applications

20

An Internet of Things Approach to "Read" the Emotion of Children with Autism Spectrum Disorder

Tiffany Y. Tang and Pinata Winoto

Wenzhou-Kean Autism Research Network, Assistive Technology Research and Development Center, Department of Computer Science, Wenzhou-Kean University, Zhejiang Province, China

20.1 Introduction

Following the discussion on the contributions of Internet of Things (IoT) in health science and management in previous chapters, this chapter describes a prototype of IoT system designed specifically to help children with autism spectrum disorder (ASD) who are notoriously known to lack some social skills. Specifically, the application is designed to help both users with or without ASD to recognize their emotion in order to facilitate social interaction. Since the effectiveness of the prototype has not been clinically tested, readers are advised to consider it as another future IoT application in health and learning management. In this chapter, we describe the design and development of such an IoT prototype.

During a social interaction, it is important for all participants to interpret correctly the emotional nature of the message being communicated in order to avoid misunderstanding. Prior researches in the past decades have pointed out the lack of emotion recognition skills and social interaction skills among children and adults with ASD and the possible causal link between the former and the latter (Harms et al., 2010, Baron-Cohen et al., 1993). Since all participants in a social interaction should be held accountable for correctly interpreting other's emotional state, the ability of neurotypical (NT) individuals to recognize the emotion of children with ASD should also be evaluated. A small number of recent works began to study these issues by empirically evaluating the emotion expressiveness of individuals with ASD (Brewer et al., 2016; Macdonald et al., 1989; Faso et al., 2015; Stagg et al., 2014). However, the research on the ability of NT individuals in "reading" the emotions of those with ASD is still rare. The study involving advanced technology is even rarer, appealing, and yet seems

Internet of Things A to Z: Technologies and Applications, First Edition. Edited by Qusay F. Hassan.
© 2018 by The Institute of Electrical and Electronics Engineers, Inc. Published 2018 by John Wiley & Sons, Inc.

to be more feasible with the overwhelming focus on and success of sensing and wearable technologies. In particular, would it be possible for a collection of sensors and wearable devices acting as "eyes" and "ears" to collectively "label" the emotion of individuals situated in a social environment engaging in natural interactions? Drawn from earlier studies on psychology that have successfully linked emotions with expressive body movements (Boone and Cunningham, 1998; De Meijer, 1989), researchers have advanced our understandings of discern emotion of NT individuals from such multimodal data, including facial expression, hand gestures, body movements, and so on (Camurri et al., 2003; Kapur et al., 2005; Gunes and Piccardi, 2007; Robinson, 2014). However, there is a lack of work to assist NT individuals to understand the emotional states of those with ASD under natural social interactions (Tang, 2016), which motivates this study.

In this chapter, we report our early attempt to construct an IoT-based natural play environment designed specifically to "read" the emotion of children with ASD. The rationale behind this play environment is that the behavioral data including children's body movements and hand gestures provide rich data for an emotion-learning algorithm (Mitchell, 2009), which is coined as an area called computational sensing (Rehg et al., 2014). In order to capture valid behavioral data, a naturalistic IoT environment was created that contains embedded sensors, toys (e.g., LEGO® toys), and other objects, combined with the facial and behavioral data captured by smartwatch, IP camera, and Kinect™ sensor to "generate" emotion labels in the provided play environment where ASD children and other NT individuals interact (Figure 20.1).

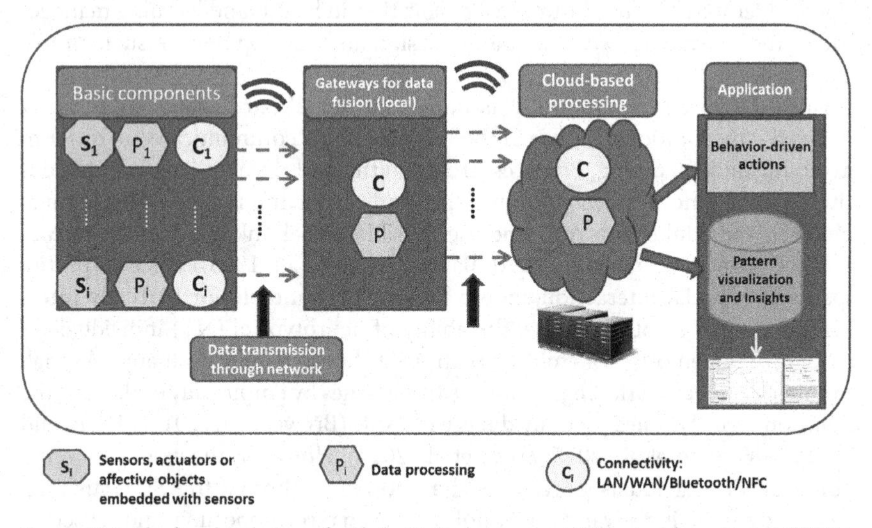

Figure 20.1 The box-level view of a typical IoT environment. (Adapted from Karimi (2013).)

20.2 Background

In this section, some background information on current research and development, as well as the challenges of technology-based intervention on autism spectrum disorder in China will be presented.

20.2.1 Current Approaches of Technology-Based Intervention on Autism Spectrum Disorder in China[1]

Thanks to the numerous impressive works on technology-based intervention (TI) (among many, Carter and Hyde, 2015; Beykikhoshk et al., 2014; Hong et al., 2015, Seo et al., 2015, Tartaro et al., 2015; Tang et al., 2015), significant positive outcomes have been achieved medically and clinically.

However, compared to the plenty of research efforts on the design and evaluation of TI on ASD in the United States and Europe, there are a few recently published works (Wang, 2015; Tang, 2016; Tang et al., 2015, 2017a, 2017b; Tang and Flatla, 2016; Winoto et al., 2016a, 2016b; Winoto, 2016) in Eastern Asia, but compared with its Western counterpart, the progress still lag far behind (Sun et al., 2015). Until now, the number of ASD population (both adults and children) in China is still undetermined (Sun et al., 2015). Indeed, very few researches had been published regarding the assessment, diagnosis, and early intervention strategies for children with ASD in China until recently (Sun et al., 2015).

20.2.2 The Challenges of Technology-Based Intervention on Autism Spectrum Disorder in China

As discussed earlier, the special education system in China is largely different from that in developed countries such as the United States in that the majority of children with autism would end up going to private educational centers to receive education. There are also a number of government-funded special education schools in each province or city, but due to the extremely limited resources, the majority of these special education schools are largely inaccessible to these children. In addition, there is a huge gap between the available number of teachers who are trained to deal with children with autism and the needed number.

1 The focus of this chapter is on the research and practices in mainland China (excluding Taiwan and Hong Kong) where the diagnosis, assessment and technology-based intervention have been far lag behind its neighboring countries and regions. For a more complete discussion of the current practices and research progress in mainland China, readers can refer to (Tang and Flatla, 2016).

Developing an affordable, portable, and personalized application that can be delivered at home has become time-pressing. Familial, social, and cultural factors could affect the participation, acceptability, and even outcomes of therapeutic approaches (Zwaigenbaum et al., 2015). Therefore, to include socially and culturally diverse populations in intervention, research is essential so as to enhance and deepen our understanding of how technology can be accepted, adapted, and deemed helpful across populations of individuals with ASD (Tang and Flatla, 2016).

20.3 Related Work

Social interaction is inherently bidirectional (Halberstadt et al., 2001), requiring individuals to recognize each other's emotion and intent that is vital for their action and behavior (Tang et al., 2017a, 2017b). In this section, previous works related to emotion recognition abilities and emotion expressivity of individuals with ASD will be discussed.

20.3.1 Emotion Recognition in Autism Spectrum Disorder

Individuals with ASD generally have a typical or delayed emotion processing (World Health Organization, 1993). Most works in this area focus on examining their abilities to perceive others' emotions via their static facial expressions (Harms et al., 2010; Baron-Cohen et al., 1993), which is typically attributed to anomalies in facial expressions (Simmons et al., 2009). However, these empirical studies have yielded mixed and even some contradictory outcomes: For example, it is argued that previous research might have underestimated the emotion recognition abilities via face among children with ASD (Peterson et al., 2015); additionally, it is revealed that ASD children are more skilled in recognizing emotion via body movements (Peterson et al., 2015). Others have failed to find any significant differences between NT and ASD groups involving basic emotion recognition tasks (Baron-Cohen et al., 1993, Baron-Cohen et al., 1997; Harms et al., 2010).

20.3.2 Emotion Expressiveness of Individuals with Autism Spectrum Disorder

Unlike the many research to study the emotion recognition ability of individuals with ASD, very few empirical studies have been reported on their emotion expressiveness (Brewer et al., 2016; Macdonald et al., 1989; Faso et al., 2015; Stagg et al., 2014). Emotion expressiveness includes the abilities to understand, mimic, and pose various emotions; of these, the first requires mental understanding of emotion and its communicative value, and the last two require facial muscle movement skills and proper feedback. Prior researches showed that children with ASD have reduced facial muscle movement during playing (Czapinski and Bryson, 2003), "atypical" looks during emotional storytelling (Grossman et al., 2013), and decreased proprioception awareness levels toward their own facial muscle movements (Weimer et al., 2001), all of which might decrease their emotion expressive abilities (Brewer et al., 2016). Taken together, it might reduce the quality of their social interaction, and negatively affect the participating NT individuals.

20.3.3 Emotion Recognition by Neuro-Typical Individuals

To date, very few literatures have reported empirical studies or technology-based design for helping NT or ASD individuals to recognize the emotions of individuals with ASD. For example, Park et al. (2012) proposed a system framework for teaching children with ASD to identify their own emotion through body language, without implementing it. A recent study discussed an early pilot study of utilizing a portable motion sensor to continuously capture the facial landmark data of children with ASD when they are watching a video; the temporal facial data will then be used to automatically generate emotion labels to inform NT individuals so as to facilitate the social interaction between NT individuals and those with ASD (Tang et al., 2017a, Tang et al., 2017b). Compared with an overwhelming number of previous studies probing the impairments of emotion recognition among individuals with ASD and implementing various computerized emotion-recognition training application for individuals with ASD, applications such as those proposed in Tang et al. (2017a, 2017b) that can assist them in informing others (especially those NT individuals) of their emotion are highly desired (Virnes et al., 2015; Tang et al., 2017a, 2017b; Tang, 2016).

20.3.4 Affective Computing, Multisensory Data Collection in Naturalistic Settings, and Ubiquitous Affective Objects

20.3.4.1 Naturalistic Settings and Ubiquitous Affective Objects

Thanks to more affordable sensory and wearable technologies, impressive improvements have been made in affective computing (Williams et al., 2015). Physiological data can be obtained easily from various wearable affect monitors (e.g., Fitbit Surge) and affordable sensors, including pressure sensors, capacitive sensors, and GSR; user's environmental information can also be collected using other sensors, such as ambient light sensors, temperature meter, PIR infrared sensor, and so on, which could provide relevant contextual information to predict users' emotion and eventually to help in mediating their effects (Williams et al., 2015). Ertin et al. (2011) designed a multisensor suit (a total of six sensors) to continuously measure a user's stress level where both body and ambient temperature data were obtained to monitor the body thermoregulatory and nervous system activation. These raw sensory data can then be transmitted via Bluetooth to a central system consisting of an algorithm to compute user emotions. Mitchell (2009) argued that various speakers' information such as their intonation, physical distance, body language, and other upper body movements can be used to characterize their interpersonal interactions.

Affective objects refer to ". . . any physical object which has the ability to sense emotional data from a person, map that information to an abstract form of expression and communicate that information expressively, either back to the subject herself or to another person" (Scheirer and Picard, 2000). This is particularly important especially for people with physical or learning disabilities (Williams et al., 2015). For the visually impaired who cannot identify emotions of others, a glove (named as *VibroGlove*) with vibrations can convey that information to the wearer (Krishna et al., 2010). Williams et al. (2015) go even further by designing a fashionable actuator-based scarf that can be used to sense the wearers' emotional state using attached sensors and then helping them to mitigate and convey their emotions to others via some actuators (e.g., LEDs). However, in the initial user testing (including one with high functioning autism), regarding whether to publicly share the user's emotion, all participants, including the one with ASD, strongly disapprove the design. As for the acceptability of group emotion broadcast, all participants are more willing despite overall reluctance except for the person with ASD and a visually impaired female participant. The results showed a strong privacy concern on broadcasting user emotion either at individual or group levels. In addition, the representation and evaluation of emotions are simplified in the study, for instance, low body temperature is associated with both "stressed" and "excited." However, ambient temperature might also have a significant impact on human being's physiological responses to daily physical activity (Tyka et al., 2009).

20.3.4.2 Sensing the Emotion from Behavioral Data Analysis

Some earlier attempts have focused on associating a specific emotion type with a set of behavioral data. Pollick et al. (2001) showed that different qualities of body movements can effectively be associated with distinct emotions. Castellano et al. (2007) proposed an emotion recognition technique based on body movement and gesture expressivity analysis by which emotional states such as anger, joy, pleasure, and sadness can be learned. Rehg et al. (2014) pointed out the drawbacks of current methods for acquiring social and communication behavioral data, and thus proposed the adoption of a so-called behavioral imaging to efficiently and effectively collect multimodal behavioral data through audio, video, and wearable sensing. Since human behaviors are multimodal in nature, these behavioral cues might provide rich information to construct user's behavior modeling. Such physiological data such as electrodermal, respiratory, and cardiovascular ones can shed light on the quality of social interactive behaviors that are not typically visible or observable to a therapist directly.

Similar to Rehg et al., 2014, Robinson (2014) suggested that emotional cues can be automatically learned from facial expressions, tone of voice, body postures, and gestures, which, if appropriately broadcasted, can assist individuals with ASD overcome their notoriously known difficulty of being understood by others.

20.3.5 The Internet of Things in Monitoring and Tracking Individuals for ASD Intervention

Although IoT has become increasingly popular in the health care industry, few published research has been reported for its adoption in ASD till recently. Sula et al. (2013) constructed a SmartBox featuring vibrator, light, smell, and sound control with an aim to provide a calming and motivating environment for autistic children when they are engaging in various learning activities. For example, the room light will be changed to adapt to the child's visual reading preferences (reading mode); a smell control to maintain pleasant room smell; vibrating the chair or bed to keep the child relax and calm. Children were asked to wear wireless body sensors to detect their body and hand movement as well.

Karimi (2013) discussed an IoT framework for remote emotive computing where such physiological variables and states were discussed for emotion data retrieval; these variables include muscle relaxation and contraction (via a pressure sensor); heart rate variability (via an on-chip two-electrode ECG), sweat (via a capacitive sensor), attitude (via an accelerometer to monitor wearer's body), and hand movements. These data are suggested to be measured in a natural environment where contextual awareness can be ensured and made meaningful for software designers to conduct behavioral analysis in the application. When an object is embedded with such sensors, it becomes affective object (Scheirer and Picard, 2000).

In summary, although most of these previous studies have, to a certain extent, effectively demonstrated the great potentials of probing the behavioral pattern of both individuals with ASD and NT individuals with more unobtrusive and efficient means, little research has been done on a wider adoption of various kinds of embedded sensors on objects or toys and on-body sensors, and how these sensing data can be combined with facial expression analysis, which motivates the research here. Figure 20.1 illustrates the functional box-level view of an IoT environment.

In the following sections, the system design and some initial analysis on the potentials of such an environment will be presented.

20.4 The Internet of Things Environment for Emotion Recognition

Driven by previous studies, an integrated IoT platform was developed and discussed in this section.

20.4.1 System Background and Architecture

To solve the privacy issues, this model is primarily designed for activities at home, school, or medical centers; and it focuses on embedding some sensors into "objects" children would play with (e.g., toys). The goal is to sense their emotion indirectly or noninvasively because individuals with ASD tend not to like wearable objects (Williams et al., 2015). Selective sensors are embedded into children's natural play environments to obtain their routine behavioral data combined with other information retrieved from monitoring devices such as Kinect and web cams to recognize the children's emotional state so as to inform their teachers, parents, or therapists.

Unlike the work by Sula et al. (2013) that requires the autistic children to wear body sensors, the current IoT system did not consider it due to observations from the numerous field testing in autistic educational centers and the feedback from parents and special education teachers. These wearable sensors are considered more intrusive to autistic children.

Figure 20.2 shows the overall system architecture, and Table 20.1 lists the emotion-related physiological data and emotion-sensitive environmental data recorded and stored in the system that follows the recommendation for remote-emotive computing in Karimi (2013). Two types of sensors have been designed to obtain information either from the human being or from the ambient environment (Figure 20.2).

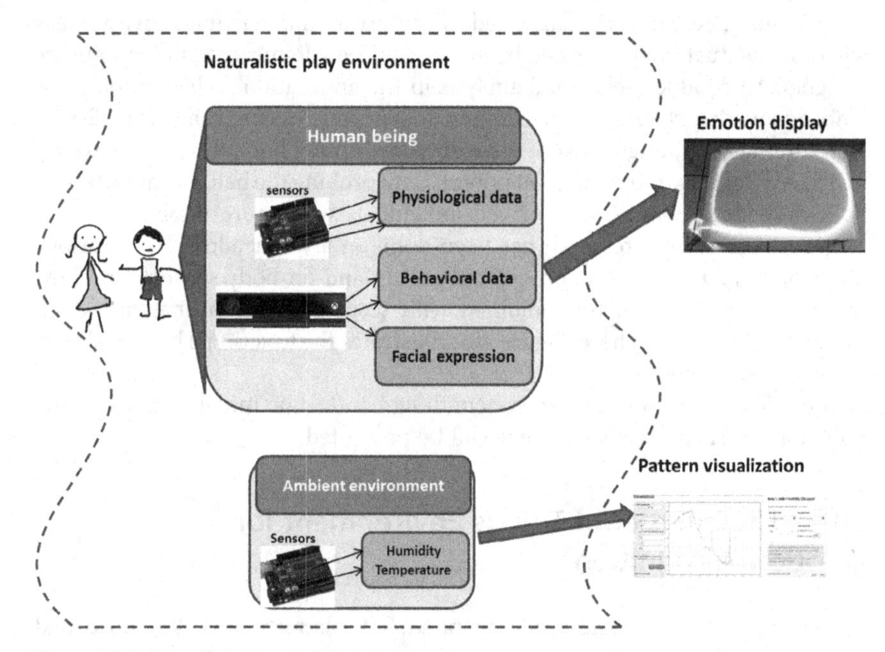

Figure 20.2 Overall system architecture.

Table 20.1 Emotion-related physiological data and emotion-sensitive environmental data in an IoT environment for emotion recognition.

Emotion-related physiological data	Sensors	Current IoT environment
Muscle relaxation	Pressure sensor	√
Muscle contraction	Pressure sensor	√
Sweat	Capacitive sensor	√
Body movement	Accelerometer	×
Heart rate variability	Wearable affect monitors	×
Emotion-related behavioral data	**Sensors**	
Attitude via body, head, and hand movements	Motion capture sensors such as RealSense™, Kinect, Leap Motion™, and Myo™ Gesture Control Armband	√
Facial expression	Motion capture sensors such as RealSense™, Kinect	√
Ambient factors contributing to emotion	**Sensors**	
Temperature	Thermometer	√
Humidity	Barometer	√
Light	Light sensor	×

In the next section, the details of the system setup, running environment, and report on some initial findings will be presented here.

20.4.2 The Naturalistic Play Environment

In our current design, there are three types of measurements used concurrently to sense users' behavioral patterns and individual and group emotions (see Table 20.1):

a) *Physiological (individual) measurement.* heart rate and perspiration (obtained via a set of Microsoft Band 2 worn by the target user)
b) *Behavioral (individual) measurement.* upper-body movements (including head and hands), gestures and motions (obtained via pressure and touch sensors), and facial expression (will be included in the future system design)
c) *Sociometers.* embedded sensors and affordable depth and RGB-B sensors (e.g., two sets of Kinect V2 sensors).

Figure 20.3 An example of a testing room consisting of two sets of Kinect sensors, a humidity + temperature sensor, two foam-like drawings equipped with capacitive sensors, a 1 m^2 synthetic grass equipped with pressure sensor (max. 40 kg), and some LEGO bricks.

Figure 20.3 illustrates the initial setup of playing environment. An IP camera was installed (above the play tables) to capture the play interactions (Figure 20.4). Figure 20.5 demonstrates a play moment captured by the IP camera; as can be seen, a player can touch the picture of a car where two capacitive sensors were installed behind its wheels (Figures 20.6 and 20.7). The collected touch-force data will be stored in the Cloud for later analysis. Currently, our design goal is to use the sensors to obtain the correlation between the level of touch force and the player's mood.

20.4.3 Sensors and Sensor Fusion

20.4.3.1 Hardware Design on Emotion and Actuation

Motivated by prior work on the link between various emotional state and biological and environmental triggers (Williams et al., 2015; Robinson, 2014, Rehg et al., 2014), we embedded a number of sensors for the purpose of collecting users' behavioral, physiological, and environmental data. And similar to the work described in Williams et al. (2015) and Tang et al. (2015), the sensors were employed in the Arduino platform due to a wide variety of compatible and affordable sensors available on the market. Table 20.2 lists the sensors used in this work. In addition, there is also a plan to make use of wearable devices to obtain users' physiological data (Microsoft Band 2).

Figure 20.4 An IP camera has been installed to capture the interactions between players and the "affective" objects.

Figure 20.5 An image captured by the IP camera shown in Figure 20.4.

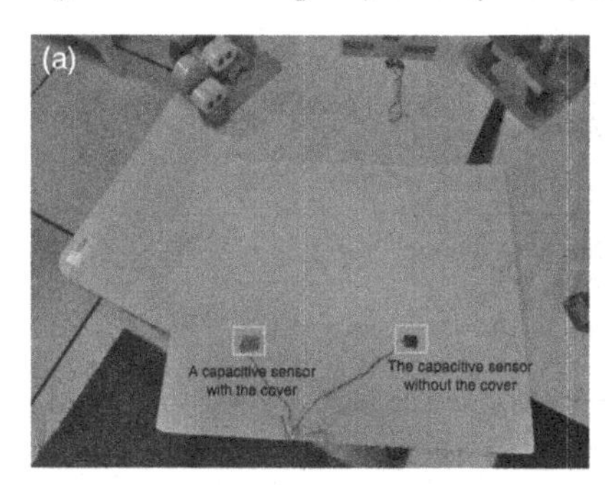

Figure 20.6 Two capacitive sensors embedded in the car wheel (a). A Windows tablet is put to visually present the touch force at real-time (b).

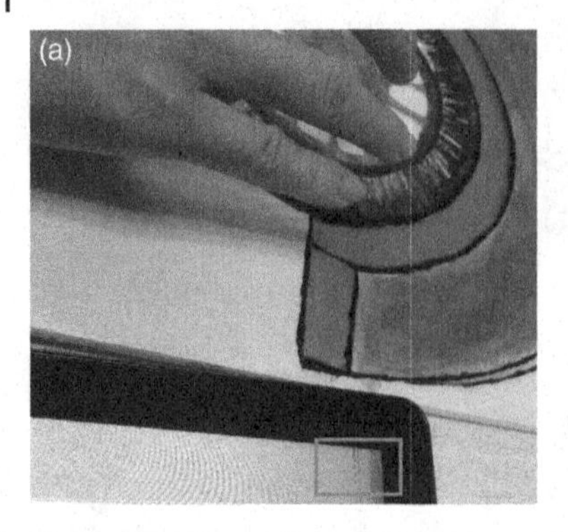

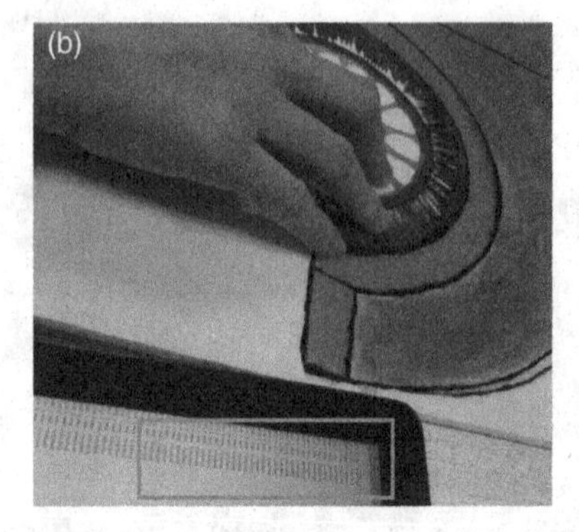

Figure 20.7 The testing of capacitive sensors, before (a) and after (b) touching the car wheel; the data bar is seen increasing on the screen.

Table 20.2 List of sensors used in current environment.

Function	Hardware components
Indoor temperature and humidity detection	(a) SEEED Grove – Temperature and Humidity sensor DHT22[a] (b) Arduino UNO R3 (a microcontroller board)[b] (c) Dragino Yun Shield V1.1[c] (*a strong* shield for Arduino Board for Internet connectivity and storage issue) (d) SEEED Base Shield V2 Grove for Arduino[d] (can be plugged into an Arduino as an expansion board)
Painting interaction	(a) SEEED Grove – I2C Touch Sensor[e] (b) Arduino Conductive Wire with Capacitive Sensing Library Supported (for sensing the electrical capacitance of the human body) (c) Arduino UNO R3 (d) Dragino Yun Shield V1.1 (e) SEEED Base Shield V2 Grove for Arduino
Step-on detection	(a) Pressure Sensor – 40 kg Pressure Sensor with HX711 AD Module[f] (b) Arduino UNO R3 (c) Dragino Yun Shield V1.1 (d) SEEED Base Shield V2 Grove for Arduino

a) http://wiki.seeed.cc/Grove-Temperature_and_Humidity_Sensor_Pro/
b) https://www.arduino.cc/en/main/arduinoBoardUno
c) http://www.dragino.com/products/yunshield/item/86-yun-shield.html
d) http://wiki.seeed.cc/Base_Shield_V2/
e) https://www.seeedstudio.com/Grove-I2C-Touch-Sensor-p-840.html
f) http://www.sunrom.com/p/loadcell-sensor-24-bit-adc-hx711

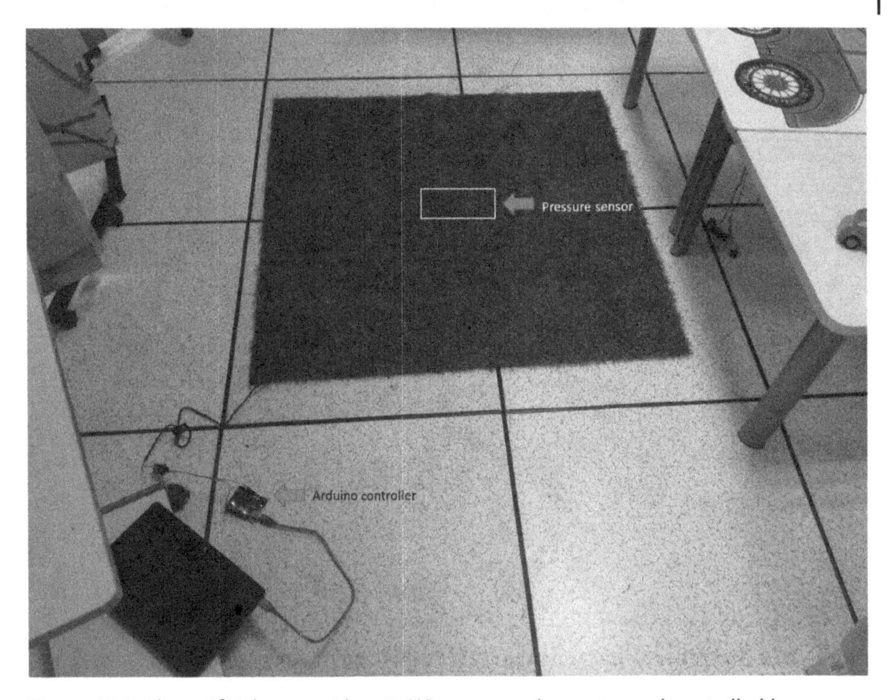

Figure 20.8 The artificial grass with a 40 KG sensor at the center and controlled by an Arduino board.

20.4.3.2 Pressure Sensors: Two Exemplary Play Scenarios

Two types of pressure sensors had been tested in the pilot studies: 40 KG pressure sensor and a capacitive one. The pressure sensor is controlled by an Arduino board (shown in Figure 20.8, at the center of the grass).

Each time the pressure sensor is activated, the self-calibration is conducted to correct the initial self-weight value. Intuitively, the initial pressure value is 0, and after the user steps on the grass (Figure 20.9), the value returned by the pressure sensor will be adjusted based on the level of pressure.

A second type of the pressure sensor has been embedded in the back of the two car wheels of a paper car toy (see Figure 20.6) that provides additional touch points for children during their play.

Figure 20.10 shows one of the typical play scenarios where a tablet is used to visualize the real-time temporal pressure values during the play. Figure 20.7 shows two testing outputs of the before and after the continuous pressures derived from users' interaction with the wheel.

Figure 20.9 The tester stepping on the grass when heading to the front.

Figure 20.10 The play scenario where the tester is observed triggering the capacitive sensors.

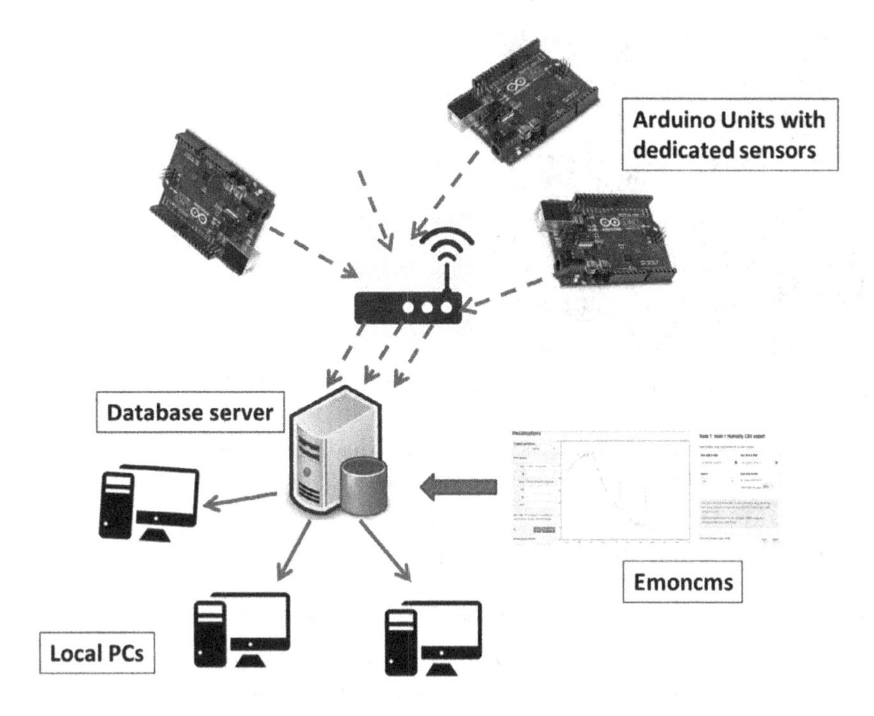

Figure 20.11 An illustration of the data transmission and network management architecture.

20.4.3.3 Data Management and Visualization for Indoor Temperature and Humidity Detection

For the purposes of managing real-time data from multiple sensors, a simple yet efficient data transmission and management system had been set up whose structure is shown in Figure 20.11.

Data capturing can be achieved by expanding a single Arduino board so that it consists of four modules: Arduino Uno, Base Shield, Yun Shield, and the sensors to serve their corresponding purposes (Figure 20.12 and Table 20.2). Data collected by the sensor will be stored at a local server via a Wi-Fi router; therefore, it will allow local computers to access and manipulate it accordingly.

Emoncms,[2] an open-source web-app for data processing and visualization of environmental data, had been deployed on the local database server. The application provides rich functions to store, visualize, and export the sensor data (Figure 20.13). It also supports JSON format making data transmission between the Arduino Yun unit and the database server more feasible and

2 https://emoncms.org/

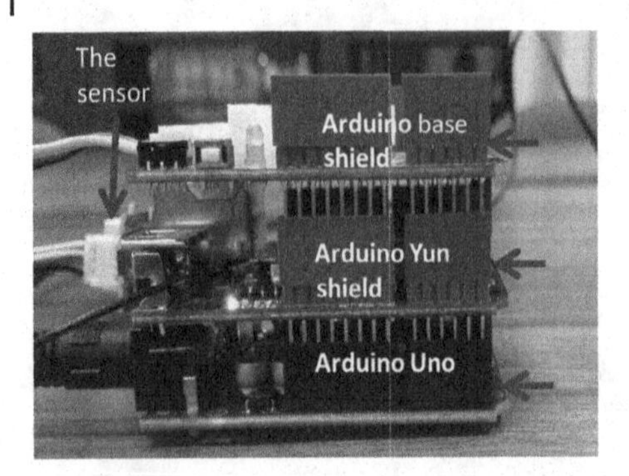

Figure 20.12 The Arduino unit at a closer look.

efficient. At present, the humidity sensor data had been tested with satisfying preliminary results (Figure 20.13).

20.5 The Study and Discussions

20.5.1 Emotion Recognition through Microsoft Kinect

So far, the capacitive sensors placed on paintings, and the pressure sensors on grass, as well as the HD Face SDK in Kinect 2.0 has been tested.

20.5.1.1 The Emotional Facial Action Coding System (EMFACS) and Kinect HD Face API

Facial Action Coding System (FACS) is a widely applied approach for objectively describing the facial animation (Ekman and Friesen, 1978). Six universal basic emotions of happiness, sadness, anger, surprise, disgust, and fear were proposed. Friesen and Ekman further proposed a methodology—Emotional Facial Action Coding System (EMFACS)—to improve the aforementioned universal emotions extraction strategy by directly analyzing the facial activities (Friesen and Ekman, 1983; Kanade et al., 2000; Mao et al., 2015) that was adopted in our current study.

In our preliminary experiment, only happy and sad emotions were studied.

Compared with the old Face Tracking API, HD Face API in Kinect V2 not only supports face *tracking* but also provides face *capture* capabilities, the latter providing high-definition raw facial data nodes for further manipulation (Kinect, 2016). Kinect V2 API natively offers 17 Action Units (AUs); all AUs are assigned with the numeric weight varying from 0 to 1, except for Jaw Slide Right, AU R4, and AU L4 whose values vary from −1 to 1.

It is known that several essential AUs in the FACS algorithm cannot be directly retrieved with the HD Face API; therefore, in our experiment, only AU L12, AU R12 (as Lip Corner Puller Left, Lip Corner Puller Right) are used to infer *happiness*, while AU L4, AU R4, AU L15, and AU R15 (as Left Eye Brow Lowerer, Right Eye Brow Lowerer, Lip Corner Depressor Left, and Lip Corner Depressor Right) are

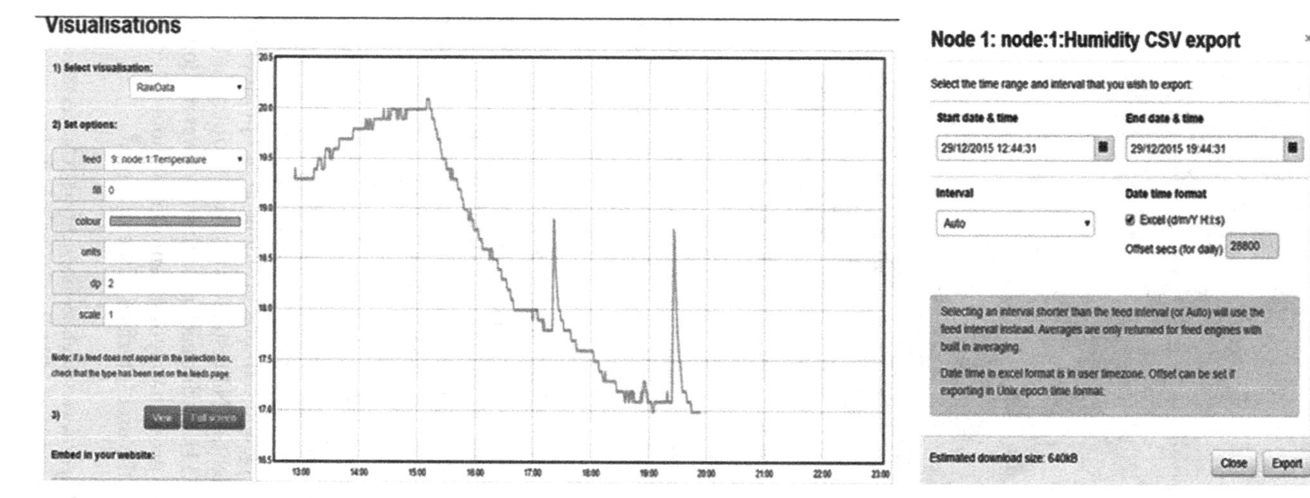

Figure 20.13 The Emoncms user interface where humidity data are visualized and shown.

Figure 20.14 The face tracking testing environment where a set of Kinect sensors is positioned to capture player's facial expressions.

applied to infer *sadness*. Additionally, some initial experiments were conducted to retrieve the proper thresholds for the different dedicated AUs and these updated thresholds were fine-tuned before the actual preliminary experiment.

20.5.1.2 Emotion Recognition: Preliminary Testing Results

Due to space limitations, an initial testing on happy and sad emotions will be reported here. During the preliminary experiment, both sitting and standing status were examined. In each type of experiment, an NT tester was instructed to play the toys laid out on the table freely where images of testing moment in the sitting status are captured by Kinect V2 sensor (Figures 20.14 and 20.15).

At present, the objective of our preliminary testing is merely to find the minimum weights used in the calculation of different AUs so that we could establish a relationship between the combinations of these AUs and the corresponding emotion labels. We tested both sitting and standing positions in our preliminary experiments. In each type of experiment, an NT tester was instructed to play the toys laid out on the table freely (Figure 20.11). During each play, the tester was randomly instructed (for testing purposes) to express a happy or sad emotion without inducing his/her moods. Figure 20.16 shows the user interface of the module.[3]

3 The testing videos can be watched at: https://www.youtube.com/watch?v=8Y17gj7zrx4&feature=youtu.be and https://www.youtube.com/watch?v =ZSyR1AEJg4s&feature=youtu.be

Figure 20.15 The face tracking testing environment where the tester can be seen sitting while playing the LEGO blocks.

20.5.2 Emotion Visualization and Broadcasting through Affective Object

After obtaining the emotion label, instead of simply returning the emotion label to the user, the emotion label was visualized and broadcasted through colored stripes (referred to as *affective objects* (Scheirer and Picard, 2000)) mainly due to the following reasons (Figure 20.2):

- Individuals with ASD tend to appear very different outwardly than what they might really be (through their behavioral and/or interactions in a social/individual environment) (Picard, 2009); thus, it is vital to inform others of their emotional state.

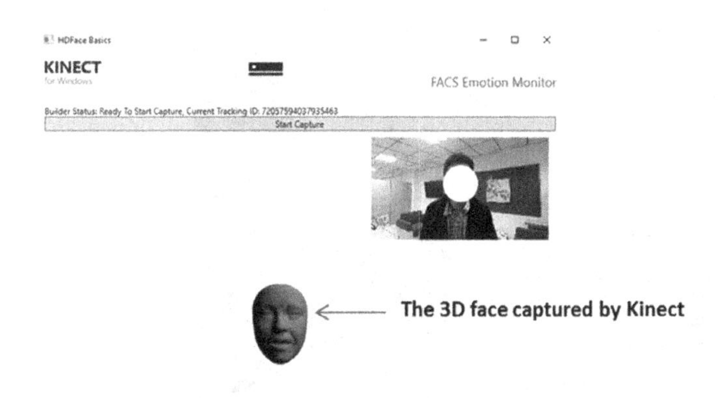

HAPPY

Figure 20.16 The Kinect HD facial capture moment.

- Previous studies have shown that many individuals prefer not to "broadcast" their emotion/mood to others due to the privacy concern except for individuals with ASD who have difficulties expressing their own emotions (Williams et al., 2015).
- Identifying and broadcasting the emotion states (of individuals either with ASD or TD) to others (again including both types) are vital for successful and effective communication (Williams et al., 2015; Picard, 2009)

In summary, instead of "announcing" the mood of the child to all the people in the environment, such emotion display through "light-up table" is considered to be more private—although it offers enough clues to other individuals to act during social interactions. The affective object designed in our experiment is a neo-pixel stripe embedded under a play table (Figure 20.17) that contains three layers: The neo-pixel stripe was positioned on top of an ordinary play table and covered with an acrylic board on which children can play with toys.

Our emotion visualization and broadcast module consists of two basic functional units: emotion tracking and responsive units, as shown in Figure 20.18.

The table is controlled through the Arduino Uno board that is linked to a local PC. Through the digital pin, the Arduino controller sends commands to the stripe to display a corresponding emotion color (see Table 20.3 for the displayed emotion color scheme used in the system (Nijdam, 2009)).

Figure 20.19 captures one of the testing moments when the player was happy in playing.

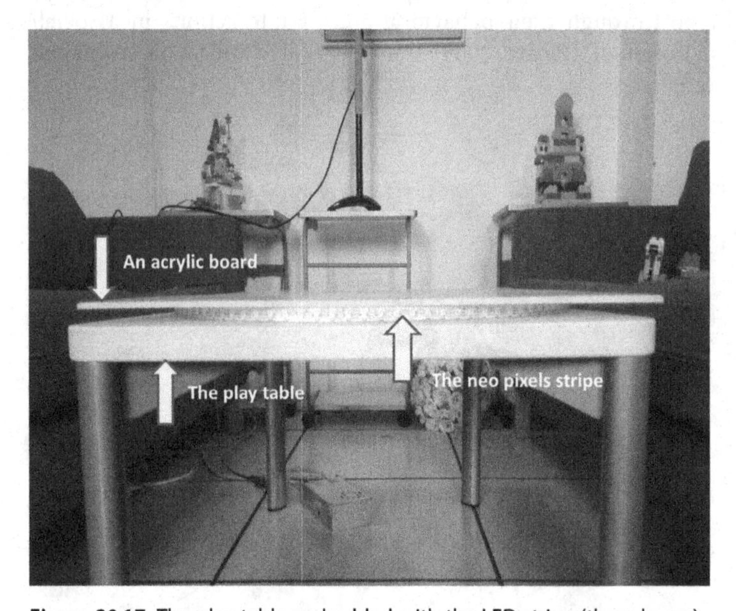

Figure 20.17 The play-table embedded with the LED stripe (three layers).

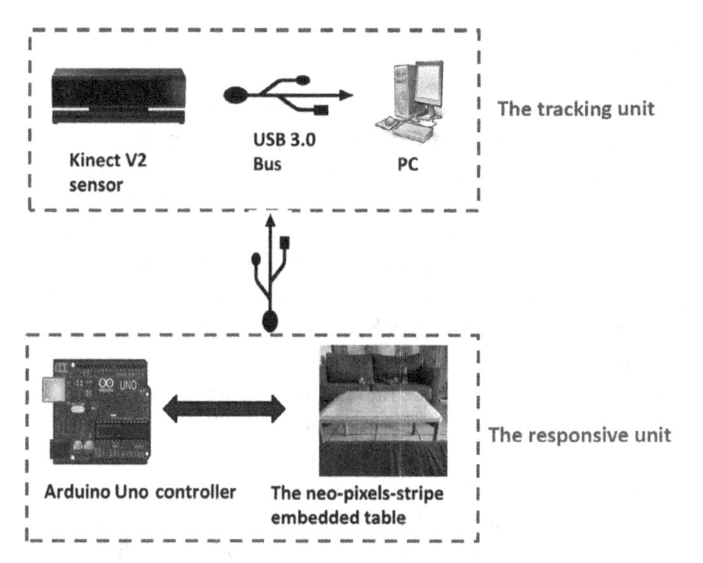

Figure 20.18 Two functional units in the emotion visualization and broadcast module.

Table 20.3 Correlation between each emotion and the displayed color.

Emotion	Corresponding color	RGB value
Happy	Orange red	(255,69,0)
Sad	Light blue	(34,219,182)
Fear	Purple	(128,0,128)
Surprise	Hot pink	(255,105,180)
Neutral	Light green	(144,238,144)

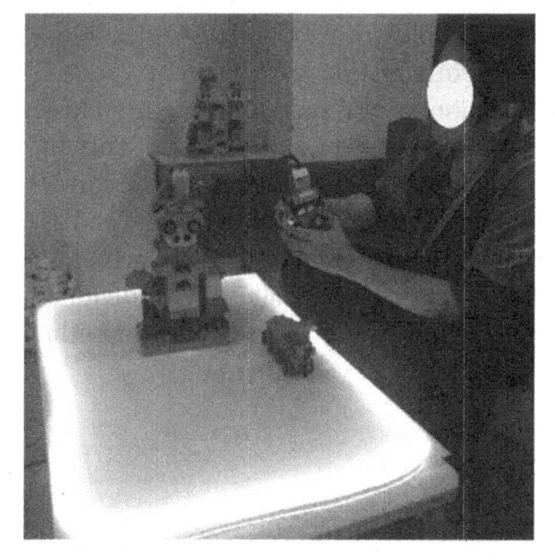

Figure 20.19 The happy moment broadcasted through the neo-pixel stripe-embedded play table.

20.6 Conclusions

In this chapter, the integration of a naturalistic multisensory IoT play environment has been presented; such an integration aims at capturing emotion-related behavioral, physiological, and ambient environment data for emotion recognition. The long-term goal of such a platform is to help NT individuals "read" the emotions of children with ASD. By adopting IoT, it may reduce direct intervention and interruption during children's playing moment; hence, children could enjoy their activities more naturally. The chapter presented the system design and offered some early insights derived from the conducted testing sessions on the potential of such environment.

At present, some sensors (including the facial data captured by the Kinect sensor) are being tested separately. Achieving data fusion (i.e., integrating data captured from the different sensors) for the purposes of generating meaningful emotional label is one of our future challenges. Meanwhile, we have also used and tested the Emotion API (part of the Microsoft Cognitive Services[4]) to label user's emotions: Although the application has been successfully experimented on NT individuals, how or whether it can be applied to children with ASD is unclear. As we have already mentioned, inferring the emotion of those with ASD is very challenging (Tang, 2016, Tang et al., 2017a, 2017b); therefore, it could be very useful if we could have a training data set obtained from children with ASD that includes their indicative emotional labels and behavioral patterns. One avenue to pursue is to collect emotional behaviors of children with ASD by inducing their emotion; another is to invite caregivers and parents living with them to rate the generated labels from natural events (e.g., playing moments). Either way, since the interest is to help caregivers and parents to better recognize their children's emotional behaviors in daily activities, the natural setting is very important here. Inferences based on other emotional behaviors (e.g., heart rate and perspiration) should be generalizable from NT individuals, since those characteristics are sympathetic nervous responses.

Moreover, there are two immediate questions that require our attention here: (1). How will the children's emotion be conveyed to others (e.g., synthetic speech, LEDs, etc.)? (2). Does the severity of disorder influence the children's emotional affect? If yes, how can this be incorporated into the existing models? Regardless of these two questions, current work has demonstrated an early effort to build an IoT-based natural play environment that would allow NT individuals to better recognize ASD children' emotion with the goal to open up another possibility to establish better social interactions between them.

4 https://www.microsoft.com/cognitive-services

Acknowledgments

The authors would like to acknowledge the financial support from the Academic Affairs Office for this project. Thanks also go to Relic Yongfu Wang for implementing the system when he was an undergraduate student at the University, and Odd Aonan Guan for his time and efforts in the experiments. We greatly acknowledge our editor, Dr. Qusay F. Hassan, for his insightful comments and efforts to help improve the readability of this chapter.

References

Baron-Cohen, S., Spitz, A., and Cross, P. (1993) Can children with autism recognize surprise? *Cognition and Emotion*, **7**, 507–516.

Baron-Cohen, S., Wheelwright, S., and Jolliffe, T. (1997) Is there a "language of the eyes"? Evidence from normal adults and adults with autism or Asperger syndrome. *Visual Cognition*, **4**, 311–332.

Beykikhoshk, A., Arandjelovic, O., Phung, D., Venkatesh, S., and Caelli, T. (2014) Data-mining twitter and the autism spectrum disorder: a pilot study. Proceedings of the ASONAM'2014, Calgary, Canada, IEEE Press, pp. 349–356.

Boone, R.T. and Cunningham, J.G. (1998) Children's decoding of emotion in expressive body movement: the development of cue attunement. *Developmental Psychology*, **34**, 1007–1016.

Brewer, R., Biotti, F., Catmur, C., Press, C., Happé, F., Cook, R., and Bird, G. (2016) Can neuro-typical individuals read autistic facial expressions? Atypical production of emotional facial expressions in autism spectrum disorders. *Autism Research*, **9**(2), 262–271.

Camurri, A., Lagerlof, I., and Volpe, G. (2003) Recognizing emotion from dance movement: comparison of spectator recognition and automated techniques. *International Journal of Human-Computer Studies*, **59**, 213–225.

Carter, E.J. and Hyde, J. (2015) Designing autism research for maximum impact. Proceedings of the ACM CHI'2015, Seoul, Korea, ACM Press, pp. 2801–2804.

Castellano, G., Villalba, S.D., and Camurri, A. (2007) Recognising human emotions from body movement and gesture dynamics. Proceedings of the ACII 2007, Lisbon, Portugal, LNCS 4738, pp. 71–82.

Czapinski, P. and Bryson, S.E. (2003) Reduced facial muscle movements in autism: evidence for dysfunction in the neuromuscular pathway? *Brain and Cognition*, **51**(2), 177–179.

De Meijer, M. (1989) The contribution of general features of body movement to the attribution of emotions. *Journal of Nonverbal Behavior*, **13**, 247–268.

Ekman, P. and Friesen, W.V. (1978) *Facial Action Coding System: A Technique for the Measurement of Facial Movement*, Consulting Psychologist Press, Palo Alto, CA.

Ertin, E., Raij, A., Stohs, N., al'Absi, M., Kumar, S., and Shah, S. (2011) AutoSense: unobtrusively wearable sensor suite for interring the onset, causality and consequences of stress in the field. Proceedings of the ACM SenSys'11, Seattle, WA, ACM Press, pp. 274–287.

Faso, D.J., Sasson, N.J., and Pinkham, A.E. (2015) Evaluating posed and evoked facial expressions of emotion from adults with autism spectrum disorder. *Journal of Autism and Developmental Disorders*, **45**(1), 1–15.

Friesen, W.V., and Ekman, P. (1983) EMFACS-7: Emotional Facial Action Coding System. Unpublished manual, University of California, California.

Grossman, R.B., Edelson, L.R., and Tager-Flusberg, H. (2013) Emotional facial and vocal expressions during story retelling by children and adolescents with high-functioning autism. *Journal of Speech, Language and Hearing Research*, **56**(3), 1035–1044.

Gunes, H. and Piccardi, M. (2007) Bi-modal emotion recognition from expressive face and body gestures. *Journal of Network and Computer Applications*, **30**(4), 1334–1345.

Halberstadt, A.G., Denham, S.A., and Dunsmore, J.C. (2001) Affective social competence. *Social Development*, **10**(1), 9–119.

Harms, M.B., Martin, A., and Wallace, G.L. (2010) Facial emotion recognition in autism spectrum disorders: a review of behavioral and neuroimaging studies. *Neuropsychology Review*, **20**(3), 290–322.

Hong, H., Gilbert, E., Abowd, G.D., and Arriaga, R.I. (2015) In-group questions and out-group answers: crowdsourcing daily living advice for individuals with autism. Proceedings of the ACM CHI'2015, Seoul, Korea, ACM Press, pp. 627–636.

Kanade, T., Cohn, J.F., and Tian, Y. (2000) Comprehensive database for facial expression analysis. Proceedings of the Fourth IEEE International Conference on Automatic Face and Gesture Recognition, Grenoble, France, IEEE Press, pp. 46–53.

Kapur, A., Kapur, A., Babul, N.V., Tzanetakis, G., and Driessen, P.F. (2005) Gesture-based affective computing on motion capture data. Proceedings of the ACII 2005, Beijing, China, pp. 1–7.

Karimi, K. (2013) The Role of Sensor Fusion and Remote Emotive Computing (REC) in the Internet of Things. Freescale Semiconductor, SENFEIOTLFWP REV 1, White Paper.

Kinect (2016) High definition face tracking. Available at https://msdn.microsoft.com/en-us/library/dn785525.aspx (retrieved October 9, 2016).

Krishna, S., Bala, S., McDaniel, T., McGuire, S., and Panchanathan, S. (2010) VibroGlove: an assistive technology aid for conveying facial expressions. Proceedings of the ACM CHI'2010, Atlanta, GA, pp. 3637–3642.

Macdonald, H., Rutter, M., Howlin, P., Rios, P., Conteur, A.L., Evered, C., and Folstein, S. (1989) Recognition and expression of emotional cues by autistic and normal adults. *Journal of Child Psychology, Psychiatry, Allied Disciplines*, **30**(6), 865–877.

Mao, Q.R., Pan, X.Y., Zhan, Y.Z., and Shen, X.J. (2015) Using Kinect for real-time emotion recognition via facial expressions. *Frontiers of Information Technology & Electronic Engineering*, **16**, 272–282.

Mitchell, T. (2009) Mining our reality. *Science*, **326**, 1644–1645.

Nijdam, N.A. (2009) *Mapping Emotion to Color*, University of Twente, The Netherlands.

Park, J.H., Abirached, B., and Zhang, Y. (2012) A framework for designing assistive technologies for teaching children with ASDs emotions. Proceedings of the ACM CHI'2012, Austin, TX, pp. 2423–2428.

Peterson, C.C., Slaughter, V., and Brownell, C. (2015) Children with autism spectrum disorder are skilled at reading emotion body language. *Journal of Experimental Child Psychology*, **139**, 35–50.

Picard, R. (2009) Future affective technology for autism and emotion communication. *Philosophical Transactions of the Royal Society B*, **364**, 3575–3584.

Pollick, F., Paterson, H., Bruderlin, A., and Sanford, A. (2001) Perceiving affect from arm movement. *Cognition*, **82**, 51–61.

Rehg, J.M., Rozga, A., Abowd, G.D., and Goodwin, M.S. (2014) Behavioral imaging and autism. *IEEE Pervasive Computing*, **13**(2), 84–87.

Robinson, P. (2014) Computation of Emotions. Proceedings of the ACM ICMI 2014, Istanbul, Turkey, pp. 409–410.

Scheirer, J. and Picard, R. (2000) Affective objects. MIT Media Laboratory Perceptual Computing Section, Technical Report No. 524.

Seo, J.H., Sungkajun, A., and Suh, J. (2015) Touchology: towards interactive plant design for children with autism and older adults in senior housing. Proceedings of the CHI 2015 Seoul, Korea, pp. 893–898.

Simmons, D.R., Robertson, A.E., McKay, L.S., Toal, E., McAleer, P., and Pollick, F.E. (2009) Vision in autism spectrum disorders. *Vision Research*, **49**(22), 2705–2739.

Stagg, S., Slavny, R., Hand, C., Cardoso, A., and Smith, P. (2014) Does facial expressivity count? How typically developing children respond initially to children with autism. *Autism*, **18**(6), 704–711.

Sula A., Spaho E., Matsuo K., Barolli L., Xhafa F., and Miho R. (2013) An IoT-based framework for supporting children with autism spectrum disorder, in Park, J., Barolli, L., Xhafa, F., Jeong, H.Y. (eds.), *Information Technology Convergence*. Lecture Notes in Electrical Engineering, vol. 253, Springer, Dordrecht, The Netherlands.

Sun, X., Allison, C., Auyeung, B., Zhang, Z., Matthews, F., Baron-Cohen, S., and Brayne, C. (2015) Validation of existing diagnosis of autism in mainland China using standardised diagnostic instruments. *Autism Research*, **9**(8), 1010–1017.

Tang, T. (2016) Helping neuro-typical individuals to "Read" the emotion of children with autism spectrum disorder: an Internet-of-Things Approach. Proceedings of the ACM IDC'2016, Manchester, UK, pp. 666–671.

Tang, T. and Flatla, D. (2016) Autism awareness and technology-based intervention research in china: the good, the bad, and the challenging. Proceedings of the Workshop on Autism and Technology: Beyond Assistance & Intervention, C in conjunction with the *ACM CHI'2016*, San Jose, CA.

Tang, T., Wang, Y. You, Y., Huang, Z., and Chen, P. (2015) Supporting collaborative play via an affordable touching + singing plant for children with autism in China. Proceedings of the ACM UbiComp/ISWC'15, Osaka, Japan, ACM Press, pp. 373–376..

Tang, T., Falzarano, M., and Morreale, P.A. (2017a) Assessment of the utility of gesture-based application for the engagement of Chinese children with autism. Universal Access in the Information Society, Special Issue on Information Society Skills. Is knowledge Accessible for All? Springer.

Tang, T., Winoto, P., and Chen, G.X. (2017b) Emotion recognition via face tracking with RealSense™ 3D camera for children with autism. Proceedings of the ACM IDC'2017, Stanford, CA, pp. 533–539.

Tartaro, A., Cassell, J., Ratz, C., Lira, J., and Nanclares-Nogués, V. (2015) Accessing peer social interaction: using authorable virtual peer technology as a component of a Group Social Skills Intervention Program. *ACM Transactions on Accessible Computing (TACCESS)*, **6**(1), 1–29.

Tyka, A., Palka, T., Tyka, A., Cison, T., and Szygula, Z. (2009) The influence of ambient temperature on power at anaerobic threshold determined based on blood lactate concentration and myoelectric signals. *International Journal of Occupational Medicine and Environmental Health*, **22**, 1–6.

Virnes, M., Kama, E., and Vellonen, V. (2015) Review of research on children with autism spectrum disorder and the use of technology. *Journal of Special Education Technology*, **30**(1), 13–27.

Wang, S. (2015) In China, the making of an app for autism. The Wall Street Journal. Available at http://blogs.wsj.com/chinarealtime/2015/05/19/in-china-the-making-of-an-app-for-autism/ (accessed October 17, 2016).

Weimer, A., Schatz, A., Lincoln, A., Ballantyne, A., and Trauner, D. (2001) "Motor" impairment in asperger syndrome: evidence for a deficit in proprioception. *Journal of Developmental & Behavioral Pediatrics*, **22**(2), 92–101.

Williams, M.A., Roseway, A., O'Dowd, C., Czerwinski, M., and Morris, M.R. (2015) SWARM: an actuated wearable for mediating affect. Proceedings of the TEI'2015, Stanford, CA, ACM Press, pp. 293–300.

Winoto, W. (2016) Reflections on the adoption of virtual reality-based application on word recognition for Chinese children with autism. Proceedings of the ACM IDC'2016, Manchester, UK, ACM Press, pp. 589–594.

Winoto, P., Chen, G.X., and Tang, T. (2016a) The development of a Kinect-based online socio-meter for users with social and communication skill impairments: a computational sensing approach. Proceedings of the IEEE International Conference on Knowledge Engineering and Applications (ICKEA'2016), Singapore, IEEE Press, pp. 139–143.

Winoto, P., Xu, N., and Zhu, A. (2016b) "Look to Remove": a virtual reality application on word learning for Chinese children with autism. Proceedings of the 2016 Human-Computer Interaction International Conference (HCII'2016), Toronto, Canada, Springer, pp. 257–264.

World Health Organization (1993) *International Statistical Classification of Diseases and Related Health Problems*, 10th edn (ICD-10), World Health Organization, Geneva.

Zwaigenbaum, L., Bauman, M.L., Choueiri, R., Fein, D., Kasari, C., Pierce, K., Stone, W.L., Yirmiya, N., Estes. A., Hansen, R.L., McPartland, J.C., Natowicz, M.R., Buie, T., Carter, A., Davis, P.A., Granpeesheh, D., Mailloux, Z., Newschaffer, C., Robins, D., Roley S.S., Wagner, S., and Wetherby, A. (2015) Early identification and interventions for autism spectrum disorder: executive summary. *Pediatrics*, **136** (Suppl. 1), S1–S9.

21

A Low-Cost IoT Framework for Landslide Prediction and Risk Communication

Pratik Chaturvedi,[1,2] Kamal Kishore Thakur,[3] Naresh Mali,[4] Venkata Uday Kala,[4] Sudhakar Kumar,[5] Srishti Yadav,[4] and Varun Dutt[5]

[1]*Applied Cognitive Science Laboratory, Indian Institute of Technology Mandi, Kamand, India*
[2]*Defence Terrain Research Laboratory, Defence Research and Development Organization, New Delhi, India*
[3]*Computer Science and Engineering Department, Thapar Institute of Engineering and Technology, Patiala, India*
[4]*School of Engineering, Indian Institute of Technology Mandi, Himachal Pradesh, India*
[5]*School of Computing and Electrical Engineering, Indian Institute of Technology Mandi, Himachal Pradesh, India*

21.1 Introduction

Landslides are uncertain geological events and they pose great dangers to life and infrastructure (Parkash, 2011). In India, especially in the Himalayan region, landslides are more frequent than any other geological phenomena causing more than 200 deaths and, on average, $82 million in damages to infrastructure yearly (Chaturvedi et al., 2017; Chaturvedi and Dutt, 2015). Because of large costs and many deaths due to landslides, there is a need to design and develop frameworks that monitor landslides and alert people before they occur. To be effective, those frameworks should possess the following features: sense soil properties and soil movement at landslide-prone sites; log-sensed data at a remote site via a cloud infrastructure; allow analyses of logged data; and alert people via mobile applications before landslides occur.

Several techniques have been proposed for monitoring landslides. Some of these techniques include the use of stereoscopic aerial photographs (Casson et al., 2003); use of satellite and unmanned aerial vehicle (UAV)-based remote sensing (Guzzetti et al., 2012; Niethammer et al., 2012); use of digital elevation models (DEM) from airborne laser altimetry data (McKean and Roering, 2004); and use of Brillouin Optical Time-Domain Reflectometry method (Dan et al., 2004). Although these techniques provide effective landslide monitoring technologies, they only scan the terrain's surface and they are expensive to use.

Internet of Things A to Z: Technologies and Applications, First Edition. Edited by Qusay F. Hassan.
© 2018 by The Institute of Electrical and Electronics Engineers, Inc. Published 2018 by John Wiley & Sons, Inc.

IoT frameworks provide a foundation for connecting sensors, actuators, and other smart technologies and thus they help to improve automation and control in various operations (Borgia, 2014). These IoT frameworks could provide alternate solutions for monitoring landslides. In the past few years, researchers have designed and developed certain IoT frameworks for monitoring landslides and alerting people (Arnhardt and Neussner, 2013; Aziz and Kamarulzaman, 2011; Ramesh, 2009). These IoT frameworks involve data collection, transmission, preprocessing, machine learning, decision-making, and information dissemination via mobile applications (Khan et al., 2012). Thus, these IoT frameworks go beyond the typical wireless sensor networks (WSNs), which consist of only sensors that collect data and transmit it wirelessly (Othman and Shazali, 2012). However, these frameworks use costly sensing components like geophones, pore pressure transducers, and tilt meters (Arnhardt and Neussner, 2013; Aziz and Kamarulzaman, 2011; Ramesh, 2009). The high cost of sensing landslides via existing IoT frameworks limits their large-scale deployment across several landslide-prone areas in the world. Thus, there is a need to design and develop IoT frameworks that are low in cost, yet efficient.

One way of reducing cost is by using microelectromechanical systems (MEMS)-based sensors (Gupta and Ahmad, 2007) as a part of existing landslide-sensing IoT frameworks. MEMS is a technology that uses advances in fabrication techniques to embed an electromechanical system on a single chip (Mehregany and Roy, 1999). In MEMS-based sensors, mechanical elements like gears and beams are added on top of electrical systems, where this electromechanical combination allows one to sense the physical world (Mehregany and Roy, 1999). IoT frameworks involving MEMS-based sensors are likely to be useful for monitoring landslides. However, sensors used for landslide monitoring in the real world need to be first calibrated and tested at the lab scale.

In this chapter, the design and development of a low-cost IoT framework for monitoring landslide is discussed. This framework involved the use of MEMS-based sensors for monitoring landslides at the lab scale. The proposed framework can monitor soil moisture and movement and generate alerts based on predefined thresholds.

21.2 Background

As mentioned earlier, several technologies have been used for monitoring landslides (Guzzetti et al., 2012; Niethammer et al., 2012; McKean and Roering, 2004; Dan et al., 2004). Although these technologies provide different methods for landslide monitoring, they can only scan terrain's surface and are expensive to use. For example, in the city of Portland, where DEMs were used for landslide monitoring, the mapping cost itself was between $400 and $600 per square mile (Weinstein, 2010). This high-mapping cost was a reason why

Portland was unable to afford a large-scale DEM implementation. Similarly, high costs and operating constraints have been associated with UAVs, which limits their usage (McKinnon, 2015).

The high cost of traditional landslide monitoring technologies calls for finding alternative technologies that are affordable and can look beneath the terrain's surface. MEMS-based sensors have been used across a wide range of applications involving healthcare, automotive, defense, and communication sectors (Crone, 2008). Recently, MEMS-based sensors have also drawn the attention of landslide researchers and IoT frameworks that use these sensors can provide promising low-cost solutions for monitoring landslides (Arnhardt and Neussner, 2013; Chaturvedi et al., 2017; Manconi and Giordan, 2016; Marciano et al., 2014; Martelloni et al., 2012). However, as the efficacy of the MEMS-based sensors has not been evaluated for landslide applications in the real world (Manconi and Giordan, 2016), more research is needed in designing and developing IoT frameworks that involve the use of these sensors. To design and develop such frameworks, it is important to first evaluate the capabilities of these sensors using controlled lab-scale simulations and then to test them in field applications. In the following section, the design and development of an IoT framework are presented where MEMS-based sensors were used at the lab scale.

21.3 System Design and Implementation

The design of the proposed landslide monitoring IoT framework consists of the following components: a sensing unit; a data-logging and thresholding unit; and an alert-generating unit (see Figure 21.1). In this section, a detailed description of each of these components is provided.

21.3.1 Sensing Unit

In the proposed framework (see Figure 21.1), MEMS-based accelerometers and moisture sensors sense soil movement (accelerations) and soil moisture, respectively. The accelerometer measures accelerations (rate of change of

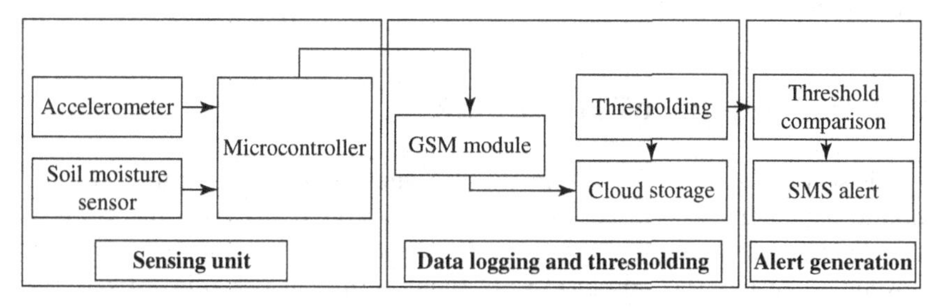

Figure 21.1 Design of the low-cost IoT framework for landslide monitoring.

velocity of an object) in three orthogonal directions (e.g., *X*, *Y*, and *Z*). When interfaced with a microcontroller, this sensor provides analog acceleration values. These analog units are converted to "m/s^2" units by using an appropriate calibration procedure. The soil moisture sensor uses the resistance property to measure water content in the soil surrounding its electrodes. Resistance is inversely proportional to soil moisture and output voltage. When the sensor is dry, a high value of resistance is recorded. This resistance value results in a low-output voltage. In contrast, when the sensor's electrodes are moist, a low value of resistance is recorded and this resistance results in a high-output voltage.

The analog values from the accelerometer and soil moisture sensors are fed to the analog pins of the microcontroller, which is directly interfaced with these sensors. The sensor values received by the microcontroller are sent to the cloud via a Global System for Mobile (GSM) module (Rahnema, 1993). GSM is used for data transmission rather than Wi-Fi because in the real world the sensing unit will be installed on a remote hill where a Wi-Fi signal will not be present. However, GSM (mobile) networks are readily available in hilly terrain, especially in the Himalayas in Himachal Pradesh, where future deployment of the framework is planned.

21.3.2 Data Logging and Thresholding Unit

The GSM module receives sensor values from the microcontroller and then it transmits them to a database on the cloud. This setup is used in controlled lab-scale experiments, where certain critical values (i.e., threshold) are determined for soil moisture and soil movement. These threshold values when breached would cause landslides.

21.3.3 Alert-Generating Unit

The alert-generating unit compares sensor values with their predefined thresholds. Once the sensor values for soil moisture or soil movement cross their threshold values, this unit generates SMS alerts concerning landslides on mobile phones. The alert-generating unit also has the provision of preregistering mobile numbers for sending SMS alerts.

21.4 Testing the IoT Framework

In this section, a lab-scale experiment involving the proposed framework for monitoring landslides is discussed. The main purpose of the experiment was to test the capability of the framework for measuring soil moisture and soil movement. Also, a second objective of the experiment was to investigate the threshold values for soil moisture and soil movement, which were likely to trigger landslides. In the experiment, a soil sample from a hill was packed on a ramp. Next, the moisture level was increased in the soil sample until a landslide was triggered on the ramp. This experiment served as a proof of concept for testing the suitability of MEMS-based sensors for landslide monitoring as well as for gauging threshold values of soil moisture and soil movement.

21.4.1 Methodology

21.4.1.1 Soil Characteristics

To investigate thresholds for soil moisture and soil movement via the framework, an experiment was performed on a ramp involving soil with certain physical properties. First, soil tests were carried out where certain soil properties were determined using the Indian Standard (IS) procedures.

These IS procedures included tests for specific gravity (IS 2720-1, 1983), particle size distribution (IS 2720-4, 1985), compaction characteristics (IS 2720-8, 1983), and direct shear test (IS 2720-13, 1986). Specific gravity is the ratio of the density of the soil sample to the density of water at a specified temperature.

The soil sample was further analyzed in terms of its percentage constituents. To do this analysis, a grain size analysis test was performed that gave percentages of gravel sizes (particle diameter > 4.75 mm), sand sizes (particle diameter between 0.075 and 4.75 mm), silt sizes (particle diameter between 0.075 and 0.0002 mm), and clay sizes (particle diameter < 0.0002 mm) in the soil sample. As shown in Table 21.1, the soil sample mostly consisted of sand and silt particles and the sample was classified as a poorly graded sand. The angles of friction were greater between the sand particles compared to that between the sand particles and the ramp material. Also, the cohesion between sand particles

Table 21.1 Physical characteristics of soil sample.

Property	Value for soil sample
Specific gravity(ratio)	2.58
Gravel size (%)	4
Sand size (%)	46
Silt size (%)	47
Clay size (%)	3
Soil classification	Poorly graded sand
Angle of internal friction between sand and sand, Φ (°)	28
Angle of internal friction between sand and ramp material, Φ (°)	5
Cohesion between sand and sand, C (kPa)	5
Cohesion between sand and ramp material, C (kPa)	3

was slightly greater compared to that between the sand particles and the ramp material. Overall, the sand particles bonded together much more compared to their bonding with the ramp material.

21.4.1.2 Lab-Scale Ramp Setup

A wooden ramp was prepared for performing an experiment with the framework (see Figure 21.2a). The dimensions of the ramp were as the following: length = 1 m; breadth = 0.3 m; and height = 0.3 m. In addition, the ramp was attached with supports on both sides so that it could be raised to a wide range of angles with the horizontal surface (between 1° and 90°; see Figure 21.2b). In the experiment, soil was packed in the ramp and moisture was added using a water sprinkler until soil moved and a landslide occurred.

As shown in Figure 21.2b, accelerometers and soil moisture sensors were placed at the base and at the top of a flexible pipe on the toe-end of the ramp. These sensors recorded soil movement and soil moisture values, respectively. One set of the accelerometer and soil moisture sensors were located at the base of the pipe and the second set of sensors were located at the top of the pipe. The pipe holding the two sensor sets was placed perpendicular to the ground level. The entire arrangement could sense soil movement and soil moisture and communicate the sensed values to the microcontroller. Subsequently, the sensor values received by the microcontroller were sent to a database via the cloud infrastructure and the GSM module.

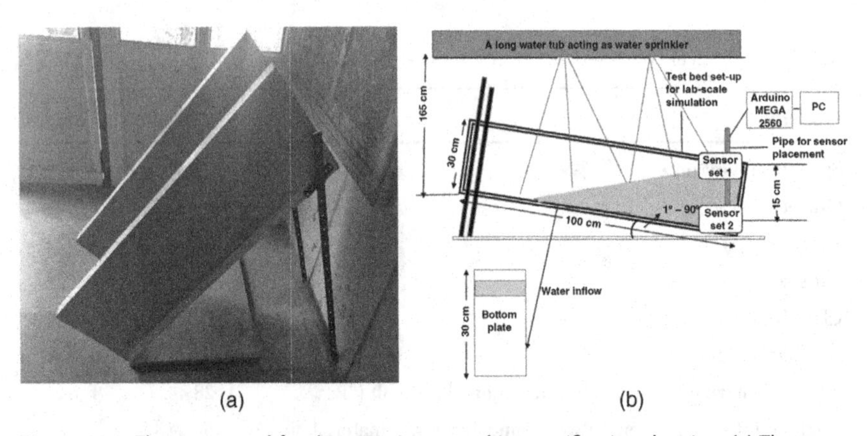

Figure 21.2 The ramp used for the experiment and its specification drawing. (a) The ramp elevated to an angle with the horizontal surface for packing a soil sample. (b) The specification drawing of the ramp showing different dimensions and placement of sensors at ramp's toe-end.

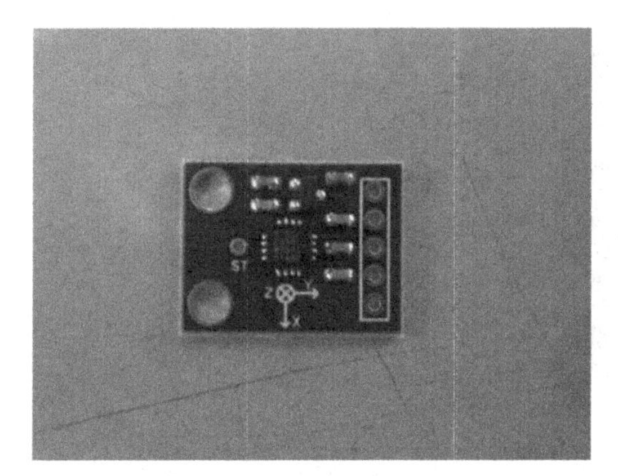

Figure 21.3 GY-61 accelerometer sensor used in the framework.

21.4.1.3 Selection of Components for IoT Framework

Two MEMS-based accelerometers and moisture sensors were placed at the base and at the top of a flexible pipe on the toe-end of the ramp (see Figure 21.2b). The specifications of the different items used in this framework are discussed as follows:

- *Accelerometer.* A GY-61 accelerometer module[1] is used in the framework (see Figure 21.3). This module measures triaxial accelerations in the range $\pm 3\,g$ (where $g = 9.8\ \text{m/s}^2$) across three orthogonal directions (see Figure 21.3). This sensor is a low-power device with an operating voltage between 1.8 and 3.6 V and an operating current of 350 μA. The accelerometer is calibrated by aligning each of its axis one by one with Earth's gravitational acceleration ($= 9.8\ \text{m/s}^2$). Thus, when the Z-axis is aligned with the Earth's gravitational acceleration, then the accelerometer gives $+1\,g$ acceleration in the Z-axis and $0\,g$ acceleration in the X- and Y-axes. For the accelerometer placed at the top of the pipe, A_x, A_y, and A_z refer to directions perpendicular to the base of the ramp (pointing upward), along the width of the ramp (from right to left), and sloping at an angle ϕ with the soil away from the ramp (ϕ is the angle that the ramp made with horizontal), respectively. Similarly, for the accelerometer at the base of the pipe, A_x, A_y, and A_z refer to directions perpendicular to the base of the ramp (pointing downward), along the width of the ramp (from left to right), and sloping at an angle ϕ with the soil toward the ramp, respectively.
- *Moisture Sensor.* For sensing the percentage of the soil moisture, a YL-69 module[2] is used in the framework (see Figure 21.4). The operating voltage for this sensor ranges between 3.3 and 5.0 V. Depending upon the level of moisture content in the close vicinity of the sensor, this sensor gives an analog value between 0 and 1023. Analog readings for soil moisture sensors in the dry state (0% moisture) and wet state (100% moisture) are 395 and 1023,

---

1 https://www.sparkfun.com/datasheets/Components/SMD/adxl335.pdf
2 https://randomnerdtutorials.com/guide-for-soil-moisture-sensor-yl-69-or-hl-69-with-the-arduino/

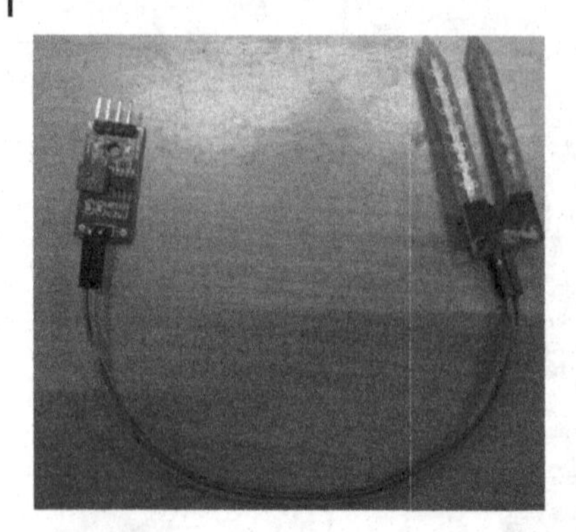

Figure 21.4 YL-69 soil moisture sensor used in the framework.

respectively. To compute the percentage moisture corresponding to an analogue sensor value, the following equation is applied:

$$\text{Moisture}\,precentage = (1023 - \text{Analog}\,value)/(1023 - 395) \times 100 \quad (1)$$

- *Microcontroller.* For reading and processing sensor values, an open-source microcontroller board, that is, Arduino MEGA 2560[3] is used (Barrett, 2013; see Figure 21.5). This microcontroller has 16 analog pins and 54 digital pins. The operating voltage and current of this microcontroller are 5 V and 500 mA, respectively.
- *GSM Module.* To transmit the sensor data to the cloud, a SIM900A GSM module[4] is used (see Figure 21.6). The GSM module is connected to the

Figure 21.5 Arduino MEGA 2560 module used in the framework.

3 https://www.arduino.cc/en/Main/arduinoBoardMega2560
4 http://www.mt-system.ru/sites/default/files/docs/simcom/docs/sim900ds/sim900-ds_hardware_design_v1.00.pdf

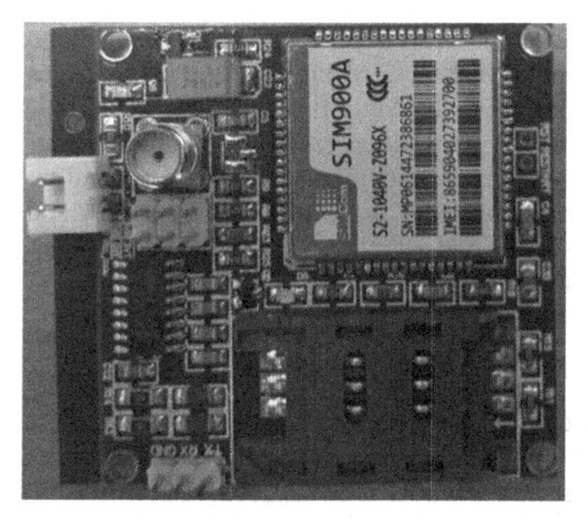

Figure 21.6 SIM900A GSM module used in the framework.

microcontroller and it can operate on four different frequencies, namely, 850, 900, 1800, and 1900 MHz. Since the GSM module consumes up to 2 A current, an external power supply is used with an operating voltage and current of 5 V and 2 A, respectively. A SIM card is used in the GSM module for establishing a wireless connection between the microcontroller and the cloud.

In Figure 21.7, a GY-61 accelerometer, a YL-69 soil-moisture sensor, and a SIM900A GSM module are interfaced with an Arduino MEGA microcontroller board. Since GY-61 module has an on-board 5 V to 3.3 V converter, the V_{CC} pin of this module is connected to the 5 V output pin of the Arduino MEGA microcontroller. Furthermore, the X, Y, and Z pins of GY-61 are connected to the analog input pins A1, A2, and A3 of Arduino microcontroller, respectively. As shown in Figure 21.7, the YL-69 module's V_{CC}, GND, and SIG pins are

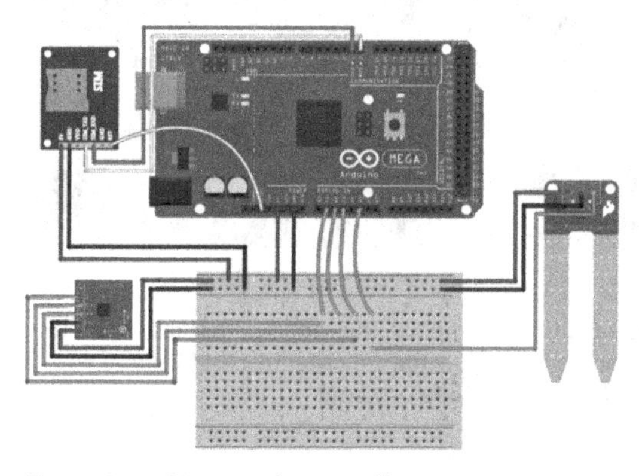

Figure 21.7 Diagram showing all components connected to the microcontroller board.

connected to the 5 V output, ground, and analog A5 pins of the Arduino microcontroller, respectively. The SIM900A GSM module uses its TTL pins to connect with the Arduino microcontroller. Since this module is responsible for transmitting sensors values, its transmitter (SIM_TXD) and receiver (SIM_RXD) pins are connected with the receiver (digital pin 0) and transmitter (digital pin 1) pins of Arduino microcontroller, respectively. Lastly, the 5 V and RST pins of the GSM module are connected with the 5 V output and RESET pins of Arduino microcontroller, respectively.

21.4.2 Data Logging and Alerts

As mentioned already, once the microcontroller receives sensor values, these values are transmitted via the GSM module to a cloud database. A PHP script is written to perform the data saving action (Rahnema, 1993). Data stored in the database is later used for analyzing thresholds for different sensors. If sensor values crossed certain predefined thresholds, then an automated landslide SMS alert is sent to registered mobile numbers using a free service called Twilio[5]. Twilio is a cloud communications platform that uses an account identification number and a mobile number to send a custom SMS to that mobile number.

21.4.3 Experimental Procedure

Soil was packed in the ramp and the ramp was elevated at an angle slightly less than the soil's critical angle. The critical angle was the minimum angle calculated from soil properties at which soil movement would occur due to the gravitational force without the addition of moisture (see Figure 21.8a). For the soil used

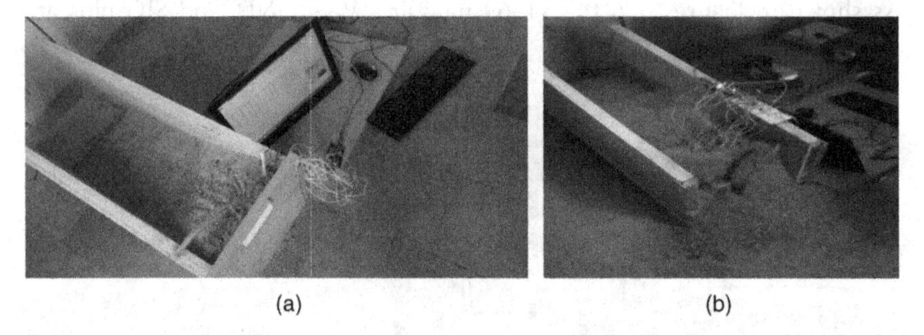

(a) (b)

Figure 21.8 Pictures of the experiment with the ramp. (a) The sand was packed inside the ramp and the ramp was raised at an angle just below the critical angle. (b) Moisture was added to the soil using a water sprinkler till a landslide occurred.

5 https://www.twilio.com/

in the experiment, the critical angle was 5° (see Table 21.1). Next, moisture was added to the soil until the soil moved and a landslide occurred (see Figure 21.8b). During the experiment, raw data from sensors were stored in a cloud database. Subsequently, sensor values were analyzed statistically to determine thresholds at which soil became loose and a landslide occurred. If sensor values crossed predetermined thresholds, then the alert-generating unit automatically sent SMS alerts to registered mobile phones.

21.5 Results

In this section, soil moisture and soil movement values recorded by different sensors are analyzed over time. Figure 21.9a–c show the plot of soil moisture, accelerometer, and cumulative displacement values, respectively, from sensors placed at the top of the pipe at the toe-end of the ramp. As seen in Figure 21.9a, the soil moisture sensor value increased and peaked at 18% when the landslide started occurring on the ramp (water was poured on the ramp at time $t = 0$ s). In addition, the changes in accelerations at the top of the pipe within the first 2 s of pouring water were the following: 0.7 to $-1\,g$ (X direction), 0.1 to $-1.5\,g$ (Y direction), and 0.15 to $-2.5\,g$ (Z direction), respectively. These accelerations and the corresponding displacements showed that the soil at the top of the pipe moved upward and outward when the landslide occurred. Furthermore, soil moisture values and acceleration values reached their peaks at the same time.

Table 21.2 shows the values stored in the cloud database, where the stored values were collected from the sensors placed on the ramp. The first column in Table 21.2 shows the date and time at which data were collected. The second column shows the soil moisture percentage recorded from the sensor at the base of the pipe. The third column shows the accelerations in the three orthogonal directions recorded by the accelerometer put at the base of the pipe. Here, acceleration values have been represented in terms of g ($1\,g = 9.8\,\text{m/s}^2$). Similarly, fourth and fifth columns show the soil moisture and acceleration values obtained from sensors placed at the top of the pipe.

In this framework, the real-time sensor values are compared with their threshold values determined from the experiment above. If the sensed values are close to the critical soil movement and soil moisture thresholds, then SMS alerts are automatically sent to the registered mobile numbers. Figure 21.10 shows an SMS alert received on a mobile phone from the framework, as well as a web-based interface for landslide monitoring.

Table 21.3 shows the cost of the components used in this IoT landslide monitoring framework for experimentation at the lab scale. The framework was assembled in the lab for only \$32. If this framework was to be deployed on a real

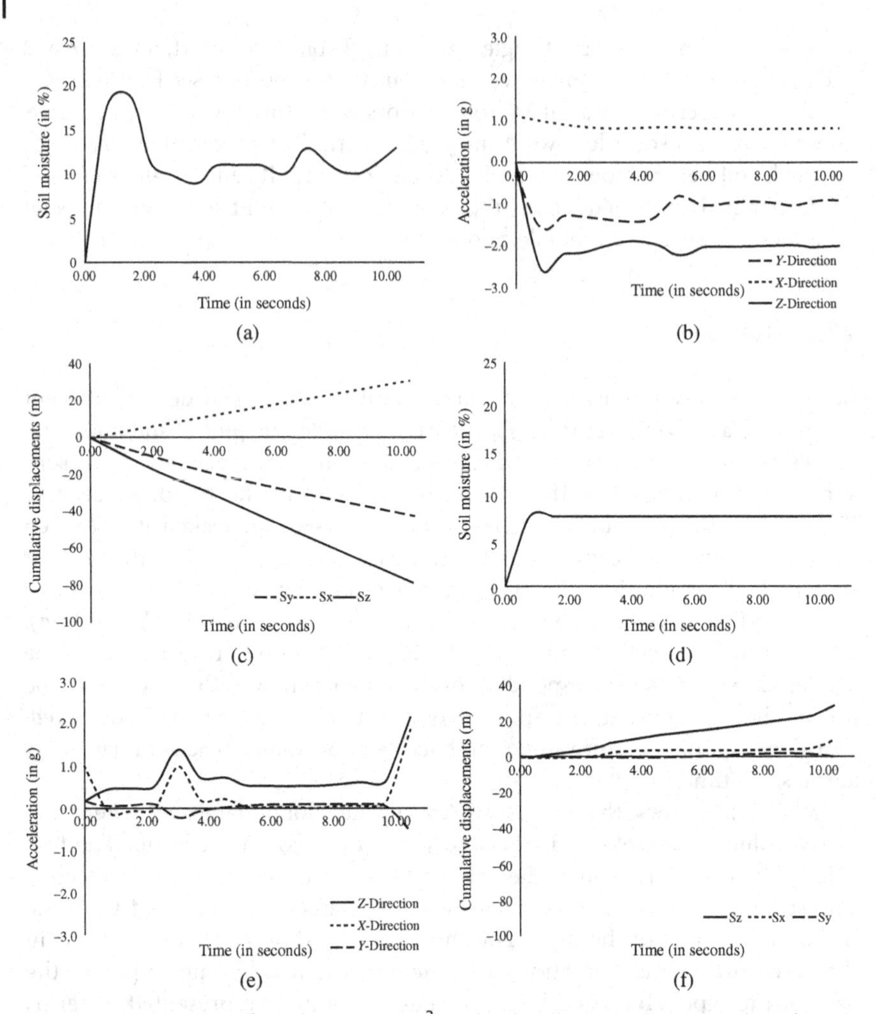

Figure 21.9 The value of accelerations (m/s^2) and soil moisture (in percentage) on the pipe at the toe-end of the ramp. (a) Soil moisture at the top of the pipe. (b) Accelerometer values at the top of the pipe. (c) Displacement at the top of the pipe. (d) Soil moisture (in percentage) at the base of the pipe. (e) Accelerometer values at the base of the pipe. (f) Displacement at the base of the pipe.

hill, then it would likely use 15 moisture sensors and 15 accelerometers at the crest-end, 10 moisture sensors and 10 accelerometers at the toe-end, 5 microcontrollers, and 2 GSM modules. The cost of this infrastructure would add up to \$400 per crest-toe pair per hill. In summary, the proposed IoT framework provides a very cost-effective landslide monitoring and alerting solution to society.

Table 21.2 Sample data obtained from sensors on the pipe at the toe-end of the ramp.

Time (in seconds)	Soil moisture at pipe's top (in %)	Accelerations at pipe's top (*X*, *Y*, and *Z*-axes) (in *g*)	Soil moisture at pipe's base (in %)	Accelerations at pipe's base (*X*, *Y*, and *Z*-axes) (in *g*)
00	0.40	1.1, −0.2, 0.13	0.5	0.98, 0.2, 0.2
01	18.0	0.96, −1.57, −2.5	8.0	−0.07, 0.08, 0.47
02	13.0	0.82, −1.29, −2.1	8.2	−0.04, 0.1, 0.50
03	15.0	0.81, −1.37, −1.9	8.3	0.21, −0.01, 0.73
04	12.5	0.81, −1.32, −1.9	8.2	0.24, 0.02, 0.75
05	14.0	0.85, −0.8, −2.25	8.4	0.04, 0.1, 0.57

21.6 Conclusions

In this chapter, a low-cost lab-scale IoT framework was proposed for monitoring landslides. MEMS-based sensors were used in this framework to measure soil moisture and soil movement on a ramp. Sensors' threshold values, derived from the experiment, were compared to real-time sensor data, and SMS alerts were sent to the registered mobile phones when sensor values exceeded their critical threshold levels.

The experimental results revealed that the base of the pipe experienced movement with only 8% moisture accumulation compared to the top of the

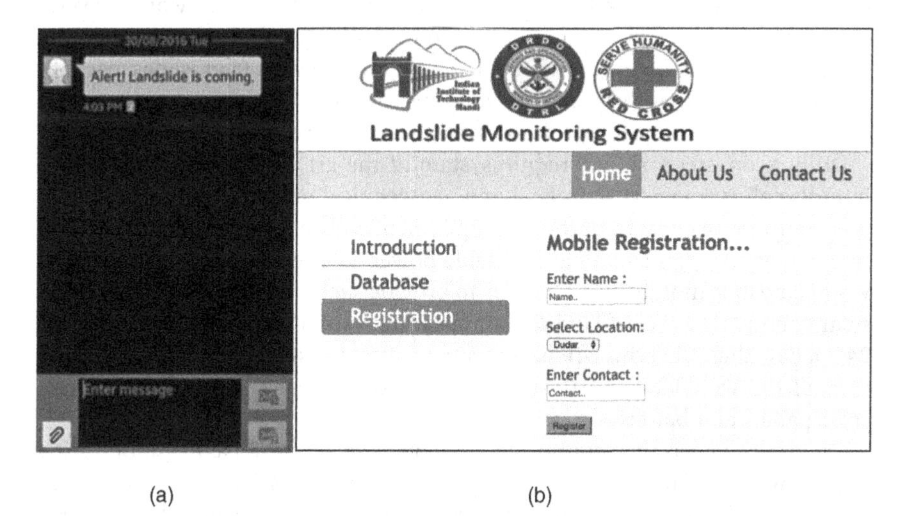

(a) (b)

Figure 21.10 An SMS alert generated from the IoT landslide monitoring framework on a mobile phone. (a) The mobile interface showing the SMS alert. (b) The web-based interface for landslide monitoring.

Table 21.3 Cost of components used in the development of the IoT landslides monitoring framework for the laboratory experiment.

Name of the component	Cost per unit (in USD)	Quantity	Total cost (in USD)
Accelerometer	3.5	2	7.0
Soil moisture sensor	3.0	2	6.0
Microcontroller board	12.5	1	12.5
GSM module	13.5	1	13.5
Breadboard + wires	3.0	1	3.0
Total			32.0

pipe with 18% moisture accumulation. One likely reason for this observation is that the soil–soil bonding at the pipe's top was much stronger compared to the soil–wood bonding at the pipe's base. Second, it was found that the moisture sensor at the pipe's base experienced little variance in its moisture value over time. This result could be due to the pouring of water from the ramp's top. Thus, it did not accumulate as much at the pipe's base as compared to the accumulation at the pipe's top (pipe's top absorbed most of the moisture present).

Although this chapter reported only a single experiment involving lab-scale landslide monitoring, the results show the potential of using MEMS-based sensors for landslide monitoring in the real world. However, before scaling up the proposed framework for monitoring landslides in the real world, several challenges need to be overcome. This includes drilling on a hill with steep slopes, requiring a self-sustaining power source, 24 h monitoring of landslides, packaging of sensors so that they can withstand the rugged hilly terrain, and the use of other groundbreaking IoT technologies that enable rapid data collection.

Drilling on steep slopes requires state-of-the-art drilling equipment and interdisciplinary expertise in civil and geotechnical areas. For this purpose, it is encouraged to engage appropriate equipment and an interdisciplinary team of researchers. A self-sustaining and stable power source will be needed for real-world landslide monitoring. One way to provide such a power source is by using solar energy. However, if landslide monitoring system stays in an "on" state for 24 h a day, then the power requirements of such a system would be very large. Overall, this would increase the monitoring cost due to the investment in several solar panels and batteries.

One way to keep the monitoring cost low is to go for a passive monitoring system, which sleeps for 10–15 min, wakes-up to record sensor values, and then sleeps again for 10–15 min. If rain occurs during the time when the system is sleeping, then the rain wakes the system from sleep via a hardware interrupt. As revealed in the experiments, rain (or water) caused an increase in soil moisture over time. However, accumulation of moisture in the soil may not necessarily trigger a landslide, where the landslide trigger is also dependent upon the type of soil and its properties. Although only one soil type was tested in the experiment reported above, both soil properties and soil moisture were considered as causal factors in the experiment. The experiment's methodology could be easily adapted to monitor other soil types.

Another challenge is related to the packaging of the MEMS-based sensors that is crucial so that these sensors do not get damaged when used in the real world. One way of ensuring sensor durability is to mount them on a flexible rubber or plastic pipe while placing them on a hill. In fact, mounting these sensors on a flexible pipe is what was done in the experiment reported.

The proposed IoT framework is a whole, cost-effective solution, which requires no communication system beyond existing 2G telephony. However, this framework is currently at a prototypical level where it is planned to be deployed for landslide monitoring on a real hill in the future. Scaling up the system across many hills will allow for investigation of other groundbreaking IoT technologies, such as 5G broadband (Osseiran et al., 2014). These technologies will have bandwidth speeds more than 100 megabits per second and will enable gathering massive amounts of sensed data from hundreds of sensors in real time.

In conclusion, overall, the proposed IoT landslides monitoring framework helps to reduce the cost of monitoring landslides in the real world. In fact, once scaled up to a real hill, the framework is expected to cost less than 4% of the cost of conventional systems that are currently used for monitoring landslides. Given that there are 200 lives lost per year in the Himalayas alone due to landslides (Parkash, 2011), the cost-effectiveness of this IoT framework is expected to be only $1.3 per head. This low cost is expected to make the framework an affordable landslide monitoring system in the future—a system that is deployable at several landslide-prone sites across the world.

Acknowledgment

This research was supported by grants from the following agencies to Varun Dutt: State Council for Science, Technology & Environment, H.P. (IITM/HPSCSTE/VD/130); Defence Terrain Research Laboratory, Defence Research and Development Organisation (IITM/DRDO-DTRL/VD/179); and, National Disaster Management Authority (IITM/NDMA/VD/184).

References

Arnhardt, C. and Neussner, O. (2013) Setup of a landslide monitoring system on the Philippine Island of Leyte near the village of Malinao (Municipality of St.

Bernard). *Landslide Science and Practice*, Springer, Berlin Heidelberg, pp. 161–167.

Aziz, N.A.A. and Kamarulzaman, A.A. (2011) Managing disaster with wireless sensor networks. 13th International Conference on Advanced Communication Technology, IEEE, 202–207.

Barrett, S.F. (2013) Arduino microcontroller processing for everyone! *Synthesis Lectures on Digital Circuits and Systems*, **8**(4), 1–513.

Borgia, E. (2014) The Internet of Things vision: Key features, applications and open issues. Computer Communications, **54**, 1–31.

Casson, B., Delacourt, C., Baratoux, D., and Allemand, P. (2003) Seventeen years of the "La Clapiere" landslide evolution analysed from ortho-rectified aerial photographs. *Engineering Geology*, **68**(1), 123–139.

Chaturvedi, P. and Dutt, V. (2015) Evaluating the public perceptions of landslide risks in the Himalayan Mandi town. *Proceedings of the Human Factors and Ergonomics Society Annual Meeting*, **59**(1), 1491–1495.

Chaturvedi, P., Arora, A., and Dutt, V. (2017) Interactive landslide simulator: a tool for landslide risk assessment and communication. *Advances in Applied Digital Human Modeling and Simulation*, Springer International Publishing, 231–243.

Chaturvedi, P., Shrivastava, S., and Kaur, P. (2017) Landslide early warning system development using statistical analysis of sensors' data at Tangni landslide, Uttarakhand, India. *Advances in Intelligent Systems and Computing*, Springer International Publishing, p. 547.

Crone, W.C. (2008) A brief introduction to MEMS and NEMS. *Springer Handbook of Experimental Solid Mechanics*, Springer, US, pp. 203–228.

Dan, Z., Bin, S., Hong-Zhong, X., Junqi, G., and Hong, Z. (2004) Experimental study on the deformation monitoring of reinforced concrete T-beam using BOTDR. *Journal of Southeast University (Natural Science Edition)*, **4**, 012.

Gupta, A. and Ahmad, A. (2007) Microsensors based on MEMS technology. *Defence Science Journal*, **57**(3), 225.

Guzzetti, F., Mondini, A.C., Cardinali, M., Fiorucci, F., Santangelo, M., and Chang, Kang-T. (2012) Landslide inventory maps: new tools for an old problem. *Earth-Science Reviews*. **112**(1), 42–66.

IS 2720-1 (1983) Indian Standard Methods of Tests for Soils. Part 1 – Preparation of Dry Soil Samples for various tests. Available at https://law.resource.org/pub/in/bis/S03/is.2720.1.1983.pdf

IS 2720-13 (1986) Indian Standard Methods of Tests for Soils. Part 13 – Direct Shear Test. Available at https://law.resource.org/pub/in/bis/S03/is.2720.13.1986.pdf

IS 2720-4 (1985) Indian Standard Methods of Tests for Soils. Part 4 – Grain Size Analysis. Available at https://law.resource.org/pub/in/bis/S03/is.2720.4.1985.pdf

IS 2720-8 (1983) Indian Standard Methods of Tests for Soils. Part 8 – Determination of Water Content-Dry Density Relation Using Heavy Compaction. Available at https://law.resource.org/pub/in/bis/S03/is.2720.8.1983.pdf

Khan, R., Khan, S.U., Zaheer, R., and Khan, S. (2012) Future internet: the internet of things architecture, possible applications and key challenges. In Frontiers of Information Technology (FIT), 2012 10th International Conference on (pp. 257–260). IEEE.

Manconi, A. and Giordan, D. (2016) Landslide failure forecast in near-real-time. *Geomatics, Natural Hazards and Risk,* 7(2), 639–648.

Marciano, J.S., Hilario, C.G., Zabanal, M.A.B., Mendoza, E.V., Gumiran, B.L., Flores, B.F., Peña, M.O., and Razon, K.H. (2014) Monitoring system for deep-seated landslides using locally-developed tilt and moisture sensors: system improvements and experiences from real world deployment. Global Humanitarian Technology Conference (GHTC), 2014 IEEE, October, pp. 263–270.

Martelloni, G., Segoni, S., Fanti, R., and Catani F. (2012) Rainfall thresholds for the forecasting of landslide occurrence at regional scale. *Landslides,* **9**(4), 485–495.

McKean, J. and Roering, J. (2004) Objective landslide detection and surface morphology mapping using high-resolution airborne laser altimetry. *Geomorphology,* **57**(3), 331–351.

McKinnon, A.C. (2015) 3D printing, drones and crowdshipping: city logistics game-changers or over-hyped curiosities. *Urban Freight and Behavior Change (URBE)* Roma Tre University, Rome, Italy.

Mehregany M. and Roy S. (1999) Introduction to MEMS, in Helvajian, H., (ed.), *Microengineering Aerospace Systems,* Aerospace Press, Los Angeles, CA.

Niethammer, U., James, M.R., Rothmund, S., Travelletti, J., and Joswig, M. (2012) UAV-based remote sensing of the Super-Sauze landslide: evaluation and results. *Engineering Geology,* **128**, 2–11.

Osseiran, A., Boccardi, F., Braun, V., Kusume, K., Marsch, P., Maternia, M., Queseth, O., Schellmann, M., and Schotten, H. (2014) Scenarios for 5G mobile and wireless communications: the vision of the METIS project. *IEEE Communications Magazine.* **52**(5), 26–35. doi: 10.1109/MCOM.2014.6815890.

Othman, M.F. and Shazali, K. (2012) Wireless Sensor Network Applications: A Study in Environment Monitoring System. Procedia Engineering, **41**, 1204–1210.

Parkash, S. (2011) Historical records of socio-economically significant landslides in India. *Journal of South Asia Disaster Studies,* **4**(2), 177–204.

Rahnema, M. (1993) Overview of the GSM system and protocol architecture. *IEEE Communications magazine,* **31**(4), 92–100.

Ramesh, M.V. (2009) Real-time wireless sensor network for landslide detection. 3rd International Conference on Sensor Technologies and Applications, IEEE, pp. 405–409.

Weinstein, N. (2010) Lasers Help Identify Potential Landslides. Available at http://djcoregon.com/news/2010/08/10/lidar-helps-identify-potential-landslides/

Glossary

5G: A new generation of wireless cellular networks is now under development; 5G networks will be five times as fast as the highest current speed of today's 4G networks (with download speeds as high as 5 Gbps—4G offering only up to a maximum of 1 Gbps).

Action Unit (AU): An observable component of facial movement adopted in the FACS system.

Actuator: A device providing the means of implementing actions in the physical world such as moving or controlling a mechanism or system.

Adaptable System: A software-based system that analyzes the results of its decisions and actions, and then adapts the control algorithm accordingly. The adaptation (learning) method is defined, but the actual algorithm changes depending on the conditions and is unpredictable.

Advanced Message Queuing Protocol (AMQP): A messaging protocol for message-oriented middleware.

Affective Object: A physical object that can sense emotional data from a person.

Analytics: The exploration, interpretation, and communication of important patterns in data.

Antifragile System: A system that evolves and develops in the changing environment.

ARIMA: Autoregressive integrated moving average is a statistical analysis model that uses differenced time series data to predict and monitor data trends.

ARM: A CPU that is based on the RISC (reduced instruction set computer) architecture developed by Advanced RISC Machines.

Asset Administration Shell (AAS): A term coined by the German initiative "Plattform Industrie 4.0" to denote a software component that forms, together with a physical asset, an Industry 4.0 component, that is, a cyber-physical system (CPS) within smart factory contexts.

Internet of Things A to Z: Technologies and Applications, First Edition. Edited by Qusay F. Hassan.
© 2018 by The Institute of Electrical and Electronics Engineers, Inc. Published 2018 by John Wiley & Sons, Inc.

Authorization: Process for granting approval to a client entity to access a resource hosted and exported by a server entity.

Autism Spectrum Disorder (ASD): A neurodevelopment disorder characterized by atypical social-communicative skills, restricted and repetitive patterns of behaviors.

Autonomous Driving System: An automation system to control the driving and steering of agriculture vehicle to move along a predetermined path based on GPS navigation autonomously.

Biocatalytic Membranes: Biohybrid artificial biochemical transformations catalyzed by enzymes or cells where a biochemical conversion is combined with a membrane separation through selective mass transport with chemical reactions.

Biometric Data: Data related to physical or behavioral attributes of the human body.

Bitcoin: The first cryptocurrency, which is a digital or virtual currency that uses cryptography for security.

Black Swan: A rare event that should not happen (and therefore is forgotten) but sometimes happen anyway.

Blockchain: A distributed database for storing a continuously growing list of linked records called blocks.

Blockchain 2.0: Blockchain implementations that extend the functionality of the Bitcoin blockchain.

Breadboard: A solderless board for building experimental models of electronic circuits.

Chip: A chip is a tiny electronic circuit on a semiconductor material.

Cloud Computing: A computing model consisting of a pool of computing resources, such as networks, servers, storage, applications, and services, instead of a local server or a personal computer. Commonly, these resources are accessed remotely and are available for free or at a fee by service providers.

Commercial Commodity Products or Commercial off-the-shelf (COTS) Products: Products bought on an open market. Typical choice criterion is its price. The supplier is not well known and the continuity of long-term delivery is unsure.

Complex System: A system containing many interacting elements. The interactions form a network with many connections and often are nonlinear. Complex systems may manifest self-organization, adaptation, and emerging properties.

Connected Buildings: Buildings that make extensive use of building management systems to control the various energy-consuming entities in the building such as HVAC, lighting, and use of sensors, networks. IoT resources are intrinsic to the concept.

Connected Lighting: Lighting systems where lights are interconnected over an IP (Internet Protocol) network; the connected lighting paradigm makes use of sensor(s); controllable and intelligent SSL sources; wired and/or wireless networks (both in-building and into the cloud); analytics as a service; and IoT principles on systems and data.

Constrained Application Protocol (CoAP): A web-based transfer protocol for IoT-constrained networks and nodes.

COPD: Chronic obstructive pulmonary disease is a collective term of chronic inflammatory lung diseases that describe a range of progressive lung diseases that may result in severe impairment of lung function. These include bronchitis, emphysema, and certain types of refractory asthma and bronchiectasis.

Crop Water Stress Index (CWSI): The characterization of crops' needs to water based on temperature measurements.

Crowdsensing: An IoT environment that allows a large population of mobile devices to measure phenomena of common interest over an extended geographic area, enabling "Big Data" collection, analysis, and sharing.

Cryptocurrency: A digital or virtual currency that uses cryptography for security.

Cyber-Physical System (CPS): The combination of a physical asset with a software component that interacts with the physical system (sensing and actuation), may offer local functionalities (digital representation, business rules, etc.), and communicates with other systems via specified interfaces.

Cyber-Physical-Social System: A system composed of tightly interacting computing elements, physical parts and natural environment, and humans.

Data Distribution Service (DDS): A publish–subscribe middleware standard for communications for real-time and embedded systems.

Decentralized App (DApp): A smart contract in Ethereum.

Demand Response (DR): Methods for reducing consumer electricity use at times of high demand.

Demand Response Management System (DRMS): A system (or collection of systems) that enable utilities to manage their DR programs utilizing an integrated system.

Denial-of-Service (DoS) Attack: Attempt to make a network node less available or completely unavailable to its intended users, typically by flooding superfluous processing requests.

Device: A hardware component with communication capabilities linking it to other computerized systems.

Diffie–Hellman Key: A shared secret encryption/decryption key created with a key agreement protocol called the Diffie–Hellman protocol.

Digital Elevation Model (DEM): A digital map of the elevation of an area on the earth.

Digital Signature: A cryptographic record of a document created with a private key.

Dispatchable (e.g., Dispatchable Generation): A description for an energy source that can be adjusted to provide the desired energy flow.

Distributed Energy Resources (DER): Small power generators typically located at users' sites where the energy they generate is principally used; examples include distributed elements such as solar/wind systems and MicroGrids.

Distributed Energy Resource Management System (DERMS): An integrated system that provides utilities with capabilities to manage and optimize distributed energy resources, automate business processes, and engage with customers.

Distributed Renewables: Renewable energy produced by local communities and private homeowners and circulated through the common grid.

Distribution Management System (DMS): A decision support system to facilitate monitoring, controlling, and optimizing the performance of the electric distribution system.

DNS Service Discovery (DNS-SD): A zero-configuration protocol that pairs services to hosts within a domain.

Edge Computing: Concept with significant overlap to fog computing. The distinction between the two is so far not totally clear in the research community. However, some sources see edge computing more human activated in contrast to fog computing, which would act fully autonomously.

Electronic Product Code (EPC): A tag standard that assigns a unique identity to every physical object at all times worldwide.

Embedded System: An embedded system is a computer system consisting of both hardware and software components, and may also include mechanical parts, designed for a specific function.

Emerging Properties: Properties that emerge in a complex system of interacting elements. The whole is greater than the sum of its parts.

Emotion Expressiveness: The ability of making recognizable representation of typical facial emotional expressions.

Energy Internet of Things (EIoT): An interconnected web of EIoT devices that can measure energy use and then regulate it.

Energy Internet of Things Device: A device that controls energy consumption, production, or storage, while having all the characteristics of an IoT device, such as internet connectivity.

Energy IoT: IoT principles, methods, and technologies focused on the electrical smart grid.

Ether: The cryptocurrency in Ethereum.

Ethereum: A Blockchain 2.0 platform.

eXtensible Markup Language (XML): A markup language that defines a set of rules for encoding documents in a format that is understandable by machines and humans.

Extensible Messaging and Presence Protocol (XMPP): A technology for real-time communication between devices.

Facial Action Coding System (FACS): An emotion recognition system based on the observable human facial movement.

Fertigation (Chemigation): The technology that enables the application of water-soluble chemical matters to farmland through irrigation.

Flying Ad Hoc Network (FANET): Mobile ad hoc networks designed for the needs of UAVs.

Fog Computing: Describes an approach in which, compared with cloud computing, IT actions are performed at the edge of the network, thus creating user or application proximity. This leads to lower service latency and denser geographical distribution, as well as improved security.

Fragile System: A system designed for a limited set of conditions, breaks under change.

Gateway: A device that compiles, transmits, and provides protocol translation between different devices.

Genesis Block: The first block in a blockchain.

Geo Mapping: To identify the quantity of agriculture treatment required of each specific site in the farmland.

Global Positioning System (GPS): A radionavigation system that provides time information and physical location to devices anywhere where there is a clear line of sight to at least three GPS satellites.

Hardware Crypto Engine or Encryption Chip: A hardware component that performs cryptographic calculations and accelerates the execution of applications that require cryptographic functions.

Hash: A cryptographic value of a fixed size calculated from some data of any size.

IEEE 802.15.4: A networking protocol for low-rate wireless personal area networks (LR-WPANs).

Independent System Operator (ISO): An electric power transmission system operator who makes decisions about the electric grid. An ISO usually is responsible for managing a grid within a specific state.

Industrial Internet of Things (IIoT): The use of IoT technologies within industrial use cases, such as smart buildings, smart grids, or smart factories. A major challenge in this context is the convergence of Operational Technology (OT) with Information and Communication Technology (ICT).

Industry 4.0: A term describing the fourth industrial revolution that targets the combination of ICT with OT in the manufacturing context based on intelligent systems in order to enhance efficiency and beyond that envisions a digital networking of the whole industrial value chain across factory and domain boundaries. The term is also known as Industrie 4.0, which was first used at the Hannover Messe in 2011.

Integration Platform: Software that connects together different application and services.

Intelligent Transport Systems: The integration of ICT with vehicles and transport infrastructure.

Internet of Flying Things (IoFT): Refers to a version of IoT that includes UAVs not only as "things," but also as infrastructure providers for IoT environments.

Internet of Things (IoT)—Long Definition: Refers to a network of physical (and virtual) objects (usually known as "things") that are identified with unique IP addresses for Internet connectivity, and also the communication among these "things" and other Internet-enabled devices and systems.

Internet of Things (IoT)—Short Definition: A global infrastructure enabling everyday objects to have network connectivity allowing them to exchange data.

Internet Protocol (IP): A fundamental communications protocol used across the Internet, home networks, and business networks.

Internet Protocol (IP) Address: An identifier assigned to every computer and other devices connected to a network.

Internet Protocol Security (IPsec): A networking protocol that is employed with IPv6 for authentication and end-to-end encryption among devices.

Internet Protocol version 6 (IPv6): Network layer protocol that may eventually replace IP (IPv4 in particular) that provides a much larger address space and mobility support.

Interoperability: The term describes a system's ability to share information and services with another system ideally based on common standards. Much of the success of IoT relies on the ability of connected devices to operate seamlessly and effectively together.

Intrusion Detection System: Entity responsible for monitoring a network or system and to detect anomalies, malevolent activities, violation of policies, and compromised nodes.

IoT Audit: The assessment and interrogation of an organization's IoT framework, policies, and procedures. IoT audits ensure that the organizations espoused policies for operation are actually implemented and enforced.

IoT Hardware Development Platform: A physical component of an IoT development kit used for implementing prototypes.

IPv6 Over Low-Power Wireless Personal Area Networks (6LoWPAN): A low-power wireless mesh network that connects every node directly to the Internet using open standards via its own IPv6 address.

Jamming Attack: Type of denial-of-service attack that targets wireless communication technologies by physically interfering with the network's operational frequencies.

Key Management: Process for generating, revoking, renewing, and distributing cryptographic key material used for secure communication or other administrative purposes.

Kinesiological Data: Data related to the movement of the human body.

Landslides: The rapid downward movement of a mass of rock, earth, or artificial fill on a slope under the influence of gravity.

Load Curve: A graph that shows the amount of electricity at each point in time that is needed to provide service for all of the consumers in a particular area.

Localization: A process of estimating the geographical location of an object.

Machine Ethics: Study of the ethical principles implemented by the machines and their relation to human ethics.

Machine-to-Machine (M2M): A broad term describing technology that allows for one connected device to communicate and exchange information with another connected device, without human assistance.

Message Digest: A hash of message.

Message Queue Telemetry Transport (MQTT): A lightweight messaging protocol for small sensors and mobile devices optimized for unreliable networks. It is useful for connections with remote locations where a small code footprint is required.

Microelectromechanical System (MEMS): A miniaturized mechanical and electromechanical system designed to carry out a specific function.

Microfluidics: Techniques based on the precise control and manipulation on fluids in geometrically constrained spaces, usually at the submillimeter scale.

MicroGrid: A localized power grid that can operate both independent from the utility electric grid but still connects to the power grid at a point of common coupling. MicroGrid owners can often sell excess power while grid is connected.

MicroGrid Control (MGC): A management control system for real-time monitoring and control of a MicroGrid, also providing historical reports, analytics, and context-based automation.

Mining: A distributed computational review process on a block before it can be linked to a blockchain.

MIPv6: A mobility management capability (protocol) used in conjunction with IPv6.

Mobile Ad Hoc Network (MANET): It is a continuously self-configuring, infrastructure-less network of mobile devices connected wirelessly.

Motes: A digital sensor node with communications infrastructure, which serves as nodes in wireless sensor networks.

Multicast DNS (mDNS): A zero configuration and infrastructure-independent protocol that provides host names to IP addresses inside local networks.

Named Data Networking (NDN): Content-centric networking where communication is driven by receivers (who are the data consumers); the focus is on the data itself, not the location where the data is stored.

Nanny State: A state where the government assumes it knows the best the needs of the citizens, cares for them, and controls their good behavior.

Near-Field Communication: A set of RFID-based protocols that enable peer-to-peer communication between a pair of electronic devices within a range of about 4 cm of each other.

Neomania: The drive to buy or implement everything new, just because it is possible.

Node: A connection point, a redistribution point, or a communication endpoint. The definition depends on the network and the protocol layer referred to. A network node is an active electronic device that is attached to a network and is capable of creating, receiving, or transmitting information over a communications channel.

Nonce: A randomly chosen value.

Nondispatchable (e.g., Dispatchable Generation): A description for an energy source that cannot be adjusted to provide the desired energy flow.

Normalized Difference Vegetation Index (NDVI): A parameter calculated from vegetation light reflection that indicates the plant health.

Object: Refers to a "thing" in IoT (in contrast to the digital and network connection shared between these systems).

Open Systems Interconnection (OSI): A reference model for how applications are managed inside a network.

Operational Technology (OT): The use of computers within industrial use cases, for example, in order to automate physical control loops by processing sensor signals and triggering actuation signals in real time.

Peak Reduction: Advanced planning that reduces spikes, or peaks, in electricity consumption.

Peakers: Small and highly dispatchable generators that are built to meet peak electricity demand on a few days each year. Peaker generators give off large amounts of greenhouse gases when used. They also add considerably to infrastructure costs, yet sit idle most of the time.

Physiological Data: Data related to the functioning of the human body, including physical and chemical reactions in any of the organs and systems that make up the human body.

PKI: Public key infrastructure, a set of roles, policies, and procedures needed to create, manage, distribute, use, store, and revoke digital certificates and manage key pairs in public key cryptography.

PKI Signature: A signature created and verified with a certified key pair.

Point of Care: Defined as when a clinician delivers health care support to a patient at the time of care. When outside the hospital, this can be either on-scene treatment or at the ambulance.

Pop-Up Satellite Archival Tags (PSATs): A PSAT is a recording device attached to marine animals for data collection, which can automatically pop up to the surface and transmit data through the satellite link.

PoW: Proof-of-Work, repeated hashing of a concatenation of a blockchain block and a new nonce until a hash with a defined difficulty is calculated.

Power-Line Communication (PLC): A set of communication protocols that use power-line wiring to simultaneously transmit both data and alternating current.

Power over Ethernet (PoE): IEEE standards (and supporting technology) to deliver low-voltage power to remote devices that utilize Ethernet wiring and protocols. The original IEEE 802.3af supported power delivery in the 15 W range (at 48 V DC); extensions have been underway to support power in the 50–70 W range (IEEE 802.3bt, Type 3) and, possibly,100 W (IEEE 802.3bt, Type 4).

Precision Agriculture (PA): A new agriculture practice to treat crops on each minimum plot separately based on their real-time needs.

Precision Livestock Farming (PLF): A modern livestock breeding system that can monitor animals and farm environment in real time, and provide quantized treatments to tend them accordingly.

(Private) Standards Consortium: A privately funded and run SSO. Especially in the ICT sector such consortia are a major driving force.

Prototype: An early model or sample of an electronic or software product built in order to test a design concept.

Public Key Cryptography: Cryptographic algorithms using a key pair, that is, a private key and a related public key. Data encrypted with one key is decrypted with the other key

Publish/Subscribe: A communication method in which, any number of computers (publishers) communicate with any number of computers (subscribers) based on an event.

Quality of Service (QoS): The concept that transmission rates, error rates, and other characteristics can be measured, enhanced, and, guaranteed.

Quick Response Code: A matrix barcode that is placed on an item to describe it and the code is readable by machines.

Radio-Frequency Identification (RFID): The application of wireless radio frequency signal in object identification, data transfer, and tracking.

Real Time Kinematic (RTK): A method to improve the GPS precision with the reference signal from a separate ground station.

Recovery Point Objective (RPO): Considers the volume of data that could be at risk. It is based on the frequency of data protection endeavors. As such it reveals the amount of data that could be retrieved and that which might be potentially lost during disaster recovery.

Recovery Time Objective (RTO): A metric that attempts to measure the time it takes to recover from any event comprising the loss of data. It also measures the time period it takes for an organization to be back to full functional capacity.

Regional Transmission Operator (RTO): An electric power transmission system operator who makes decisions about the electric grid. An RTO usually is responsible for managing a grid that covers several states.

Representational State Transfer (RESTful): A set of predefined and uniform set of stateless operations to allow requesters to access and manipulate textual representations of Web resources.

Request/Response: A communication method in which, a computer requests a data, and the other computer replies with the requested data.

Resident: A person who occupies or lives in a house.

Resistor: An electrical component that limits the flow of electric current in a circuit.

RFID Reader: A device that employs either radio waves or nonpropagating electromagnetic fields to transfer data between itself and an RFID tag.

RFID Tag: A transponder that responds to interrogation signals from RFID readers.

Robust System: A system that resists to change, in a certain limit.

Routing Protocol for Low-Power and Lossy Networks (RPL): An IPv6 routing protocol for resource-limited devices.

Scalability: The ability of a computing process to be used in a range of capabilities.

Secure Communication: Exchange of messages among a set of network nodes that fulfills a number of security requirements, including message confidentiality, message integrity and authenticity, and replay prevention.

Self-Calibration: The ability of a device or system to undergo self-diagnosis in order to revert its predetermined operational parameters back to factory-defined baselines.

Self-Driving Automation. A technical solution where the operator no longer has any responsibility for safe operation of a vehicle, and is not expected to monitor road conditions or take control at any point during the trip.

Semantics: A concept to encode meanings separately from application code and data and content files.

Sensor: A device that detects events or changes in its environment and sends information to other electronic components.

Service: A software component enabling interaction with computing resources through a well-defined interface, often via the Internet.

Smart Buildings: Commercial buildings where building management systems (BMSs), and more specifically IoT-based BMSs, are used to manage in real time, often remotely, and often using cloud-based capabilities, a number of building-related functions, principally power consumption, and HVAC, but more recently also lighting control, access control, and surveillance.

Smart Cities: A concept that tries to create a more intelligent city infrastructure by using modern information and communication technologies. Smart cities propose a more flexible adaptation to certain circumstances, more efficient utilization of resources, higher quality of life, more fluid transportation, and so on. This may be achieved through networking and integrated information exchange between humans and things.

Smart City: A city that make broad use of information and communications technologies in general and IoT technologies in particular to improve the delivery of services to the residents, such as improved traffic management, improved access to real-time conditions, infrastructure monitoring (power, water, sewer), and surveillance.

Smart Connected Home: A residence equipped with sensors, systems, and devices that can be controlled and monitored, typically via the Internet. It may exhibit some form of "intelligent" logic for activity recognition and allowing it to perform some automated actions on behalf of the residents.

Smart Contract: A computer program that encompasses contractual terms and conditions that enable the verification, negotiation, or enforcement of a contract.

Smart Controls: Devices connected to the Internet and the electric grid that use machine learning to make intelligent decisions on energy use for the owners of energy consuming devices.

Smart Cow/Chicken/ . . . Farm (IoT Cow/Chicken/ . . .): The application of the IoT technologies in the precisely farming practice of various livestock, which can significantly improve the productivity and welfare of animals.

Smart Device: A device featuring sensors and "intelligent" logic allowing it to operate and communicate with other connected devices and users possibly in an autonomous way.

Smart Factories: A term used to describe manufacturing facilities in which machines, products, processes, and factories themselves are networked with each other. Based on self-organizing cyber-physical systems (CPS) and the digital representation of all aspects of the production process (the so-called digital twins), the facility allows for dynamic reconfigurations and can efficiently build products of lot size one. The factory is fully digitalized using IoT technologies to converge the exiting Operational

Technology (OT) with modern Information and Communication Technology (ICT).

Smart Grid: An advanced digital infrastructure with bidirectional communications capabilities for transmitting information, controlling equipment, and distributing electrical energy. It is an electrical grid that incorporates Information and Communication Technology-enhanced energy resources, including smart meters, smart appliances, and renewable energy resources.

Smart Health Care Ecosystem: The interrelated components that are connected via the IoT to support patient-centered healthcare systems, including human participants (patients, healthcare providers, and other stakeholders); autonomous and subordinate devices (computers, smartphones, sensors); and the related applications, interfaces, and data.

Smart Home: A home equipped with several intelligent systems, like access control, monitoring cameras, heating and lighting optimization, and so on. In this way, processes within a home can be monitored and controlled automatically to optimize quality of life, cost, security, and environmental impact.

Smart Manufacturing: The integration of ICT into manufacturing processes to enable proactive and intelligent manufacturing decisions in dynamic environments.

Smart Meter: A device that not only measures the total consumption of electricity in the home or office but can also communicate with in-home displays to inform the consumer how much energy they are using as a function of time (over the day, week, or month).

Social Engineering: A means of infiltration/exfiltration of any valuable resource by taking advantage of the human factor in a chain of security. The tendency here is to trick somebody into breaking or bypassing standard security operations.

Software-Defined Network (SDN): A computer networking approach that allows administrators to programmatically control network behavior dynamically via open interfaces.

Soil Moisture: The water contained within the soil pores.

Solid-State Lighting (SSL): Lighting that uses semiconductor light-emitting diodes (LEDs) or organic light-emitting diodes (OLED) as sources of illumination in place of the traditional electrical filaments.

Stakeholder: An individual or group of users with an interest or that may be affected by an organization, strategy, or project.

Standard: A document that provides requirements, specifications, guidelines, or characteristics that can be used consistently to ensure that materials, products, processes, and services are fit for purpose.

Standardization: The process of negotiating and, eventually, agreeing upon a specification as "standard."

Standards Developing Organization (SDO): An SSO that is in some way "officially" recognized. Examples include ITU (a UN organization) and the European Standards Organizations (through EU Regulation).
Standards Setting Organization (SSO): An organization that provides a platform for standardization.
Supervisory Control and Data Acquisition (SCADA): An industrial automation control system in wide use.
Supply Curve: A graph that shows the amount of electricity at each point in time that is provided by the available power sources in a particular area.
Swarm Intelligence: Collective behavior of decentralized, self-organized systems, natural or artificial, like a flock of birds or a set of interacting software-based artificial agents.
Syndromic Surveillance: Advance detection of certain health indicators that are used in the public health system for disease outbreak forecast.
System on a Chip (SoC): An integrated circuit (i.e., "chip" or "IC") that contains all parts of a computer or other electronic system.
Tag: A label or other physical object used to identify the physical entity to which it is attached.
TCB Measurement: Hashes of all TCB components.
TCP/IP: The IP suite is the networking conceptual model that describes a set of communications protocols used on the Internet and similar computer networks.
Technical Revolution: A disruptive change in the technology that has a profound impact on human life, from harnessing fire and introducing agriculture to smartphones and the Internet of Things. Every step marks a point of no return. The frequency of steps is getting higher and the duration of change is getting shorter.
Terminal: An interface in which a user can type and execute commands.
Thing: In the term "IoT," the word "thing" denotes a physical entity (in contrast to the digital and network connection shared between these systems).
Threshold: The magnitude or intensity of a parameter that must be exceeded for a certain phenomenon or condition to occur.
Time Stamp: The current time of an event that is recorded by a computer.
TM-Coin: TCB Measurement-Coin, a protocol for the trustworthy management of TCB measurements in IoT devices.
TOR: The onion router, an anonymizing network.
Transport Layer Security (TLS): A protocol that provides data integrity and privacy to communicating applications.
Trusted Computing Base (TCB): The set of hardware, firmware, and/or software components that ensure the security of a computer system.
Ubiquitous Code (uCode): A numeric identification system that uniquely identifies physical objects and places in the real world.

Ubiquitous Computing: A connected world in which computers are seamlessly integrated into their surroundings and easily and constantly accessible, without traditional interfaces, such as screens and keyboards.

Ultra-Wide Bandwidth (UWB): A wireless technology that uses low-energy transmissions to provide high-bandwidth communications over a wide radio spectrum.

Uniform Resource Identifier (URI): A unique sequence of characters in the form of a web link that refer to an abstract or physical resource.

Unmanned Aerial Vehicles (UAV): Also known as a drone, a UAV is an aircraft without a human pilot aboard.

Unmanned Aircraft System (UAS): The set of elements that support a UAV flight, for example, ground control station, propellers, and rescue system.

Value-Added Service (VAS): In the telecommunication industry, a service beyond basic services, for example, voice calls; also used to describe services in IoT systems offered on top of the basic services, such as computing resources and storage, which are commonly offered by today's cloud providers.

Variable Energy Source: A description for an energy source that fluctuates minute by minute in terms of the amount of energy it provides.

Variable Rate Technology (VRT): The technology that enables devices to precisely control the rate of materials applied to the farmland in real time.

Wearables: Smart electronic devices that can be worn on the body as accessories or implants. Often, wearable technology is utilized to quantify a physical process (e.g., heartbeat monitoring) or to augment human capabilities.

Web of Things (WoT): A term describing software architectural models that allow physical objects to be part of the World Wide Web.

Wide Flying Ad Hoc Network (Wide-FANET or WFANET): Refers to a FANET composed by remote sub-FANETs connected via a cloud-based secure tunnel.

Wireless Sensor Network (WSN): A wireless network comprising of spatially distributed autonomous sensors (motes), employed to monitor physical or environmental conditions.

Zigbee: An IEEE 802.15.4-based specification for wirelessly connected devices. Its key features include low-power short-range duplex communication and interoperability.

Author's Biography

Abdullah Abuhussein is a Ph.D. candidate in the Computer Science Department of the University of Memphis. He has a B.Sc. degree in Computer Science (1999) and an M.Sc. degree in Information Systems (2003). His research involves the challenges and solutions of cloud computing security and privacy. He currently works as a faculty member in the Information Systems Department at St. Cloud State University, USA.

Faisal Alsubaei is a Ph.D. student in the Computer Science Department of the University of Memphis. He has a B.S.Eds. degree in Computer Science from King Abdulaziz University, Saudi Arabia, and an M.Sc. degree in Computer Science from RMIT University, Australia. His research interests include security and privacy in IoMT and cloud computing.

Kalinka Regina Lucas Jaquie Castelo Branco has Master's degree in Computer Science from University of São Paulo (1999) and Ph.D. in Computer Science from University of São Paulo (2004). She is currently an Associate Professor of the Institute of Mathematics and Computer Science (ICMC-USP), working in the department of Computer Systems. She has experience in computer science, with emphasis on computer networks, security, embedded systems, and distributed computing.

Internet of Things A to Z: Technologies and Applications, First Edition. Edited by Qusay F. Hassan.
© 2018 by The Institute of Electrical and Electronics Engineers, Inc. Published 2018 by John Wiley & Sons, Inc.

Willie Brown, Jr. is an Assistant Professor in the Department of Engineering and Aviation Sciences at the University of Maryland Eastern Shore, Maryland. He earned his Ph.D. in Business Administration with a specialization in Homeland Security Leadership and Policy (Aviation Safety and Security) from Northcentral University and a Master's degree in Software Engineering from Embry-Riddle Aeronautical University. Brown also received dual degrees from Elizabeth City State University, North Carolina with a Bachelor of Science in Aviation Science, and a Bachelor of Science in Computer Science. In addition, Brown is a Federal Aviation Administration (FAA) licensed private pilot. He earned a Management Development Graduate Continuing Education Certification from Harvard University and completed the Leadership Management Program from the Federal Emergency Management Agency. As a member of several aviation professional organizations, he has received numerous awards, most notably from the North Carolina Department of Transportation and the Office of Naval Research/National Aeronautics and Space Administration. Brown serves as the Vice President of Administration of the Eastern North Carolina Institute of Electrical and Electronics Engineers (IEEE) for Geoscience and Remote Sensing Society. Brown has authored/coauthored publications in both national and international peer-reviewed journals with conference proceedings involving aviation and engineering education.

Joseph Bugeja received his B.Sc. degree in Computer Science and Artificial Intelligence from the University of Malta, in 2005, and M.Sc. degree in Information Security from Royal Holloway University of London, in 2011. Since August 2015 he is a Ph.D. student at the University of Malmö working in the department of Computer Science and the Internet of Things and People (IoTaP) research center. His research activities are mainly focused on cybersecurity and privacy aspects of IoT applications, in particular smart connected homes. Joseph also has more than 10 years of full-time experience working professionally in the software industry, occupying development, leadership, consultancy, and managerial roles specializing in Web technologies and Information Security.

Pratik Chaturvedi is a scientist at Defence Terrain Research Laboratory, Defence Research and Development Organisation, India and a senior research scholar at the Applied Cognitive Science Laboratory, Indian Institute of Technology Mandi. He is a Member of Institute of Electrical and Electronics Engineers. His current research interests include Landslide Risk Assessment and Management, AI, and Cognitive Science.

Ibibia K. Dabipi's research interests include computer security and network management, parallel computing and algorithms development, performance evaluation of computer networks, optimization of transportation networks, and economic analysis of transportation facilities. Dabipi holds Ph.D. in Electrical Engineering (1987) and a Master of Science in Electrical Engineering (1981), Louisiana State University, Louisiana; Bachelor of Science in Electrical Engineering and Bachelor of Science in Physics/Mathematics (1979) from Texas A&I University, Texas. Dabipi was the chairman, Department of Engineering and Aviation Sciences, University of Maryland Eastern Shore. Prior to coming to University of Maryland Eastern Shore, he was the Interim Chairman, and Chairman, Electrical Engineering Department, Southern University, Louisiana. His experiences include working at Bell Communications Research and AT&T Bell Labs as a member of technical staff during the summers of 1984 through 1987. He has authored/co-authored many technical articles for publications and presentations. He is currently a professor of Electrical Engineering in the Engineering and Aviation Sciences Department.

Paul Davidsson is Professor of Computer Science at Malmö University Sweden. He received his Ph.D. in Computer Science in 1996 from Lund University, Sweden. Davidsson is the Director of the Internet of Things and People Research Centre at Malmö University. His research interests include the application of agent technology, simulation, artificial intelligence, information systems, and data mining. Current application areas include transport and energy systems. The results of this work have been reported in more than 180 peer-reviewed scientific articles published in international journals, conference proceedings, and books. Davidsson has been member of the program committee for more than one hundred scientific conferences and workshops.

João Vitor de Carvalho Fontes received his Bachelor' degree in mechatronics engineering and M.Sc. degree in mechanical engineering from the University of São Paulo in 2012 and 2015, respectively. He is currently working toward a Ph.D. degree at the same university. His current research interests include multibody system, robotics, and control.

Varun Dutt is an assistant professor and principal investigator at the Applied Cognitive Science Laboratory, School of Computing and Electrical Engineering, Indian Institute of Technology Mandi. He is also a Senior Member of Institute of Electrical and Electronics Engineers. His current research interests include AI, Cognitive Science, and HCI.

Jeanette Eriksson holds a Ph.D. in Software Engineering from Blekinge Institute of Technology, Sweden. Her dissertation, "Supporting the cooperative design process of end-user tailoring," emphasizes the need for collaboration between different roles (stakeholders) such as end users, administrators, and software developers in order to achieve sustainable systems. Eriksson's research has addressed tailorable business systems, sensor games for emotion regulation, games for well-being, and wearables for managing health and chronic disease. She has worked with health care-related research for several years and has collaborated with a range of different health care institutions, including hospitals, retirement homes, and senior day care centers. Eriksson is leader of the application area Smart Health in the Internet of Things and People Research Center (IoTaP RC, http://iotap.mah.se/). The IoTaP Research Center collaborates with over 30 industrial partners to conduct research that contributes to the development of Internet of Things (IoT) technologies and applications that are both useful and usable. In particular, the Research Center emphasizes the importance of including "people" as a part of IoT systems.

Akaa Agbaeze Eteng obtained a B.Eng degree in Electrical/Electronic Engineering from the Federal University of Technology Owerri, Nigeria in 2002, and a M.Eng degree in Telecommunications and Electronics from the University of Port Harcourt, Nigeria in 2008. In 2016, he obtained a Ph.D. in Electrical Engineering from Universiti Teknologi Malaysia. Currently, he is a lecturer at the Department of Electronic and Computer Engineering, University of Port Harcourt, Nigeria. His research interests include wireless energy transfer, radio frequency energy harvesting, and wireless powered communications.

Lucas Finco has years of experience working with utilities on managing the demand side of their business with data analytics, forecasting, and planning. He is a proven innovator, providing state-of-the-art ideas and analyses that are enabling a new, smarter electric grid to emerge. He continues to innovate new concepts and valuation methods that are opening up possibilities for new energy paradigms. Mr. Finco has a B.S. in Applied Math, Physics, & Engineering, an M.A. in Physics, and an MBA in Finance & Entrepreneurship.

Bernard Fong received his B.S. degree in electronics from the University of Manchester Institute of Science and Technology in 1993, and Ph.D. in health technology from the University of New South Wales in 2005. Since July 2017, he is an M.Phil. student in clinical emergency medicine at Auckland University of Technology where he previously worked as a tenured faculty member between 2001 and 2007 prior to joining Hong Kong Polytechnic University as an associate professor. He is currently serving as the chair for the System Biology and Biomedical Systems technical committee under the IEEE Systems Council and the General Chair for the 1st IEEE International Workshop on System Biology and

Biomedical Systems to be held in March 2018. His research interest is broadly in the areas of assistive care, medical devices, public health, rehabilitation engineering, and telemedicine.

A.C.M. Fong is currently with Western Michigan University. He holds four degrees in computer science and electrical engineering. His research interests include machine learning, knowledge discovery, and applications of these to consumer and industrial fields. He has published about 190 papers in leading journals and conference proceedings such as *IEEE Transactions on Knowledge and Data Engineering, IEEE Transactions on Evolutionary Computation, IEEE Transactions on Industry Applications, IEEE Transactions on Affective Computing,* and *IEEE Transactions on Multimedia.*

Virginia N.L. Franqueira is currently a senior lecturer at the University of Derby, UK. Prior to that, she held a lecturer position at the University of Central Lancashire (UK), a postdoc research position at the University of Twente (NL), and worked as an information security consultant (UK). She received her Ph.D. in Computer Science from the University of Twente, The Netherlands, in 2009, and her M.Sc. from the Federal University of Espirito Santo, Brazil. She is a member of the IEEE Computer Society and of the British Computer Society, and a fellow of the Higher Education Academy. Her topics of research interest include different aspects of cybersecurity and digital forensics.

Mário Marques Freire earned a 5-year B.Sc. degree in electrical engineering and a 2-year M.Sc. degree in systems and automation in 1992 and 1994, respectively, from the University of Coimbra, Portugal. He obtained his Ph.D. degree in electrical engineering in 2000 and the habilitation title in computer science in 2007 from the University of Beira Interior, Portugal. He is a full professor of computer science at the University of Beira Interior, which he joined in the fall of 1994. When he was an M.Sc. student at the University of Coimbra, he was also a trainee researcher for a short period in 1993 in the Research Centre of Alcatel-SEL (now Alcatel-Lucent) in Stuttgart, Germany. His main research interests fall within the broad area of computer systems and networks, including network forensics and Internet traffic classification, security and privacy in computer systems, peer-to-peer networks, and cloud systems. He is the coauthor of seven international patents, the coeditor of eight books published in the Springer Lecture Notes in Computer Science book series, and the author or coauthor of about 120 papers in refereed international journals and conferences. He serves as a member of the editorial board of *ACM SIGAPP Applied Computing Review*, as an associate editor of the Wiley journal *Security and Communication Networks*, and as an associate editor of the Wiley *International Journal of Communication Systems*. In the past, he served as an editor of *IEEE Communications Surveys and Tutorials* (2007–2011) and as a guest editor of two feature topics in *IEEE Communications Magazine* (2008) and of a special issue of the Wiley *International Journal of Communication Systems* (2009). He served as a technical program committee member for several Institute of Electrical and Electronics Engineers international conferences and is cochair of the track on Networking of ACM SAC 2017. Mario Freire is a chartered engineer by the Portuguese Order of Engineers, a member of the Institute of Electrical and Electronics Engineers Computer Society and Engineers Communications Society, and a member of the Association for Computing Machinery.

Daniel Happ received his M.Sc. degree in computer science in 2013 from Free University of Berlin, Germany. He is now a Ph.D. candidate at the Telecommunication Networks Group at Technical University of Berlin. His research interests include large-scale cloud-connected sensor networks, publish/subscribe messaging, and fog computing.

Pedro Ricardo Morais Inácio was born in Covilhã, Portugal, in 1982. He holds a 5-year B.Sc. degree in mathematics and computer science and a Ph.D. degree in computer science and engineering, obtained from the University of Beira Interior (UBI), Portugal, in 2005 and 2009, respectively. The Ph.D. work was performed in the enterprise environment of Nokia Siemens Networks Portugal S. A. through a Ph.D. grant from the Portuguese Foundation for Science and Technology. He has been a professor of computer science at UBI since 2010, where he lectures on subjects related to information assurance and security, programming of mobile devices, and computer-based simulation, for graduate and undergraduate courses, namely, the B.Sc., M.Sc., and Ph.D. courses in computer science and engineering. He is an instructor of the UBI Cisco Academy. He is an Institute of Electrical and Electronics Engineers senior member and a researcher at the Instituto de Telecomunicações.

Andreas Jacobsson is an Associate Professor in Computer Science at Malmö University in Sweden. Jacobsson received his Ph.D. in Computer Science in 2008 at Blekinge Institute of Technology, and in 2016 he qualified for the title of Docent in Computer Science at Malmö University. His research interests include the theory and application of technologies for interoperability, trust, privacy, and security in Internet-based information systems, more recently in Internet of Things settings, such as pervasive smart living spaces and smart homes. The results of his work have been documented in numerous peer-reviewed scientific articles published in international books, journals, and conference proceedings. Jacobsson has also coauthored two books on trust, security, and privacy in Internet-based information systems, which have been picked up as course literature in several courses at various universities in Scandinavia. Jacobsson has also acted as conference session chair, program committee member, and reviewer for several research journals, conferences, and workshops in his fields. Moreover, Jacobsson is a member of the Internet of Things and People Research Center, where he currently leads a research project on intelligent privacy support for smart homes.

Kai Jakobs joined Computer Science Department of RWTH Aachen University in 1985. He holds a Ph.D. in Computer Science from the University of Edinburgh and is a Certified Standards Professional.

His activities and research interests focus on ICT standards and the underlying standardization process. Over time, he has (co)-authored/edited 20+ books and published 200+ papers.

Kai is Vice President of the European Academy for Standardisation (EURAS) and founder/editor-in-chief of the *International Journal of Standardization Research.*

Venkata Uday Kala is an assistant professor and coordinator at the Construction Material laboratory, School of Engineering, Indian Institute of Technology Mandi. He is also an Associate Member of American Society of Civil Engineers. His current research interests include Geotechnical Engineering, Landslide: Monitoring-Mitigation-Management, Geo-Bio-Technology, and Unsaturated Geotechnics.

Jonny Karlsson is currently a lecturer and researcher at Arcada University of Applied Sciences, Helsinki, Finland. He received his Ph.D. degree in 2016 at the Open University, UK. He has authored/coauthored 29 publications in networking, network security, information security, and applied cryptography.

Sudhakar Kumar is a project associate at the Applied Cognitive Science Laboratory, School of Computing and Electrical Engineering, Indian Institute of Technology Mandi, India. He completed his Bachelor of Technology in Electronics & Communication Engineering from Shri Mata Vaishno Devi University Katra, India. He is a FOSS enthusiast and his research interests are Internet of Things, Wireless Sensor Networks, and Embedded Systems.

Chee Y. Leow obtained his B.Eng. degree in Computer Engineering from Universiti Teknologi Malaysia (UTM) in 2007. Since July 2007, he has been an academic staff in the Faculty of Electrical Engineering, UTM. In 2011, he obtained a Ph.D. degree from Imperial College London. He is currently a senior lecturer in the faculty and a member of the Wireless Communication Centre (WCC), UTM. His research interest includes but not limited to wireless relaying, MIMO, physical layer security, convex optimization, communications theory, near-field wireless charging and 5G.

C.K. Li obtained his B.Sc. (Eng) with first class honors, M.Sc. and Ph.D. in UK in 1976,1978, and 1983, respectively. He has 40 years academic and industrial experience. Li is a Senior Member of the IEEE and Fellow of the IET and HKIE. He is currently CTO of the Add-Care Ltd. His current interest is in biomedical engineering and signal optimization.

Naresh Mali is a senior research scholar at School of Engineering, Indian Institute of Technology Mandi. He is also a Student Member of American Society of Civil Engineers. His current research interests include Geotechnical Engineering and Landslide: Monitoring-Mitigation-Management.

Manuel Meruje was born in Castelo Branco, Portugal, in 1992. He started his studies in Computer Science and Engineering in 2010 at Universidade da Beira Interior. Some of his main interests are Software Development, Internet-of-Things, and Information Security. In 2012 he discovered the Arduino and Raspberry Pi platforms and started to develop some small projects with both systems. In 2013 he joined Instituto de Telecomunicações, where he was able to develop skills related to his main interests. He received his Bachelor's degree in Computer Science and Engineering from Universidade da Beira Interior in 2014, and his Master's degree in Computer Science and Engineering from Universidade da Beira Interior in 2016. The Master's project consisted in the development of a solution that allowed the monitoring of medication boxes using wireless sensors and Android devices. He is currently learning about new technologies while developing personal projects.

Daniel Minoli, Principal Consultant, DVI Communications, has published 60 well-received technical books and 300 papers and has made 85 conference presentations. He has many years of technical-hands-on and managerial experience in planning, designing, deploying, and operating secure IP/IPv6-, VoIP, telecom, wireless, satellite, and video networks for global Best-in-Class carriers and financial companies. Over the years, Minoli has published and lectured extensively in the area of M2M/Internet of Things (IoT), network security, satellite systems, wireless networks, IP/IPv6/Metro Ethernet, video/IPTV/multimedia, VoIP, IT/Enterprise Architecture, and network/Internet architecture and services. Minoli has taught IT and Telecommunications courses at NYU, Stevens Institute of Technology, and Rutgers University.

Benedict Occhiogrosso is a cofounder of DVI Communications. He is a graduate of New York University Polytechnic School of Engineering. Occhiogrosso's experience encompasses a diverse suite of technical and managerial disciplines, including sales, marketing, business development, team formation, systems development program management, procurement and contract administration budgeting, scheduling, QA, and technology operational and strategic planning. As both an executive and a technologist, Occhiogrosso enjoys working and managing multiple client engagements as well as setting corporate objectives. Occhiogrosso is responsible for new business development, company strategy, as well as program management. Occhiogrosso also on occasion served as a testifying expert witness in various cases encompassing patent infringement and other legal matters.

Daniel Fernando Pigatto is Bachelor in Computer Sciences (2009). He holds M.Sc.. (2012) and Ph.D. in Computer Sciences (2017) from the University of São Paulo (USP). He is currently Adjunct Professor at Federal University of Technology of Paraná (UTFPR), Curitiba, Brazil. Pigatto was visiting researcher at the University of the West of England, UK (2015). His main topics of research are Critical Embedded Systems, Computer Networks, and Network Security.

Alex Sandro Roschildt Pinto received his Master's degree in Computer Science in 2003) and Ph.D. in Electrical Engineering in 2010, both from Federal University of Santa Catarina (UFSC), Brazil. Currently, he is Assistant Professor at the Department of Engineering—Blumenau of the Federal University of Santa Catarina. He has experience in the following areas: embedded systems, wireless sensor networks, and Internet of Things. He has published more than 50 international papers (journals, conferences, and book chapters) and has also achieved three IEEE awards.

Göran Pulkkis is currently a project researcher at Arcada University of Applied Sciences, Helsinki, Finland. He received his doctoral degree in 1983 from Helsinki University of Technology, Finland. He has authored/coauthored approximately 100 publications on performance evaluation of computer systems, computer architecture, networking, network security, information security, and applied cryptography.

Sharul Kamal Abdul Rahim obtained his first degree from University of Tennessee, USA, majoring in Electrical Engineering, graduating in 1996, M.Sc. in Engineering (Communication Engineering) from Universiti Teknologi Malaysia (UTM) in 2001, and Ph.D. in Wireless Communication System from University of Birmingham, UK in 2007. Currently, Dr. Sharul is a Professor at the Wireless Communication Centre, Faculty of Electrical Engineering, UTM, Skudai, Malaysia. His research interests include antenna design, RF and microwave systems, reconfigurable antennas, beamforming networks, smart antenna systems, and antennas for wireless energy transfer. He is also a Senior Member of IEEE Malaysia Section, Corporate Member of Institute of Engineer Malaysia (MIEM), Member of the Institute of Electronics, Information and Communication Engineers (IEICE), and Member of the Eta Kappa Nu Chapter (International Electrical Engineering Honour Society, University of Tennessee). He has published a number of technical papers, including journal articles, book chapters, and conference papers.

Hejamadi Raghav Rao is the AT&T Distinguished Chair in Infrastructure Assurance and Security at the University of Texas at San Antonio College of Business. He also holds a courtesy appointment as full professor in the UTSA Department of Computer Science. He graduated from Krannert Graduate School of Management at Purdue University. His interests are in the areas of management information systems, decision support systems, e-business, emergency response management systems, and information assurance. He has chaired sessions at international conferences and presented numerous papers. He has also coedited four books, including *Information Assurance Security and Privacy Services* and *Information Assurance in Financial Services.* He has authored/coauthored more than 200 technical papers, of which more than 125 are published in archival journals.

Shahid Raza is Lab Manager and Senior Researcher at the Security Lab of RISE SICS AB in Stockholm, Sweden, where he has been working since 2008. His research interests include but are not limited to security and privacy in IPv6-connected IoT, interconnection of computing clouds and IoT, WirelessHART, the smart grid, and storage security in constrained environments. Shahid is an associate editor of the premier *IEEE IoT Journal* and a TPC member of a number of targeted IoT conferences/workshops. His research work has been published in renowned international journals and conferences. In addition to participating in different EU (NobelGrid, CALIPSO, CONET, etc.) and Swedish projects (SeC-Things, PROMOS, etc.), Shahid has recently won a competitive research grant from the Swedish funding agency VINNOVA under the Strategic Innovation Program for the Internet of Things, where he is the project leader. This Certificate Enrollment for Billion of Things (CEBOT) project is primarily aimed to equip IoT devices with digital identities. Shahid has completed his industrial Ph.D. from RISE SICS AB in Stockholm and the Mälardalen University in Sweden. He also holds a Master of Science degree from the KTH Royal Institute of Technology, Stockholm.

Mariana Rodrigues has a Bachelor's degree in Electric Engineering (2007) and an M.Sc.. in Computer Science & Technology, both from Federal University of Itajubá (UNIFEI). She is currently a Ph.D. student at University of São Paulo (USP), Brazil. Her main areas of interest are embedded systems, Internet of Things, data structures, industrial automation, and telecommunications.

Jason M. Rosenberg is an IT Auditor with the Treasury Inspector General for Tax Administration. He has a Bachelor's degree in Accounting and an MBA in Information Assurance. His areas of research interest include digital forensics, fraud prevention and detection, IT auditing, and information assurance.

Nancy L. Russo is a professor in the department of Computer Science and Media Technology and the Internet of Things and People Research Center at Malmö University. She holds a Ph.D. in Management Information Systems from Georgia State University and has held positions at Northern Illinois University, Miami University, University College Cork (IE), and Slobomir P University (BiH). Her research has addressed the adoption, diffusion, customization, and use of information system development methodologies in a variety of contexts as well as the factors related to successful implementation of information systems. In addition to journal and conference papers, she has coauthored several books and book chapters on these topics. She has also studied gender issues in technology and the impact of social media on information systems development and service quality. Her current research addresses the design, development, and use of systems within the Internet of Things environment, with a particular focus on health-related applications that focus on the patient.

Musa Gwani Samaila is currently pursuing his Ph. D. in Computer Engineering at the Department of Computer Science, University of Beira Interior, Covilhã, Portugal, with research interests in Internet of Things and embedded systems security. He was a lecturer at the Department of Electrical and Electronic Engineering Technology, Federal Polytechnic Bauchi, Nigeria, from 2001 to 2009. He is an assistant Chief Engineer at the Centre for Geodesy and Geodynamics, National Space Research and Development Agency, Toro, Bauchi, Nigeria. He has authored/co-authored a number of peer-reviewed scientific articles published in both national and international journals, conference proceedings, and books. He is a corporate member of the Nigerian Society of Engineers, and is registered with the Council for the Regulation of Engineering in Nigeria.

Detlef Schoder is a Professor and chairs the Department of Information Systems and Information Management at the University of Cologne, Germany. He was appointed as a reviewer to the German Parliament's Lower House for e-commerce issues and was a consultant to the European Commission. Recently, he became one of three revisers of the German Research Foundation (DFG), Germany's largest science foundation. He was also a Visiting Scholar at Stanford University, the University of California at Berkeley, and the Massachusetts Institute of Technology, USA. His research interests and projects focus on data-centric products, processes, and business model innovations, in particular deploying techniques from the fields of machine learning and artificial intelligence. Professor Schoder has been recognized in many prestigious competitions in these fields. He was granted a patent on "Individualized printed newspapers." Professor Schoder was ranked among the top 250 management researchers in the German-speaking region by Handelsblatt, a prominent German business newspaper.

Sajjan Shiva is a Professor in the Computer Science Department of the University of Memphis. His research involves computer architecture/distributed and parallel processing, computer system security, software engineering/process improvement/software testing, artificial intelligence/expert systems, and modeling and simulation.

John S. Shu is a Ph.D. candidate at the Department of Information Systems and Cybersecurity of University of Texas San Antonio. He has a Bachelor's degree in Computer Sciences and Mathematics, and a Master's degree in Computer Sciences with a concentration in Software Security. His area of interest in research includes information security, Big Data analytics, cloud and mobile computing, and information healthcare systems.

Jan Sliwa is affiliated to the Department of Engineering and Information Technology of the Bern University of Applied Sciences, Switzerland. For over 30 years, he developed software in languages ranging from Assembler to Java, from industrial control to Web/Database applications. In recent years, his activity field has been related to medical applications. He is an author of several papers and book chapters regarding the technical, social, and ethical aspects of reuse of medical data and data distribution in the IoT systems. Coming from IT, he tries to understand the nondigital world. He sees his mission in connecting people with different backgrounds and mentalities. He is particularly interested in languages and cultures: Germanic, Romanic, Slavic, and East Asian.

James Smith is Professor of Interactive Artificial Intelligence at the University of the West of England (UWE). He has an MA in Electrical Sciences from Cambridge University and an M.Sc.. in Communicating Computer Systems and a Ph.D. in Computer Science, both from UWE. He has published extensively on many aspects of theoretical and applied artificial intelligence, sits on the editorial boards of the journals *Evolutionary Computation* (MIT Press/ACM) and *Memetic Computing* (Springer).

Tiffany Y. Tang is an assistant professor at the Department of Computer Science, Wenzhou-Kean University. Her recent research focuses on affective computing, sensing, and wearable technologies for children with autism spectrum disorder (ASD) and other special needs, human factors for smart recommendations, and artificial intelligence in education. She has previously served as the editor to special issues in Expert Systems: *The Journal of Knowledge Engineering* (Wiley) and *Journal of Educational Computing Research* (SAGE). She currently serves as a coeditor of the Special Issue on Leveraging Wearable and Sensing Technologies for Assistive Learning Environment in Special Education of *Journal of Educational Computing Research.* She was the cochair of the track for PACIS'2014 and PACIS' 2016 on Human and Social Factors in Decision Support and Recommendation Systems. In May 2015, she cofounded the Autism Research Network in Wenzhou as the city's first and one of China's few dedicated research and development centers for individuals with autism. She has coauthored conference papers published in *ACM Ubicomp, ACM IDC, HCII* and journal papers on technology-enabled intervention for individuals with autism.

Kamal Kishore Thakur is an undergraduate student pursuing his bachelor's degree in Computer Engineering from Thapar Institute of Engineering and Technology, Patiala and a student intern at the Applied Cognitive Science Laboratory, Indian Institute of Technology Mandi. He has worked with various Web Technologies and has a PHP Certification. Apart from this, he is passionate about Artificial Intelligence, Cloud Computing and Deep Learning.

Marco Tiloca is a Senior Researcher at the Security Lab of RISE SICS AB in Stockholm, Sweden. He received the Bachelor's degree and the Master's degree (cum laude) in Computer Engineering from the University of Pisa, Italy, in 2006 and 2009, respectively. He received the Ph.D. degree in Computer Engineering from the University of Pisa in 2013, with a focus on network and communication security in Wireless Sensor Networks. His research interests are in the field of network security and include security in the Internet of Things, secure group communication, key management, denial-of-service attacks, and simulative evaluation of network attacks. Results of research activities in these areas have produced several publications in renowned international journals and conferences. Tiloca has been involved in the FP7 European Collaborative Project SEGRID as leader of the work package "Novel Security Solutions" and in the EIT-Digital High Impact Initiative ACTIVE as leader of the Task "Security and Privacy Modules and APIs." He was also involved in the FP7 European Network of Excellence CONET and the FP7 European Integrated Project PLANET.

Shambhu Upadhyaya is the director of UB's Center of Excellence in Information Systems Assurance Research and Education (CEISARE) and Professor of Computer Science and Engineering at the State University of New York at Buffalo. His research focuses on protecting the US infrastructure from cyber threats. This includes electric power grids, transportation systems, financial networks, military assets, and water supplies. Upadhyaya has conducted research sponsored by the Air Force Research Laboratory, DARPA, Intel Corp., the National Science Foundation, the Department of Defense, and other organizations. He has authored/coauthored about 280 articles in refereed journals and conferences in these areas.

Magnus Westerlund is the programme director of the master programme in big data analytics and a researcher at Arcada University of Applied Sciences, Helsinki, Finland. He has completed his doctoral studies at Åbo Akademi University, Turku, Finland. His current research topics are found in the convergence area of distributed clouds that are enabled by distributed ledger technology and the application of intelligent agents. He has long experience from industry in various software development positions and published extensively on information system issues arising from regulatory activities in the EU.

Alexander Willner is the head of the Industrial Internet of Things (IIoT) Center at the Fraunhofer Institute for Open Communication Systems (FOKUS) and the head of the IIoT research group at the Technical University Berlin (TUB). He is working with his groups in applying standard-based machine-to-machine (M2M) and Internet of Things (IoT) technologies to industrial domains. With a focus on moving toward the realization of the edge/fog computing paradigm based on semantic annotations and the orchestration of microservices, particular attention is being paid to the fields of application Smart Manufacturing and Industrie 4.0.

Prior research positions include the University Bonn, he holds an M.Sc. and a Ph. D. (Dr.-Ing.) in computer science from the University Göttingen and the Technical University Berlin, respectively. His research interests are in semantics-enabled distributed information systems, linked data, the Graph of Everything, communication middleware, and service-oriented architectures. He is active in relevant standardization activities and gives a corresponding lecture at the Technical University Berlin.

Pinata Winoto is an assistant professor in computer science at Wenzhou-Kean University. He received his Ph.D. degree from University of Saskatchewan. His current research interests are recommendation systems, multiagent systems, assistive technology, and edutainment. He has published over 60 high-quality conference papers in *IEEE/ACM AAMAS, ACM Ubicomp, ACM IDC, ACM CHI, ACM CSCW,* and so on; book chapters published by Springer and CRC Press, and journal papers in *Journal of Educational Computing Research, Expert Systems with Applications, New Generation Computing,* and so on. His early research on cross-domain and emotion-aware recommendation systems has been widely cited. He is currently the Founding Director of the Computer Science Playground and the Assistive Technology Research & Development Center at Wenzhou-Kean University.

Srishti Yadav is a student intern at the Applied Cognitive Science Laboratory, Indian Institute of Technology Mandi. She is an engineering graduate in Electronics and Communication and she has experience in working on Embedded Technology, IoT, Computer Vision, and Quadrotors. Passionate about working in the field of embedded technologies, she sees herself working with farsighted technologies that make an impact on the society.

Dr. Lei Zhang received his Ph.D. Degree in Electrical Engineering in 2011 from the University of Nevada, Las Vegas. Currently, he is working as an associate professor in the Department of Engineering and Aviation Sciences, University of Maryland Eastern Shore. His research focuses on advanced computer architecture, Internet-of-Things, deep learning, and robotics.

Index

Internet of Things A to Z: Technologies and Applications, First Edition. Edited by Qusay F. Hassan.
© 2018 by The Institute of Electrical and Electronics Engineers, Inc. Published 2018 by John Wiley & Sons, Inc.

b

e

g

Printed in the United States
By Bookmasters